Mine Water

ENVIRONMENTAL POLLUTION

VOLUME 5

Editors

Brian J. Alloway, *Department of Soil Science, The University of Reading, U.K.*
Jack T. Trevors, *Department of Environmental Biology, University of Guelph, Ontario, Canada*

Editorial Board

Mine Water

Hydrology, Pollution, Remediation

by

Paul L. Younger

University of Newcastle
United Kingdom

Steven A. Banwart

University of Sheffield
United Kingdom

and

Robert S. Hedin

Hedin Environmental
Pittsburgh, USA

KLUWER ACADEMIC PUBLISHERS
DORDRECHT / BOSTON / LONDON

Library of Congress Cataloging-in-Publication Data is available.

ISBN (hardback) 1-4020-0137-1
ISBN (paperback) 1-4020-0138-X

Published by Kluwer Academic Publishers,
P.O. Box 17, 3300 AA Dordrecht, The Netherlands.

Sold and distributed in North, Central and South America
by Kluwer Academic Publishers,
101 Philip Drive, Norwell, MA 02061, U.S.A.

In all other countries, sold and distributed
by Kluwer Academic Publishers,
P.O. Box 322, 3300 AH Dordrecht, The Netherlands.

Cover photo: A derelict "valley dyke" tailings pond at the abandoned Mount Wellington Tin Mine, Cornwall, UK. The mine closed before this pond could be used as intended, and it now serves only as a seasonal source of acidic pollution to the Carnon River. The orange high water mark is controlled by decant through the lower of the two redundant draw-off towers; the burgundy low water mark reflects evaporation and seepage through the dam (which is composed of waste rock). Notice the losing battle of Mother Nature to re-establish a healthy fauna: this book is about how we can help Nature to overcome water pollution at mine sites world-wide. (Photo: P.L. Younger).

Printed on acid-free paper

Printed and bound in Great Britain by MPG Books Limited, Bodmin, Cornwall.

TABLE OF CONTENTS

FOREWORD

The authors of this book bring together fifty years of experience in the field of mine water, with a broad and multidisciplinary perspective. In reading these chapters, as someone with forty years of personal experience in the same field, I have found myself both reliving experiences and learning much from the ideas expressed by these authors. The latter will, I suspect, be the case for all readers of this work.

To best appreciate the importance of this book it is important to note that, for many years, the literature on mine water, in all of the aspects covered here, was extremely sparse. This is in no small measure due to the fact that mining companies, for obvious reasons, have not been especially keen to air their "problems" in public: and water has, for a variety of reasons, often been one of the most serious problems faced by mining operations.

In more recent times, through the inputs of universities and research institutes, the development of solutions to water problems in mining has been blessed with major advances in science and technology, which have in turn opened the door to a much wider dissemination of knowledge and experience. Now, with works such as this book, tools of great utility are being made available to society in general and the mining sector in particular. These tools are no longer the sole preserve of the specialist, but are increasingly accessible to those who must cope daily with mine water problems, using the best technology available, deployed within the most appropriate socio-economic framework.

This is a book which offers practical solutions, and it will be of great help to many professionals who respond to these challenges day-by-day, developing solutions which are both economically and environmentally acceptable. I believe that these professionals will find that the balance between theory and practice in these pages has been struck at just the right level.

Because of the particular background of the three authors, coal mining features prominently (though by no means exclusively) in this book. Many of the insights are equally applicable to mining for metalliferous, industrial and/or ornamental minerals. Since no two mines are ever identical, in solving the complex water problems that beset many mines, there is no better guide than the experience gained from both successes and failures in practice, examples of which abound in this book.

Material in the book is helpfully grouped into three thematic areas: hydrochemistry, groundwater engineering and treatment technologies. The extensive bibliography which supports all three themes will be found to be an excellent resource for the reader seeking further insights into specific topics.

In this connection, without a doubt, the authors have taken full advantage of the information available both in academic journals and in the proceedings of international conferences, which are the meeting place of both practitioners and researchers in this field. I would like to make special mention here of the efforts of the International Mine Water Association (of which I was one of the Co-Founders in Granada (Spain) in 1988), which through its Congresses, Symposia and its journal *Mine Water and the Environment*, continues to provide the single greatest source of literature on the full range of mine water issues.

Most useful for newcomers to the issue of mine waters is the introductory presentation of the various types of mine workings, both underground and on the surface, and of mine waste disposal, with consideration of their effects on their surroundings. Both active mining and closure are considered, supported by highly illustrative figures.

In discussing mine water chemistry, the text combines profundity with a clear presentational style, making it readily intelligible by non-specialists. The coverage embraces all of the key concepts, taking into account the various factors which govern pollutant release, the understanding of which is the key to predicting in advance the hydrochemical changes which may accompany the opening and working of a mine. Such a predictive capacity is particularly important in relation to the mobility of metals, which governs both the solution chemistry and contamination potential of mine waters and related leachates.

The basic concepts of hydrology, especially subsurface hydrology, are the key to developing models of ground water flow and solute transport, both of which are profoundly affected by changes in drainage patterns arising from the excavation of mine voids. The measurement techniques needed to fully characterize the most important hydrological parameters are described. Given their particular sensitivity to mining operations, the interactions between surface and subsurface hydrological systems are given special emphasis. This emphasis underlies the discussion of the movement of water within the mine workings themselves, with all the problems associated with sudden, dangerous inrushes, perennial inflows requiring continuous drainage, and/or the influence of high pore pressures on slope and floor stability.

The description of mine drainage strategies will undoubtedly serve as a good source of reference for those who must cope with such problems, both in mine voids and in other excavations made for civil engineering applications, whatever their purpose.

The possible environmental impacts of mine drainage waters, which today more than ever we must address within the broadest social context, are described along with practical measures for their mitigation. The coverage extends beyond the mine voids themselves to embrace mine waste depositories, including waste rock piles, tailings dams, and backfilled waste in the former voids. And all are considered 'from cradle to grave', that is from the first opening of a mine to the post-abandonment period.

Linked closely with the theme of hydrochemistry, the text provides in-depth coverage of treatment technologies for polluted mine waters. These technologies are currently advancing rapidly, thanks to the work of many researchers in those countries which have more advanced environmental consciousness, finally addressing some of the most persistent challenges arising from mining activities. The active and passive treatment chapters in this book offer the reader an ample overview of the state-of-the-art, which will allow them to choose the most appropriate technologies for their particular application. So rapid is the pace of innovation in this field, providing ever-deeper insights, that the specialist is forever under pressure to keep abreast of the latest developments.

To conclude, then, you have in your hands one of the most comprehensive and much-needed syntheses on the theme of mining, water and the environment ever to be produced. Hopefully it will soon be translated into other languages, especially Spanish, so that many more people may benefit from the profound knowledge of the authors, and from the great effort which they have brought to bear in the preparation of this book.

Professor Rafael Fernández Rubio
Chair of Hydrogeology, Madrid School of Mines.
Emeritus President of the International Mine Water Association

PREFACE

This book has been a long time in the making. Firstly it has drawn on the best part of 50 man-years of collective experience in hydrogeology, ecology, geochemistry and engineering (mining, environmental, civil). Secondly, it has been written in parallel with our 'day jobs', in which the three of us grapple with research and consultancy exigencies in the field of mine water management and the remediation of polluted mine waters. Although this has resulted in repeated delays, testing the patience of our editors beyond all reasonable bounds, we hope that the frequent doses of raw experience which pepper the book make the waiting worthwhile. For we have written this book primarily in the hope that it might serve the needs of the hard-pressed engineer or site manager, at least as a depository of useful guidance, if not as a source of inspiration in those awful moments when it is realized that some very difficult decisions should have been made yesterday! We especially hope that this text may prove useful to the mining engineer who has just discovered an unanticipated water management issue, or to the environmental engineer charged with developing a remedial program for an active or abandoned mine site.

Since two of us practice our engineering in the role of University teachers, we also hope that this volume will also be found instructive by students in environmental sciences, geology, hydrology, environmental engineering and mining engineering. While we can hardly expect the book to become a course text, given that 'mine water management' modules are hardly common, we believe that the scientific principles and insights which we present (especially in Chapters 2, 3 and 5) have the potential to find much wider application than the particular context which we address here.

The reader will soon discover that the book skews its overall coverage in favor of mine water hydrology (Chapter 3) and passive treatment of polluted mine waters (Chapter 5). While we believe that our coverage of mine water chemistry (Chapter 2) and active treatment (Chapter 4) includes much that is new, or is at least presented in a new (and hopefully refreshing) manner, there is already a substantial, coherent literature addressing many of the details of those topics. By contrast, mine water hydrology and passive mine water treatment are not nearly so well served by the existing literature, and therefore merit the emphases which we accord them in this volume.

As with all works of such length and complexity, we owe major debts of gratitude to numerous colleagues. While it is increasingly common practice to avoid naming any individuals for fear of offending those not mentioned specifically, we reckon we can trust our friends not to be such sensitive flowers. Moreover, we think our closest friends and colleagues would agree with the composition of the following list of colleagues who have made particularly significant contributions, both by critically reviewing early drafts and contributing in myriad ways to the development of the concepts and practices presented in this book:

- Russell Adams, Lesley Batty, Tom Curtis, Andy Aplin, Mick Jones, Charlotte Nuttall, Karen Johnson, Catherine Gandy, Jerry Kuma, Julia Sherwood, Gholam Karami, Andy Large, Will Mayes, Enda O'Connell, Marian Díaz Goebes and Marcos Arce (University of Newcastle, UK)
- Sean Burke, Katy Evans, Stephanie Croxford, John Cripps and David Lerner (University of Sheffield, UK)
- Bob Kleinmann, Bob Nairn and George Watzlaf (National Energy Technology Laboratory, USA)
- Aidan Doyle (Mining Institute) (to whom we are particularly grateful for the many excellent photographs he provided for the illustration of Chapters 1 and 3).
- Adam Jarvis, David Laine, Adrian England and Keith Whitworth (IMC Ltd)
- Keith Parker, Ian Burns, Dave Stafford and Albert Schofield (UK Coal Authority)
- Julia Sherwood, Steve Brooker, Dave Griffiths, John Taberham, Jim Wright, John Aldrick, Wayne Davies, and Paul Edwards (Environment Agency, England & Wales)
- Colleagues from a wide range of other organizations: Chas Brookes Alan McRae and the late Terry Jeffreys (Quaking Houses Environmental Trust); Richard Coulton (Unipure Environmental); William Pulles (Pulles Howard and de Lange, S Africa); Alan Lowdon (NuWater Ltd); Gia Destouni and Maria Malmström (Royal Institute of Technology, Stockholm); Bill Dudeney (Imperial College); Bo Strömberg (Swedish Nuclear Regulatory Agency); Carlos Ayora (CSIC, Barcelona); Jorge Loredo, Fernando Pendás Fernández and Alumdena Ordoñez Alonso (Escuela de Minas, Oviedo, Spain); Rafael Fernández Rubio (Universidad Politécnica de Madrid); Mike Cambridge, Jim Gusek and Clive Hallett (Knight Piésold); Christian Wolkersdorfer (TU-Bergakademie Freiberg, Germany); Nick Robins and Steve Dumpleton (British Geological Survey); Gus Spirit, Bob Sargent and David Holloway (SEPA); David Banks (Holymoor Consultancy); Ron Cohen (Colorado School of Mines); Jaap Smit (Mintek, South Africa); Colin Dudgeon (University of New South Wales); Rob Bowell (Steffen Robertson and Kirsten).

We wish to particularly note the profound influence of the late Werner Stumm (EAWAG, Switzerland). His enormous contribution to the quantitative treatment of natural water chemistry provides a critical foundation for the applications described in this book.

The most important debt of gratitude is owed to our families for their unending support and understanding, especially to our wives, Louise, Jane and Beth.

Finally, we record our gratitude for the finance provided by our communities, through publicly-funded research, and by numerous mining and environmental management companies and regulatory agencies who have funded our work over the years.

CHAPTER ONE

MINING AND THE WATER ENVIRONMENT

1.1 MINING, SOCIETY AND THE ENVIRONMENT

It is difficult to over-state the importance of mining to the development and sustenance of modern society (Shepherd, 1993). The appropriation of natural earth resources for human use is apparently instinctive, and operates at every level of social organisation: from the hunter-gatherer constructing a temporary shelter with loose boulders, to an international corporation extracting and processing many tonnes of host rock to obtain a few grams of precious metal. The extraction and processing of minerals is a prerequisite for the lifestyle of all advanced societies, to the extent that a blanket opposition to all mining is a difficult position for even the most dedicated of anti-mining pressure groups to sustain[1]. The all-pervading influence of mining on contemporary urban life is readily appreciated by the simple exercise of examining your surroundings and reflecting on the

---

Evidence of mining on all sides.

As I sat in my study writing this, I was able to compile the following list of mining-derived items in the surrounding objects, without so much as leaving my chair:

Building fabric: Bricks from fired clay; mortar from limestone; roof tiles from slate (with lead and zinc flashing); window glass from pure silica sand.

Screws and nails in all furniture: Steel, derived from iron, and various other metals in alloy. (Steel production also involves the use of limestone and fluorite).

Step ladders: Aluminum, derived from bauxite extraction.

Electricity supply: Baseload from uranium-based nuclear energy, peaks from coal-burning. (Also implicated in the existence of most other objects).

Computer: Electrical components (wires etc) derived from copper and other metals; microchips based on quartz and other silica minerals.

Electric lighting: Filaments of tungsten, glass bulbs derived from pure silica sands.

Street lighting outdoors: Filaments of sodium; lamp-posts of steel and other alloys.

Road surface: Road aggregate derived from a local basaltic intrusion.

Vehicles: Bodywork of light alloys; chassis of steel; brake and clutch linings zinc-based, windows from silica sand.

Rock and mineral collection: An array of specimens from mines all over the world clutters my window-sill and shelf space.

---

[1]For web-sites of current anti-mining campaigns world-wide, access the following URL: http://www.moles.org/

origins of the various materials you see. Above are the results of undertaking this exercise in the author's study.

Notwithstanding the reliance of urbanised society on mining, controversy often attends the pursuit of mineral wealth. The main source of such controversy historically has been the dangers which mining has all-too-often posed to the life and limb of miners. In virtually all mining situations, the health and safety of miners can be at risk from rock falls and the inhalation of dust. In deep mines[2] inadequate ventilation can result in:

• asphyxiation, if the air becomes deficient in oxygen, and/or
• devastating explosions, fuelled by ignition of methane and airborne coal dust.

Sudden inrushes of water are a further hazard particularly relevant to this book; some notable examples are recounted in Chapter Three.

Since the 1960s, the possible environmental impacts of mining have grown increasingly controversial, to the point that environmental objections are often the principal obstacle to the development of new mines or the extension of existing sites (Beynon *et al.*, 2000). The increase in controversy over the environmental impacts of mining is probably due to a combination of factors, including the following:

• The general heightening of environmental awareness in all sectors of society in the 1980s and 1990s, which was arguably the most significant development in socio-political philosophy in the 20th Century.
• The exhaustion of resources in long-established mining districts, leading to the pursuit of new prospects in hitherto pristine regions, including emblematic conservation areas such as the Amazonian rainforest (e.g. Cleary and Thornton, 1994), and remote areas which had been established as inviolable reserves for indigenous peoples.
• The increase in surface mining, which differs from deep mining in being generally more visually obtrusive, and in temporarily precluding other land uses in the worked area. For this reason, it is particularly environmentally contentious (Beynon *et al.*, 2000). Until the development of suitably large excavation equipment in the 1930s (Grimshaw, 1992), nearly all large-scale mining was carried out underground, and had relatively limited impacts on pre-existing surface land uses.
• A number of spectacular incidents of pollution from active and abandoned mines, which have occurred in all continents in recent years (UNEP, 1996; ICOLD, 2001).

Indeed long-term problems of water pollution from abandoned mines have become so frequent and extensive in some parts of the world that some State governments have gone so far as to ban any new mines which are predicted to require water treatment in perpetuity after closure (at the time of writing, this is true in New Mexico and Quebec; Dr D Chambers, Centre for Science in Public Participation, Montana, and G Ferris, Houston, BC, *personal communications*, 2001).

In parallel with the growth in environmentally-based objections to mining, the industry has evolved to establish environmental management systems at the heart of development,

---

[2]A glossary in Appendix I defines all of the mining terminology used in this book

operation and restoration activities (Warhurst, 1999). This has resulted in increasing global uniformity in environmental control strategies in the larger international mining corporations. The picture is generally less rosy in the SME[3] and 'informal' mining sectors (Cleary and Thornton, 1994). Many SME mining companies operate marginally economic mines, some of which are of considerable size, having been formerly worked by the predecessors of the current large international corporations. Under-capitalised and marginalised from the growing environmental management culture of the trans-national companies, SME and informal sector mines often use antiquated mining and processing techniques which have long been shown to be incompatible with environmental protection. Perhaps the most striking example of this phenomenon is the use of mercury by the informal '*garimpeiro*' gold mining sector in the Brasilian Amazon (e.g. Cleary and Thornton, 1994; Meech *et al.*, 1997), and by similar informal and SME sector mines in the Andes of Bolivia, Peru, Ecuador and Colombia (e.g. Grosser *et al.*, 1994), and in the gold fields of Ghana.

In most cases of mine development, it should be possible to achieve a consensus between the developers and other interested parties on the most appropriate means to strike a balance between the socio-economic benefits of mining and the mitigation of unavoidable environmental impacts. This book is in large measure concerned with these means in relation to the water environment. While water is undoubtedly the single greatest pathway and receptor for mining-related contaminants, it is important to appreciate that mining can seriously impact other key compartments of the natural environment. Some of the major possible non-aquatic impacts include:

- Direct disruption of the landscape (and thus pre-existing soils, drainage, and ecology) by surface mining, by land subsidence above deep mine workings, and by surface disposal of waste rock and tailings.
- Air pollution through dispersal of dust, engine exhausts and process flue gases from mines and mineral processing plants.
- Disturbance of surrounding habitats by noise and vibrations from blasting and haulage activities.

The investigation and mitigation of these non-aquatic impacts is beyond the scope of the present volume; the interested reader is advised to consult the recent syntheses of Sengupta (1993), Ripley *et al.* (1995) and Azcue (1999).

1.2 SCOPE OF THIS BOOK

1.2.1 WHAT'S COVERED?

This book concerns the impacts of mining on the water environment. It is intended as a practical book, enabling the reader to:

---

[3]small and *medium*-sized enterprises

- Investigate mine water hydrology and chemistry
- Recognise and understand the causes and impacts of mine water pollution
- Develop appropriate remedial strategies for pollution and related problems arising from mine waters.

The book is based on the state-of-the-art in mine water science and engineering. However, it is deliberately *not* an exhaustive, erudite review of the vast international literature on "acid mine drainage". There are two reasons for this. Firstly, admirable reviews of the causes of acid mine drainage (Evangelou, 1995a) and its biological impacts (Kelly, 1988, 1999) have recently been published. Secondly, compilation of a complete review of all other works concerning acid mine drainage would be a task of Pharaonic proportions, and would likely result in a dense text of little direct appeal to practitioners who must develop solutions to specific mine water problems. Instead of presenting a thorough literature review, therefore, our approach is to direct the reader to understand and apply what we see as the more directly useful elements of the global canons of mine water experience. Clearly, the directions we give are based on our own, necessarily subjective judgement and experience, and the reader must decide for themselves whether the perspective we offer is applicable to their particular problems. However, between us we have experience of mine water investigation and remediation in five continents, and expertise spanning mining geology, chemistry, ecology, hydrology and the three most relevant engineering disciplines (civil, environmental and mining). Our experience encompasses the metals, industrial minerals and coal sectors. There is, if anything a slight bias towards the coal sector in the origins of the examples used in the text, because:

(i)   All three authors currently reside in some of the historically most productive coal-producing regions in the world, and
(ii)   The coal sector arguably has the most coherent and extensive track-record in mine water hydrology and remediation to date, at least in Europe and the eastern USA.

In order to make the book intelligible and useable in "stand-alone" mode, we include substantial quantities of theoretical and documentary information in relation to:

- Aqueous chemistry (Chapter Two), with particular emphasis on the meaning and significance of acidity and alkalinity. A robust understanding of acidity and alkalinity is the key to successful mine water remediation, and yet few texts include sufficient exposition on these topics to give the reader fluency in practice
- Groundwater engineering (Chapter Three), with special reference to the ways in which mining alters the behaviour of natural strata.
- Treatment technologies (Chapters Four and Five). As our focus is on long-term "sustainable" solutions to polluted mine water discharges, we emphasise passive treatment methods (Chapter Five) to a greater extent than conventional active treatment (Chapter Four); however, we include sufficient information on the latter to allow the reader to select and develop outline designs for active treatment systems.

Apart from a few practical guidelines, the book does not provide exhaustive coverage of the nature and assessment of biological impacts of mine water pollution. The reader seeking detailed treatment of this theme is directed to existing literature such as Scullion and Edwards (1980), Kelly (1988, 1999), Jarvis and Younger (1997, 2000) and Azcue (1999).

1.2.2 WHO SHOULD READ THIS BOOK?

In writing the text, we have imagined ourselves to be addressing a mixed, technically oriented readership including:

- environmental scientists and engineers, particularly those involved in making environmental impact assessments and/or regulatory decisions in mining districts, and those involved in the design of remediation schemes
- hydrologists, hydrogeologists, geologists, chemists, civil engineers etc who encounter mine drainage problems in their work
- mining engineers who need to get to grips with environmental water management issues
- scientists and engineers interested in analogous problems of flow and solute transport in karst terrains, in and around deep waste repositories, and in other redox boundary zones etc

1.2.3 BEYOND THIS BOOK

The intensity of the environmental problems encountered at many active and abandoned mine sites around the world ensures that research and development in this field is vigorous. Advances in science and technology in relation to mine water are reported in a wide variety of journals, conference proceedings and web sites. Some of those which we have found especially useful include:

International Journals	**Publisher**
Applied Geochemistry	Pergamon
Mine Water and the Environment	Springer
Ecological Engineering	Elsevier
Journal of Contaminant Hydrology	Elsevier
Environmental Geology	Springer
Engineering and Environmental Geoscience	Geological Society of America
Science of the Total Environment	Elsevier
Quarterly Journal of Engineering Geology and Hydrogeology	Geological Society of London

Conference Proceedings
International Mine Water Association
International Conference on Acid Rock Drainage
International Water Association
International Conference on Tailings and Mine Waste (held every January since 1994 in Fort Collins, Colorado, with Proceedings being published by Balkema).

Web Resources
Up-to-date links to web sites containing information on mine water pollution are maintained by the authors at the following URL: *http://www.minewater.net.* A brief listing of some of the more useful web-sites is given below:

Organisation	Web address
The European Commmission's PIRAMID project (Passive *in-situ* remediaton of acidic mine drainage).	*www.piramid.org*
Enviromine (international information network on mining and the environment)	*www.infomine.com/technology/enviromine*
Mine Environment Neutral Drainage (MEND) – Canadian mine drainage research network)	*http://mend2000.nrcan.gc.ca*
International Network on Acid Prevention	*http://www.inap.com.au*
International Council on Metals and the Environment (ICME)	*www.icme.com*
The UK Coal Authority's mine water remediation programme	*www.coal.gov.uk/environment*
The US EPA's mining home page	*http://cfpub1.epa.gov/npdes/home.cfm?program_id=8*
United Nations Environment Programme (UNEP) mineral resources forum	*www.natural-resources.org*
Australian Centre for Mining Environmental Research (ACMER)	*www.acmer.co.au*
National Mine Land Reclamation Center (USA)	*http://www.nrcce.wvu.edu/nmlrc*
Mining Environmental Management Magazine	*www.miningenvironmental.com*
Sweden's Mitigation of Mine Waste Impacts (MiMi) R&D programme	*http://www.mimi.kiruna.se*
Hedin Environmental Inc.	*www.hedinenv.com*

1.3 OVERVIEW OF MINING AND MINE WASTES

1.3.1 PURPOSE OF OVERVIEW

This section is intended as a brief overview of mining and related practices, insofar as a knowledge of these will assist non-specialists in:

- understanding the nature of mined systems (for the purposes of conceptualising them hydrologically and geochemically), and
- obtaining a basic mining vocabulary to expedite discussions with mining professionals. A full glossary of mining and related specialist terminology used in this book is given in Appendix I.

It should be noted that this Chapter makes no pretence to be a manual of mining engineering. Readers seeking formal, quantitative advice on excavation, ventilation and routine drainage operations in active mines should consult standard texts such as Hartman (1987), Brady and Brown (1993) and Hoek *et al.* (2000). Surveying practices in the coal mining industry are described in great detail by NCB (1977), which makes a very useful starting point for the interpretation of mine plans. Valuable information on historic mining practices can be found in old engineering manuals, such as those of Pamely (1904) and Mason (1951), and in the thorough review of ancient mining by Shepherd (1993).

In the sections which follow we consider how deep mining and surface mining alter natural strata, and how the storage/disposal of mine wastes (a common activity to both deep and surface mining, albeit on different scales) creates sediments with hydraulic characteristics distinct from those of the parent rocks.

1.3.2 DEEP MINING

"Deep mining" is simply mining underground, and hence a deep mine is any mine in which the miner and/or his machinery work beneath a cover of soil or rock. Of the various activities involved in deep mining, two have particularly profound implications for the movement of mine water during and after mining:

1. The development and maintenance of underground access (for humans, machinery, ancillary services and ventilation).
2. The extraction of run-of-mine ore or coal, and associated underground handling of waste rock.

Variations in the way these two activities are undertaken result in a complex range of patterns of underground voids, with implications for the short- and long-term hydrological behaviour of mined systems.

Access to deep mines is achieved by tunnels varying in inclination from horizontal to vertical (Figure 1.1). Although examples of every conceivable angle could be cited, the most common forms of deep mine access are:

- shaft – a vertical or sub-vertical (>70°) tunnel connecting the workings to the ground surface (Figures 1.2 and 1.3).
- adit – an essentially horizontal tunnel from the ground surface to adjoining workings (Figure 1.4), particularly common in hilly areas.
- decline – an inclined tunnel from ground surface to the workings (Figure 1.5).

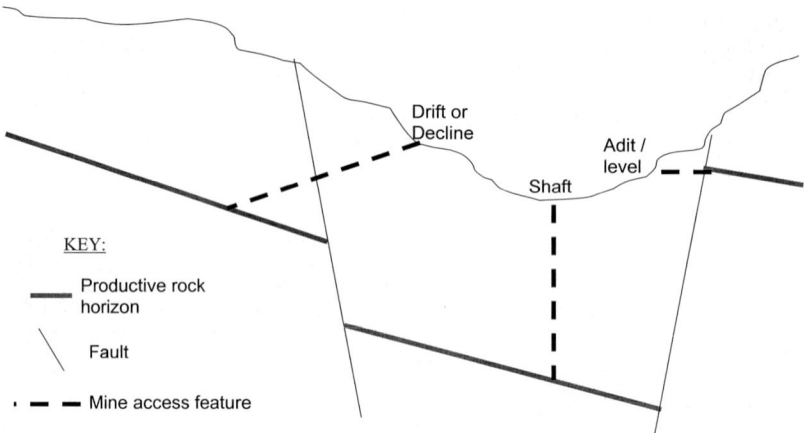

Figure 1.1: A summary of the principal means of access to underground workings (after Younger and Adams, 1999).

Shafts are constructed for a variety of purposes, and thus in a corresponding variety of plan forms and diameters. The type of shaft which most commonly springs to mind is the archetypal "main shaft" (Figure 1.2), which is typically used for a variety of purposes, including:

- "man-riding" (i.e. access for personnel)
- transport of materials, such as timber and steel roof supports, explosives etc
- drawing of run-of-mine product to bank
- access for electrical cables and the pipe ranges associated with dewatering operations
- ventilation

Such a multi-purpose shaft will typically be of large diameter (commonly in the range 5 m to 8 m) and may be circular or rectangular in plan form (Figure 1.2). In most countries, it is a legal requirement for every deep mine to have at least two means of access/egress to the deep workings. This is to ensure a secondary escape route in case of mechanical problems or blockage at the principal access route, and/or roof-falls etc within the workings. The maintenance of a reliable ventilation system is also greatly facilitated by the presence of two deep mine access routes; otherwise the downcast and upcast ventilation streams must be carried in the same shaft/adit, separated by partitions (brattices) running the full height of the shaft. It is therefore common to find two major shafts in close proximity (often within 50 m of each other) at modern mine sites (Figure 1.3).

Because the long-term viability of the mine will depend on the unhindered use of the main shafts, they are generally engineered to an extremely high-quality specification, and their immediate subsurface surroundings (for a radius of 100 m or more) will often be left unworked (save for penetration by main access roadways), so that collapse of old workings cannot threaten the structural integrity of the shaft. The cylindrical body of unworked

Figure 1.2: A view down one of the main shafts of an underground coal mine in Durham (UK), showing the spacious dimensions and highly engineered lining, designed for permanence. The shaftsmen are riding in a 'kibble', a large bucket used for maintenance and inspection work. (Photo: A. Doyle).

Figure 1.3: A typical pit-head scene, in this case at Boldon Colliery, Durham (UK), in 1950, showing two closely adjoining shafts, marked by the two headframes. The twin shafts provide between them a complete ventilation circuit for the mine and two independent means of egress from the underground workings. (Photo: National Coal Board, salvaged by A. Doyle).

Figure 1.4: The portal of a typical, long-established drainage adit: the Blackett Level, Allendale Town, Northumberland. At the time this photograph was taken (1999) this adit had been successfully dewatering the nearby Pb-Zn mines for more than 125 years. (Photo: A. Doyle).

Figure 1.5: The portal of a 1-in-4 decline (Frazer's Hush Dib, North Pennine Orefield, UK). This decline produced fluorite, brought to the surface ('bank') by means of tubs on rails hauled by cable to a standing engine at bank. (Photo: A. Doyle).

strata surrounding a major shaft is termed a "shaft pillar". With high specification liners and intact shaft pillars, major shafts can be confidently expected to remain open long after closure, unless deliberately capped and/or back-filled.

Besides such major shafts, several types of smaller, single-purpose shaft are frequently encountered, especially in older workings. For instance, counter-weight shafts are small diameter (2 m to 4 m), shallow (typically < 50 m) shafts constructed to allow the descent of a weight on the end of a cable, as an essential part of the operations of certain 18th and 19th Century steam-driven winding mechanisms. More common small shafts are the air shafts which are usually found at regular intervals along the line of most old adits. As their name suggests, air shafts are constructed to allow natural ventilation of the adit. Because they were not regularly used for man access or materials haulage, these air shafts are typically as narrow in diameter as was economically feasible, given contemporary sinking technology. For instance, diameters of as little as 1.5 m to 2 m are common amongst the 18th and 19th Century hand-dug air-shafts in Europe. In more recent times, air shafts have often been replaced altogether by large-diameter boreholes (250–750 mm). Some of the older air shafts were deliberately constructed such that surface water could fall into them (Sopwith, 1833), with the purpose of improving ventilation by physical entrainment of air and by thermal convection promoted by the contrast in temperature between the air and the water. Shafts used in this way were commonly termed "waterblast shafts".

Interactions between air and falling water were perfected in certain lead mine shafts in England, in which water was poured down the shafts to gain sufficient kinetic energy to work a hydraulic compressor at the foot of the shaft, from which compressed air was transmitted in pipelines to the working faces where it powered mining machinery (Smith, 1923, pp 88–89).

In the early history of mining in most parts of the world, the first shafts are shallow, informally constructed "bell-pits", so named because of their shape in vertical profile (Figure 1.6). These are instinctively constructed under primitive mining conditions: when a valuable deposit is followed in from outcrop in a simple adit, it is soon discovered that there are limits to the span width of an unsupported roof before collapse occurs. Extending down-dip from the first line of collapses, a shallow shaft will be sunk and the mineral worked radially from it until the limit of roof stability is reached once more. In cases where the roof is self-supporting over large spans, the maximum radius of working around each bell-pit shaft may be dictated by the maximum distance to which a shovel of ore may be thrown by the miners. In this manner, clusters of neighbouring bell-pits may be excavated over a number of years in areas of shallow economic reserves. For instance in one complex of Neolithic flint mines in eastern England, more than 360 bell-pits have been discovered (Holgate, 1991). Similar complexes of mediaeval bell-pits are widespread near seam outcrops in the coalfields of Europe, albeit they are almost always unmapped, and frequently only discovered during opencasting or other excavation projects. In the New World, bell-pits are less common, though they are to be expected wherever mining was initiated by the informal sector (as in the "Gold Rush" zones of the western USA and Brasil). To this day, bell-pits are under construction in remote areas of active informal-sector mining, such as the upper tributaries of the Amazon in Perú and Ecuador, and in Diwalwal, Mindanao (Philippines; Vidal, 1998).

Not all shafts reach daylight. A significant number of shafts are constructed entirely underground, to provide access to the more remote zones of a dipping ore-body. An

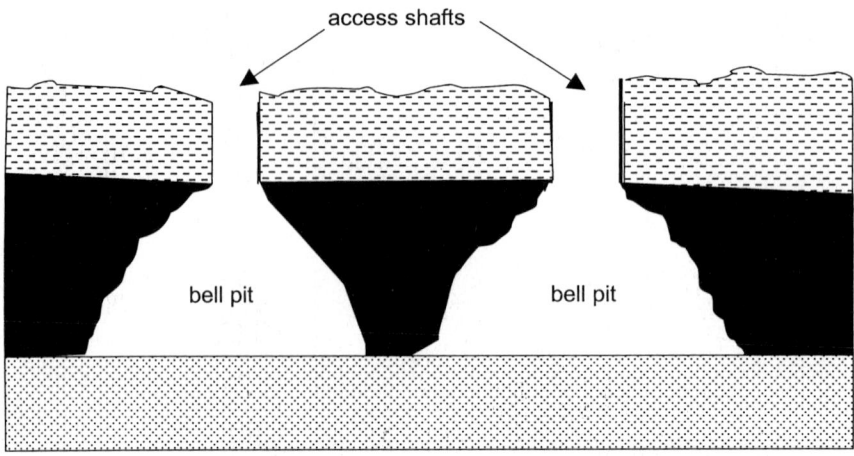

Figure 1.6: Schematic cross-section of typical bell pits. These can be of various diameters depending on local ground conditions, but are rarely more than 8 m wide at the widest point.

underground shaft is usually termed a "staple shaft"[4] (Figure 1.7). They often represent important hydraulic connections between otherwise separate worked horizons in a given mine.

Shafts have four principal drawbacks as mine access features:

(i)   the high cost of sinking, which is often more than half of the total capital cost of establishing a deep mine.

(ii)   in stratiform deposits, little or no payable rock will be extracted during shaft sinking.

(iii)   the full weight of the mined rock must be taken by the winding gear cables. This necessitates the use of very high-specification winch plant, which is expensive to purchase and maintain. The limits of winding technology serve as the principal limitation on the depth of shafts. At the present level of technology, the deepest single shaft in the world is limited to 2700 m (South Shaft, Western Deep Levels Gold Mine, far west Rand, South Africa).

(iv)   water in workings accessed only by shafts must always be pumped out of the mine, for it cannot flow out by gravity.

Where their installation is feasible, *adits* have the potential to mitigate all of these drawbacks: in stratiform deposits, they can often be driven in the productive horizon, so that revenue generated during construction can defray the cost of excavation. Moreover, in horizontal or gently-inclined adits, much of the weight of the mined rock can be borne by the strata (via rails in most cases), making the haulage plant concomitantly less expensive

---

[4]"Staple" is pronounced to rhyme with "apple"

Figure 1.7: A view up a typical staple shaft in an underground coal mine in the Great Northern Coalfield (UK). The landing of the next highest mine level can be clearly seen. The timber work partitions the shaft for different uses (occasional haulage in the central sector, and pipe ranges and power lines in the sector behind the timber framework on the right). (Photo: A. Doyle).

to install and maintain. Finally, for waters entering the workings above adit-level, simple gravity drainage can be used for dewatering. The revenue savings associated with replacing pumping with gravity drainage can be so great that it makes economic sense to drive drainage adits through many kilometres of country rock to under-drain workings hitherto accessed only by shafts. For instance, the adit shown in Figure 1.4 was driven for more than 7 km in the mid-19th Century to dewater a group of highly productive lead mines. This adit, the Blackett Level, drains around 0.2 $m^3.s^{-1}$ from these mines, and has per-manently lowered the water table beneath them by as much as 180 m. The cost saving which this represents was so substantial that the mines served by the Blackett Level remained in production long after falling lead prices had forced the closure of most other mines in the region (Younger, 1998a).

In all but the most mountainous of orefields, mining will eventually extend below the lowest feasible level for gravity-drainage by means of adits. In such cases, existing adits are often maintained as routes for waters pumped from below, as they at least offer the potential to save the cost of pumping against the greater head needed to raise water all the way to the shaft collar.

In terms of geometry, adits are relatively simple, being long tunnels of fairly constant proportions. The excavated diameters of adits have tended to become larger over the centuries. For instance, Sopwith (1833) describes the typical dimensions of early 19th Century adits as "3 feet [0.91 m] wide at the bottom, gradually widening to middle height, where it is $3\frac{1}{2}$ or 4 feet [1.07 or 1.22 m] wide, and from thence it has an arched form to the top, which is from 6 to 7 feet [1.82 to 2.13 m] high". By these standards, the Blackett Level (Figure 1.4) is of fairly generous proportions for its era. Haulage in such old adits was typically by pony, with small hardy Galloways being the favoured breed. Nevertheless, where adits were driven in thick beds of payable rock, adits were correspondingly more spacious. For instance, in the iron ore mines of Cleveland, UK, the Main Seam was up to 5 m thick, and the roofs of adits driven in this horizon were so high that haulage was accomplished using shire horses. Modern adits in most settings are comparable to the roomy adits of Cleveland, with both the median width and maximum roof clearance often being around 3 m. This provides sufficient room for haulage of tubs by means of loco-motives, conveyor belts or by rope haulage to a standing engine adjacent to the portal.

In certain circumstances, particularly where stratiform ore-bodies have a significant structural dip, the most economic mine access may be a decline (Figures 1.1 and 1.5). Where the angle of dip of the decline can be less than about 6°, it should be possible to work the mine using free-travelling vehicles (usually diesel powered). Conveyor belts can handle steeper gradients, but cable-hauled rail tubs are likely to be the best option where the dip steepens beyond about 10°. Beyond these considerations, declines hardly differ in form from ordinary circum-horizontal adits.

Having obtained subsurface access via a shaft, adit or decline, the next major access feature of any deep mine is the "level" or major roadway (Figures 1.8 and 1.9). These are simply underground tunnels, radiating away from the shaft or main adit. As shown in Figure 1.8, the largest levels may have sufficient space for two pairs of locomotive rails. This means that they can easily have diameters in excess of 8 m. These larger diameter levels are also relatively uncluttered with fittings, and are engineered for permanence. As the major levels are traced in-bye (i.e. towards the working areas), more modest district roadways ("main gates") are entered. In these district roadways, the diameter of the arch girders will usually decrease once more, to 4 m or less. Furthermore, with a shorter access life, it is possible for the roof and wall supports to be less permanent in nature (Figure 1.9) and therefore less expensive. Conveyor belts and other equipment associated with zones of active extraction can begin to use up much of the cross-sectional area of a district roadway (Figure 1.9).

Having traced the major access routes of a deep mine towards the working zone, we must now consider the various extraction techniques. There are two principal classes of extraction technique: Where the productive zone is essentially *stratiform* (i.e. disposed in layers), and the dip is relatively low (less than, say, 15°) then some of the classic "*horizontal extraction methods*" will be applicable. If a stratiform productive zone is steeply-dipping (as in classic hydrothermal vein/lode deposits), or if an ore-body is of irregular shape though with significant vertical extension, then extraction most commonly proceeds by means of the various "*stoping*" techniques.

Horizontal extraction methods are of two principal types:

1. *Bord-and-pillar* workings (also commonly called "room-and-pillar" or "pillar-and-stall")
2. *Longwall*

Figure 1.10 shows extracts from mine plans typical of these two types of workings. The notes which follow will explain how the clear differences between the two plan extracts arise.

Bord-and-pillar workings, also known as room-and-pillar and pillar-and-stall, were the principal means of extraction of European stratiform deposits for many centuries, and remain the principal technique of deep mining for coal in the USA (Hartman, 1987). In this method (Figures 1.10a and 1.11), extraction roadways (termed "bords", or "rooms") are excavated on an approximately rectilinear plan, leaving substantial "pillars" of unworked ore/coal to support the roof. Typical dimensions of modern bord-and-pillar operations in coal are (Hartman, 1987):

Bord (room) widths: 6 m to 9 m (larger openings require auxiliary roof support)
Pillar widths: 9 m–30 m.

These dimensions have been derived from centuries of experience in Carboniferous coal-bearing strata world-wide. Adjustments to these recommendations will be necessary in practice, depending on roof strata competence etc. In the halite mines of Cheshire, UK, for instance, bords and pillars are usually both 30 m–35 m wide, and 8 m high. The precise dimensions of structurally stable bords and pillars in a given geological setting are functions of the mechanical state of the strata. Bord-and-pillar is one of the prime examples of a "supported" method of extraction, in that the pillars support the roof by distributing loads such that strains in the roof strata remain elastic in nature, i.e. the strata are prone to sag and bend rather than fracture. Where the spans of roof beams exceed a critical threshold (which varies with rock type and stress history), the mode of deformation will switch from elastic bending to brittle fracture. Figure 1.12 illustrates the consequences of inappropriate sizing of bords and pillars, with fracturing at the edges of each pillar leading to complete plugging of bords with sandstone from above. In the example shown in Figure

Figure 1.8: Example of a major deep mine "level", in this case a junction of two main locomotive roadways in a deep coal mine (Westoe, north-east England). This locality (known as "Cathedral Junction") is the point at which the access from the foot of the main shaft diverges towards the two principal working districts in the mine. The roof support is engineered for permanence: steel arch girders with intervening steel cross-plates. The large pipe entering the right-hand branch of the level at head-height carries compressed air for ventilation. Most of the other continuous "pipes" in this view are electricity cables, with the exception of two pipe ranges conveying pumped mine water: the lowermost narrow pipe, hanging on chains on the left-hand wall, serves a local pump, whereas the large pipe at floor level on the right (partly obscured by spare materials) is a major pipe-range from the working districts. (Photo: A. Doyle).

1.12, the collapse was apparently so rapid and instantaneous that it did not even disturb the stratification of the subsiding sandstone.

Adequately designed and mined bord-and-pillar workings can be expected to achieve extraction rates on the order of 48% in Carboniferous coal, though rates as high as 60% are claimed for fully-mechanised bord-and-pillar workings in the USA (Hartman, 1987). To attain rates in excess of 60%, it is necessary to extract the pillars, which for obvious reasons is undertaken during the retreat from a district of workings, heading back towards the mine entrance. Pillar extraction can increase coal yields to as much as 90% of reserves. However, it inevitably induces brittle fracture in the overlying strata, which collapse into

Figure 1.9: District roadway ("main gate"). Closer to the working face than in Figure 1.8, this is a major route by which coal is brought out from the working face to the locomotive haulage roadways, by means of the conveyor belt in the left-hand side of the view. (Photo: A. Doyle).

the void in a manner analogous to that shown in Figure 1.12, forming a mass of brecciated roof material which is generally termed "goaf" or "gob"[5] .

One of the key advantages of bord-and-pillar is that it can operate in strata dipping as steeply as 30° (Brady and Brown, 1993); at such dips most longwall methods fail to be viable. For this reason, bord-and-pillar retains a specialist niche even in mining regions where longwall is now the norm. Another advantage of bord-and-pillar is economic: capital investment is relatively modest compared to longwall. This is precisely why the technique remains so popular in the USA, where most mining is undertaken by relatively small, private sector companies with limited investment capital. In Europe, where state-owned mining companies have been common until recent decades, the initial investment associated with longwall (which is, over the long-term, cheaper and more efficient than bord-and-pillar) has not been such a disincentive.

Longwall workings of various types have been recorded since the early 1800s. However, widespread uptake of fully mechanised longwall extraction was essentially a mid-20th

---

[5]Both of these terms are derived from the Welsh word '*ogof*', meaning cave

(a) Bord-and-pillar workings

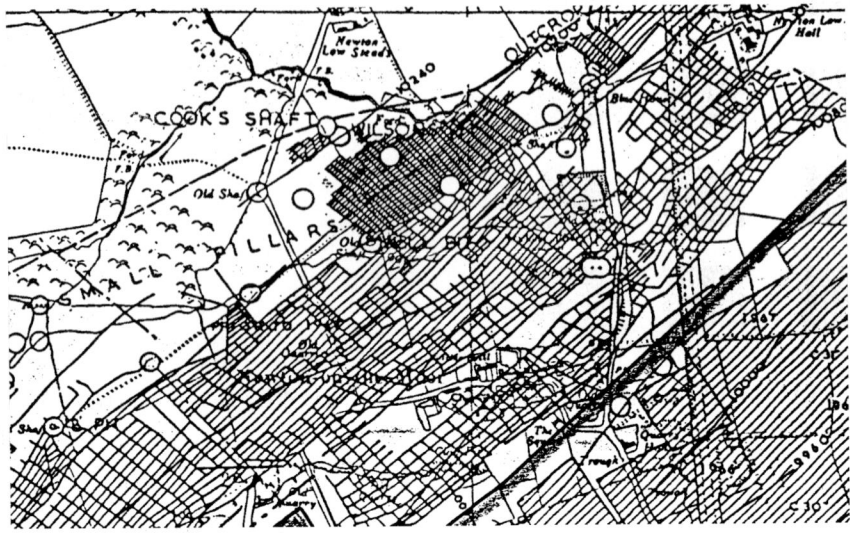

(b) Longwall workings

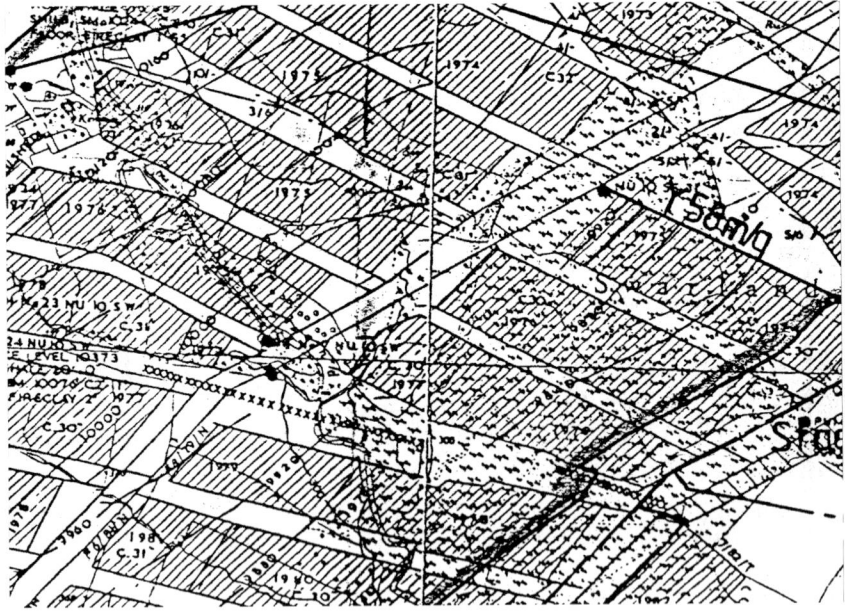

Figure 1.10: Mine plan extracts (a) bord-and-pillar workings (rectilinear network of roadways) (b) modern longwall panels (large rectangular shaded zones). (Whittle Colliery, Northumberland, UK). (Source: M. Foster).

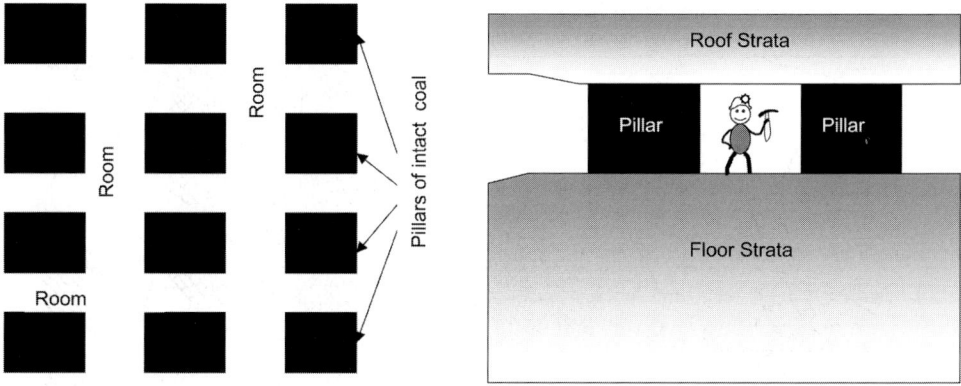

Figure 1.11: Sketch showing the general layout of bord-and-pillar workings (after Younger and Adams, 1999).

Figure 1.12: Two old mine roadways dating to the late 19th Century, which have subsequently closed entirely by plug failure of the overlying sandstone. (High Main Seam, Morrison North Pit, Co Durham (UK)). These were exposed during working of the Chapman's Well Opencast Mine in the early 1990s. (Photo: A. Witcomb).

Century development. It is now the most common deep mining technique for coal in Europe and South Africa (accounting for more than 80% of total production) but has only begun to penetrate the US industry since the 1960s, where it still accounts for little more than 15% of deep-mined coal. Modern longwall workings offer the possibility of extraction rates of 80% or more in a single phase of mining. Longwall working is undertaken in rectangular sub-divisions ("panels") of the coal seam. The layout of longwall panels in a typical deep coal mine operation is shown in Figures 1.10b and 1.13. Each longwall panel will be up to 250 m wide, and as much as 1 km in length. Where ground conditions are difficult, it may be necessary to reduce these panel dimensions, such that mines can effectively end up with a *"shortwall"* mining method, with panels as little as 40 m wide by 200 m long. (Further consideration of possible hydrological motives for a switch from longwall to shortwall are discussed in Section 3.6.4).

Within each longwall panel, a drum shearer passes back and forth along the working face, trimming the coal as it goes (Figure 1.14). If the shearer track is initially operative in close proximity to the main haulage roadway in the mine, so that the face moves ever further from the main roadway at each pass of the shearer, then the face will be termed an "advance face". The contrary configuration is shown in Figure 1.13a, in which the main gate and tail gate roadways on either edge of the panel were driven out to the full extent of the planned working before shearing commenced, so that the face retreats towards the main haulage roadway as shearing continues. In this configuration, the face is termed a "retreat face". Under ideal conditions, retreat faces are preferable to advance faces, for a number of reasons including the following:

• in a retreat face, the driving of the main gate and tail gate will have revealed at an early stage any faults or wash-outs that might affect the economic viability of the panel.
• advance faces have been known to hole unexpectedly into unrecorded old workings filled with water, causing lethal inrushes (e.g. Calder, 1973).

As the coal is sheared away and the face retreats, the shearer track is automatically moved to keep the shearer in contact with the seam. Simultaneously, the powered face supports move to ensure that the shearer is shielded from falling rocks. In the area behind the face supports, from which the coal was recently sheared, the roof is left unsupported. Once the span of unsupported roof strata exceeds the limits of elastic deformation, brittle fracturing sets in and the roof strata collapse into the void from which the coal has been extracted, forming goaf (Figure 1.13a). For this reason longwall is classified as a "caving" method of mining, which invariably induces brittle deformation of the overlying strata.

Stoping
Stoping is the name given to mining in the vertical plane, which produces vertically extensive voids known as "stopes". At its simplest, stoping may be little more than the application of bord-and-pillar techniques to steeply inclined, non-coal strata (in which case it is often termed "stope-and-pillar"; Hartman, 1987). Given the variations in ore grades of hydrothermal vein deposits etc, there is usually a strong incentive to make the pillars coincide with patches of low-grade ore. Hence pillars in stoping operations are often smaller, and of less regular plan-form than pillars in coal workings. This means that single pass mining (for pillar removal is rare in stope-and-pillar operations) can yield 60% to 80% recoveries.

(a) An individual longwall panel (retreat working mode)

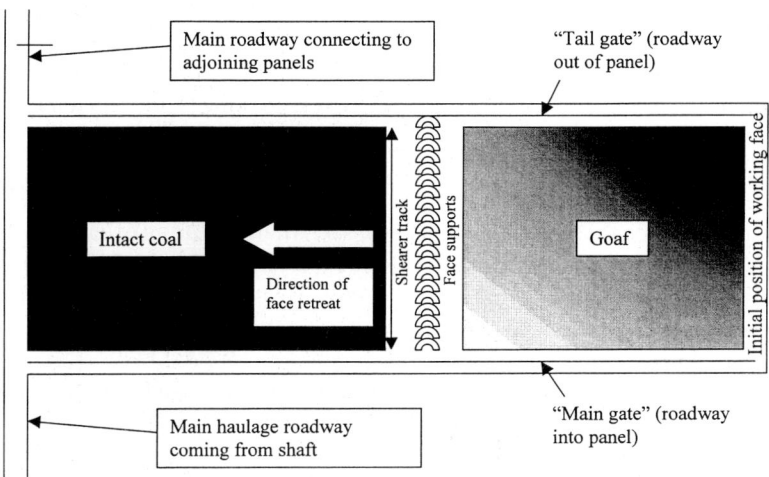

(b) Layout of a series of panels in relation to main haulage and shaft access.

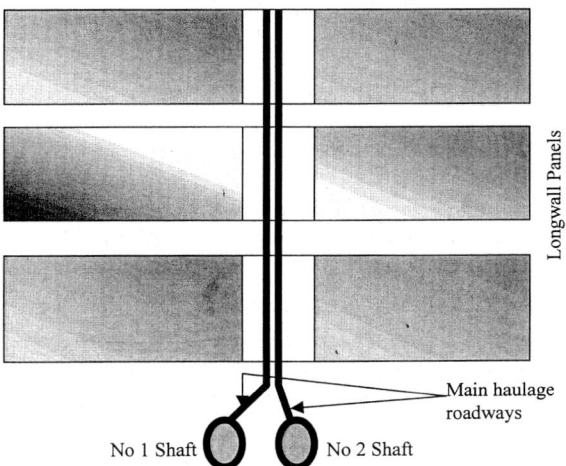

Figure 1.13: Sketch showing the general layout of longwall workings (after Younger and Adams, 1999).

Most stoping is carried out by working upwards from a major access roadway. Stoping upwards in this manner is termed "overhand stoping". Overhand stoping techniques have the great benefit that gravity can be used to transport ore from the working face to collection points below. The alternative, "underhand stoping", in which the extraction proceeds downwards from a major access roadway is less attractive, as ore must be raised in defiance of gravity in cramped conditions. Consequently, little underhand stoping is undertaken these days.

Figure 1.14: Close-up view of a longwall coal face, showing the drum shearer (circular feature on the right) which is being adjusted by the miner working within hydraulic supports. In the foreground is the Deputy (supervisor) with a flame lamp at the ready to check for gas, for the working face is generally the most prolific source of methane in a working coal mines. (Photo: A. Doyle).

Modern overhand stoping methods (Figure 1.16) involve blasting the current roof of the stope, so that ore falls through the void below (formed by previous extraction) to accumulate at the stope foot. The foot of the stope is accessed from a major haulage roadway by means of draw-points (short roadways driven into the vein), via which the broken ore can be "mucked out", i.e. excavated and loaded onto the mine haulage system, which will take it to the surface. Specific approaches to overhand stoping are classified on the basis of how the blasting work is undertaken. In "shrinkage stoping", the process is as follows:

(a)   The roof of the stope is blasted and the shattered rock falls into the stope.
(b)   Enough of the shattered rock is mucked out at the draw-points so that the pile of shattered rock "shrinks" sufficiently for miners to gain access to the roof of the stope by walking on top of the rock pile.
(c)   The miners on top of the rock pile set charges for the next blast, then withdraw without firing the charges.
(d)   More rock is mucked out of the stope to make room for that which will fall when the shots are fired.
(e)   return to step (a) above.

Shrinkage stoping is well suited to situations in which the ore grade varies rapidly over short distances, as the miners are able to adapt the blasting patterns to leave pillars where grades are low. Where grades are more uniform, a more mechanised approach to overhand stoping may be pursued, which is typically some variant of "sub-level open stoping". Sub-level stoping is undertaken as follows (Figure 1.15):

(a)   The main access roadway communicates with the vein not only via the draw-points, but also via small shafts known as "manway raises", from which sub-levels (i.e. road-ways intermediate in elevation between the main access roadways) are driven off at intervals into the vein.
(b)   Miners access the vein via the sub-levels and drill blast holes in a radial pattern around the sub-level. These holes are charged, and then fired at the end of each shift.
(c)   Broken ore which falls under gravity into the stope is mucked out via the draw-points.
(d)   Drilling and firing of blast holes is then repeated all the way back along the sub-level to the manway raise over the succeeding days, leaving an empty stope behind.

Such sub-level open stoping has the advantage over shrinkage stoping that mucking out occurs in one cycle per shift, which lessens confusion and increases production efficiency.
   Further, less common, stoping techniques include (Brady and Brown, 1993):

•   Cut-and-fill stoping, in which waste rock or tailings are back-filled into the empty stopes, thereby providing support for the stope walls and minimising the problems inherent in surface disposal of tailings and waste rock (see Section 1.3.5). This is particularly appropriate where the ore is of high economic value and the country rock is relatively weak, and thus prone to collapse if left unsupported.
•   Vertical crater retreat (VCR) stoping, also known as "longhole open stoping", in which the blast holes are sufficiently long that they can reach from one main access level to another without the need for miners to enter the evolving stope.

Temporal persistence of deep mine voids
Where the strata are competent and the mine has been well-designed and excavated, major mine roadways, and indeed bord-and-pillar workings, can remain open virtually indefinitely. Where spans are suitably short, the rate of elastic deformation of roof strata can be so slow that convergence of the roof and floor, or of the walls, occurs only over geological time. It is for this reason that mine workings as old as 4500 years remain open today (Holgate, 1991; Shepherd, 1993). In cases where void roof spans are only marginally in excess of those necessary to ensure long-term persistence of quasi-stable elastic strain, brittle failure such as that shown in Figure 1.12 can be expected to occur eventually (Taylor *et al.*, 2000). When void roofs finally collapse, then where the overburden is less than 30 m thick, or less than ten times the thickness of the worked seam, void migration can result in land subsidence (Waltham, 1994), typically in the form of a circular "crown hole".
   In certain cases, it is conceivable that roof spans which were originally stable can become unstable due to mass wasting of the pillars. This can occur, for instance, where flooding of coal workings leads to slaking of mudstones which immediately underlie the seam, prompting slippage and collapse of the pillars (e.g. Smith and Colls, 1996).

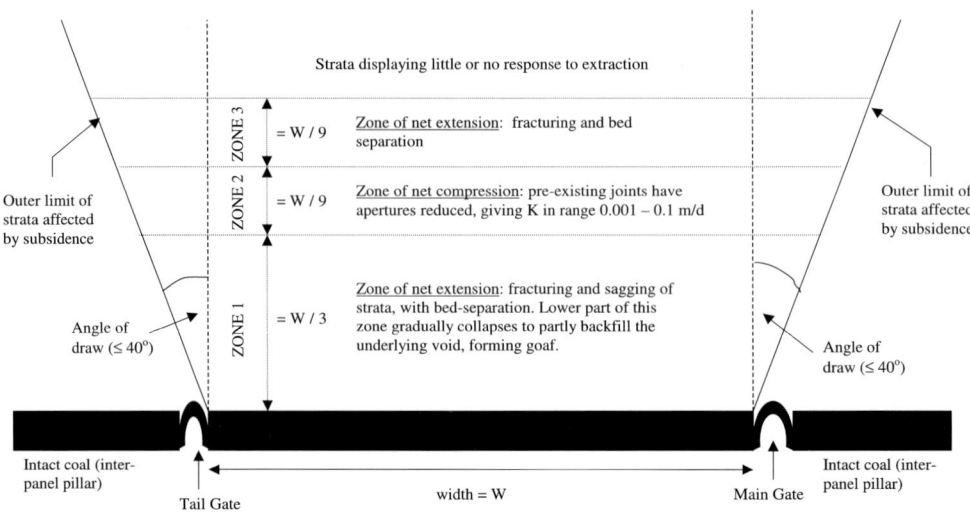

Figure 1.15: Schematic cross-section showing the deduced zones of strata deformation above a longwall panel. (After NCB, 1975, and Younger and Adams, 1999).

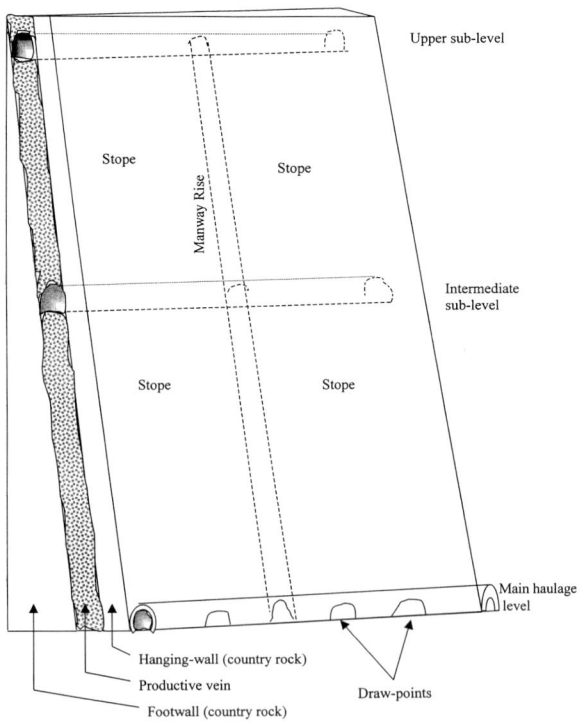

Figure 1.16: Diagrammatic cross-section of overhand stoping method for working sub-vertical mineral veins (after Younger and Adams, 1999).

Mines in evaporitic strata might also be prone to collapse if pillars are winnowed by dissolution in mine waters during flooding of the mine (J. Lamont-Black, University of Newcastle, *personal communication*, 1999). Alternatively, where roof strata are thinly bedded and prone to spalling, the void may migrate upwards even though the pillars remain intact.

Mine roadways can also close up due to floor heave. This is particularly common where plastic clays underlie a competent worked horizon. The transmission of loads from the roof beams through the pillars can result in elastic deformation of the floor strata, which then bulge up into the roadway and fill the void. Bords in which floor heave has occurred were historically termed "crept bords".

As stopes are generally excavated in very competent country rock, they can remain open long after the cessation of mining, at least at moderate depths where confining pressures are modest. Published data on the rock mechanics of abandoned stopes are scarce (Brady and Brown, 1993). However, *a priori* reasoning suggests that the degree and rate of stope closure will increase with the depth of working. Data from some of the world's deepest mines corroborate this notion. For instance, Figure 1.17 shows the results of some direct strain measurements at different depths within gold mines of the Kolar Goldfield, southern India, which worked near-vertical sheet orebodies in a hard rock (metamorphic) terrain. Stopes have been excavated to total depths slightly in excess of 3000 m. In a relatively shallow setting, at 560 m depth in the Nundydroog Mine, a total of 7 mm of stope closure was observed over a 33 month period (Figure 1.17a). This was accounted for by extensional deformation within both the footwall and the hanging-wall, most of which took place within 6 m of the stope walls. At much greater depths, the degree and rate of stope closure are correspondingly greater. At 2400 m depth in the Champion Reef Mine, monitoring over a period of 26 months following stope excavation revealed a total wall convergence on the order of 80 mm (Figure 1.17b). However, as shown in Figure 1.17b, the rate of closure declined exponentially over time. Assuming no fundamental change in the deformation process over time, then judging from the later observed rates of convergence, complete closure of this 8 m-wide sub-vertical stope could take more than 1000 years. This surprisingly slow convergence rate is explicable only by the sub-vertical disposition of the stopes. By contrast, low-angle stopes (45° from the horizontal) at depths of up to 3700 m in the world's deepest mine (Western Deep Levels gold mine, South Africa) are not expected to remain open for more than five years following abandonment.

Effects of deep mining on surrounding strata
Movements in rocks surrounding mine voids have been studied for many years, mainly in relation to problems of land surface subsidence induced by longwall mining of flat-lying coal seams (e.g. NCB, 1975; Aston and Whittaker, 1985; Whittaker and Reddish, 1989; Hartman, 1987). Relatively few studies have considered the issue of greatest interest in the context of mine water management, namely the degree to which the deformation of strata around underground workings affects their hydraulic properties and hydrogeological behaviour. Notable exceptions include the works of Aston and Whittaker (1985), Booth and Spande (1992), Liu *et al.* (1997) and Zipper *et al.* (1997). Full consideration of their findings is reserved to Section 3.5.3; in the paragraphs which follow a conceptual overview of some of the processes which occur around mined voids is presented.

a

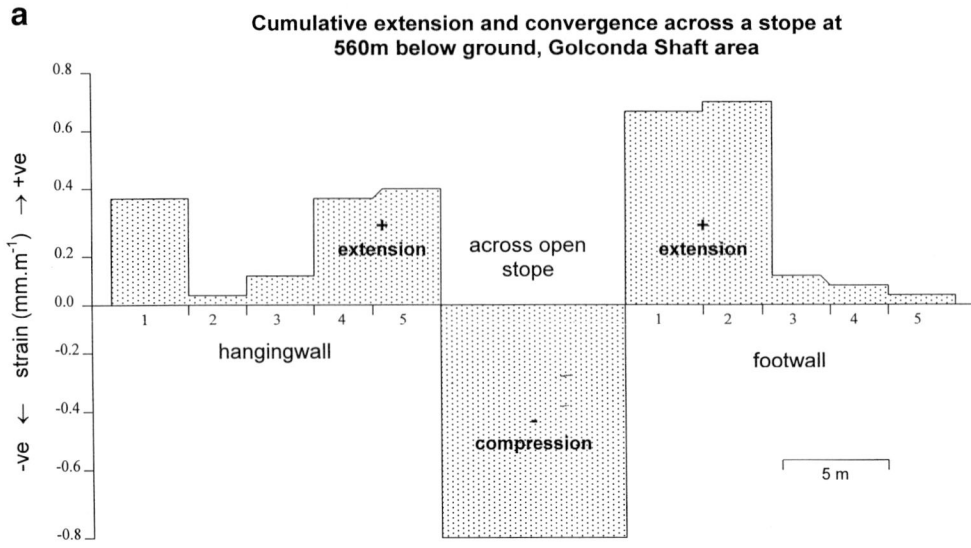

Cumulative extension and convergence across a stope at 560m below ground, Golconda Shaft area

b

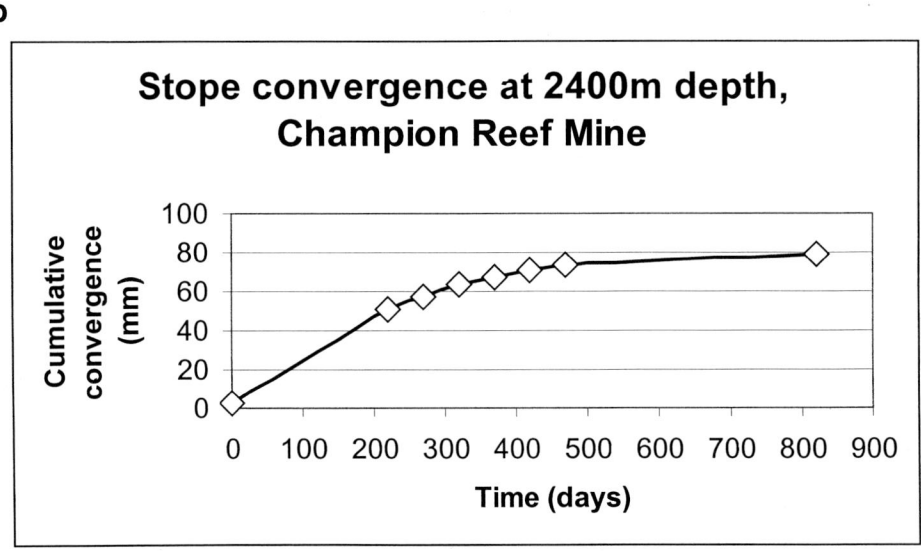

Figure 1.17: Examples of gradual closure of stopes at two different depths in mines of the Kolar Gold Field, Mysore, southern India (adapted from unpublished data collected by Whittaker, 1968): (a) Total cumulative strain across an open stope over a 33 month monitoring period, (Golconda Shaft area, 560 m depth). All measurements were made using borehole strain gauges in roadways. The hangingwall and footwall measurements were made in cross-cut roadways driven at right-angles to the strike of the gold-bearing reef. The measurements made 'across open stope' were made in a roadway developed in (and following the strike of) the stoped gold-bearing reef. (b) Cumulative closure curve for an open stope at 2400 m depth, made with a tubular total strain gauge installed across the stope.

In all but the most tectonically active areas, the subsurface stress regime can reasonably be assumed to be in quasi-steady state prior to the commencement of deep mining, at least in relation to the time-scales of human interest. The opening of voids in previously intact strata perturbs the subsurface stress regime, and induces strains in the rocks surrounding the void (Brady and Brown, 1993). The rate of deformation in these rocks tends to be most rapid in the first few months after extraction, and then tails off exponentially over the following months, so that only subtle adjustments of the strata continue in the long term. In practical terms, this is the basis for the generalisation that subsidence attributable to longwall mining generally makes its presence felt within two years of extraction. (NCB, 1975). There is an abundance of data which demonstrate the veracity of this generalisation. Apparent exceptions to the rule, in which land subsidence occurs long after mining has ceased, invariably prove on closer examination to be due to collapse of voids which had been left open at mine closure, and which have eventually exceeded their elastic limits.

The collapse of worked longwall panels to form goaf induces strains in the overlying strata which are relatively predictable. Figure 1.15 is a composite diagram which summarises the orthodox view of zones of deformation above a longwall panel. The geomechanical theory and observations upon which this summary is based are presented by NCB (1975), with substantial further development by Garritty (1980, 1982), Whitworth (1982), Aston and Whittaker (1985), Whittaker and Reddish (1989) and Brady and Brown (1993). The key feature to note on Figure 1.15 is the existence of a zone of net *compression* sandwiched between the two zones of net *extension*. The occurrence of a zone of compression is not universally accepted (e.g. Whitworth, 1982). However, observed increases in piezometric head just before subsidence occurred above two longwall mine faces in Illinois (Booth *et al.*, 1999) provide compelling evidence for the existence of compressional stresses. Furthermore, subsidence management projects which have been predicated on the occurrence of compression within subsidence profiles have been highly successful (e.g. Orchard, 1975; Whittaker and Reddish, 1989).

Where stopes have been developed in competent country rock, they are likely to remain open in the long-term, and hence little deformation of the surrounding strata is possible. In strata where stopes do collapse (Figure 1.18), deformation of the hanging-wall strata is expected to be extensional, by means of brittle fracture, whereas the footwall strata would be expected to remain under elastic strain. The hydrological implications of this pattern are discussed briefly in Section 3.5.3.

1.3.4 SURFACE MINES

A "surface mine" is any mine in which the miner and/or his machinery work in an excavation open to the skies. Surface mining has been undertaken at small-scale since the earliest days of mining. For instance, reliably-dated Neolithic and Bronze Age open-pit workings and "hushes" (open cuts made by hydraulic mining methods – see Appendix I) are known from several European countries (Shepherd, 1993). However, these ancient surface mines are generally little more than scrapings on vein outcrops, generally less than 100 m in width and rarely more than 10 m in depth (O'Brien, 1996). Where ancient miners contended with depths of more than a few metres, overburden removal became too onerous and bell-pits or other simple deep mines were excavated.

(a)   **(b)**

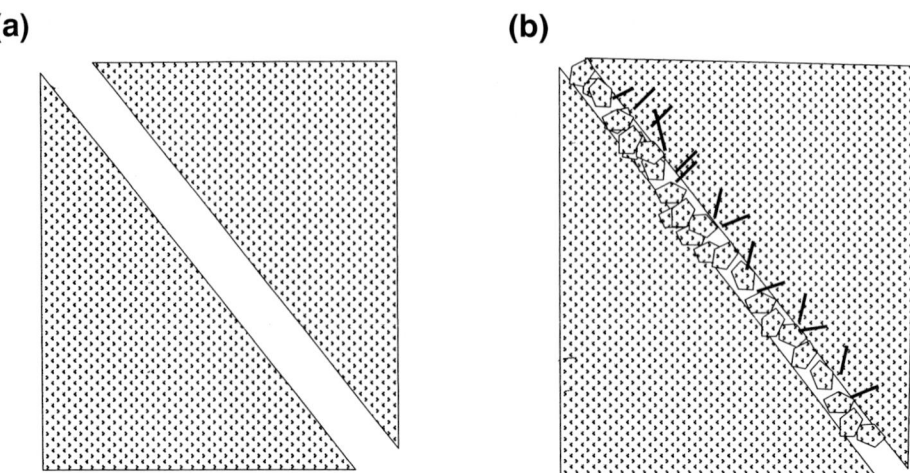

Figure 1.18: Diagrammatic cross-section showing collapse of a stope. (a) Open stope immediately after extraction is finished. (b) Closure of stope by a combination of extension in the hanging-wall and footwall (as in Figure 1.17a,b), with associated fracturing and spalling of the hangingwall.

The first large-scale surface mine of modern proportions was the famous Kimberley Mine in the Free State, South Africa[6]. At Kimberley, an open-pit mine exploited a diamond-bearing igneous intrusion of vertical cylindrical shape. Between its discovery in 1871 and its exhaustion in August 1914, some 22.5 million tonnes of rock had been excavated from the Kimberley Mine pit, yielding more than 14.5 million carats (2.7 tonnes) of diamonds. The open-pit is 500 m in diameter and was excavated manually, reaching an astonishing 215 m in total depth. This compares closely with the maximum depth attained in the highly mechanised modern surface coal mines of Europe (245 m at Westfield in Scotland; Grimshaw, 1992). The Kimberley Mine void remains to this day (Figure 1.19), and is now known affectionately as 'the Big Hole'.

For minerals less valuable than diamonds, large-scale surface mining did not become a viable proposition until modern heavy-duty draglines and dumper trucks (Figure 1.20) first became widely available in the 1940s. Since then, world-wide uptake of surface mining has burgeoned, so that at the start of the 21st Century, surface mines account for more than 80% of global mineral production.

There are three principal activities involved in all surface mining:

1.   The stripping of overburden (i.e. the excavation of non-economic deposits which overlie the ore or coal)
2.   Mining of the ore or coal
3.   Restoration and/or abandonment of the mine void.

---

[6]The information which follows was obtained from the Kimberley Mine Museum, South Africa

Figure 1.19: The Big Hole, Kimberley, South Africa: the world's first open-pit mine of modern proportions, albeit largely excavated before the advent of large-scale earth-moving equipment. The cylindrical shape of the Hole reflects the form of the igneous intrusion which hosted the diamonds. The total depth, at 215 m, is belied by the partially flooded state of the void at present. The void forms a 'through flow' pit lake (see Figure 3.42a; Section 3.11.1). (Photo: Kimberley Mine Museum).

Surface mines are classified according to how these three activities are undertaken. The classification used below follows that of Hartman (1987).

Open-pit mines are surface mines in which the overburden is removed to a disposal area, and the mineral is worked from stepped horizontal benches (Figure 1.21). These benches generally vary in width from around 18 to 45 m, and are typically separated by faces between 9 and 30 m tall. Inactive faces are normally blasted back, or allowed to collapse under controlled conditions, until they form "aprons" with batters (i.e. slope gradients) of around 2:1.

Open-pit mines are typically developed in situations where there is little overburden and little gangue in the orebody, so that a significant shortfall of waste rock and/or tailings precludes easy restoration of the void by back-filling. Many limestone and road aggregate surface mines fall in this category. Very large open-pit mines of this type are currently operational in Brasil (Figure 1.21), working very large iron orebodies in which the ratio of overburden to ore is as low as 1:10. Open-pit mines may also be developed where the extracted material is of relatively low grade (i.e. only a few percent by weight of the sought

Figure 1.20: A typical scene in a surface coal mine (opencast mine), with coal being loaded by face shovels into haulage trucks. This photograph was taken in January 2001 in the last working section of the San Victor opencast mine, the largest (1300 ha) and longest-lived mountain top surface coal mine in Spain. (Photo: P. Younger, with the permission of HUNOSA).

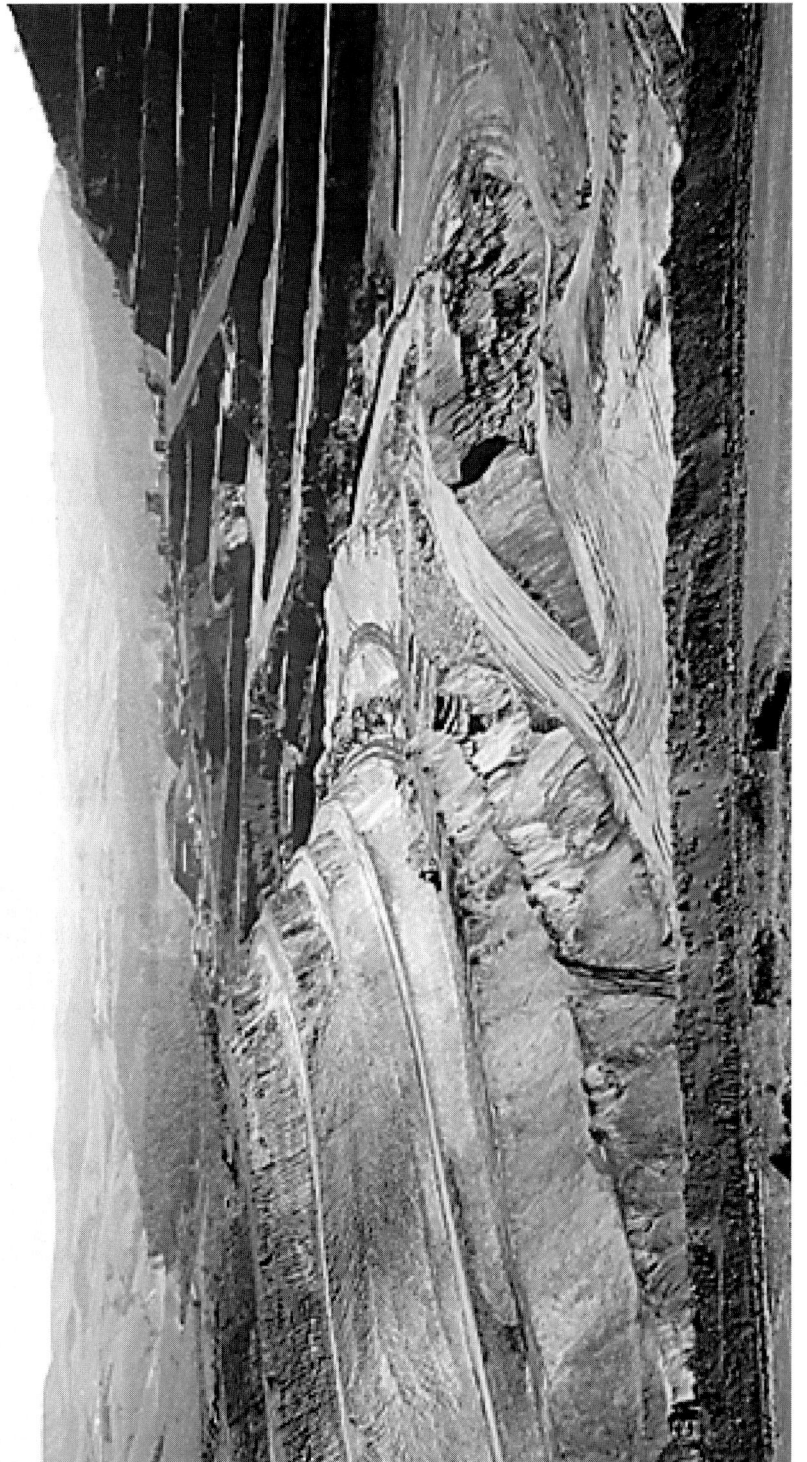

Figure 1.21: A typical hard-rock open-pit mine, in this case working Precambrian Banded Iron Formation (Mina do Pico, Minas Gerais, Brasil, in 1997). Note the typical layout of benches and haulage roads. (Photo: P. Younger, with permission of MBR).

ore), so that gangue material obtained during mineral processing would be sufficiently voluminous to back-fill the void, if only the mineral processing plant were located close enough to the void that back-filling would not be prohibitively expensive.

Because of the shortage of overburden for back-filling, many abandoned open-pit mines are left (or specifically reclaimed) to form lakes. These pit lakes differ radically from natural lakes in having much lower diameter-to-depth ratios (Miller *et al.*, 1996); in other words, natural lakes are usually much wider than they are deep, whereas the converse is typically true of pit lakes. This has important limnological consequences, particularly in relation to seasonal stratification (in terms of temperature, density and salinity), which in turn affects water quality dynamics (see Section 3.11.1).

The term *Quarry* is a synonym for "open-pit mine" in Australia, South Africa, Britain and Canada. However, in US usage the term quarry is used to refer quite specifically to surface mines which produce dimension stone (i.e. facing stone used in prestigious building projects).

Strip mines or *opencast mines* are surface mines working bituminous coal. They differ from open-pit mines in one major respect: with typical overburden to coal ratios on the order of 20:1, it is not economic to remove all of the stripped overburden ("spoil") to disposal areas. Consequently the spoil is "cast" (i.e. back-filled immediately after stripping) directly onto mined-out surfaces behind the advancing workings (Figure 1.22). The term "opencast" reflects this "casting" of spoil during working. The term "strip mine" reflects another facet of the mining process: the fact that the coal is won, and therefore the spoil is cast, in long, narrow strips parallel to the high-wall (which is the artificial cliff of unworked strata into which the workings gradually advance; Figure 1.22).

Opencast mines are generally developed as "area mines", where an area of ground for which a mining permit has been obtained is systematically worked for its coal. The typical sequence of events is as follows (Figures 1.20 and 1.22):

(a)   Stripping of topsoil, and careful emplacement of it in a pre-defined storage area.
(b)   Stripping of unconsolidated overburden ("drift") to rockhead. This overburden is typically tipped around the perimeter of the void to form a bund (known as "baffle banks"), which will serve a number of useful purposes such as minimisation of dust dispersal from the site and shielding from view of workings which might be deemed unsightly.
(c)   Sinking of one or more "box-cuts" (Figure 1.23), which are rectangular excavations sunk into the shallowest of the target seams.
(d)   Lateral working from the box-cuts to create the first strip excavation, and thus establish the initial high-wall position.
(e)   Repeated working of strips into the high-wall, with casting of spoil on worked-out strips behind.
(f)   After coaling is complete (which will require around 6 years for the average site in Pennsylvanian/Stephanian Coal Measures), the spoil is regraded, the unconsolidated overburden from the bund placed over the spoil, and the topsoil brought back from the storage area, spread and re-vegetated (Rimmer and Younger, 1997).

The observant reader will have noticed the introduction of the word "bituminous" into the description of strip/opencast mines as surface coal mines. This is because there are many

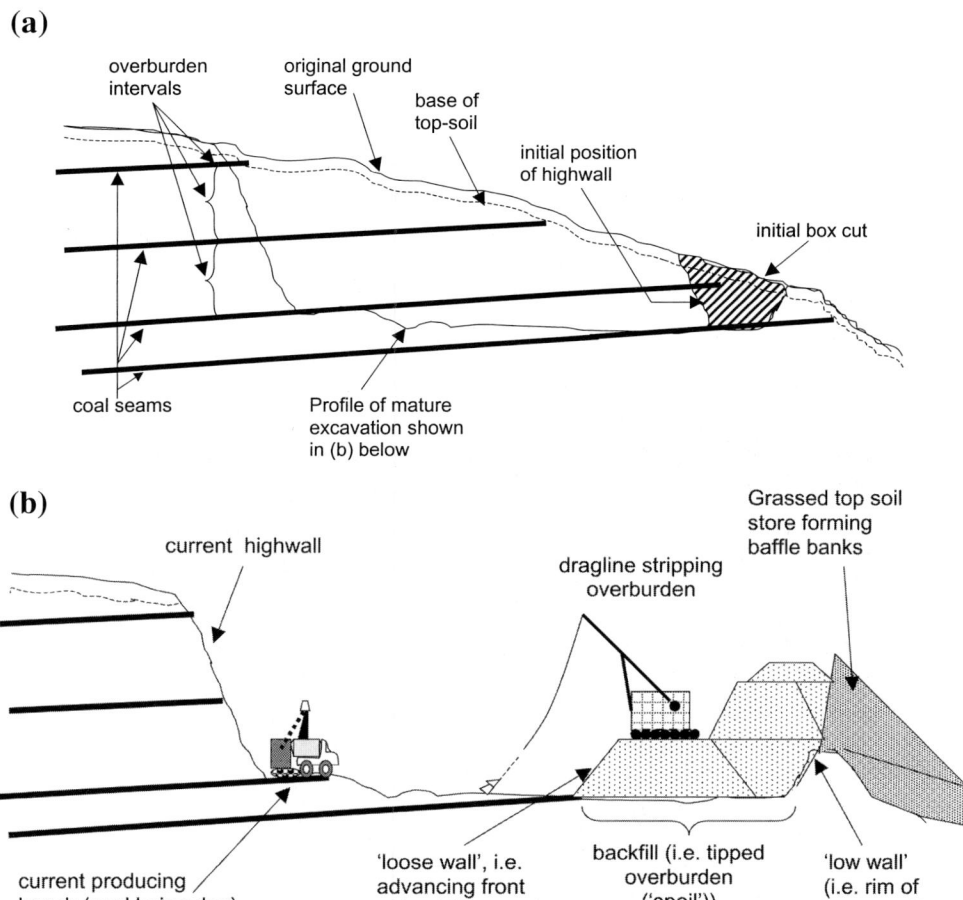

Figure 1.22: Schematic section through an opencast mine (a) pre-working land surface, soil zone and position of first box cut. (b) Mature working, with top-soil stored in baffle banks (which serve to screen the site from the exterior, preventing unsightly views into pit and helping to minimize dust dispersal), overburden tipped in spoil banks re-shaped to form the advancing 'loose wall', an active working bench and the slowly retreating 'highwall'.

areas of the world where lignite (brown coal of very low rank) is worked in areas almost devoid of overburden. With little overburden available for back-filling worked-out voids, most surface mines working lignite conform to the "open-pit" definition rather than the "opencast" definition. Germany is particularly famous for its brown coal mines. After exhaustion, many of these have been allowed to flood, forming pit lakes (Geller *et al.*, 1998). While the water quality in the German lignite pit lakes is generally good in the west of the country, problems arise in the east where the lignite was subject to diagenesis by marine waters, and is consequently rich in acid-generating pyrite (see Geller *et al.*, 1998). Acidic pit lakes in abandoned lignite open-pits are also known in Spain (Laine, 1998) and Greece (N. P. Nikolaidis, University of Connecticut, *personal communication*, 2000).

Figure 1.23: A view across the main box-cut of an open-pit gold mine in Brasil in 1997, showing rectangular sinkings to pay-zones, which will subsequently be extended laterally and transversally as long as the grade remains high. (Photo: P. Younger, with permission of CVRD).

Hydraulic mining was once a popular surface method of working unconsolidated materials containing alluvial diamonds, gold, and valuable base metals etc. Everyone is familiar with the romantic image of the lonely prospector panning for gold in the streams of California. Gold panning is the simplest form of hydraulic mining: using the power of running water to separate ore from country rock and gangue. The *garimpeiro* miners of the Brasilian Amazon use similar techniques to this day, although the romance is diminished by the devastation of watercourses through extensive dredging and silt washing, and by the toxic spillages of mercury into the tributaries of the Amazon (Cleary and Thornton, 1994). During the Californian gold rush, substantial workings for alluvial gold were worked by high pressure hydraulic jetting. (A particularly vivid, though perhaps not entirely accurate, reconstruction of these workings forms the setting for the 1985 Clint Eastwood movie, *Pale Rider*).

In Europe, hydraulic mining on an industrial scale appears to have been pioneered by the Roman army, working in the mountains of northern Spain. Both alluvial and *in situ* skarn-type deposits of gold in this region were found to be sufficiently soft that they were amenable to hydraulic mining. The Romans directed the construction of extensive leat systems along the flanks of mountains. The leats conducted water to reservoirs above the mountain passes, whence it was used to erode large open cuts in the hillsides (Lewis and Jones, 1970). From the 17th Century onwards, a similar technique enjoyed widespread application in upland Britain (Cranstone, 1992). This technique involved first impounding surface runoff waters on the edges of upland plateaux, and then releasing them suddenly in torrents which scoured the hillsides below, creating deep channels known as "hushes" (Figure 1.24). At the foot of each hush, fragments of lead ore were deposited in distinct bands (gravitational sorting being inherent in the process), from which they were readily recovered by hand-picking. Inspection of the bed of the hush would reveal the locations of rich veins suitable for further working.

While hydraulic mining is predominantly a surface mining technology, it is worth noting that the technique has also seen limited use in deep mines working soft coal deposits (Hartman, 1987). The technique has some advantages over conventional shearing and blasting, most notably in the suppression of dust and the inhibition of fire and dust explosion hazards. However, it is unlikely that subsurface hydraulic mining will ever become widespread, not least because of the problems of water management which the method entails.

Effects of surface mining on surrounding strata
Surface mining severely affects the worked strata, reducing them to spoil or tailings with management problems of their own (Section 1.3.7). However, the effects of surface mines on surrounding strata would generally be expected to be less marked than in the case of deep mines. This is because unworked strata surround a surface mine in two dimensions (beside and below), whereas they surround deep mines in three (beside, below and above). Nevertheless, the principle that extraction of mineral can induce extensional movements in adjoining strata is just as applicable to a surface mine as to a deep mine. In the case of an open-pit, the anticipated pattern is that a "halo" of extensional deformation will surround the outer perimeter of the zone of extraction (Figure 1.25). This is frequently manifest in the field in the form *en echelon* tension fractures, sometimes with considerable apertures, on benches immediately behind high-walls. These fractures can

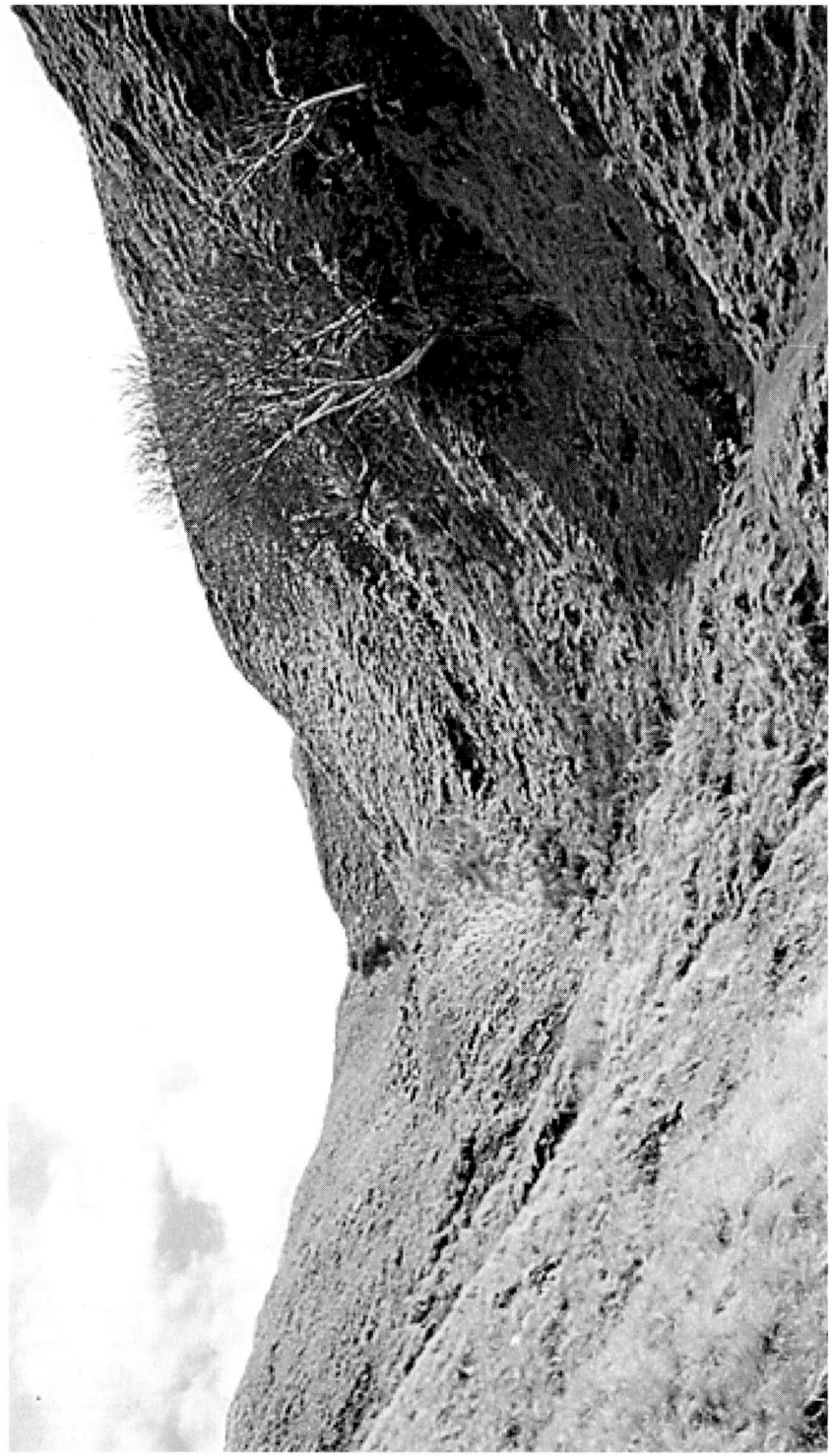

Figure 1.24: A typical hush in the North Pennines, UK. The 'hush' is the entire valley in the field of view. This valley is entirely artificial (it lies obliquely to the natural surface drainage), having been formed by hydraulic mining ("hushing"), probably in the 17th and 18th Centuries. The hushing was carried out by successively impounding and releasing large volumes of water in an earth dam located just above the break in slope at the top of the hush. (Photo: P. Younger).

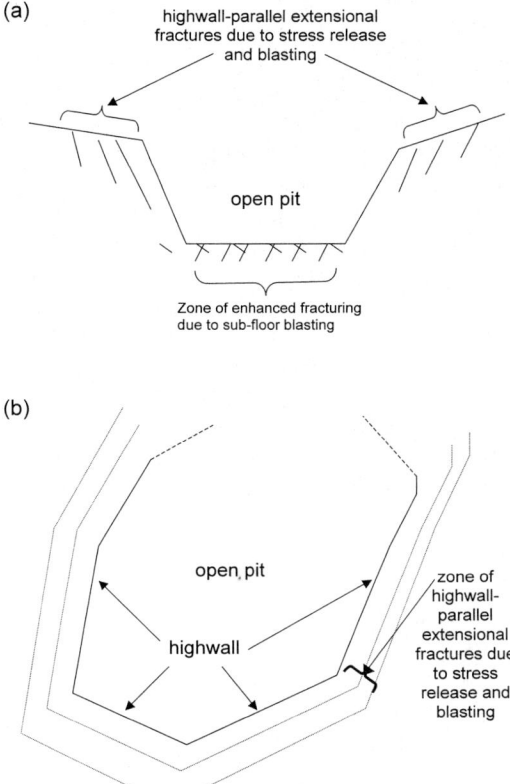

(a)

highwall-parallel extensional
fractures due to stress release
and blasting

open pit

Zone of enhanced fracturing
due to sub-floor blasting

(b)

open, pit

zone of
highwall-
parallel
extensional
fractures due
to stress
release and
blasting

highwall

Figure 1.25: Schematic diagram of the 'halo' of extensional fracturing which commonly surrounds an open-pit mine. (a) in cross-section, showing sub-floor zone as well as fracturing behind the highwalls (b) in plan, showing halo of highwall-parallel fractures.

represent a slope stability hazard in open-pit mining, for they can facilitate the toppling of large blocks of rock from otherwise sound faces. There is a need for further research in rock mechanics before the nature and extent of such extensional fracturing behind high-walls will be amenable to generalisation. Some of the hydrogeological implications of the phenomenon are discussed in Section 3.5.3.

1.3.5 HYBRID MINING METHODS

There are a number of specialist mining techniques which do not fall neatly into the "deep mining vs surface mining" classification. These include auger mining, *in situ* solution mining, *in situ* leaching, underground coal gasification and coal bed methane production.

Auger mining involves the horizontal drilling of large-diameter (typically 0.5 m) boreholes into the outcrop of a coal seam, which may be exposed in a natural cliff or more commonly in an opencast high-wall (Figure 1.26). The auger drills resemble giant corkscrews, and they retrieve coal in lumps large enough for conventional processing.

Figure 1.26: Auger mines in the two leaves of the Dundas Seam (KwaZulu-Natal Coalfield, South Africa). The concrete plugs mark the portals of horizontal borings which won coal beneath the overlying cliff face without the need to develop human access underground. (Photo: P. Younger).

Auger holes serve to increase the yield of an open-cast mine beyond that obtainable within the limits of excavation specified in mining permits.

In situ solution mining is a technique solely applicable to salt and sulfur deposits which are amenable to dissolution in water. It is undertaken by drilling boreholes into the target strata and then injecting cold or hot water. Cold water is injected to extract potassium and other alkali metals from evaporite mineral deposits by simple congruent dissolution. Hot water is injected to melt sulfur in appropriate strata, which is then transported to the surface in a stream of cooler water, aided by rising bubbles of air (Hartman, 1987). This method of sulfur winning is known as the "Frasch Technique". While both methods of *in situ* solution mining have the capacity to form subsurface voids, the cold water injection techniques are the more prolifc of the two in this regard, and can cause land subsidence. For instance, Gómez de las Heras and Rivadeneira de Vega (1999) document subsidence craters of several tens of metres in diameter above the Polanco solution mine in northern Spain.

In situ leaching also involves borehole injection of fluids, but in this case the fluid is a powerful solvent which preferentially dissolves a metal or other substance of economic interest (Hartman, 1987). Pumping of the resultant solution to surface allows extraction of the metal. The solvent used depends on the target deposit, as follows:

Mineral deposit to be dissolved	Solvent
Copper sulfide materials	Acid ferric sulfate
Copper oxide minerals	Sulfuric acid
Uranium	Hydrogen peroxide
Gold	Sodium cyanide

Generally, *in situ* leaching removes so little of the rock to which it is applied that surface subsidence should not be an issue. Of greater concern is the risk of ground water pollution by the leaching solvents. It is clear that the solvents listed above are substances which would not normally be consented for discharge into the subsurface, let alone into the saturated zone of an aquifer[7]. However, for *in situ* leaching to work at all, the target deposit must be in the saturated zone, must be reasonably permeable, and must be confined above and below by low-permeability strata. To obtain a permit for *in situ* leaching, therefore, it will be necessary to establish a convincing case that the leaching solvent will not be able to migrate to other aquifers or surface water courses. Despite these limitations on its use, *in situ* leaching currently accounts for around 4% of global mineral production. Nevertheless, cases are known in which former *in situ* leaching operations which were considered environmentally secure at the time of permitting subsequently gave rise to surface water contamination. In at least one case, surface water pollution arose after open-pit mining exposed strata that had previously been subject to *in situ* leaching (Miller *et al.*, 1996).

Underground coal gasification (UCG) and *coal bed methane* (CBM) extraction are two emerging technologies for the exploitation of coal seams which are inaccessible by conventional mining (Hartman, 1987; Coal Authority, 2000). Both techniques involve drilling boreholes from the surface into coal seams at depths of hundreds of metres. The boreholes

---

[7]see Chapter 3 for relevant definitions

are usually arranged in pairs, at either end of a target area of coal seam. In UCG operations, one of the boreholes is used to ignite the seam (using electrodes, explosives or gas burners; Hartman, 1987) and then sustain the fire by injection of oxygen. The second borehole is used to extract the gaseous products of the burning, and exploit them for energy supplies. In CBM developments, the seam is not ignited; rather hydraulic or pneumatic fracturing of the seam is undertaken between the boreholes, and suctions applied to one or more boreholes to drain naturally occurring methane from the coal (Clarke, 1996).

Experiments and small-scale UCG and CBM operations have been undertaken in many countries since the 1950s. While these met with sufficient technical success to foster full-scale implementation in the USSR and China, neither technology proved economically viable at more than a handful of sites in western Europe and North America. A productive CBM field in Alabama is one of the few western success stories to date (Hartman, 1987; Clarke, 1996). On the other hand, after expensive UCG experiments in central England in the 1950s, subsequent open-casting of the trial site revealed that the areas of coal seam that had actually been gasified ranged in size from a few square metres to no more than 1000 m^2 (Grimshaw, 1992, p. 71). The reason for such failures lay in the fact that the seams were accessed only by vertical boreholes, which have miniscule interface areas with the target seams. This problem could be overcome by the use of directional drilling technologies which were originally pioneered in the oil-industry. With this technology, holes can be drilled vertically until they reach the target seams, then switch to lower angles to allow them to remain within the seams for tens of metres (Figure 1.27). Until about 1995, such technologies were prohibitively expensive for coal mining and civil engineering applications, but a proliferation of low-cost systems since then is currently fostering renewed interest in UCG and CBM. European field trials are focusing on the reliability of directional drilling from the surface into seams 600 to 1200 m below ground. In seam, the boreholes will be virtually horizontal, and will be completed with diameters of about 215 mm (Coal Authority, 2000). If drilling trials prove successful, trial UCG burns will be undertaken. If these trials overcome the problems encountered with non-directional drilling in the 1950s, the way will be open for more widespread uptake of UCG and CBM. It may be that UCG and CBM eventually become *preferred* options to deep mining.

1.3.6 MINE WASTES

It has been estimated that more than 70% of all the material excavated in mining operations world-wide is waste (Hartman, 1987). At the start of the 21st Century, more than 99% of all mine waste rock is being generated by surface mines, which is out of proportion to their share of global mineral production (80%). This correctly implies that the deep mining sector is relatively efficient in terms of waste management. There are two main reasons for this:

- Since deep mining generally involves little stripping of overburden, little of this material ends up as waste rock *sensu stricto*, even though considerable quantities will end up as goaf wherever caving methods of extraction are used.
- The high costs of haulage of run-of-mine product in deep mines provide a strong incentive for disposing of waste rock in the underground voids. In many cases, waste rock can be used underground to make packs (artificial pillars) to provide additional roof support.

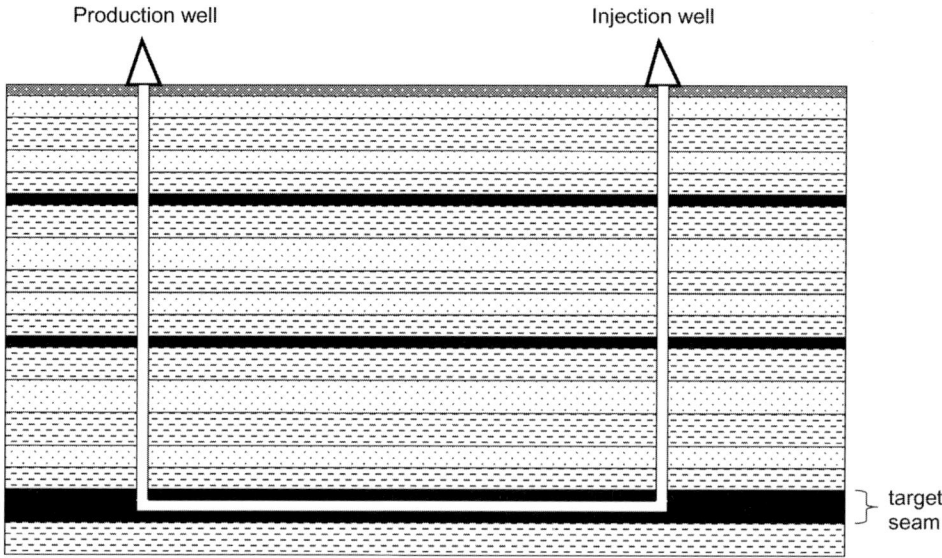

Figure 1.27: Directionally drilled boreholes, with vertical sections linked to a horizontal in-seam section, suitable for coalbed methane stimulation or underground coal gasification.

Mine wastes can be generally classified into two major categories (Figures 1.28 and 1.29):

- *Waste rock* or *Spoil.*
- *Tailings* (also known as "finings" in the coal sector).

As shown in Figure 1.28, waste rock can arise both before and during mineral extraction, and also during mineral processing. By contrast, tailings arise solely in the course of mineral processing.

While mineral extraction has been discussed in some detail above, we have not yet considered mineral processing techniques. A comprehensive review of mineral processing has been published by Wills (1992). Because of the availability of this review, and because it is not really necessary to understand all of the details of mineral processing in order to appreciate the environmental significance of tailings, few details are given here. Suffice it to say that the process of transforming run-of-mine ore into a marketable commodity generally involves the following sequence of activities:

1. Primary screening (i.e. large-scale sieving) of the run-of-mine rock to remove large blocks of country rock.
2. Crushing of the ore to a sufficiently fine grain size that virtually all grains of ore mineral will be free of gangue.
3. Separation of the ore and gangue minerals.

Waste rock and tailings both arise during this sequence, with early phases generating most of the waste rock, and the final step releasing tailings (Figure 1.28). Genetic considerations

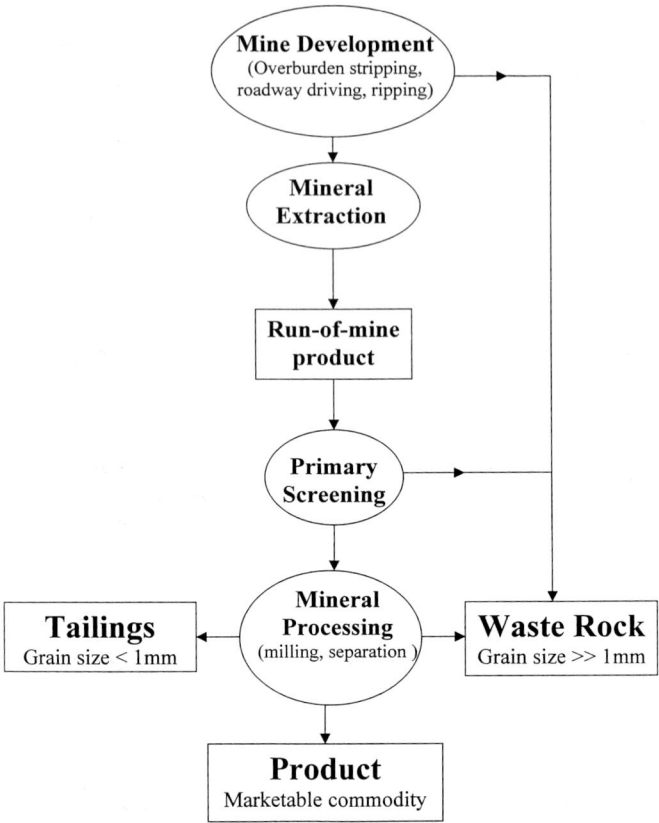

Figure 1.28: A classification of mine wastes according to their origins within the mining and mineral processing cycle.

aside, the fundamental distinction between waste rock and tailings is one of grain size, with most tailings being finer than 1 mm in diameter, and most waste rock being considerably coarser than this (Figure 1.29).

The differences between waste rock and tailings may be summarised as follows:

Waste rock	Tailings
Predominantly coarse-grained (1 mm–50 mm)	Predominantly fine-grained (<1 mm)
Moderately reactive if sulfidic	Highly reactive if sulfidic
Moderate to high permeability en masse	Low permeability en masse
Generally tipped dry	Generally deposited from flowing water

The contrasts in granulometry between different waste rock materials, and between waste rock and tailings, are highlighted in Figure 1.29. The figure gives examples of grain-size distributions for waste rock materials from:

- A "hard rock" mining setting (the Gibraltar Copper Mine, British Columbia; Poulin *et al.*, 1996)

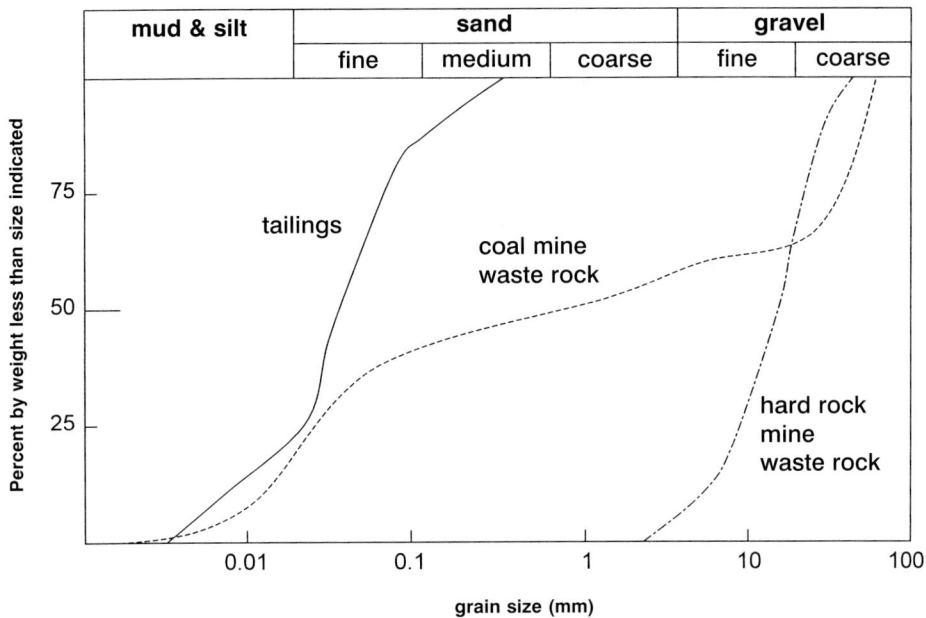

Figure 1.29: Median grain-size distribution curves for typical tailings, coal mine waste rock and waste rock from a typical 'hard rock' metalliferous mining setting. (Plots for tailings and hard rock mine waste rock after Poulin et al., 1996; plot for coal mine waste rock is a median profile for the entire UK, after NCB, 1973).

• deep coal mine waste rock piles in the UK (the median of a national database of analyses; see NCB, 1973, and Bell, 1996)

Also shown is the plot for milled tailings from the Gibraltar mine site (Poulin et al., 1996), which compares closely with the median plot of the compilation of tailings grain-size analyses presented by Vick (1983).

Examination of Figure 1.29 shows that the three different mine wastes differ considerably from one other. It is immediately obvious that the coal mine waste rock is far more variable in grain size than the "hard rock" waste (which represents shattered quartz diorite country rock). Indeed the grain size distribution for the colliery waste rock shows a distinctly bimodal distribution, with the smallest 40% of the material being less than 0.06 mm, while the largest 40% of clasts are all greater than 10 mm in diameter. This strongly bimodal grain-size distribution has been observed wherever waste rocks from coalfields of Upper Carboniferous age have been studied (e.g. NCB, 1973; Groenewold and Bailey, 1979; Bell, 1996). It can be ascribed to the disintegration of shale clasts, which are usually present in large quantities in colliery spoils. The shale lumps weather very readily during handling, releasing their constituent fine-grained clay minerals to form mud. It is likely that most of the weathering of shale clasts is by simple physical abrasion during excavation, transport and tipping (Bell, 1996). However, sloughing of clay particles from wetted clast surfaces is commonly observed under saturated conditions; this process may account for some of the exacerbated settlement of coal mine waste rock piles which is sometimes observed when they are flooded (Norton, 1983).

The "hard rock" plot in Figure 1.29 trends to an upper-bound grain size of around 50 mm, which is no doubt an artefact of the use of standard-sized crushing and screening plant at the two sites (i.e. 2 inch grizzlies). Turning to the contrast between tailings and waste rock, it is evident that the upper 50-percent of the tailings curve falls in a grain size range which is one to two orders of magnitude finer than the equivalent portions of the hard rock waste rock curve. The same applies to the lowermost 50-percentile grain size range. However, the strongly bimodal distribution of grain size in the colliery waste rock results in it having an even greater proportion of the very finest particles than the copper mine tailings.

There are at least two general lessons to be drawn from this examination of Figure 1.29:

(i)   source rock lithology is at least as important as processing technology in determining the granulometry of waste rock and tailings
(ii)   the specific outcome of processing a particular ore is difficult to predict accurately.

Consequently, it is recommended that site-specific granulometric measurements be made whenever mine wastes are the subject of crucial management decisions.

Disposal of waste rock
Where waste rock cannot be back-filled into the mine void, or else exported from the mine site for use as bulk fill in construction projects, it will generally be disposed of in a *waste rock pile* (or "spoil heap"). Waste rock piles are generally formed by loose tipping ("end tipping"), from wagons or conveyor belt systems. The process of end-tipping results in the development of significant variability in the sediment fabric (Figure 1.30). Gravity sorts end-tipped wastes as they roll down the face of the heap (which typically conform to angles of repose of around 40°). Large blocks and cobbles roll to the foot of the slope while finer sediment remains near the top of the slope (Figure 1.30a). This phenomenon has been widely observed (e.g. Winczewski, 1979; Groenewold and Rehm, 1982; Hawkins, 1994; Newman *et al.*, 1997). Repeated end-tipping results in the accumulation of inclined layers of systematically graded rock fragments. If the flank of a waste rock pile is excavated, it often reveals an internal structure akin to that shown diagramatically in Figure 1.30b (see also Winczewski, 1979; Newman *et al.*, 1997, and Wilson *et al.*, 2000). This has significant hydrological implications, as will be seen in Section 3.9.1.

Most waste rock is emplaced without deliberate compaction. In deep mine spoil heaps, which are typically formed by end-tipping from conveyor belts, compaction under the weight of the spoil itself may be the only operative compaction process. In strip mines, however, where heavy machinery repeatedly tracks across tipped waste rock, accidental compaction in excess of that due to the weight of the rock itself can be expected. Studies of waste rock settlement in back-filled strip mine voids have revealed exponential decreases in the rate of settlement over time. A typical example is shown in Figure 1.31. This shows that the bulk of the settlement can be expected to occur within the first decade after back-filling is complete. Consideration of the straight line relationship which results when the data are re-plotted with time on a logarithmic axis shows that the overall rate of settlement is such that total settlement will still be less than 2% of the total fill height after 150 years. While saturation of the coal mine waste rock with water can

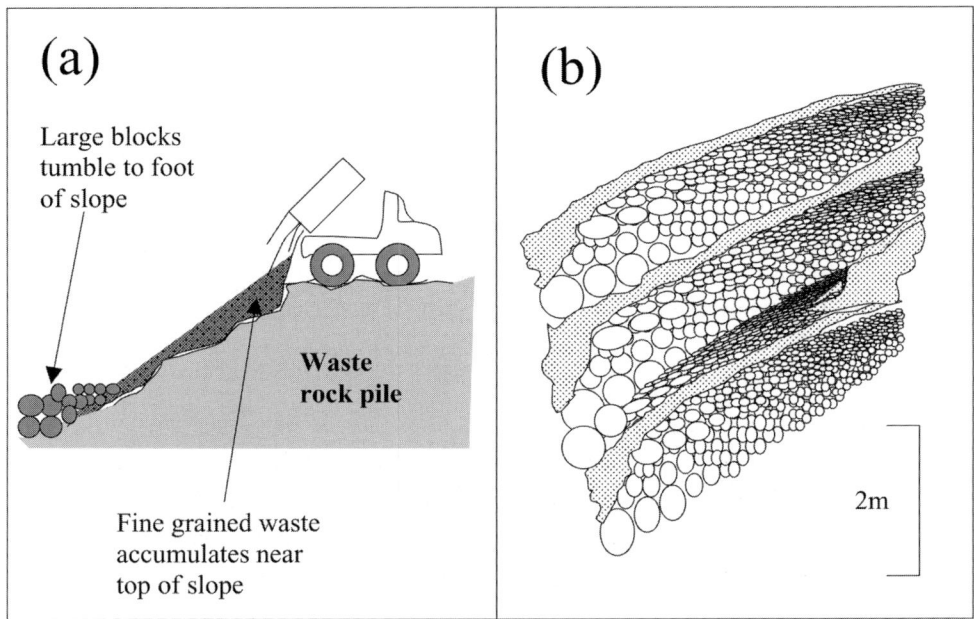

Figure 1.30: (*See next page for Figure 1.30c*) End-tipping as the source of heterogeneity in mine waste rock. (a) End-tipping, resulting in sorting of the large and small fragments as they roll down the tip face. (b) Schematic diagram showing the fabric of the waste rock accumulated from end-tipping. (c) The profoundly contrasting physical properties of intact strata and waste rock backfill. This is a view from the highwall of Pegswood Moor Farm Opencast Coal Site, Northumberland (UK), in August 2000. On the left is the loose-wall (see Figure 1.22b) comprising end-tipped overburden; on the right are the intact strata in the sidewall. The sorting of clasts in the loose-wall (in accordance with (a) above) is clearly evident, and the contrast in the style and magnitude of heterogeneity in permeability and porosity between the loose-wall and the sidewall needs no amplification. (Photo: P. Younger, with permission of H J Banks Ltd).

exacerbate settlement to a limited extent (Norton, 1983), settlement of waste rock piles is not sufficiently severe to be a major issue in itself. However, because strip mines are back-filled progressively as mining continues, there is generally some scope for differential settlement between adjacent back-fill strips (cf Figure 1.22), and between the body of fill and the intact strata behind the high-wall, at least in the early years after site abandonment. The shallow closed depressions which can develop on top of waste rock piles by such differential settlement can affect site drainage, and are therefore of potential hydrological significance (see Section 3.9.1, and Groenewold and Rehm, 1982).

Revegetation of waste rock piles is now common practice in the mining industry world-wide. Considerable research has been undertaken into the best way to stimulate plant growth and pedogenic (soil-forming) processes on reclaimed spoil (e.g. Rimmer and Younger, 1997). However, until recently far less attention was paid to ensuring that the drainage of reclaimed waste rock piles was arranged such that it minimised leachate generation in the long term. Consequently, many visually attractive spoil heaps which were reclaimed in the 1970s are now giving rise to polluting dsischarges of acidic, metalliferous leachate (see for instance Younger *et al.*, 1997).

Figure 1.30c: (*See previous page for Figure 1.30a&b and caption*).

Disposal of tailings

Until the early 20th Century, virtually all tailings were simply released into the nearest watercourse (e.g. Pirrie *et al.*, 1997). Where mining continued for many centuries, this practice sometimes led to rivers and estuaries becoming so choked with tailings that they became no longer navigable (Pirrie and Camm, 1999), which could be counter-productive to mining in places where the finished product was taken to market by boat. Added to this complaints from riparian owners downstream from mineral processing plants, and the

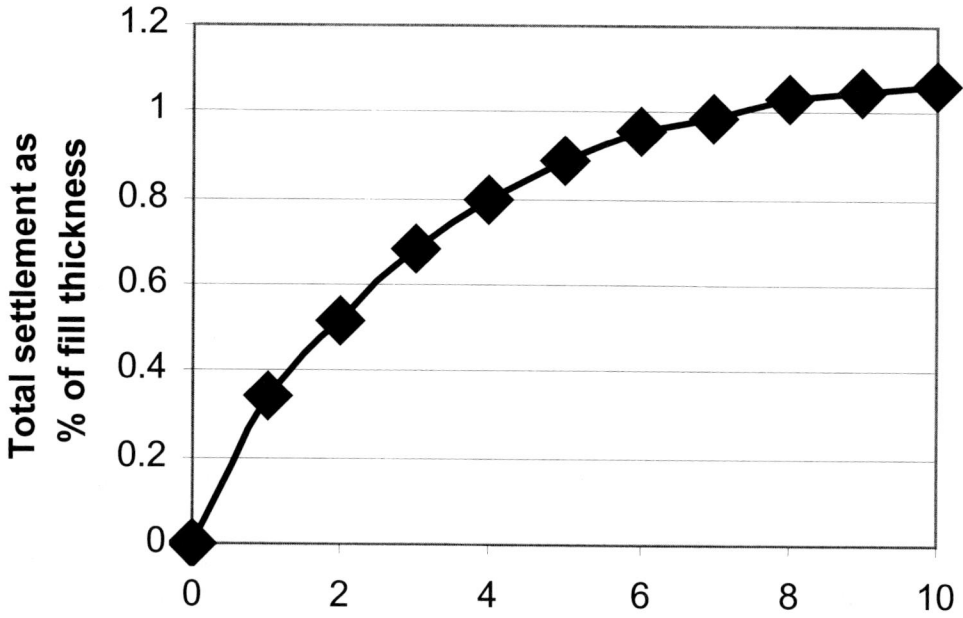

Figure 1.31: Settlement of opencast back-fill: an example from ten years of monitoring at the former Chibburn Opencast Coal Site in Northumberland (UK). Note the approximately exponential decline in the rate of settlement, so that little detectable settlement is occurring beyond the first decade after completion of the back-filling operation. (Previously unpublished data of Kilkenny, 1968).

first impulses for the development of alternative tailings disposal methods are obvious. The emergence of environmental consciousness in the mid-20th Century provided the final and greatest impulse.

By the 1940s, the disposal of tailings in purpose-built sedimentation lagoons had become common practice in the mining industry world-wide. These sedimentation lagoons are usually called "tailings dams" or "tailings dykes" (Vick, 1983), though the term "slimes dam" is also used, particularly in South Africa (e.g. Blight, 1997). Despite the global uptake of tailings dam technology, significant SME and informal sector mining operations in remote areas continue the antiquated practice of tailings disposal to streams, resulting in the same problems of watercourse clogging which used to bedevil European and North American mining districts in the 19th Century.

Tailings dam technology is deceptively simple (Vick, 1983; UNEP, 1996). There are two main types of tailings dam – ring dykes and valley dams – and three methods of dam construction. *Ring dykes* are constructed in flat-lying areas, and consist of a polygonal dyke (i.e. a bund) entirely enclosing the sedimentation lagoon. *Valley dams* are tailings dams which are constructed by impounding a natural valley, or part of a natural valley. Three sub-types of valley dam are recognised (UNEP, 1996):

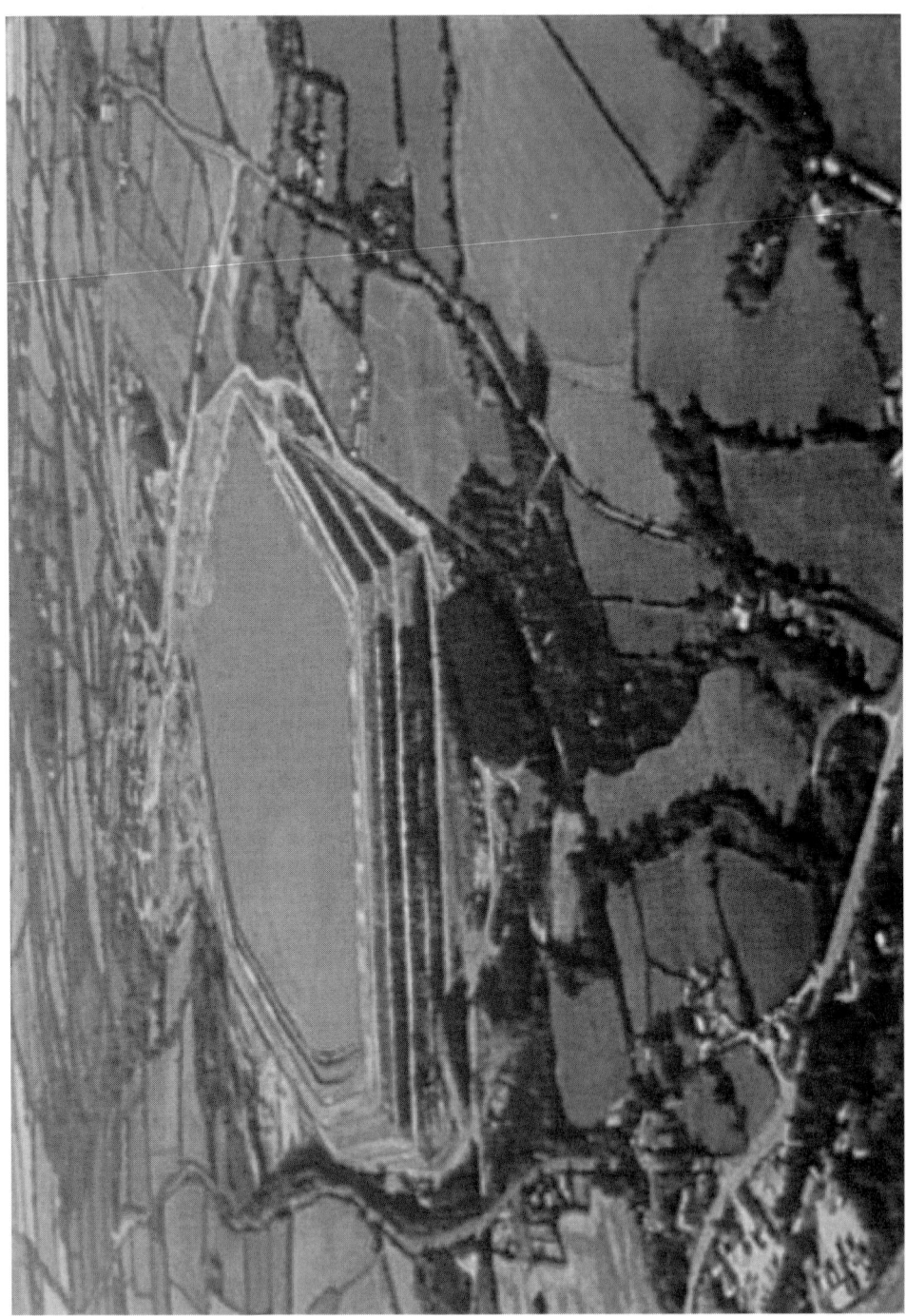

Figure 1.32: A typical, large, well-constructed tailings dam: the Clemows Valley Tailings Dam, Cornwall, UK. (Photo: Environment Agency, UK). Note the stepped nature of the dyke, recording successive periods of raising of the embankment in line with the original design of Knight Piésold Ltd. The dam has been subject to ongoing testing and inspection throughout its operational life and beyond.

- cross-valley dams, in which the entire width of a natural valley is impounded
- valley side dams, in which a dam is constructed on one flank of the valley, using the natural ground to form one side of the lagoon.
- valley bottom dams, which are constructed where a valley is too wide or too flat-lying for a cross-valley dam to be effective. In these circumstances, part of the width of the valley is impounded, and a further dyke parallel to the valley side is constructed to enclose the impoundment.

In practice, tailings dams which were essentially valley dams in their early stages grow above the surrounding ground surface in later phases of construction to become ring-dykes. The tailings dam shown in Figure 1.32 is a case in point.

Tailings dams are usually constructed from waste rock and the tailings themselves. Except in the case of large valley dams, the construction of a tailings dam never occurs in a single phase of activity. Rather the dyke is raised successively higher in phases as mining progresses, with more waste rock becoming available for dyke construction and ever more tailings arising for disposal. Three methods of construction are used (Figure 1.33):

- *Upstream* construction involves the raising of the dam ever inwards (Figure 1.33a), with each successive rampart of the dyke being founded on part of the former lagoon surface. This has the inherent disadvantage that dense material (i.e. the coarse waste rock/tailings which form the dyke) is being emplaced onto less dense, finer-grained tailings. Depending on the nature of the tailings, this can result in the toe of each successive rampart being a line of structural weakness.
- *Downstream* construction (Figure 1.33b) avoids the structural ambiguities of upstream construction by constructing the dyke rampart ever outwards, so that it is always founded upon well-drained dense material. The gain in structural integrity is at the expense of increasing land take throughout the life of the mine. It also depends upon an increasing supply of materials suitable for dyke construction throughout the period of use, which is often contrary to the realities of mine development, since most overburden tends to be excavated in the early years of a mining operation.
- *Center-line* construction (Figure 1.33c) is a compromise between the upstream and downstream techniques in both method and merit. In this technique, sequential raising of the dyke is accomplished without any lateral movement of its center line, despite this entailing the founding of parts of the dyke rampart on fine-grained tailings (Figure 1.33c). To compensate for this problem, excess waste rock is added to the downstream face of the dyke. This achieves structural stability nearly comparable to that of downstream-construction dykes, but uses considerably less material.

Further details of the design of tailings dams are provided by Vick (1983), while Klohn (1979) provides a comprehensive discussion of design procedures and operational interventions to minimise problems of seepage through the dykes.

World-wide there are thousands of tailings dams constructed in accordance with the above principles, the vast majority of which function according to design, preventing environmental pollution. Nevertheless, a few spectacular tailings dam failures over the last few decades have called into question the structural integrity of at

(a) upstream method

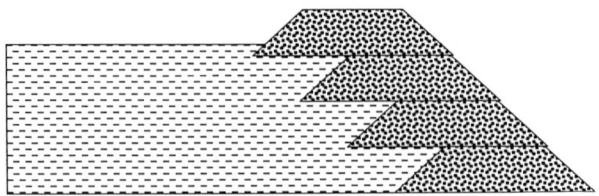

(b) downstream method

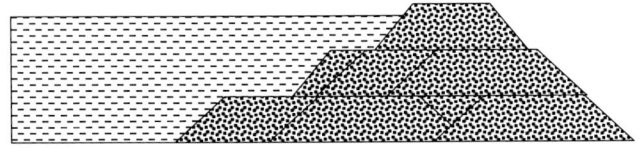

(c) center-line method

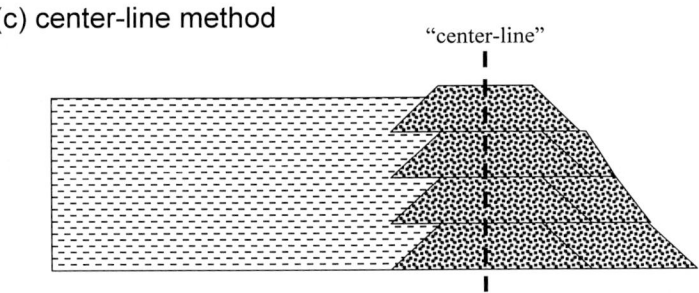

Figure 1.33: Schematic cross-sections illustrating the three principal strategies of tailings dam construction, i.e. (a) the upstream method (b) the downstream method and (c) the centre-line method. In all three sketches, the coarse stipple represents coarse discard material used to give structural strength to the dyke, while the horizontal dashed line ornament signifies tailings. (Adapted from Vick, 1983).

least some of these facilities. Table 1.1 below summarises some of the more devastating tailings dam failures. Although comparisons are not easy to make, it seems that tailings dams fail more than ten times more frequently (*pro rata*) than conventional water reservoirs.

Several world-wide surveys of tailings dam failures have concurred that the root cause of the apparently elevated rate of tailings dam failure is *not* poor design, but rather stems from inadequate implementation of the original design (UNEP, 1996; ICOLD, 2001). This arises because, unlike other dams, tailings dams are constructed gradually

Table 1.1: Summary of some of the most notorious and/or disastrous tailings dam failures

Site name	Year	Direct fatalities	Environmental impact	Reference for further information
El Cobre (Chile)	1965	~200	Not recorded.	UNEP (1996)
Bafokeng (South Africa)	1974	12	Polluted a 45 km reach of the Kwa-Leragane River and the Vaalkop reservoir.	Blight (1997)
Stava (Italy)	1985	268	Extensive impacts on river and floodplain ecology.	Chandler and Tosatti (1995)
Merriespruit (South Africa)	1994	17	Limited pollution of lake sediments.	Blight (1997)
Omai (Guyana)	1995	0	Minor fish kill in Omai River.	UNEP (1996); Narayan (1998)
Río Porco (Bolivia)	1996	0	Fish kills recorded through 300 km of river.	Edwards (1996); García-Guinea, & Huascar, 1997; Macklin *et al.* (1996)
Aznalcóllar (Spain)	1998	0	Severe contamination of 40 km reach of a river of high conservation value.	WWF International (1998); Grimalt *et al.* (1999)
Baia Mare (Romania)	2000	0	Severe contamination of 2000 km of the Danube and its tributaries, resulting in massive fish-kills (thousands of tonnes)	UNEP (2000); Lorber and Erhart-Schippek (2000)
Sebastião das Águas Claras (a.k.a. Macacos), Nova Lima, Minas Gerais (Brasil)	2001	5	A 6 km reach of the Córrego Taquaras (a tributary of the Río das Velhas) was buried up to 15 m deep in a torrent of red mud, which engulfed and uprooted trees. Threatened 70% of the water supply of the city of Belo Horizonte.	Local press reports (*Hoje em Día*, Belo Horizonte, Brasil)

over many years or even decades, so that the site operators in the later phases of construction are unlikely to be the same personnel as those who originally designed the dam many years previously. How best the scope for human error can be "designed out" of tailings dam construction is still a matter of active debate within the mining industry.

From the perspective of this book, tailings dam failures are particularly interesting because the common causative agent in all majors failures has been water (UNEP, 1996; ICOLD, 2001). The two most common ways in which water promotes dam failure are:

- through erosive down-cutting of the dyke if flows exceed spillway capacities (Smith and Connell, 1979; ICOLD, 2001), or
- by promoting slippage due to liquefaction or sapping of the dyke core by seepage waters (Robinson and Toland, 1979; Smith and Connell, 1979; UNEP, 1996).

Wet conditions downstream of a failed dyke also greatly exacerbate the movement of destructive mudflows of tailings (Blight, 1997).

As with waste rock piles, drainage of leachate from both active and abandoned tailings dams can be a significant source of surface water pollution, as well as a potential source of structural instability in the long term.

1.4 IMPACTS OF MINING AND MINE WATERS ON THE WATER ENVIRONMENT AND INFRASTRUCTURE

Though tailings dam failures are amongst the more dramatic causes of water pollution related to mining, they are by no means the most common or widespread. Far more pervasive on a global scale are the impacts that arise due to the alteration of regional surface and ground water flows by mining. In this section, a brief summary is presented on the negative impacts of mining and associated water management practices on the environment and engineered infrastructure; details of the *processes* behind these impacts are reserved to the following two chapters. Here we are concerned with the *issues* posed by mine waters, leaving the explanation of the processes to Chapters Two and Three, and descriptions of possible remedial actions to Chapters Four and Five.

Before proceeding, it is worth recalling that mine waters are not all bad news. Some mine waters have *positive* impacts on the water environment. For instance, examples of the following positive impacts have been catalogued by Banks *et al.* (1996, 1997):

- a considerable number of mine waters are of sufficiently good quality that they can be used for potable supply (Banks *et al.*, 1996), and many more are of sufficiently good quality to supply industrial or irrigation demands (e.g. Reddy *et al.*, 2000).
- pumped mine water discharges have been known to provide valuable dilution to rivers which would otherwise be heavily polluted by sewage.
- dissolved metals in some waters are sufficiently valuable that they have been extracted at a profit.
- some iron-rich ("ferruginous") mine waters have been deliberately mixed with raw waters entering water treatment works, to take advantage of the flocculating powers of the iron as it forms ferric hydroxide.

An interesting positive impact of mine water discharge relates to the RAMSAR-designated Blesbokspruit Wetland system (Gauteng Province, South Africa), which owes its very existence to mine waters pumped from Grootvlei Gold Mine and other workings (Wood and Reddy, 1998). On the other hand, the quality of the pumped mine water has given rise to some ecological concerns (Institute for Water Quality Studies, 1997; Wood and Reddy, 1998), illustrating that "good" and "bad" can be very relative terms in mine water management. Leaving aside positive impacts of mine waters, mining can have positive impacts on waters in adjoining aquifers. For instance deep mining by caving techniques has been shown to cause beneficial increases in the permeability of overlying sandstone aquifers in Illinois (Booth *et al.*, 1998). No doubt other positive examples could be catalogued at no great effort. However, few readers will be reading this book in order to address non-problematic mine waters; hence our focus on the negative impacts of mine waters in the following pages, while potentially misleading in terms of the overall picture, is nevertheless more germane to the needs of problem-solvers.

Mining impacts on the water environment arise from five distinct phases of the mining life-cycle:

(i)   The mining process itself
(ii)   Mineral processing operations
(iii)   The dewatering which is undertaken to make mining possible
(iv)   Seepage of contaminated leachate from waste rock piles and tailings dams
(v)   Flooding of workings after extraction has ceased
(vi)   Discharge of untreated waters after flooding is complete.

Of these six, the first is usually the least important, and the last is certainly the most important in the long-term.

The mining process itself affects the water environment principally through disruption of pre-existing hydrological pathways. Underground mining normally has relatively subtle impacts on the surface water environment. An exception to this generalisation is shown in Figure 1.34, in which a stope which daylights in the bed of a stream is seen to drain the entire flow into the underground workings. Similarly, but on a smaller scale, propagation of fractures due to mining subsidence into the bed of a stream in north-east Greece led to so much infiltration that the stream dried up in summer months (Grapes and Connelly, 1998). More subtly, subsidence due to longwall coal mining was found to temporarily alter the distribution of pool, riffle and cascade habitats in a mountain stream in Utah (Sidle *et al.*, 2000). In terms of groundwater systems, the working of longwall faces at considerable depth below shallow aquifers present in overburden strata has been shown to result in temporary, but locally pronounced, declines in the water table within the shallow aquifers (Booth and Spande, 1992; Zipper *et al.*, 1997). In some cases water levels will recover after the subsidence has ceased (Booth and Spande, 1992), while in other cases the lowering of the water table will be permanent (Booth *et al.*, 1998).

The excavation of an open-pit surface mine inevitably removes a large volume of rock which previously enclosed subsurface flow pathways. Indeed, the removal of unsaturated zone water storage which an open-pit working represents has been one of the key elements of controversy surrounding a number of surface limestone mines in south-west England (Hobbs and Gunn, 1998). Less subtly, slope failure due to opencast mining above old longwall workings led to the diversion of a major river into an active pit, forming a 100 ha lake in the void, which led to the suspension of mining at the site for almost a decade

Figure 1.34: A stream cascading into an open stope which has been driven (probably unintentionally) all the way to daylight. The scene is at an abandoned base metal sulphide mine in central Norway. (Photo: D. Banks).

(Hughes and Clarke, 2001). In general, however, direct impacts of the mining process *per se* on the water environment tend to be localised and of limited magnitude (e.g. Wardrop *et al.*, 2001). There is one very practical reason for this, which Kesserû (1995) has pointed out: the miner and the water resources manager share a common interest in avoiding the ingress of fresh water into a mine void; the water manager's loss of resource is the miners increase in nuisance. With both parties suitably motivated, efforts are generally made wherever feasible to minimise the direct interaction of mine voids and water resources.

Water pollution arising from *mineral processing operations* has been a cause of complaint for centuries. Possibly the earliest recorded example dates from 1556 in Germany, as recounted by Georgius Agricola in his monumental treatise on metal mining:

> "… when the ores are washed, the water which has been used poisons the brooks and streams, and either destroys the fish or drives them away …"
> (Agricola, *De Re Metallica*, as translated by Hoover and Hoover, 1950, p.8)

As discussed in the preceding section, such problems led to the development of tailings dam technology, and to the introduction of effluent treatment systems using the unit processes described in Chapter Four. Nevertheless, the problem still remains in areas of unregulated informal sector mining such as Amazonia, where mercury releases from gold ore processing cause widespread environmental damage (e,g, Thornton and Cleary, 1994).

The impacts of mine dewatering vary from negligible impacts in the case of small surface mines (e.g. Streetley, 1998; Wardrop *et al.*, 2001) to major depression of the water table and consequent impacts on connected surface water systems over very large areas (e.g. Younger, 1993a). For instance, the dewatering of major coalfields has been shown to depress regional water tables over contiguous areas of around 1,100 km^2 in Poland (Różkowski, 1997) and in excess of 2500 km^2 in north-east England (Younger, 1993a; 1998b). The consequences of water table depression due to mine dewatering can include:

1. Decreased flows in streams, wetlands and lakes which are in hydraulic continuity with groundwater.
2. Lowering of the water table in the vicinity of water supply or irrigation wells, leading at least to an increase in the pumping head (and therefore in pumping costs), if not to the complete drying-up of wells.
3. Land subsidence, either due to compaction fine-grained sediments (especially silts and clays), or due to collapse of voids in karstic terrains as buoyant support is withdrawn.
4. Surface water or ground water pollution, if the pumped mine water is of poor quality and is discharged to the natural environment without prior treatment.

In many cases, these impacts will have been anticipated well before they actually occur, and mining companies will take steps to mitigate them (e.g. Wardrop *et al.*, 2001). For instance:

1. Compensation flows might be added to sensitive surface waters, probably by treating the pumped mine water and piping it to appropriate locations where it can be discharged to advantage.

2. Water abstracted by dewatering pumps can be locally re-injected into the subsurface to support the water table in environmentally sensitive areas.
3. Alternative water supplies might be provided for communities or individuals whose wells dry up.
4. Mine waters can be treated prior to discharge.

There are very few published accounts of demonstrable negative impacts of mine dewatering in the modern literature. One possible explanation would be an endemic unwillingness to air such problems. However, this is a highly unlikely explanation, since no similar reticence has prevented the publication of literally thousands of papers on acid mine drainage problems. Rather, it is suspected that the scarcity of literature on dewatering impacts reflects two factors:

(i)   the success of the mitigation measures outlined above, which are now common-place in the mining industry worldwide, and
(ii) the fact that mines which intercept serious quantities of water will rarely remain in production for long.

Nevertheless, a few examples of demonstrable negative impacts of mine dewatering can be cited.

1. Examples of mine dewatering decreasing flows and water levels in rivers, lakes and aquifers have been reported recently from Estonia, Poland and Greece. Mine dewatering in an oil shale field in Estonia was found to intercept as much as 40% of the total annual flow in the catchment of the Purtse river (Ratsep and Liblik, 2000). A hydrochemical study of a surface limestone mine near Cracow, Poland, has revealed that dewatering is inducing the ingress of contaminated river waters from the Vistula river (Motyka and Postawa, 2000). Elsewhere in Poland, dewatering of a deep mine working copper is causing significant lowering of ground water levels in overlying limestone and sandstone aquifers (Bochenska et al., 2000). In Greece, the interactions between dewatering operations at an open-pit lignite mine and water levels in an ecologically important lake were investigated by a programme of hydrogeological characterisation. It was concluded that, while water would certainly be induced to flow from the lake to the mine, the quantities of water involved would be so small in comparison to the overall water-budget of the lake that no negative environmental impacts would ensue (Koumantakis and Dimitrakopoulos, 1999).
2. Anecdotal accounts of the drying-up of wells due to the dewatering of mines are legion. However, in practice investigations of actual instances are often inconclusive, at least to the level of proof necessary for successful litigation. For instance, in a recent detailed study of groundwater levels in an area of irrigated agriculture surrounding an open-pit limestone mine in northern Australia, Dudgeon (1997) found that it was difficult to reliably and objectively distinguish between water table lowering due to mine dewatering, and that due to a prolonged drought in the region.
3. Land subsidence due to mine dewatering (as opposed to mineral extraction) is rarely recorded, but in some karstic limestone/dolomite terrains where deep mining is exten-sive, severe problems have been reported. For instance in South Africa in the early

1970s so much subsidence followed the dewatering of a dolomite aquifer to facilitate continued deep mining for gold that entire towns had to be evacuated (Forth, 1994). More recently, in China, widespread subsidence problems have arisen due to dewatering of deep mines working coal, lead, zinc, iron and gold within (or in close stratigraphic proximity to) karstic limestone strata (Li and Zhou, 1999).

4. Surface water pollution due to discharge of poor quality pumped mine waters was once common in many mining districts. Nowadays, governmental regulations in most developed economies oblige mine owners to treat such waters prior to discharge. Nevertheless, where treatment is prohibitively expensive, problems still arise. For instance, discharge of highly saline pumped mine waters led to serious pollution of surface waters in central England (e.g. Lemon, 1991). In the former Kent coalfield of south-east England, leakage of lagoons receiving saline pumped mine waters led to extensive contamination of a major public supply aquifer (Headworth et al., 1980). Similar saline waters pumped from the mines of the Upper Silesian coalfield in Poland (Lebecka et al., 1994), and from some of the gold mines of South Africa, are so radioactive that they lead to radiation hazards in the receiving watercourses. Also in South Africa, controversy surrounds the pumping of saline mine waters into the Blesbokspruit wetland (Institute for Water Quality Studies, 1997), even though the same waters are largely responsible for the existence of the wetland in the first place (Wood and Reddy, 1998).

Seepage of contaminated leachate from waste rock piles and tailings dams is a significant cause of surface water pollution in many mining districts. Numerous instances are known where waste rock piles which were thoroughly re-vegetated several decades ago continue to release acidic leachates, emanating from shallow water table systems perched within the spoil (e.g. Younger et al., 1997). Drainage of leachate through the unlined bases of old tailings dams is also known to give rise to pollution of both surface waters and ground waters (Manzano et al., 1999; Johnson et al., 2000). Besides seepage releases, unreclaimed spoil can give rise to highly polluted surface runoff during storms (e.g. Bayless and Olyphant, 1993). Numerous examples of pollution by spoil heap/tailings dam leachates and runoff will be mentioned in the forthcoming chapters, and hence the issue is not discussed further here.

The *flooding of abandoned mine workings* after mining has ended has long been dubbed "water table rebound" (e.g. Henton, 1979, 1981), or "groundwater rebound" (e.g. Robins, 1990; Smith and Colls, 1996; Sherwood and Younger, 1997). The process of rebound in deep mines commonly results in a marked deterioration in the quality of mine waters (e.g. Cairney and Frost, 1975; Younger, 1993a, 1998b,c, 2000a,b). The reason for this is the sudden dissolution of so-called "acid-generating salts" (AGS), which is the name given to an array of secondary minerals formed by the weathering of pyrite above the water table (Bayless and Olyphant, 1993; Younger, 2000a,b). The AGS accumulate within the dewatered workings wherever oxidising pyrite is out of reach of flowing water. As the workings flood, all the AGS between the deepest level of working and the rest water table position will eventually be dissolved as the water surface reaches them. Younger (2000a) has noted that this commonly results in a ten-fold increase in the concentrations of contaminants (particularly iron) within the mine waters. Similar deteriorations in quality are sometimes observed following the initial saturation of some strip mine back-fill materials after restoration (e.g. Marsden *et al.*, 1997; Younger, 2000c), and during the flooding of open-pit mines to form pit lakes (Geller *et al.*, 1998). Such deterioration in water quality has important environmental consequences in the subsequent phase in the mine life-cycle, when the abandoned workings overspill to the surface environment (see below).

Besides these chemical changes, the rebound process can also prompt physical changes in the mined system. At least four scenarios have been documented:

1. Subsidence as open voids are eroded, either by rapidly flowing water winnowing support pillars, or perhaps more commonly by pillar collapse as the wetting of seat-earths (i.e. weak mudstones which often underlie coal seams) leads to slaking and a loss of strength. For instance, surface subsidence features which appeared in the recently-abandoned Leicestershire Coalfield (central England) in the early 1990s have been ascribed to the collapse of an 150-year old barrier pillar when it was inundated by rising mine waters (Smith and Colls, 1996).
2. Fault re-activation has been recorded during the flooding of deep mine workings in a number of European countries (Donnelly, 2000). The increase in pore pressures which occurs during rebound can reduce the frictional resistance to movement in extensional fault planes, leading to a temporary increase in seismicity, sometimes with associated property damage.
3. Rising ground levels of up to 25 cm have actually been recorded above an area of actively-flooding workings in the Limburg coalfield of the southern Netherlands (Bekendam and Pottgens, 1995). While the exact mechanism is not known, it is suspected to involve the re-hydration of swelling clays in the sequence, and an increase in buoyant support of open voids.
4. Development of high pressure gas pockets has been directly observed by miners who have been temporarily trapped in high points within deep mine workings, many metres below the water table, during catastrophic flooding incidents (e.g. Llewellyn, 1992; Younger and LaPierre, 2000). In one recent case (Younger, 1999; Younger and LaPierre, 2000), audible explosions originating at many hundreds of metres' depth in actively-flooding tin mine workings were ascribed to pneumatic fracturing of the crowns of open stopes as the pressure of gas trapped by rising mine waters exceeded the strength of the rock mass. The perceived sequence of events is illustrated diagrammatically in Figure 1.35.

The violent venting of air or other mine gases trapped in high pressure pockets during rebound is only one aspect of a more general process of acceleration of mine gas emissions during rebound. Low density mine gases such as methane will tend to behave like air: they will either vent readily from the workings if the pathways are available, or else they will accumulate under increasing pressure in high spots in the workings. High density mine gases, by contrast, will tend to form an invisible blanket "floating" on the surface of the rising mine water. This is the expected behaviour of some important hazardous mine gases, such as radon and stythe (CO_2-rich gas). The rising water table therefore tends to drive mine gases before it. This is a point worth remembering for anyone contemplating exploring old workings which do not receive forced ventilation. A tragic case which occurred in Northumberland, UK, in 1995 illustrates the dangers. In this instance a man working in a small factory was overcome by carbon dioxide when he

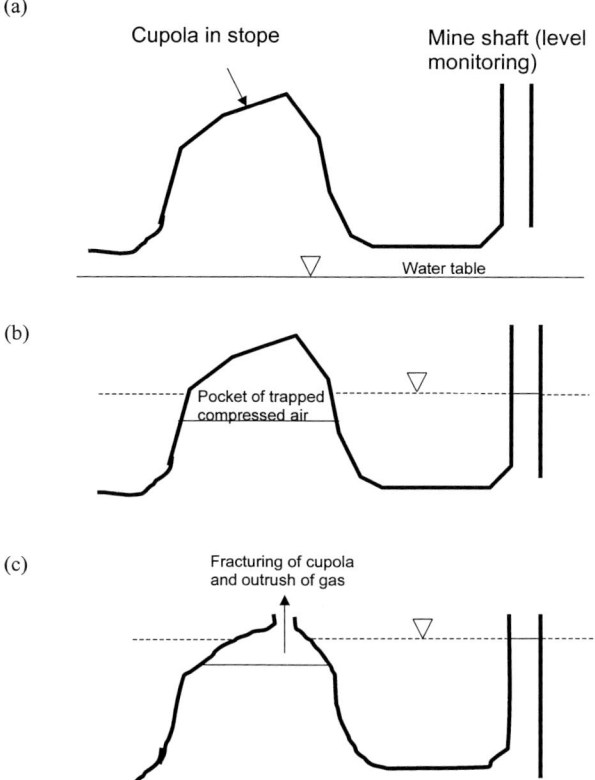

Figure 1.35: Possible process of pneumatic fracturing of a cupola in a large open stope, as is postulated to have occurred during mine water rebound in South Crofty Tin Mine, Cornwall (UK), in late July 1998 (after Younger and LaPierre, 2000).

entered a well-frequented basement area, which happened to adjoin an old drift portal (Burrell and Friel, 1996). The hitherto unprecedented carbon dioxide accumulation in this area has been ascribed to a local rise in the water table in the old workings.

One indirect impact of the rebound process is that, as long as it continues, previously-pumped waters will no longer be released to the surface environment. Where these pumped waters had been useful (for instance in providing dilution for sewage effluents entering the same rivers), then this temporary loss of surface flows can be significant (Banks *et al.*, 1996).

The *discharge of untreated mine waters after the flooding of workings* can lead to:

- Surface water pollution
- Pollution of over-lying aquifers

- Localised flooding, and
- Over-loading and clogging of sewers

As the first of these impacts is one of the major themes of this book (see Chapters Two and Three), there is little point in multiplying examples of the phenomenon here. With regard to the other three impacts:

(i)   Pollution of over-lying aquifers by upward migration of mine water has been recorded far less frequently than might be imagined. Indeed only one unequivocal case has been published, in which part of a public-supply dolomitic aquifer in north-east England was polluted when underlying coal mines were allowed to flood in the late 1970s (Younger and Adams, 1999).

(ii)   Localised flooding has affected residential, agricultural and industrial land in at least four UK coalfields (Younger and Adams, 1999), wherever new mine water discharges have emerged onto previously dry land. In many cases the flooding is a side affect of the following process:

(iii)   Over-loading of sewers occurs when the volume of water routed into a sewer exceeds the design flow. In some cases, high-volume mine water discharges have so overwhelmed the capacity of sewers that surface flooding has resulted (e.g. Younger, 2000c). Even if the mine water flows are not excessive, the clogging of the sewers with ferric hydroxide (ochre) precipitates can reduce the effective diameter of the sewer sufficiently that water backs-up and flooding occurs. This has been observed in Scotland (Younger, 2000c) and in Cleveland, UK (Younger, 2000d).

1.5 SUMMARY

The world-wide expansion in mining in the 20th Century released vast quantities of mineral wealth for human use. At the same time, the possible environmental impacts of mining grew ever more controversial, for a number of reasons:

- The general heightening of environmental awareness in all sectors of society
- The expansion of mining into hitherto untouched regions such as the Amazonian rainforest
- The switch in mining methods in the middle of the 20th Century from inconspicuous deep mining to visually obvious surface mining. Unlike deep mining, surface mining at least temporarily precludes other land uses in the worked area.
- A number of spectacular "own goals" by the mining industry between 1985 and 2000 (UNEP, 1996, 2000), in the shape of major incidents of pollution, in some cases accompanied by hundreds of fatalities, due to the failure of tailings dams, which are, ironically, constructed as pollution-prevention systems.

In order to develop appropriate, long-term measures to minimise (or remediate) the environmental damage caused by mining it is necessary to understand something of the processes by which deep mining and surface mining affect the ground. Deep mining can be undertaken by "supported" or "caving" methods; if the latter are used, subsidence of the strata overlying the workings is inevitable. Such subsidence can affect water levels in overlying aquifers (e.g. Booth and Spande, 1992; Booth *et al.*, 1998) and can also affect the dynamics of surface water courses (e.g. Sidle *et al.*, 2000). In extreme cases, open mine voids can propagate to the surface, capturing the flow of surface streams (Figure 1.34). Where "supported" methods of deep mining are used, deep mine voids can remain open for millennia, providing high-permeability pathways through the subsurface. On the other hand, old mine voids can suddenly collapse, either as a result of a belated switch from elastic to brittle deformation in the roof strata, or due to changes involving water movement. For instance, if flooded workings are drained, the removal of buoyant support can lead to roof collapse. On the other hand, if previously dry working are flooded, slaking of incompetent strata can lead to the collapse of support pillars (e.g. Smith and Colls, 1996).

Surface mining involves the wholesale excavation and (in many cases) back-filling of the strata which enclose valuable minerals. The back-filled waste rock can be expected to have radically different hydraulic properties from the unmined parent rock. The process of back-filling itself introduces marked heterogeneities in the grain size distributions of the waste rock (Figure 1.30), which can be expected to have important consequences for subsurface drainage (Newman *et al.*, 1997). The same considerations apply to waste rock piles associated with deep mines.

Tailings dams are purpose-built sedimentation lagoons which capture the fine-grained fraction of the waste stream resulting from mineral processing operations. They are typically constructed over many years, using waste rock to form the outer dyke. The loss of continuity in personnel during the life of some tailings dams is considered to be the underlying cause for some of the more spectacular cases of tailings dam failure in recent years (UNEP, 1996). In other words, the design of tailings dams is generally adequate; failures often arise due to a mis-match between the original design and the details of the construction actually implemented years later.

Negative impacts of mining on the water environment arise in at least six ways:

(i)   From the mining process itself (as mentioned above)
(ii)   From mineral processing operations (tailings dam failures being a case in point)
(iii)   From dewatering activities
(iv)   As pollution by spoil heap/tailings dam leachates and runoff
(v)   During the flooding of workings after mining has ceased
(vi)   By discharge of untreated waters after flooding is complete.

The latter is the most significant of all these impacts in the long term, and is therefore accorded particular attention in the chapters which follow.

CHAPTER TWO

MINE WATER CHEMISTRY

2.1 SOURCES OF MINE WATER CONTAMINANTS AND THEIR NATURAL ATTENUATION

The minerals and coal that represent economically valued ores are largely chemically stable under *in situ* geological conditions. When excavated and exposed to the atmosphere, however, these solid phases become chemically unstable. Coal and hydrocarbons can be oxidized through combustion to gain useful energy. The sulfide minerals associated with metal ores and present as sulfur contamination in coal will spontaneously dissolve when in contact with water. The *chemical weathering* of sulfide minerals represents a series of linked geochemical and microbiologically-mediated reactions through which contaminants are released from ore and mine waste into the hydrological cycle and become mobile and thus bioavailable as potentially toxic solutes.

Release of *metal ions* results from the weathering of sulfide ores containing minerals such as sphalerite ($ZnS(s)$, galena ($PbS(s)$) and arsenopyrite ($FeAsS(s)$). *Acidity* is primarily released during the weathering of pyrite ($FeS_2(s)$) in solutions containing dissolved oxygen. Subsequent *precipitation* of iron under oxic conditions produces *iron oxyhydroxide* minerals. These precipitates are generically termed *ochre* which is visibly deposited as yellow to red-brown coatings on solid surfaces within effluent channels at mined sites.

Metal ions, acidity and ochre all represent significant environmental hazards to freshwater resources. Environmental plans for decommissioning of active mines and environmental risk assessment of abandoned sites must consider the strength of the contaminant source (contaminant flux), the intensity of contamination (contaminant concentrations), the duration of the source (contaminant lifetime) and natural attenuation processes. In addition, hydrological transport pathways to sensitive receptors such as valued surface waters and groundwater abstraction zones must be considered. Figure 2.1 shows a schematic diagram of a mine workings showing contaminant sources, possible receptors, transport pathways and possible attenuation processes.

This chapter focuses on chemical aspects of contaminant behavior. Although the aqueous chemistry of mine water contaminants is relatively well understood, there remains a great deal of active research into detailed descriptions of processes that control contaminant release and attenuation (see reviews by Evangelou and Zhang, 1995; Nordstrom and Southam, 1997; and Ledin and Pedersen 1996). The chemical principles presented in the following sections provide a relatively simple mass balance approach to making initial assessments of contaminant loads and lifetimes and natural attenuation processes and their duration. Knowledge of how to balance chemical reactions and an ability to calculate and apply atomic and molecular formula weights is assumed. A concentration scale of

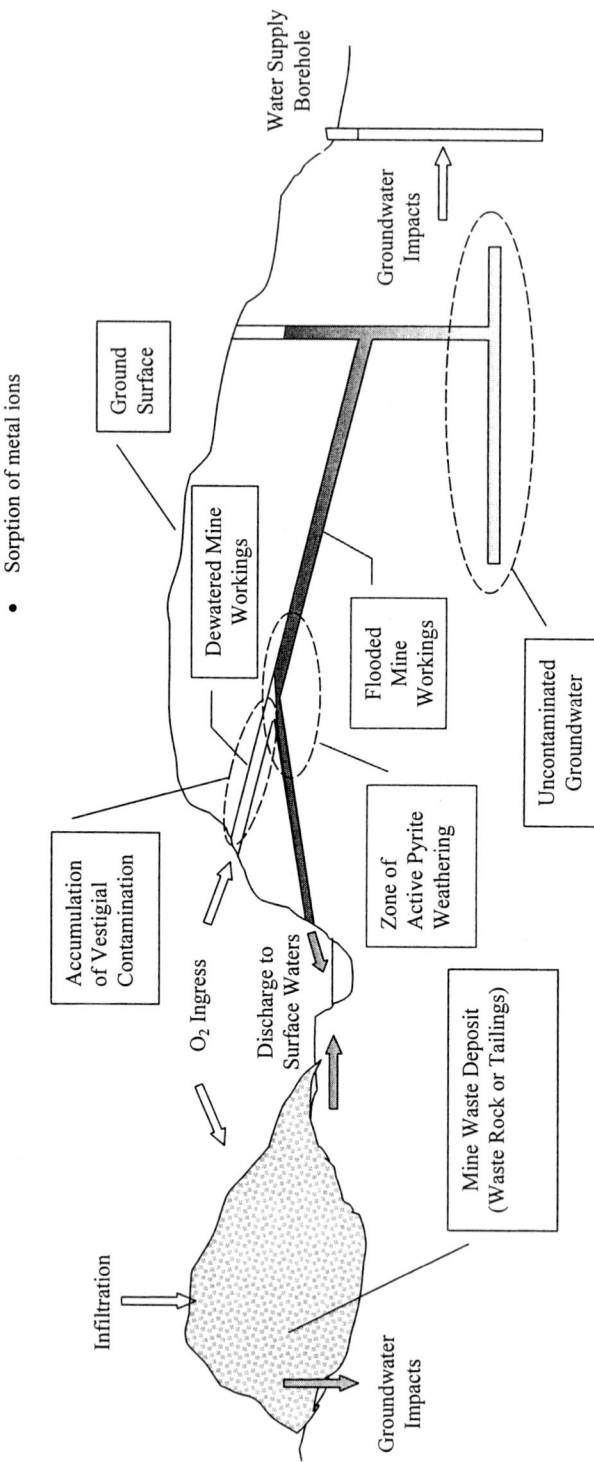

Figure 2.1: Diagram of mine workings where oxidative weathering of sulphide minerals produces acidity, metal ions and ochre as significant environmental hazards via transport through surface water and groundwater to sensitive receptors such as abstraction points for stream water, groundwater boreholes (wells) and stream biota. Active weathering occurs where oxic waters are in contact with sulphide-bearing rock. In the absence of active transport of contaminants, secondary weathering products can accumulate contaminants as solid precipitates in hydraulically unsaturated zones within surface workings or around dewatered tunnels.

moles l^{-1} is used with conditional (determined for stated ionic strength and temperature conditions) equilibrium constants when applying the thermodynamic law of mass action. Correction of thermodynamic data for the effects of changes in ionic strength and temperature is beyond the scope of this chapter and the reader is referred to a standard text on water chemistry (see for example Stumm and Morgan,1996, sections 3.4 and 3.5). Other concepts are introduced as needed.

Ultimately, the focus will be on applying these principles in order to answer the following questions.

1. Will there be a problem?
2. How severe will the problem be?
3. How long will the problem last?

2.1.1 MINERAL WEATHERING AS A SOURCE OF MINE WATER CONTAMINATION

Table 2.1 lists the relative abundance of rock forming minerals for several rock types that are representative for coal and metal ore mining. As for many mined areas, such data demonstrate the low contribution of sulfide minerals to the total mass of geological material present. It is the small fraction of sulfide minerals, usually only a few weight per cent in the ore, coal or mine waste, that has the potential to create significant environmental degradation that can last over decades or even centuries. In order to quantify contaminant loads from mine sites, and estimate their duration, it is necessary to understand which minerals contribute to the contamination.

The following chemical reactions represent processes that produce acidity, ochre and soluble metal ions when sulfide minerals dissolve in the presence of dissolved oxygen. These weathering processes are examples of redox (or *electron transfer*) reactions. Worked Example 2.1 illustrates how to write redox reactions for aqueous solutions.

Transfer of electrons to dissolved O_2 from Sulfur and Iron in the mineral phases, transforms S and Fe into soluble forms. The flow of electrons that occurs during these reactions can be harnessed by bacteria as an energy source to grow and to maintain cell functions. In most cases, microbial activity plays an important role in how fast these redox reactions proceed. Detailed models of the reaction mechanism and reaction rates generally must consider a number of linked geochemical and biochemical reaction steps. The reactions listed below only represent the average stoichiometry of overall weathering processes. Such stoichiometric descriptions are useful for assessing contaminant mass balances but provide no information on factors that influence weathering rates. These rates and the environmental factors that affect them are discussed in more detail in the next section.

Pyrite weathering that produces sulfate also releases soluble ferrous iron (Fe^{2+}) and acidity which is represented by production of protons in equation (2.1). The following reactions are also relevant for the weathering of the mineral marcasite, which has the same composition of pyrite ($FeS_2(s)$), but with a different crystallographic structure.

$$2FeS_2(s) + 7O_2(aq) + 2H_2O \rightarrow 2Fe^{2+} + 4SO_4^{2-} + 4H^+ \qquad (2.1)$$

Table 2.1: Examples of mineralogical composition as volume % of mine rock

Mineral Group	Mineral Names	[1]Coal Measure Mudrocks, North America	[1]Coal Measure Mudrocks, UK	[2]Coal Measure Sandstone, UK	[3]Biotite Gneiss with Mica Schist, Sweden
Sulfide	Pyrite FeS_2	1.0	2.5		0.57
Primary Silicates	Quartz SiO_2	25	16.5	55–62	24
Feldspars	Albite $NaAlSi_3O_8(s)$	1.0	0.5	24–41	13
	Anorthite $CaAl_2Si_2O_8(s)$				6
	K-Feldspar $KAlSi_3O_8(s)$				24
Micas	Biotite $KMg_{1.5}Fe_{1.5}AlSi_3O_{10}(OH)_2$				8
	Muscovite $KAl_2(AlSi_3O_{10})(OH)_2$				10
Iron Oxides	Goethite α-$FeOOH$		3	1–4	
	Hematite Fe_2O_3				
Carbonates	Calcite $CaCO_3$	2.0	4.5		0.1
	Dolomite $CaMg(CO_3)_2$				
	Magnesite $MgCO_3$				
	Ankerite $Ca_2MgFe(CO_3)_4$				
	Siderite $FeCO_3$				
Clay Minerals	Smectite $Ca_{0.5}(Al_{1.5}Mg_{0.5}Si_4)O_{10}(OH)_2$	70	74		
	Illite $K_{0.7}(Al_{1.4}Fe_{0.6})(Al_{0.6}Si_{3.4})O_{10}(OH)_2$				
	Mixed Layer Illite – Smectite				
	Kaolinite $Al_2Si_2O_5(OH)_4$				
	Chlorite $(Mg_5Al)(Si_3Al)O_{10}(OH)_2$				
Coal	[2]The sulfure content of UK coal is 0.8 – 4 weight % as organic S, framboidal pyrite and occasionally as massive pyrite				

[1]Taylor (1988), [2]Eden *et al.* (S. Yorkshire Coal Fields; 1957;), [3]Strömberg and Banwart (1994)

Worked Example 2.1 Writing Redox Reactions

The following reaction represents reduction of the charge on Oxygen atoms from 0 for dissolved O_2 (since elemental oxygen has no net charge) to 2– in H_2O (deduced by the fact that protons have a charge of 1+ and water itself has no net charge), due to transfer of electrons. This is written as the half-reaction for *reduction* of O_2.

$$O_2(aq) + 4e^- + 4H^+ \rightleftharpoons 2H_2O$$

The following half-reaction represents reduction of sulfate to hydrogen sulfide gas.

$$SO_4^{2-} + 8e^- + 10H^+ \rightleftharpoons H_2S(g) + 4H_2O$$

These reactions can also be written as reverse reactions for *oxidation* of water and hydrogen sulfide, respectively, where the charge on O or S atoms is increased due to loss of electrons. The *electroneutrality condition* dictates that net charge including that from electrons cannot accumulate in aqueous solution; i.e. the bulk solution has zero net charge. During redox reactions, conservation of electrons is maintained by necessarily coupling simultaneous oxidation and reduction of separate elements. This principle is illustrated by combining half cell reactions to give a coupled reduction-oxidation (redox) reaction. The oxidation of hydrogen sulfide gas by dissolved O_2 is written as follows, where the number of moles of each reactant and product in the oxygen half-reaction is multiplied by two in order to conserve the eight electrons released by sulfur oxidation.

$H_2S(g)$ oxidation	$H_2S(g) + 4H_2O$	$\rightleftharpoons SO_4^{2-} + 8e^- + 10H^+$
$O_2(aq)$ reduction	$2O_2(aq) + 8e^- + 8H^+$	$\rightleftharpoons 4H_2O$
$H_2S(g)$ oxidation by $O_2(aq)$	$H_2S(g) + 2O_2(aq)$	$\rightleftharpoons SO_4^{2-} + 2H^+$

A feature of many redox reactions in aqueous solution is the transfer of protons with electrons in order to maintain charge balance. In natural waters, many redox reactions can impact the acid-base balance. The oxidation of pyrite by molecular oxygen to produce free acidity is a classical example.

If sufficient dissolved oxygen is present, or if solutions can be oxygenated by contact with the atmosphere, the dissolved ferrous iron will be oxidized to ferric iron (Fe^{3+}), consuming acidity in the process.

$$2Fe^{2+} + \tfrac{1}{2}O_2 + 2H^+ \rightarrow 2Fe^{3+} + H_2O \tag{2.2}$$

A much greater net production of acidity occurs because ferric iron can react further to precipitate as iron oxyhydroxide forming ochre (equation 2.3), or by reacting further with the pyrite to produce more acidity and ferrous iron (equation 2.4). Because the forward and reverse reactions are relatively rapid for the precipitation and dissolution of ferric hydroxide, compared to the residence times of discharge water in mine workings, mine waters often achieve solubility equilibrium with these minerals. This is denoted by writing Equation (2.3) as both a forward and reverse reaction at equilibrium.

$$2Fe^{3+} + 6H_2O \rightleftharpoons 2Fe(OH)_3(s) + 6H^+ \tag{2.3}$$

$$14Fe^{3+} + FeS_2(s) + 8H_2O \rightarrow 2SO_4^{2-} + 15Fe^{2+} + 16H^+ \tag{2.4}$$

The ferrous iron produced in reaction (2.4) can then be re-oxidized by available dissolved oxygen, perpetuating the cycle represented by reactions (2.2) – (2.4). If dissolved oxygen becomes depleted, reaction (2.4) can proceed to completion yielding predominantly ferrous iron in solution.

Metal sulfides other than pyrite will not necessarily produce acidity, but will release soluble metal ions to solution. Some examples are given below where the corresponding sulfide mineral name is listed to the left of each weathering reaction.

Sphalerite	$ZnS(s) + 2O_2(aq) \rightarrow Zn^{2+} + SO_4^{2-}$	(2.5)
Galena	$PbS(s) + 2O_2(aq) \rightarrow Pb^{2+} + SO_4^{2-}$	(2.6)
Millerite	$NiS(s) + 2O_2(aq) \rightarrow Ni^{2+} + SO_4^{2-}$	(2.7)
Greenockite	$CdS(s) + 2O_2(aq) \rightarrow Cd^{2+} + SO_4^{2-}$	(2.8)
Covellite	$CuS(s) + 2O_2(aq) \rightarrow Cu^{2+} + SO_4^{2-}$	(2.9)
Chalcopyrite	$CuFeS_2(s) + 4O_2(aq) \rightarrow Cu^{2+} + Fe^{2+} + 2SO_4^{2-}$	(2.10)

It is clear from these weathering reactions that water infiltrating a mine workings will accumulate solutes as minerals dissolve. As acidity is released to solution during pyrite weathering, the pH will drop. There are other minerals however, that can consume acidity as they dissolve, thereby providing natural attenuation of the acidity produced by pyrite weathering, helping to buffer the pH. This is a critical aspect of assessing mine water contamination since the solubility and therefore the mobility and bioavailability of metal ions is strongly pH dependent. Metal ions generally have increasing solubility with decreasing pH. Natural attenuation of acidity (see Section 2.1.4) and pH buffering (see Section 2.2) turn out to be critical controls on the environmental behavior of metal ions that are released in mine water discharges.

2.1.2 THE WEATHERING KINETICS OF PYRITE AND RELATED SULFIDE MINERALS

Kinetic rate expressions are mathematical equations that describe how quickly pyrite oxidizes. Table 2.2 summarizes rate expressions relevant to the weathering of pyrite and marcasite. These equations show how this reaction rate depends on the amount of pyrite reacting and on the concentration of reactants such as dissolved oxygen and dissolved Fe^{3+}. The reader is also referred to excellent research publications by Holmes and Crundwell (2000) and Pfeifer and Stubert (1999) for discussions on factors that influence pyrite weathering rates and the application of rate laws to describe the reaction rates. Before delving into the equations in Table 2.2, it is useful to review a few basic aspects of mineral weathering reactions. Because mineral weathering reactions take place at the interface between the solid phase and the aqueous phase, reaction rates are proportional to the amount of mineral surface area that is reacting with solutes such as Fe^{3+} and $O_2(aq)$. This is reflected by the fact that weathering rates obtained in laboratory studies are often reported in units such as mole $m^{-2} s^{-1}$. Multiplying this rate by the amount of wetted mineral surface area reacting per litre of solution (A_w, $m^2 l^{-1}$) would yield a rate expressed in units of mole s^{-1}. It is clear in this case, that if the amount of reacting surface area is increased by a factor of 10, the rate of pyrite weathering would also increase by a factor of 10.

Table 2.2: Rate expressions for the oxidation of $FeS_2(s)$ and ferrous iron

Kinetic Reaction Process	Reference

$FeS_2(s)$ oxidation by dissolved $O_2(aq)$

$$2FeS_2(s) + 7O_2(aq) + 2H_2O \rightarrow 2Fe^{2+} + 4SO_4^{2+} + 4H^+$$

Rate expression: $-\dfrac{d[O_2]}{dt} = kA_w[O_2(aq)]^{0.5}$ mole L^{-1} s^{-1} 1

$k = 3 \pm 2 \times 10^{-8}$, mole$^{0.5}$ m^{-2} s^{-1}
A_w: reactive surface area, m^2 L^{-1}
Apparent activation energy for abiotic reaction: $E_a = 56.9$ kJ mole^{-1} 2
Microbial catalysis: evidence for 25- to 34-fold rate increase.   3

$FeS_2(s)$ oxidation by dissolved $Fe(III)$

$$FeS_2(s) + 14Fe^3 + 8H_2O \quad 15Fe^{2+} + 2SO_4^{2+} + 16H^+$$

Rate expression: $-\dfrac{d[Fe^{3+}]}{dt} = kA_w[Fe^{3+}]^{0.6}$ mole L^{-1} s^{-1} 4

$k = 3.0 \pm 2.0 \times 10^{-5}$, mole$^{0.4}$ L$^{0.6}$ m^{-2} s^{-1}
A_w: reactive surface area, m^2 L^{-1}
Apparent activation energy for abiotic reaction: $E_a = 92$ kJ mole^{-1} 5
Microbial catalysis: possible catalysis; no conclusive evidence.   3

Oxidation of aqueous $Fe(II)$ species by $O_2(aq)$

Dissolved species:	(0)	$2Fe^{2+} + \frac{1}{2}O_2 + 2H$	$\rightarrow 2Fe^{3+} + H_2O$
	(1)	$2Fe(OH)^+ + \frac{1}{2}O_2 + 2H^+$	$\rightarrow 2Fe(OH)^{2+} + H_2O$
	(2)	$2Fe(OH)_2(aq) + \frac{1}{2}O_2 + 2H^+$	$\rightarrow 2Fe(OH)^{2+} + H_2O$
Adsorbed species*:	(3)	$2XFeOFe^+ + \frac{1}{2}O_2 + 2H^+$	$\rightarrow 2XFeOFe^{2+} + H_2O$
	(4)	$2XFeOFeOH + \frac{1}{2}O_2 + 2H^+$	$\rightarrow 2XFeOFeOH^+ + H_2O$

Rate expression: $-\dfrac{d[Fe(II)]}{dt} = \Sigma k_i[Fe(II)]_i[O_2]$ mole $L^{-1}s^{-1}$ 6

Fe^{2+}	$\log k_0 = -5.1$	$XFeOFe^+$	$\log k_3 = 0.7$
$FeOH^+$	$\log k_1 = 1.4$	$XFeOFeOH$	$\log k_4 = 0.7$
$Fe(OH)_2(aq)$	$\log k_2 = 6.9$		

Microbial catalysis: up to a factor of 10^6 for oxidation of Fe^{2+}, depending on nutrient   3, 7
status, temperature and pH.

Equilibrium Reactions for Formation of Reacting Aqueous Species

i	Species i	Formation Reaction	Formation constant LogK, 25°C, I = OM	
0	Fe^{2+}	-	-	-
1	$FeOH^+$	$Fe^{2+} + H_2O \rightleftharpoons FeOH^+ + H^+$	-9.5	6
2	$Fe(OH)_2(aq)$	$Fe^{2+} + 2H_2O \rightleftharpoons Fe(OH)_2(aq) + 2H^+$	-20.6	6

Table 2.2 continued overleaf

Table 2.2 continued

3 XFeOFe$^+$	Fe^{2+} + XFeOH $\rightleftharpoons$ XFeOFe$^+$ + H$^+$		−1.99	8**
4 XFeOFeOH	Fe^{2+} + XFeOH + H$_2$O $\rightleftharpoons$ XFeOFeOH + 2H$^+$		−8.39	8**
O$_2$(aq)	O$_2$(g) $\rightleftharpoons$ O$_2$(aq)		−2.90 (mole L^{-1} atm^{-1}) 6	

1. Summarized in Strömberg and Banwart (1994).
2. McKibben and Barnes (1986).
3. Reviewed in Nordstrom and Southam (1997).
4. Rimstidt and Newcombe (1993).
5 Wiersma and Rimstidt (1984).
6. Wehrli (1990).
7. Singer and Stumm (1970).
8. Zhang *et al.* (1992).
*XFeOH represents unoccupied adsorption sites
**Determined at I = 0.1 M, 25°C.

The specific surface area (m^2 g^{-1}) of minerals is highly variable. It depends on crystal morphology, surface texture, porosity and degree of aggregation with other minerals in a particular geological environment. This makes it very difficult to predict how fast a weathering reaction will proceed, even if the ratio of mineral mass to solution volume is known. The point to emphasize is: A$_w$ is usually a very uncertain parameter.

The constant of proportionality in the rate expression is the *rate constant*. This parameter, although characteristic for the reaction of a particular mineral with specific solutes, does vary with the geological conditions in which the mineral formed; i.e. the same type of mineral sampled from different settings could have different *kinetic reactivity*. The standard errors listed for the values of the rate constants in Table 2 reflect this variability in reactivity. The rate constant also varies with temperature according to the Arrhenius equation (Equation 2.11), and depends directly on the activation energy of the reaction being studied (E$_a$, J mole^{-1}). R is the universal gas constant (8.314 J mole^{-1} K^{-1}) and T is the absolute temperature in Kelvin. The subscripts 1 and 2 refer to two different conditions of temperature. The value of k at T$_2$ can be calculated if the rate constant at temperature T$_1$ is known.

$$\frac{k_{T1}}{k_{T2}} = Exp \left[\frac{E_a}{R} \left(\frac{1}{T_1} - \frac{1}{T_2} \right) \right] \tag{2.11}$$

An apparent activation energy can be determined by assessing reaction rate constants calculated from experimental data determined at different temperatures, and then fitting the Arhenius equation to the data with E$_a$ as an adjustable parameter. Table 2.2 lists published values of E$_a$ for pyrite weathering reactions.

The weathering rate is also generally dependent on the amount of oxidant available to react with the pyrite surface. The rate expressions are expressed as power laws where the rate is proportional to the concentration of a particular reactant raised to the power *n*. The value of the exponent *n* is the *reaction order*. Mineral weathering is said to be *first-order* in surface area, since the rate is proportional to (A$_w$)$^{1.0}$. Rate laws for mineral weathering are often *fractional-order* in dissolved reactants; i.e. the reaction order is less than 1 (Hering and Stumm,1990).

Table 2.2 lists the rate laws for the reactions represented by equations (2.1), (2.2) and (2.4). The rate laws for abiotic oxidation of pyrite by dissolved oxygen and dissolved ferric ion are fractional order with respect to these oxidants, with $n = 0.5$ and $n = 0.6$, respectively. There is also evidence that the reaction of pyrite with dissolved oxygen is also fractional order with respect to protons; the rate being proportional to $[H^+]^{0.11}$ (Williamson and Rimstidt, 1994). This dependence states that a drop in pH from 7 to 2 corresponds to a 0.5 log unit drop in rate which corresponds to a factor of about 3. This dependence is neglected in Table 2 since application to assessing natural environments generally includes other uncertainties such as mineral surface area which are far greater than this effect of pH.

So what controls the rate of pyrite or marcasite oxidation? A key point is that the oxidation of pyrite by Fe^{3+} is generally faster than oxidation by dissolved $O_2(aq)$; i.e. reaction (2.4) above is faster than reaction (2.1). The rate-limiting step in the overall pyrite weathering process is therefore reaction (2.2), which provides a continuous supply of Fe^{3+} to react with pyrite. Although this abiotic chemical reaction is rapid at neutral pH and above, it is very slow under acidic conditions. However, at low pH, acidophilic bacteria can thrive, and can catalyze reaction (2.2), increasing the reaction rate by up to a factor of 10^6 (Singer and Stumm, 1970). In this case, pyrite weathering can accelerate significantly over rates that occur at neutral pH. First, this occurs because the rate limitation on the production of Fe^{3+} as an oxidant is removed by microbial catalysis. Second, the oxidation of pyrite by Fe^{3+} is enhanced by the high solubility of $Fe(OH)_3(s)$ at low pH; i.e. the weathering rate is accelerated by increased Fe^{3+} concentration.

The abiotic rate expressions can be useful to assess the limits of these rates. However, it can be expected that reaction rates will depend on ecological factors that influence microbial activity; i.e. the composition of the indigenous microbial consortia, temperature, pH and nutrient status of the aqueous environment.

This section has outlined the theoretical framework for describing the weathering of sulfide minerals, and has also given some indication of the factors upon which the rate of $FeS_2(s)$ weathering depends. The following section discusses proposed methods for controlling reaction rates as a method of pollution prevention. Application of weathering rates in the assessment of mine water pollution is covered in Section 2.2.

2.1.3 BIOLOGICAL AND CHEMICAL METHODS FOR POLLUTION PREVENTION

The rate laws in Table 2.2 demonstrate that reactive surface area, oxidant concentration and microbial activity are key controls on the rate of sulfide mineral oxidation. Inhibition of these reactions can occur through either chemical or microbiological effects. Strategies for chemical inhibition focus on adding solutes that block the reactive surface area by forming surface precipitates; and thus prevent access to oxidants and release of weathering products (Evangelou and Zhang, 1995; Roy and Worral, 1999). Strategies for microbiological inhibition seek to add bactericides that decrease microbial activity and thus slow reaction rates (Kleinmann, 1980; McCready, 1982; Stichbury *et al.*, 1995; Lortie *et al.*, 1999).

A known chemical inhibitor is orthophosphate (PO_4^{3-}) which forms sparingly soluble iron(III) phosphate mineral phases. Such phases are known to be effective in lowering pyrite weathering rates, if the phosphate phase can be induced to precipitate directly on the pyrite surface. The main problem is that pyrite can reduce Fe(III) to Fe(II)(aq) species, thus preventing formation of the Fe(III)-phosphate precipitate. This has been circumvented in trial systems by dosing pyrite with hydrogen peroxide as a highly reactive oxidant for Fe(II) on the pyrite surface, followed by immediate addition of phosphate (Evangelou, 1995b). A further complication is that the surface precipitate has decreasing stability with decreasing pH. Other types of surface precipitates, particularly silica coatings, remain stable at low pH and may be more effective under conditions of fluctuating mine water quality (Vandiviere and Evangelou, 1998). An example is the process of silica micro-encapsulation which renders sulfide minerals and metal precipitates relatively inert through surface precipitation of silica coatings and is currently applied to treat mine water discharges containing reactive suspended particulates (Mitchell *et al.*, 2000).

A common problem of bactericide treatments to inhibit pyrite weathering is that potentially potent inhibitors are themselves pollutants and should not be introduced into the environment. Exceptions include some anionic surfactants and heterocyclic mercaptans and road salt. The detergent sodium lauryl sulfate has been demonstrated in laboratory and field trials to be effective, at least in the short term (weeks-months) in reducing contaminant production from pyritic rock waste (Kleinmann, 1980). Lab studies and field trials of 2,5 dimercapto-1,3,4-thiadazole and 5-amino-1,3,4-thiadiazole-2-thiol suggested that microbial activity could be arrested in tailings deposits for periods of up to one year (Lortie *et al.*, 1999). The activity of the best-known iron-oxidizer, *Thiobacillus ferroxidans*, is also known to be inhibited in saline environments (McCready 1982; McCready and Krouse, 1982). Laboratory studies of salt layers together with clay caps demonstrate that pyrite oxidation in underlying layers can be effectively inhibited. This is achieved by preventing ingress of $O_2(g)$ through the clay cover, coupled with microbial inhibition due to saline pore waters arising from the salt.

For practical application of either chemical or microbiological inhibition, there are common problems of 1) delivering effective inhibitors to areas in the mine environment where active sulfide weathering occurs and 2) maintaining effective inhibitor loads in open hydrological systems due to flushing. Potential applications generally suppose a solid reservoir of inhibitor with slow release. For phosphate or silica precipitates, mineral sources of P and Si incorporated with mine waste may provide slow release to pore waters, thus delivering the inhibitor to sub-surface flow paths in contact with oxidizing sulfide minerals. Slow release Si resins for environmental applications are also commercially available. Trials with anionic detergents have utilized a slow release polymer matrix impregnated with the inhibitor (Kleinmann, 1980; Kleinmann *et al.*, 1981), while road salt can be applied as part of a multi-layer capping system.

Although the principles of chemical and microbiological inhibition are outlined above, any potential application will require a number of potentially complicated site-specific issues to be addressed. Regulatory issues related to release of compounds to the environment may prevent application. Thorough testing at bench and pilot scale will be required, as will demonstration of long-term effectiveness and financial feasibility. A good control of transport processes including $O_2(g)$ diffusion and water flow at any site is presumed as a prerequisite to effective performance.

Some of these issues can be circumvented if $O_2(g)$ diffusion control and hydraulic control methods are applied. Both wet (ponded water) and dry covers (compost, organic soils, clay covers, etc.) are commonly used to prevent ingress of $O_2(g)$, while clay caps and other impermeable covers can minimize infiltration of water to mine environments with pyritic rock deposits. These methods are discussed in Section 3.12.

2.1.4 MINERAL WEATHERING AS A SOURCE OF NATURAL ATTENUATION

The weathering of many carbonate and silicate minerals consumes acidity and helps buffer pH. Worked example 2.2 calculates the amount of acidity produced by pyrite weathering in a mine waste deposit, and compares it with the pH of the discharge in order to demonstrate the role of natural attenuation. Dissolution of calcite is a classical example of a weathering process that consumes acidity. This is represented by the following reaction that consumes protons and releases calcium and bicarbonate ions to solution. Calcite generally dissolves sufficiently rapidly to maintain a mine water discharge in solubility equilibrium with the mineral if it is present.

$$CaCO_3(s) + H^+ \rightleftharpoons Ca^{2+} + HCO^{3-} \tag{2.12}$$

In addition to calcite dissolution, weathering of aluminosilicate minerals consumes acidity, although rates of dissolution are much slower than for calcite. Some examples of weathering reactions for primary silicate minerals are listed below.

Feldspars
K-feldspar
$$KAlSi_3O_{8(s)} + H^+ + 9/2H_2O \rightarrow 2H_4SiO_{4(aq)} + 1/2Al_2Si_2O_5(OH)_{4(s)} \tag{2.13}$$

Plagioclase, including:
Anorthite
$$CaAl_2Si_2O_{8(s)} + 2H^+ + H_2O \rightarrow Ca^{2+} + 2H_4Si_{4(aq)} + 1/2Al_2Si_2O_5(OH)_{4(s)} \tag{2.14}$$

Albite
$$NaAlSi_3O_{8(s)} + H^+ + 9/2H_2O \rightarrow Na^+ + 2H_4SiO_{4(aq)} + 1/2Al_2Si_2O_5(OH)_{4(s)} \tag{2.15}$$

Micas
Biotite
$$KMg_{3/2}Fe_{3/2}[AlSi_3]O_{10}(OH)_{2(s)} + 7H^+ + 1/2H_2O \rightarrow K^+ + 3/2Mg^{2+} + 3/2Fe^{2+} +$$
$$2H_4SiO_{4(aq)} + 1/2Al_2Si_2O_5(OH)_{4(s)} \tag{2.16}$$

Muscovite
$$KAl_2[AlSi_3O_{10}](OH)_{2(s)} + H^+ + 3/2H_2O \rightarrow K^+ + 3/2Al_2Si_2O_5(OH)_{4(s)} \tag{2.17}$$

The weathering of iron carbonate ($FeCO_3(s)$) and ferroan dolomite minerals ('ankerite') (($Ca,Mg,Fe)CO_3(s)$) also initially consumes acidity. It can subsequently be released upon oxidation of the ferrous iron and precipitation of ochre resulting in no net loss or gain of acidity.

$$FeCO_3(s) + H^+ \rightleftharpoons Fe^{2+} + HCO_3^- \qquad\qquad (2.18)$$

$$Fe^{2+} + \tfrac{1}{4}O_2 + 5/2H_2O \rightarrow Fe(OH)_3(s) + 2H^+ \qquad\qquad (2.19)$$

2.1.5 CONTAMINANT MOBILITY: pH AS A MASTER VARIABLE

As sulfide minerals dissolve, the metal ions that are released to solution can remain soluble or can be precipitated, particularly as oxyhydroxide minerals such as $Zn(OH)_2(s)$, $Al(OH)_3(s)$ and $Fe(OH)_3(s)$. The solubility of such phases is critical since they affect the mobility of metal ions in water. Precipitated solids will accumulate within transport pathways, while ions will move with the water flow. Figure 2.2 shows the dependence of the solubility of some metal hydroxide minerals on pH. The increasing solubility with decreasing pH is a common feature of such mineral phases.

Table 2.3 emphasizes this feature of contaminant chemistry by showing hydrochemical analyses from a variety of mine water discharges. The data demonstrate that low pH discharges are generally characterized by relatively high concentrations of dissolved metals while circumneutral pH conditions are associated with much lower concentrations. A critical control on the solubility and thus mobility and bioavailability of metal ions is therefore the pH of the discharge. Although pH is linked to acidity production from the weathering processes outlined above, weathering of silicate and carbonate minerals must also be considered. These minerals act as potentially important sources of neutralizing capacity for the natural attenuation of acidity. Such buffering processes stabilize the drop in pH that results from pyrite weathering.

The metal ions released by sulfide weathering can also precipitate as sulfate-, carbonate- and in some cases silicate- mineral phases. Some examples include minerals that immobilize Ferric Iron (Jarosite: $KFe_3(SO_4)_2(OH)_6(s)$), Ferrous Iron (Melanterite: $FeSO_4 \cdot 7H_2O(s)$), Aluminum (Alunite: $KAl_3(SO_4)_2(OH)_6(s)$), Lead (Anglesite: $PbSO_4(s)$; Cerrusite: $PbCO_3(s)$) and Copper (Malachite: $Cu_2(OH)_2(CO_3)(s)$). Formation of sulfate-bearing minerals is favored where extensive pyrite oxidation has produced high sulfate concentrations. In the absence of significant acidity consumption by weathering of calcite or silicate oxides, high sulfate concentrations will also be associated with low pH. Due to the high solubility of oxyhydroxide minerals at low pH, these conditions often prevent their precipitation which would otherwise limit metal ion concentrations to values below the saturation limits where sulfate-bearing minerals precipitate.

Cited situations (Langmuir,1997) where Alunite-Jarosite minerals are likely to occur include evaporative concentration of pore and capillary waters such as those occurring within the walls of dewatered underground workings or in hydraulically-unsaturated pyritic waste rock deposits. These precipitates represent residual contaminant loadings of metal ions and acidity which can be re-mobilized if hydraulic flushing occurs such as during mine water rebound subsequent to mine closure (Younger,1997a).

2.2 ACIDITY, ALKALINITY AND pH

Because acidity is a key contaminant in mine water, and because metal ion mobility is critically dependent on pH, it is necessary to clearly understand how pH depends on

Worked Example 2.2 Comparing Acidity Production and Discharge pH

The following hydrochemical data for drainage from a mine waste deposit shows sulfate release as well as copper, resulting from weathering of pyrite and traces of chalcopyrite in the ore rock. If there is no natural attenuation, the amount of acidity in the discharge should correspond to the protons produced by pyrite weathering. The amount of acidity produced by pyrite weathering can be estimated from hydrochemical data.

Hydrochemical Analyses of Discharge Water from a Copper Mine Waste Deposit (mg L^{-1}).

pH	3.8	SO_4^{2-}	1310	Ca^{2+}	185	Mg^{2+}	57
Al_T	75	Cu^{2+}	19	Fe_T	2	Mn_T	12
Na^+	46	K^+	17	Si_T	19	Cl^-	27

Note: The subscript T refers to total dissolved concentration

Step 1. Calculate the molecular weight of sulfate and list the atomic weight of Copper.

Element	S	O	Cu
Atomic Weight	32.066	15.9994	63.546

Formula Weight (FW) for SO_4^{2-}; 1 mole of S and 4 moles of O:

FW = (1)(32.066) + (4)(15.9994)   = 96.064   g mole^{-1}

Step 2. Calculate molar concentration of Sulfate and Cu in discharge.

$[SO_4^{2-}]$ $= 1310 \div 96.064$ $= 13.64$ mmole L^{-1}

$[Cu]$ $= 19 \div 63.546$ $= 0.299$ mmole L^{-1}

Step 3. Calculate sulfate release from pyrite; accounting for sulfate released from chalcopyrite (2 S for each Cu in $CuFeS_2(s)$).

$[SO_4^{2-}]_{Py} = [SO_4^{2-}]_T - 2[Cu^{2+}]$

$[SO_2^{2-}]_{Py} = 13.64 - 2(0.299) = 13.04$ mmoles L^{-1}

Step 4. Calculate protons released from pyrite weathering (Eq. 2.1–Eq. 2.3 in text); 4 protons are produced when pyrite dissolves to produce sulfate and ochre. Note that very little dissolved Fe is produced, indicating that most Fe released by pyrite weathering precipitates within the deposit as ochre.

$[H^+] = 4(13.04)$ $= 52.17$ mmoles L^{-1}

$= 0.05217$ moles L^{-1}

Step 5. Calculate pH from expected proton concentration.

$pH = -\log[H^+] = -\log(0.05217)$

$\cong 1.3$

The observed proton concentration in solution, pH = 3.8, is more than 2 pH units (2 orders-of-magnitude in proton concentration) above the calculated value. This is because dissolution of carbonate and silicate minerals neutralizes much of the acidity produced by pyrite weathering. Further evidence for natural attenuation is given by the relatively high concentrations of cations (Ca^{2+}, Mg^{2+}, Na^+, K^+) in the discharge that are released by carbonate and silicate weathering. At acid pH, total acidity is approximately equal to the proton concentration, although this is not always the case. The relationship between acidity and pH is presented in section 2.2.

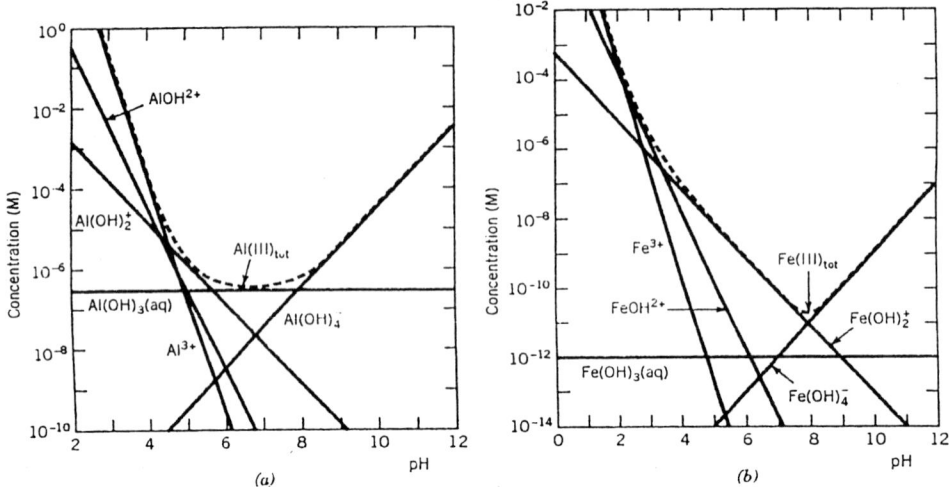

Figure 2.2: The pH dependence of solubility for aluminium- and iron-hydroxide minerals (from Stumm and Morgan, 1996). Note that below neutral pH, solubility of these phases increases with decreasing pH. This emphasises the role of pH as a master variable to control the solubility of metal ions and thus their mobility. Metal ions such as Fe^{3+} react with hydroxide ions to form hydrolysis complexes such as $FeOH^{2+}$. The straight lines that are plotted show how concentrations of individual metal species such as Fe^{3+}, $FeOH^{2+}$, $Fe(OH)_4^-$, etc. depend on pH. The dashed line in each plot is the sum of concentrations for all species and thus corresponds to the total dissolved concentration of Fe or Al for solutions in solubility equilibrium with these minerals.

Table 2.3: A comparison of metal ion concentrations for mine water discharges (Banks *et al.*, 1997). Concentrations in units of mg L^{-1}.

Site Description	pH	$[SO_4^{2-}]$	[Fe]	[Al]	[Mn]	[Zn]	[Cu]
Pyrite Mine	2.5	5110	1460	84.21	3.05	0.94	0.16
Coal Spoil Tip	2.7	1077	179	27.5	3.2	1.3	0.168
Abandoned Coal Mine	3.6	1044	101.3	17.3	4.02	0.221	0.007
Coal Spoil Tip	3.7	77	1.6	1.8	0.40	0.49	0.014
Abandoned Coal Mine	4.2	1554	180	< 0.5	6.1	0.061	
Coal Spoil Tip	5.5	146	287	0.97	5.2	0.05	< 0.007
Abandoned Coal Mine	6.3	83	4.9	0.078	0.36	0.048	0.005
Abandoned Coal Mine	6.3	210	10.6	< 0.045	1.26	< 0.007	< 0.007
Metal Mine	6.5	124	14.9	0.132	2.10	0.029	
Coal Spoil Tip	6.8	1327	18.6	< 0.045	2.0	< 0.007	
Pumped Coal Mine	7.1	690	0.63			0.056	
Pumped Coal Mine	7.1	1170	5.8			0.034	
Pumped Coal Mine	7.3	380	5.0			0.030	
Abandoned Coal Mine	7.9	176	0.097	< 0.01	0.138	< 0.005	0.0007
Pumped Coal Mine	8.2	7.4	< 0.01	< 0.02	0.004	0.055	< 0.005

acidity and alkalinity. The relationship is not trivial. A rigorous definition requires review of some basic concepts from aqueous chemistry. We start by defining strong acids and bases (Section 2.2.1), and subsequently describe weak acids and how their dissociation to release protons depends on pH (Section 2.2.2). The relationship between alkalinity and pH for unpolluted waters can then be defined rigorously in terms of the acid-base equilibria of the weak acids that constitute the carbonate buffer system (Section 2.2.3). The models that we develop are useful for predicting how pH in unpolluted receiving streams will change as a mine water with known acidity is discharged to them.

In Section 2.2.4 we consider other weak acids associated with mine water discharges. In particular we describe how metal ions contribute to mine water acidity. This requires extending the concepts of acidity and alkalinity to a more general framework; the "proton condition". This framework is applied in later chapters that consider acid neutralization within the context of remediation technology. The final two sections on acidity and pH describe how acidity can be stored as soluble mineral salts (Section 2.2.6), and describe the impact of biota on mine water pH (Section 2.2.7). Subsequent chapters build on these section to describe how contamination from mineral salts accumulates in rising groundwater during mine water rebound, and how processes such as microbial sulfate reduction can be harnessed to remediate mine water discharge quality.

2.2.1 STRONG ACIDS AND BASES: DEFINING ACIDITY AND ALKALINITY

A general definition of acidity is "the excess of strong acid over strong base in an aqueous solution". The acidity of a mine water corresponds to the amount of strong base that must be added in order to raise the pH to a specified value. Alternatively, the alkalinity is defined as the amount of strong acid that must be added to lower the pH to the same defined value.

Operationally, the threshold value is often defined as pH 4.5, the endpoint of the bromcresol green–methyl red indicator. If a mine water sample has a pH below 4.5 it is net acidic, and if the pH is above 4.5 it is net alkaline. In practice, alkalinity is determined by adding a few drops of the indicator to a known volume of mine water and then incrementally adding strong acid until the color change indicates that the pH has been lowered to pH 4.5. This titration provides a quantitative empirical measurement of the alkalinity of the sample in units of equivalents per litre. Worked Example 2.3 presents calculations of alkalinity on the basis of titration data. Acidity determinations are somewhat more complicated since addition of strong base to an acidic mine water sample can cause very rapid local precipitation of $Fe(OH)_3(s)$ or $Al(OH)_3(s)$ as the base mixes with the sample. Dissolution of the precipitate is not necessarily immediate and errors to the titration are introduced since precipitation and dissolution of these hydroxides add and remove protons in solution (Equation 2.3).

Before the acidity and alkalinity of a mine water can be related to the pH, some additional concepts are needed. In particular it is necessary to understand what differentiates "strong" acids and bases from "weak" ones. Strong acids and bases are compounds that dissociate completely when dissolved in water, releasing protons or hydroxide ions respectively, in the process. For natural waters, strong acids include H_2SO_4, HNO_3 and HCl, while strong bases include $NaOH$, KOH, $Mg(OH)_2$ and $Ca(OH)_2$. Because these compounds dissociate completely, the amount added can be determined by

---

Worked Example 2.3 Calculating Alkalinity from Titration Data

Alkalinity is determined empirically by acidimetric titration. A 50 ml. sample of mine water with circumeutral pH is titrated with 0.01 Molar (Mole L^{-1}) H_2SO_4. The acid is added drop-wise with constant stirring until the indicator color change occurs. The total amount of acid added is 3.70 mL. Calculate the alkalinity as equivalents per litre of mine water. The term "equivalents" is used to count the equivalent number of moles of H^+ associated with a mole of acid; i.e. 2 molar equivalents of H^+ per mole of H_2SO_4.

Step 1. Calculate the moles of acid added from H_2SO_4.
$\qquad$ 3.70 mL = 3.70×10^{-3} L
$\qquad$ moles of acid = (Litres of acid added) × (concentration of acid in moles L^{-1})
$\qquad$ = $(3.70 \times 10^{-3})(0.01) = 3.70 \times 10^{-5}$ moles

Step 2. Calculate the charge equivalents of protons added.
$\qquad$ 1 mole of sulfuric acid = 2 charge equivalents of protons
$\qquad$ protons added = $2(3.70 \times 10^{-5})$
$\qquad\qquad$ = 7.40×10^{-5} equivalents

Step 3. Calculate the mine water alkalinity in equivalents per Litre.
$\qquad$ Alk $\qquad$ = (equivalents of protons) ÷ (volume of sample)
$\qquad\qquad$ = $(7.40 \times 10^{-5}) \div (0.050) = 1.48 \times 10^{-3}$ eq. L^{-1}

---

the amount of "acid anion" (SO_4^{2-}, NO_3^-, Cl^-) or "base cation" (Na^+, K^+, Mg^{2+}, Ca^{2+}) in solution. Equations (2.20) and (2.21) mathematically define acidity (Acy) and alkalinity (Alk) according to the excess of strong acid or base in solution. Each mole of H_2SO_4 dissociates to produce two charge equivalents of protons, and each mole of $Ca(OH)_2$ and $Mg(OH)_2$ dissociate to provide two equivalents of hydroxide ion. The concentrations of the respective acid anion and base cations must therefore be multiplied by two. The square brackets refer to the molar concentration scale with units of mole litre^{-1}.

$$[Acy] = 2[SO_4^{2-}] + [NO_3^-] + [Cl^-] - [Na^+] - [K^+] - 2[Mg^{2+}] - 2[Ca^{2+}] \qquad (2.20)$$

$$[Alk] = [Na^+] + [K^+] + 2[Mg^{2+}] + 2[Ca^{2+}] - 2[SO_4^{2-}] - [NO_3^-] - [Cl^-] \qquad (2.21)$$

$$[Acy] = -[Alk] \qquad (2.22)$$

2.2.2 WEAK ACIDS: THE CO_2–H_2O BUFFER SYSTEM

Unlike strong acids, weak acids do not dissociate completely. Equation (2.23) shows the dissociation of carbonic acid.

$$H_2CO_3 \rightleftharpoons HCO_3^- + H^+ \qquad (2.23)$$

A common shorthand for this reaction represents the left hand side as "carbonic acid" with an asterisk. This is because the equilibrium for the reaction $CO_2(aq) + H_2O \rightleftharpoons H_2CO_3$ lies far to the left and true carbonic acid (H_2CO_3) is therefore present in negligible concentrations compared to dissolved $CO_2(aq)$). The dissociation constant for the deprotonation of $H_2CO_3{}^*$ reflects both the hydration of $CO_2(aq)$ and the deprotonation of true H_2CO_3 to form $HCO_3{}^-$ (Equation 2.24).

$$H_2CO_3{}^* \rightleftharpoons HCO_3{}^- + H^+ \quad -logK_1 = 6.35 \ (25°C, I = 0M) \quad (2.24)$$

The protonated form $H_2CO_3{}^*$ is the acid; donating protons to solution. The deprotonated form $HCO_3{}^-$ is the *conjugate base*. The total concentration of inorganic carbon is equal to sum of the concentrations of the acid and conjugate base. The pK value of the dissociation constant is defined as $pK_1 = -logK_1$ and reflects the pH value where the concentration of acid equals the concentration of its conjugate base. For carbonic acid in equation (2.24), the concentration of acid is greater than the conjugate base if the pH is below pH 6.35. If the pH is higher than pH 6.35, then the concentration of conjugate base is higher. Carbonic acid is an example of a "weak" acid because it does not dissociate completely and exists partially as the protonated acid and partially as its conjugate base.

The bicarbonate ion ($HCO_3{}^-$) can also deprotonate further to donate another proton to solution and yield carbonate ion ($CO_3{}^{2-}$) as its conjugate base.

$$HCO^{3-} \rightleftharpoons CO_3{}^{2-} + H^+ \quad -logK_2 = 10.3 \ (I = 0M, 25°C) \quad (2.25)$$

At pH values above pH 10.3, carbonate ion dominates, while at pH values between 10.3 and 6.3, bicarbonate ion dominates. Figure 2.3 shows plots of the concentration of carbonic acid, bicarbonate ion and carbonate ion as a function of pH, for solutions which are open and closed to carbon dioxide gas. By combining the thermodynamic law of mass action, described below, with the mass balance for inorganic carbon, it is possible to calculate concentrations of carbonic acid and its conjugate bases. This is developed for the particular case of mine water discharges that are in contact with the atmosphere; i.e. open to carbon dioxide.

The principle of thermodynamic mass action can be applied to a general stoichiometric reaction: $aA + bB + cC + \ldots \rightleftharpoons nN + pP + rR + \ldots$, with an equilibrium constant K_{eq} that has been corrected for temperature and the ionic composition of the solution. The brackets indicate units of molar concentration (moles L^{-1}) for reactants (A,B,C, etc) and products (N,P,R,etc.).

$$K_{eq} = \frac{[N]^n[P]^p[R]^r\ldots}{[A]^a[B]^b[C]^c\ldots} \quad (2.26)$$

Although there are a large number of solutes which act as weak acids, these can often be neglected in natural waters except for carbonic acid which forms when carbon dioxide dissolves in water.

$$CO_2(g) + H_2O \rightleftharpoons H_2CO_3{}^* \quad -logK = 1.47 \ (25°C, I = 0 \ M) \quad (2.27)$$

For the dissolution of $CO_2(g)$ in water, and the dissociation of the resulting carbonic acid, the following thermodynamic mass action laws follow from equations (2.27), (2.24) and

(a)

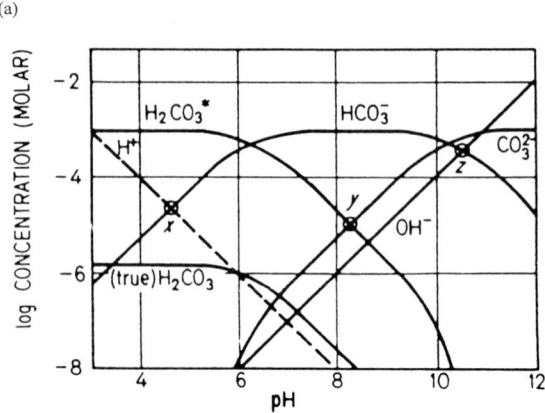

(b)

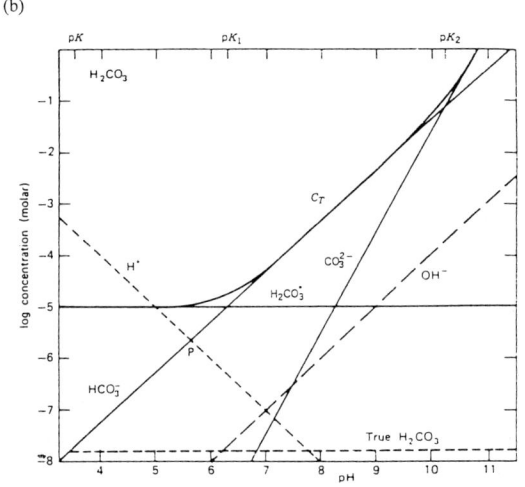

Figure 2.3: Plots of concentrations for individual inorganic carbon species as a function of pH for aqueous solutions that are (a) closed and (b) open to carbon dioxide (from Stumm and Morgan, 1996). Deep aquifers that are not in contact with the unsaturated zone, and with no biogenic gases or gas production from sedimentary carbon, would be closed to $CO_2(g)$. Surface waters receiving mine water discharges are in contact with atmospheric $CO_2(g)$ and are therefore open to $CO_2(g)$. The envelope of total carbon concentration (C_T) plotted in (b) represents the solubility of inorganic carbon as a function of pH. For any single pH value, the ratios of bicarbonate and carbonate ion concentrations to the carbonic acid concentration are the same in both figures and are dictated by the equilibrium described in the text. For an open system, the carbonic acid concentration remains constant with pH, determined by equilibrium at a fixed $PCO_2(g)$. In a closed system the carbonic acid concentration decreases as pH increases, since more carbon is converted to bicarbonate and carbonate ions, with no possibility of dissolving more carbon dioxide from a gas phases. The assumption of open or closed conditions can be tested by assuming open conditions and calculating the expected $PCO_2(g)$ from the equilibrium equations presented in the text. Waters that are substantially undersaturated with respect to atmospheric $PCO_2(g)$; i.e. calculated $PCO_2(g)$ at least 2–3 orders-of-magnitude lower than $PCO_2(g)$ atm, can be characterised as closed systems.

(2.25), respectively. Note that gases have concentrations expressed as a partial pressure with units of atmospheres (atm). For example, the earth's atmosphere is 78% by volume $N_2(g)$, and has a total pressure of 1.0 atm. The atmospheric partial pressure of $N_2(g)$ is therefore $PN_2(g) = 0.78$ atm. The partial pressure of carbon dioxide in the atmosphere is approximately $PCO_2(g) = 3.55 \times 10^{-4}$ atm.

$$K = 10^{-1.47} = \frac{[H_2CO_3^*]}{PCO_2(g)} \qquad (25°C, I = 0\ M) \qquad (2.28)$$

$$K_1 = 10^{-6.35} = \frac{[HCO_3^-][H^+]}{[HCO_3^*]} \qquad (25°C, I = 0\ M) \qquad (2.29)$$

$$K_2 = 10^{-10.33} = \frac{[CO_3^{2-}][H^+]}{[HCO_3^-]} \qquad (25°C, I = 0\ M) \qquad (2.30)$$

The total concentration of inorganic carbon is defined by mass balance.

$$C_T = [H_2CO_3^*] + [HCO_3^-] + [CO_3^{2-}] \qquad ,moles\ L^{-1} \qquad (2.31)$$

Some simplifications arise from inspection of Figure 2.3. The median pH value between pK_1 and pK_2 defines the point in a hypothetical titration where $[H_2CO_3^*] = [CO_3^{2-}]$. This occurs at pH = 8.3. This is termed an "equivalence point" in the titration. It defines the pH range beyond which the concentrations of particular acids and conjugate bases no longer contribute significantly to alkalinity or acidity.

For the carbonic acid buffer system, the following assumptions can be applied. For pH > 8.3, $[H_2CO_3^*]$ is much lower than the concentration of bicarbonate and carbonate ion and can be neglected in the mass balance. For pH < 8.3, $[CO_3^{2-}]$ can be neglected in the mass balance. In the respective pH ranges, these assumptions yield the following simplified expressions by combining the mass action laws with the mass balance for carbon. By neglecting the concentration of carbonate ion, Equation (2.32) defines the mass balance for total carbon, solved for the carbonic acid concentration.

$$pH < 8.3 \qquad [H_2CO_3^*] \cong C_T - [HCO_3^-] \qquad (2.32)$$

This expression can be substituted into Equation (2.29) in order to eliminate $[H_2CO_3^*]$ and allow Equation (2.33) to be solved for bicarbonate concentration, yielding Equation (2.34).

$$pH < 8.3 \qquad K_1 = \frac{[HCO_3^-][H^+]}{(C_T - [HCO_3^-])} \qquad (2.33)$$

$$pH < 8.3 \qquad [HCO_3^-] = \frac{K_1 C_T}{[H+] + K_1} \qquad (2.34)$$

A second expression for the concentration of carbonic acid can be obtained by solving the mass balance for bicarbonate concentration (Equation 2.35), eliminating $[HCO_3^-]$ in the mass action law (Equation 2.36) and solving for $[HCO_3^*]$, yielding Equation (2.37).

$$pH < 8.3 \qquad\qquad [HCO_3^-] \cong C_T - [H_2CO_3^*] \qquad\qquad (2.35)$$

$$pH < 8.3 \qquad\qquad K_1 = \frac{(C_T - [H_2CO_3^*])[H^+]}{[H_2CO_3^*]} \qquad\qquad (2.36)$$

$$pH < 8.3 \qquad\qquad [H_2CO_3^*] = \frac{[H^+]C_T}{[H^+] + K_1} \qquad\qquad (2.37)$$

By analogy, expressions for bicarbonate and carbonate concentrations in the region pH > 8.3 can be obtained by assuming that $[H_2CO_3^*] \cong 0$.

$$For\ pH > 8.3 \qquad\qquad [CO_3^{2-}] = \frac{K_2 C_T}{[H^+] + K_2} \qquad\qquad (2.38)$$

$$[HCO_3^-] = \frac{[H^+]C_T}{[H^+] + K_2} \qquad\qquad (2.39)$$

The concentration of carbonate ion can also be directly related to pH and the concentration of carbonic acid by combining equations (2.24) and (2.25). The reactions are added together like algebraic equations with the reaction arrows as an equal sign. All reactants are added together on the left side of the combined reactions and all products are added together on the right. Bicarbonate ion cancels out since it is both a reactant and a product in the combined reaction. The logarithm of the equilibrium constant for the combined reaction is the sum of the logarithms of the two added reactions.

$$H_2CO_3^* \rightleftharpoons HCO_3^- + H^+ \qquad -logK_1 = 6.35 \qquad\qquad (2.40)$$

$$HCO_3^- \rightleftharpoons CO_3^{2-} + H^+ \qquad -logK_2 = 10.33 \qquad\qquad (2.41)$$

$$H_2CO_3^* + HCO_3^- \rightleftharpoons HCO_3^- + CO_3^{2-} + 2H^+ \qquad -logK_2' = 6.35 + 10.33 = 16.68 \quad (2.42)$$

As mentioned, HCO_3^- can be eliminated since it appears on both sides of the reaction. The combined reaction and the corresponding thermodynamic mass action law are now defined.

$$H_2CO_3^* \rightleftharpoons CO_3^{2-} + 2H^+ \qquad -logK_2' = 16.68\ (25°C, I = 0\ M) \qquad\qquad (2.43)$$

$$K_2' = 10^{-16.68} = \frac{[CO_3^{2-}][H^+]^2}{[H_2CO_3^*]} \qquad\qquad (2.44)$$

$$[CO_3^{2-}] = \frac{K'_2 = [H_2CO_3^*]}{[H^+]^2} \qquad (2.45)$$

For a water that is not in contact with $CO_2(g)$; i.e. the atmosphere or another source, C_T is constant. Inorganic carbon can neither be gained nor lost. This is termed a "closed" system to $CO_2(g)$. If the system is "open" to $CO_2(g)$, the carbon dioxide can be dissolved from the gas phase to increase dissolved inorganic carbon, or can be lost by degassing from the solution. C_T is not constant for an open system, and is seen to increase with increasing pH in Figure 2.3b. Worked examples 2.4 and 2.5 calculate the distribution of carbonic acid, bicarbonate and carbonate ions for mine waters that are open and closed, respectively, to $CO_2(g)$. These calculations do not account for the possible dissolution or precipitation of calcite or other carbonate phases. Since the objective here is to understand how the pH of an unpolluted stream will change when receiving mine water discharges, we can focus on the solution chemistry. Treatment processes such as lime addition, or geochemical modelling of interstitial waters in sedimentary strata containing carbonate minerals, must necessarily consider interactions that add or remove dissolved inorganic carbon due to interactions with minerals. These concepts are applied as needed in subsequent chapters on mine water treatment and remediation.

2.2.3 BUFFERING MINE WATER QUALITY: THE RELATION BETWEEN ACIDITY AND PH

A complete description of the acid-base balance for mine waters must also take account of proton acidity and hydroxide alkalinity, that contributed directly by H^+ and OH^- ions. The relationship between protons and hydroxide ions is described by the dissociation of water.

$$H_2O \rightleftharpoons H^+ + OH^- \qquad -logK_w = 14 \ (25°C, I = 0 \ M) \qquad (2.46)$$

To obtain a relationship between pH and acidity or alkalinity, the concentrations of weak acids, their conjugate bases, protons and hydroxide ions must all be related to the earlier definition of [Alk] and [Acy] based on the excess of strong acids or bases. This is done by considering that the electroneutrality condition must hold for all aqueous solutions, and that it can be defined by setting the charge equivalents of all cations equal to that of all anions.

$$[Na^+] + [K^+] + 2[Mg^{2+}] + 2[Ca^{2+}] + [H^+] =$$
$$- 2[SO_4^{2-}] - [NO_3^-] - [Cl^-] + [HCO_3^-] + 2[CO_3^{2-}] + [OH^-] \qquad (2.47)$$

By combining equation (2.47) with the earlier definitions of acidity and alkalinity as the excess of strong acid or base (Equations 2.20–2.21), these two chemical capacities can be defined in terms of the conjugate bases of carbonic acid and protons and hydroxide ions. Note that alkalinity is identically equal to $-[Acy]$.

$$[Alk] = [Na^+] + [K^+] + 2[Mg^{2+}] + 2[Ca^{2+}] - 2[SO_4^{2-}] - [NO_3^-] - [Cl^-]$$
$$= [HCO_3^-] + 2[CO_3^{2-}] + [OH^-] - [H^+] \qquad (2.48)$$

Worked Example 2.4 Mine Waters Closed to Carbon Dioxide

Closed $CO_2(g)$ systems are solutions that cannot exchange carbon dioxide gas with a gas phase; for example, deep groundwaters without contact with the atmosphere and in the absence of $CO_2(g)$ production from biodegradation. This case is seldom encountered in mine water pollution since the environmental problems of interest generally arise due to exposure of mine workings and mine waste to the atmosphere. The problem is relevant to many groundwater problems however, where atmospheric and biogenic $CO_2(g)$, or gas evolution from lithological strata with sedimentary carbon can be neglected. In this case, a laboratory determination of total dissolved inorganic carbon (TIC) in a groundwater sample can be used with a pH measurement to calculate the carbon speciation in the groundwater. For example, if TIC = 74 mg L^{-1} and the *in situ* pH of the water is 7.8, the equations presented in the text can be used to calculate the concentration of carbonic acid, bicarbonate ion and carbonate ion.

Step 1. Calculate the molar concentration of TIC
   The atomic weight (A_w) of Carbon is 12.011 grams $mole^{-1}$
   TIC = 0.074 g L^{-1}
   C_T = TIC ÷ A_w = 0.074 ÷ 12.011 = 6.2×10^{-3} moles L^{-1}

Step 2. Examine the pH range; in this case, pH < 8.3, and select the necessary expressions for the concentration of each species.

$$[H_2CO_3^*] = \frac{[H^+]C_T}{[H^+] + K_1}$$

$$[HCO_3^-] = \frac{K_1C_T}{[H^+] + K_1}$$

Step 3. Calculate the proton concentration from the pH value.
   $[H+] = 10^{-pH} = 10^{-7.8} = 1.58 \times 10^{-8}$ moles L^{-1}

Step 4. Calculate species concentrations using C_T and $[H^+]$ values; $K_1 = 10^{-6.35}$.
   $[H_2CO_3^*] = 2.12 \times 10^{-4}$ moles L^{-1}
   $[HCO_3^-] = 5.99 \times 10^{-3}$ moles L^{-1}

Step 5. Check mass balance.
   $C_T = 2.12 \times 10^{-4} + 5.99 \times 10^{-3} = 6.20 \times 10^{-3}$ moles L^{-1}

Step 6. Carbonate ion concentration can also be checked because the carbonic acid is known.

$$[CO_3^{2-}] = \frac{K_2'[H_2CO_3^*]}{[H^+]^2} = 1.77 \times 10^{-5} \text{ moles } L^{-1}$$

This values agrees with the assumption that $[CO_3^{2-}] << [HCO_3^-], [H_2CO_3^*]$ for pH<8.3.

$$[Acy] = 2[SO_4^{2-}] + [NO_3^-] + [Cl^-] - [Na^+] - [K^+] - 2[Mg^{2+}] - 2[Ca^{2+}]$$
$$= [H^+] - [HCO_3^-] - 2[CO_3^{2-}] - [OH^-] \qquad (2.49)$$

Inspection of Figure 2.3 shows that for circumneutral pH, bicarbonate ion is the dominant form of inorganic carbon in solution. This explains why alkalinity titrations, which strictly speaking determine [Alk] and not $[HCO_3^-]$, can often be used as an analytical determination of bicarbonate ion concentration. There are two limiting conditions. One is that no

---

Worked Example 2.5 Mine Waters Open to Carbon Dioxide Gas

For a solution in equilibrium with a fixed partial pressure of $CO_2(g)$, the mass action law for dissolution of $CO_2(g)$ (equation 2.28) shows that $[H_2CO_3^*]$ is constant with changes in pH; fixed only by the value for $PCO_2(g)$. The concentrations of the other species can then be calculated if $[H_2CO_3^*]$ is known. A mine water with pH 7.1 is discharged from an abandoned workings into a receiving stream. The partial pressure of $CO_2(g)$ in equilibrium with the stream is assumed to be atmospheric; $PCO_2(g) = 10^{-3.45}$ atm. The equations presented in the text can be used to calculate the concentrations of carbonic acid, bicarbonate ion and carbonate ion.

Step 1. Write relevant mass action laws.

$$K = 10^{-1.47} = \frac{[H_2CO_3^*]}{PCO_2(g)} \qquad (25°C, I = 0 \text{ M})$$

$$K_1 = 10^{-6.35} = \frac{[HCO_3^-][H^+]}{[HCO_3^*]} \qquad (25°C, I = 0 \text{ M})$$

$$K_2' = 10^{-16.68} = \frac{[CO_3^{2-}][H^+]^2}{[H_2CO_3^*]} \qquad (25°C, I = 0 \text{ M})$$

Step 2. Solve each expression respectively for $[H_2CO_3^*]$, $[HCO_3^-]$ and $[CO_3^{2-}]$.

$$[H_2CO_3^*] = KPCO_2(g)$$

$$[HCO_3^-] = \frac{K_1[H_2CO_3^*]}{[H^+]}$$

$$[CO_3^{2-}] = \frac{K_2'[H_2CO_3^*]}{[H^+]^2}$$

Step 3. Calculate the concentration of $[H_2CO_3^*]$.
$$[H_2CO_3^*] = (10^{-1.47})(10^{-3.45}) = 10^{-4.92} = 1.20 \times 10^{-5} \text{ moles L}^{-1}$$

Step 4. Calculate the concentrations of bicarbonate and carbonate ion.
$$[H^+] = 10^{-7.1} = 7.94 \times 10^{-8} \text{ moles L}^{-1}$$
$$[HCO_3^-] = 6.75 \times 10^{-5} \text{ moles L}^{-1}$$
$$[CO_3^{2-}] = 3.94 \times 10^{-8} \text{ moles L}^{-1}$$

---

other weak acids or their conjugate bases are present in significant concentrations. The other condition is that the pH is not sufficiently high or low such that $[H^+]$, $[CO_3^{2-}]$ or $[OH]^-$ become large enough that they contribute significantly to $[Alk]$. The next section will illustrate how metal ions such as Al^{3+} also act as weak acids and need to be considered in mine water applications.

Some rules of thumb for the relative concentrations of protons, hydroxide ions, carbonate and bicarbonate ions are as follows. At neutral pH, proton and hydroxide ion concentrations are near 10^{-7} M and the ratio between bicarbonate and carbonate ion concentrations is more than 10^3. The concentration of the weak acid decreases by one order of magnitude for every pH unit above the pK value. For every 1 unit decrease in pH below the pK value, the concentration of the conjugate base decreases by 1 order-of-magnitude.

If 100% refers to the concentration of conjugate base at pH = pK, then it will have 10% of this concentration at pH = pK – 1, 1% at pH = pK – 2, 0.1% at pH = pK – 3, etc. Thus, $[CO_3^{2-}]$ should be about 10^3 times less at pH 7 than at pH 10.3 because pH 7 is approximately 3 pH units lower than the value for $pK_2 = 10.3$. By the same reasoning the concentration of the protonated species decreases in a similar way at pH values above the pK. For example at pH 7, $[H_2CO_3^*]$ should be somewhat more than 10% of its concentration at pH = 6.3, since the pH is slightly less than 1 pH unit above the value for pK_1.

How does the pH of a receiving stream change as a mine water discharges into it? Under the assumptions of an open $CO_2(g)$ system and for pH < 8.3, the concentration of bicarbonate ion can be related to the carbonic acid concentration by Equation (2.29). Introducing these assumptions into the definition of alkalinity (Equation 2.47) yields Equation (2.50) which relates alkalinity and pH in the range pH < 8.3. Solving this quadratic equation for $[H^+]$ gives equation (2.51). These equations could just as easily be written in terms of [Acy] with the corresponding changes of sign reflecting [Acy] = – [Alk]; i.e. equation (2.52).

$$pH < 8.3 \qquad [Alk] \cong \frac{K_1[HCO_3^*]}{[H^+]} - [H^+] \qquad\qquad \text{moles L}^{-1} \qquad (2.50)$$

$$pH < 8.3 \qquad [H^+] = \frac{-[Alk] + \sqrt{[Alk]^2 + 4K_1[H_2CO_3^*]}}{2} \qquad \text{moles L}^{-1} \qquad (2.51)$$

$$pH < 8.3 \qquad [H^+] = \frac{[Acy] + \sqrt{[Alk]^2 + 4K_1[H_2CO_3^*]}}{2} \qquad \text{moles L}^{-1} \qquad (2.52)$$

Figure 2.4 shows the resulting theoretical *titration curve* which plots pH as a function of alkalinity for a water that is open to atmospheric carbon dioxide. This plot illustrates several key concepts and implications for the quality of surface waters that receive mine water discharges.

- pH, acidity and alkalinity are related through the $CO_2(g)$–H_2O buffer system.
- Neutral pH waters are buffered by bicarbonate alkalinity down to about pH 6.
- Further small additions of acidic discharge may result in large drops in pH.
- Precipitous drops in pH can have serious implications for aquatic life and for the behavior of other contaminants such as metal ions.
- Changes in pH from mixing of acidic and alkaline waters can be calculated by treating [Acy] and [Alk] as conservative quantities.

First of all, it is important to note that pH does not have a simple relationship to acidity or alkalinity. pH is almost independent of [Alk] above the pK_1 of carbonic acid, but drops precipitously with very small changes in [Alk] below pH 6. This behavior stresses the role of bicarbonate ions to *buffer pH* against additions of acid when an acidic discharge is released into receiving waters that are otherwise at circumneutral pH. In other words, the

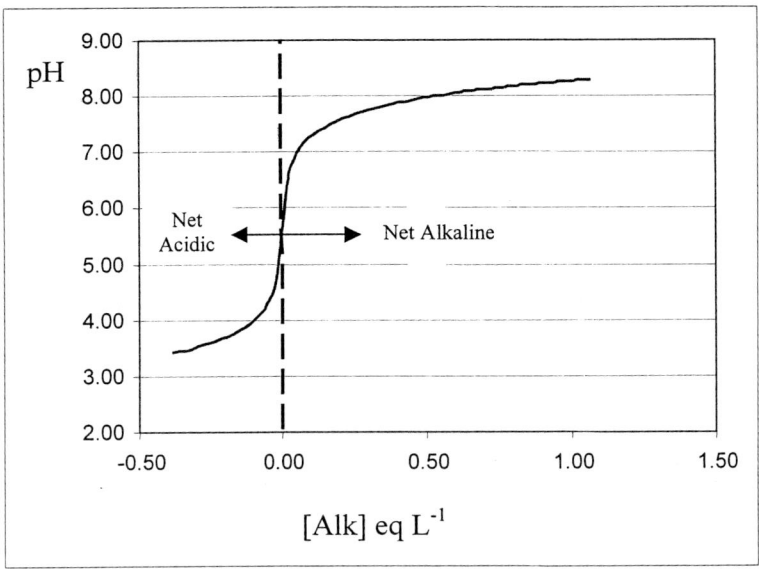

Figure 2.4: Theoretical titration curve calculated by considering the relationship between –log(H⁺) and [Alk] or [Acy] for an open system at atmospheric PCO₂(g). This plot represents a model for the pH dependence expected for an otherwise uncontaminated surface water as increasing amounts of acidic mine water are discharged into it; i.e. decreasing [Alk].

protons in the acidic discharge will be consumed by protonating HCO_3^- to $H_2CO_3^*$, rather than remaining as proton acidity in solution. According to Figure 2.4, a 90% drop in alkalinity (a decrease of ~1 meq L^{-1}) causes of drop of pH from 8 to about 6, an increase of only $\sim 10^{-6}$ moles L^{-1} in proton concentration. On the other hand, below pH 6, further small additions of acid can easily cause a pH drop to values below 4; i.e. the solution is unbuffered. Also noting Figure 2.2, it must be emphasized that loss of alkalinity as buffer capacity in a discharge will have possibly catastrophic consequences on metal ion solubility and mobility, due to the large drop in pH. Alkalinity and pH are often referred to as *master variables* in water chemistry, due to their importance in dictating changes in pH and thus the behavior of other solutes such as metal ions.

It is also important to remember that acidity and alkalinity can be assessed from major cation and anion analyses of mine waters (Equations 2.20 and 2.21). If these ions are conserved during mixing of a mine water discharge with a receiving stream, then the net acidity or alkalinity of the mixed waters downstream can be assessed *a priori*. If a volume of mine water V_M (L) with acidity $[Acy]_M$ (eq L^{-1}) is mixed with a volume of receiving water V_R (L) with alkalinity $[Alk]_R$ (eq L^{-1}), then the net acidity or alkalinity $[Alk]_M$ of the mixed water can be calculated by equation (2.53), and its pH determined from equation (2.51) or (2.52).

$$[Alk] = \frac{(V_M)(-[Acy]_M) + (V_R)([Alk]_R)}{V_M + V_R} \qquad \text{eq. } L^{-1} \qquad (2.53)$$

Worked Example 2.6 illustrates application of mixing calculations for acidity and alkalinity and prediction of pH changes in receiving waters due to mine water discharges. Such relatively simple calculations allow prediction of how a discharge will impact the pH of receiving waters; for example when considering discharge consent or the necessary degree of treatment before discharge.

By analogy, the use of [Alk] and [Acy] as conservative parameters that can be used to determine the net alkalinity or acidity of mixed waters, can be extended to geochemical and biological processes that produce and consume acidity. If calcite dissolution produces alkalinity at a faster rate than pyrite oxidation produces acidity, then the mixed waters emanating from a mine workings where both processes occur, will be net-alkaline. Alternatively, if calcite is not present or is depleted, and pyrite weathering produces acidity at a greater rate than alkalinity is produced from other attenuation processes such as aluminosilicate weathering, the mixed waters within the mine workings will be net-acidic. These principles are critical to understanding how the relative rates of competing processes dictate the acidity and thus pH of a discharge, and define a framework within which net-acidic or net-alkaline discharge are associated with a rigorously defined range of pH values. The following section expands on this relationship between net acidic or alkaline waters and pH; extending it to include other weak acids and bases, and processes other than mineral weathering that influence the acid-base balance of mine waters.

What is the exact threshold pH where acidity and alkalinity are zero? Solving Equation (2.51) or (2.52) for the condition [Acy] = [Alk] = 0 and fixing $[H_2CO_3^*]$ by equilibrium with atmospheric $PCO_2(g)$ defines the theoretical titration endpoint pH for acidimetric titration of bicarbonate alkalinity.

$$[H^+] = (K_1[H_2CO_3^*])^{1/2} = 2.32 \times 10^{-6} \qquad \text{moles } L^{-1} \qquad (2.54)$$

$$pH = -\log(2.32 \times 10^{-6}) = 5.64 \qquad (2.55)$$

Mine waters with pH > 5.64 are thus defined as net-alkaline while those with pH < 5.64 are defined as net acidic.

This broad classification of mine waters has profound practical implications since Figure 2.4 shows that the master variable pH drops dramatically across the transition from net alkaline to net acidic conditions. How does the theoretical endpoint of pH = 5.64 relate to the operational endpoint given by the bromcresol green–methyl red indicator endpoint of pH = 4.5? Applying Eq. (2.32) with pH = 4.5 yields [Alk] = -3.15×10^{-5} eq L^{-1}. For a solution originally near pH = 8 with [Alk] $\sim 10^{-3}$ eq L^{-1}, this corresponds to a slight over-estimate of [Alk]; about 3%.

2.2.4 THE PROTON CONDITION: CONSIDERING OTHER WEAK ACIDS IN MINE WATERS

The concepts of alkalinity and acidity can be extended to a general consideration of all weak acids and bases that might be present in a mine water discharge. In particular, the hydrolysis and precipitation of Aluminium and Ferric Iron ions need to be considered. The solubility products (Equations (2.56, 2.57), see Stumm and Morgan, 1996, p. 979) for these minerals depends strongly on the age and crystallinity of the precipitate. Freshly

Worked Example 2.6 Calculating the Effect of Mine Water Discharge on the Alkalinity and pH of a Receiving Stream.

A mine water discharge of $20 \, L \, s^{-1}$ and pH 3.61 enters a stream with an upstream discharge of $120 \, L \, s^{-1}$, pH 8.3 and $[HCO_3^-] = 10^{-3}$ mole L^{-1}. Calculate the alkalinity and pH of the mixed waters downstream of the discharge. Subsequently estimate what mine water discharge rate would yield pH 6 in the mixed waters downstream.

Step 1. Calculate acidity of mine water discharge.

   Note: for pH $<<$ 5.56, $[H^+] >> [HCO_3^-]$, $[CO_3^{2-}]$ and $[OH^-]$.

   $-\log[H^+] = pH$
   $[H^+] = 10^{-pH} = 10^{-3.61}$
   $[Acy] \cong [H^+] = 10^{-3.61} = 2.45 \times 10^{-4}$ eq. L^{-1}

Step 2. Calculate the alkalinity of the mixed water downstream of the discharge.

$$[Alk] = \frac{(Q_M)(-[Acy]_M) + (Q_R)([Alk]_R)}{Q_M + Q_R} \quad \text{eq. } L^{-1}$$

where: Q_M = mine water discharge $(L \, s^{-1})$, Q_R=stream discharge $(L \, s^{-1})$, $[Alk]_R$ is upstream alkalinity (eq L^{-1}), $[Alk]$ is downstream alkalinity, $[Acy]_M$ is acidity of mine water (eq L^{-1}).

$$[Alk] = \frac{(20)(-2.45 \times 10^{-4}) + (120)(10^{-3})}{20 + 120} = 8.22 \times 10^{-4} \text{ eq } L^{-1}$$

Step 3. Calculate pH. Note that $[H_2CO_3^*] = 1.20 \times 10^{-5}$ mole L^{-1} for equilibrium with atmospheric $PCO_2(g)$ and $K_1 = 10^{-6.3}$ for carbonic acid.

   For pH < 8.3:

$$[H^+] = \frac{-[Alk] + \sqrt{[Alk]^2 + 4K_1[H_2CO_3^*]}}{2}$$

$$[H^+] = \frac{1}{2} \left[\left(-8.22 \times 10^{-4}\right) + \sqrt{\left(8.22 \times 10^{-4}\right)^2 + (4)(10^{-6.3})(1.20 \times 10^{-5})} \right]$$

   $[H^+] = 7.32 \times 10^{-9}$ mole L^{-1}
   $pH = -\log(7.32 \times 10^{-9}) = 8.14$

Step 4. Estimate what mine water discharge rate would give pH = 6.0.

$$[Alk] \cong \frac{K_1[H_2CO_3^*]}{[H^+]} - [H^+]$$

Worked Example 2.6 continued overleaf

Worked Example 2.6 continued

$$[Alk] \cong \frac{10^{-6.3} (1.20 \times 10^{-5})}{[10^{-6}]} - 10^{-6} = 5.01 \times 10^{-6} \text{ eq L}^{-1}$$

$$Q_M = \frac{Q_R([Alk]_R - [Alk])}{[Alk] + [Acy]} = \frac{(120)(10^{-3} - 5.01 \times 10^{-6})}{5.01 \times 10^{-6} + 2.45 \times 10^{-4}}$$

$$Q_M = 478 \text{ L s}^{-1}$$

Step 5. Review assumptions.
The calculation assumes that the carbonic acid system is in equilibrium with atmospheric carbon dioxide. The equation relating pH and alkalinity holds only for pH values below 8.3; hence the upstream conditions are just at the limit of this assumption. It is also assumed that no acid or base is added or removed by chemical reactions during the mixing; i.e. alkalinity is treated as a conservative parameter. Under this assumption, in fact, it is very easy to determine the answer to Step 4, once the acidity of the discharge is calculated in step 1. The discharge acidity is ¼ of the alkalinity upstream. Hence, a pH near 5.56 corresponding to zero net-alkalinity would result from a discharge that provides approximately $4Q_R$, or around 480 L s^{-1} to the river.

---

precipitated hydroxides are amorphous but age over periods of weeks and months to more cystalline and less soluble forms. Section 2.3 focuses on hydrolysis reactions and their role in the pH dependence of mineral oxide solubility.

$$Al(OH)_3(s) + 3H^+ \rightleftharpoons Al^{3+} + 3H_2O \qquad 8.11 < logK_{so} < 10.8 \qquad (2.56)$$

$$Fe(OH)_3(s) + 3H^+ \rightleftharpoons Fe^{3+} + 3H_2O \qquad -1 < -logK_{so} < 5.0 \qquad (2.57)$$

The hydrolysis and solubility behavior indicates that these metal ions act as weak acids, releasing protons as they hydrolyze and as they precipitate. Proton release by hydrolysis initiates at around pH 3 for Fe^{3+} and around pH 5 for Al^{3+}. Inspection of Figure 2.2 shows that if waters with lower pH contain significant concentrations of these metal ions, increasing pH to these threshold values will cause precipitation of these metal ions until solubility equilibrium is reached; i.e. around 10^{-6} mole L^{-1}. The loss of dissolved metal ions from solution by precipitation will be accompanied by a corresponding 3-fold release of protons. Dissolved Al^{3+} and Fe^{3+} thus represent real contributions to acidity that must be neutralized if pH is to be increased into the neutral range.

Table 2.4 lists a number of weak acids that play a role in the chemistry of mine waters. The strength of each acid is denoted by the pK value associated with loss of one proton to form the conjugate base. The acidity or alkalinity can now be generally defined as the amount of strong base or strong acid that is needed to raise or lower the pH of the water to any defined pH; i.e. the selected *reference proton state*. The reference proton state defined above is pH = 5.64, the acidimetric titration endpoint for carbonic acid at atmospheric PCO$_2$(g). A more general definition of acidity and alkalinity can be developed by considering the *proton condition*; the balance of all weak acids and conjugate

Table 2.4: Weak acids and conjugated bases in mine waters

Weak acid and conjugate base	Equivalents of Alkalinity for a Proton Condition of pH = 5.64	pK (25°C, I = 0M)
Dissociation of Water $H_2O \rightleftharpoons OH^- + H^+$	0	$K_w = 14$
Carbonate Ion $HCO_3^- \rightleftharpoons CO_3^{2-} + H^+$	2	10.3
Bicarbonate Ion $^*H_2CO_3 \rightleftharpoons HCO_3^- + H^+$	1	6.3
Organic acids with phenol groups $R\text{-}OH \rightleftharpoons R\text{-}O^- + H^+$	1	~9
Organic acids with carboxyl groups $R\text{-}COOH \rightleftharpoons R\text{-}COO^- + H^+$	−1	~5
Ammonium Ion $NH_4^+ \rightleftharpoons NH_3 + H^+$	1	9.3
Aluminium hydrolysis $Al^{3+} + 3H_2O \rightleftharpoons Al(OH)_3(s) + 3H^+$	−3	[a]5.0
Ferrous Iron Oxidation and Ferric Iron Hydrolysis		
$Fe^{2+} + \frac{1}{4}O_2(aq) + 2\frac{1}{2}H_2O \rightarrow Fe(OH)_3(s) + 2H^+$	−2	[a]3.05
$Fe^{3+} + 3H_2O \rightleftharpoons Fe(OH)_3(s) + 3H^+$	−3	[a]3.05
Manganese Oxidation and Hydrolysis $Mn^{2+} + \frac{1}{4}O_2(aq) + 1\frac{1}{2}H_2O \rightarrow MnOOH(s) + 2H^+$	−2	[b]2.7
$Mn^{2+} + \frac{1}{2}O_2(aq) + H_2O \rightarrow MnO_2(s) + 2H^+$	−2	[b]~O

[a]The listed pK values correspond to the first hydrolysis constants. Although the first hydrolysis product strictly corresponds to release of 1 equivalent of acidity, the onset of hydrolysis is defined by these pK values. Inspection of Figure 2 shows that at the corresponding pH values the associated equilibrium concentrations of Al^{3+} and Fe^{3+} are on the order of 1 µM. Raising pH to these values would correspond to production of 3 equivalents of acidity for every mole of metal ion precipitated.
[b]The listed values correspond to the pH values above which the solubility of the metal ion is less than 10^{-6} M when in equilibrium with atmospheric $PO_2(g)$.

bases with respect to the reference proton state. All weak acids with a pK value below the reference proton state contribute to the acidity, while the conjugate bases of all weak acids with a pK value above the reference pH will contribute to the alkalinity.

Table 2.5 lists the charge equivalents of protons released by the weak acids upon alkalimetric titration to the reference pH, and the equivalents of protons consumed by each conjugate base upon acidimetric titration to the reference pH. For example, when a sample of net alkaline mine water is titrated to determine alkalinity, each carbonate ion in the sample is protonated twice as the pH is lowered to pH 5.64, thus contributing two

Table 2.5: The proton condition

Net Acidity = (equivalents of all weak acids with pK < pH_{ref})
           − (equivalents of all conjugate bases with pK > pH_{ref})

Reference Proton State pH_{ref} = 5.64

Bases contributing to [Alk]	Equivalents of alkalinity released by acidimetric titration to pH_{ref}
OH^-	1
CO_3^{2-}	2
HCO_3	1
$R-O^-$	1
NH_3	1

Acids contributing to [Acy]	Equivalents of acidity released by alkalimetric titration to pH_{ref}
H^+	1
R-COOH	1
Al^{3+}	3
Fe^{3+}	3
Fe^{2+}	2
Mn^{2+}	2

$$[Acy] = [R\text{-}COOH] + 3(Al^{3+}) + 3[Fe^3] + 2[Fe^{2+}] + 2[Mn^{2+}] + [H^+]$$
$$-[OH^-] - 2[CO_3^{2-}] - [HCO_3^-] - [R\text{–}O^-] - [NH_3]$$

Reference Proton State pH_{ref} = 9.5

Bases contributing to [Alk]	Equivalents of Alkalinity
OH^-	1
CO_3^{2-}	1

Acids contributing to [Acy]	Equivalents of Acidity
R-OH	1
$H_2CO_3^*$	1
R-COOH	1
Al^{3+}	3
Fe^{3+}	3
Fe^{2+}	2
Mn^{2+}	2
NH_4^+	1

Net Acidity
$$[Acy] = [H_2CO_3^*] + [R\text{-}OH] + [HN_4^+] + [R\text{-}COOH] + 3(Al^{3+}) + 3[Fe^{3+}] + 2[Fe^{2+}] + 2[Mn^{2+}]$$
$$+ [H^+] - [OH^-] - [CO_3^{2-}]$$

equivalents of alkalinity for each mole of carbonate ion. Likewise, in a sample of net acidic mine water, each Al^{3+} ion that precipitates as $Al(OH)_3(s)$ will release 3 protons to solution when the pH is raised to pH 5.64.

The definition of acidity as the amount of strong base that must be added to raise the pH to a defined threshold dictates that raising the pH to 5.6 requires (for example) sufficient lime addition to neutralize the free H^+ and also the acidity that is produced when the solubilized Al^{3+} and Fe^{3+} precipitate. Inspection of equations (2.2) and (2.3) also indicates that each mole of ferrous ion in solution or stored in secondary mineral precipitates represents 2 equivalents of acidity. This is produced upon oxidation of ferrous iron and precipitation as ferric hydroxide (1 equivalent of protons is consumed upon oxidation to Fe^{3+}, while 3 are produced upon precipitation as $Fe(OH)_3(s)$). This concept can likewise be extended to oxidation of manganese.

$$Fe^{2+} + \tfrac{1}{4}O_2(aq) + 2\tfrac{1}{2}H_2O \; (Fe(OH)_3(s) + 2H^+ \qquad (2.58)$$

$$Mn^{2+} + \tfrac{1}{4}O_2(aq) + H_2O \rightarrow \tfrac{1}{2}Mn_2O_3(s) + 2H^+ \qquad (2.59)$$

The definition of acidity and alkalinity based on the proton condition will change if the reference proton state is changed. For example, chemical treatment of an acidic mine discharge by high pH treatment for precipitation of metal ions may require that pH is raised to near pH = 9.5. In this case, all weak acids with a pK < 9.5 contribute to acidity, and the conjugate bases of all weak acids with a pK > 9.5 contribute to alkalinity. This includes all Fe^{3+}, Al^{3+}, Fe^{2+} and Mn^{2+} in solution. The corresponding expression for the net acidity is listed in Table 2.5. This acidity corresponds to the amount of base which must be added to raise the pH to 9.5.

2.2.5 VESTIGIAL AND JUVENILE ACIDITY

Acidity produced by active weathering of pyrite that is transported by advective flow from the zone of weathering is termed *juvenile* acidity. This term arises because the combination of active weathering and transport will cause recently produced contamination to manifest itself relatively quickly where the flow emerges from a mine workings. In Figure 2.1, the active zone of pyrite oxidation at the water table is an example of where juvenile acidity is produced.

Vestigial acidity develops in hydraulically unsaturated void space that is in contact with the atmosphere but not actively flushed by water flow (Younger,1997a). This void space includes the walls and hydraulically unsaturated rock around dewatered underground workings, and the pore space in mine waste deposits situated above the water table. Because these voids are not actively flushed, sulfide weathering products accumulate within the pore waters; acidity as H^+ and as metal ions such as Al^{3+} and Fe^{3+}.

Under persistently oxic conditions that drive weathering of pyrite and other sulfide minerals, metal ions and sulfate will continue to accumulate in pore waters until the solubility limit for metal sulfate and hydroxy-sulfate minerals is reached. As sulfide weathering continues, these secondary weathering products accumulate as mineral precipitates. Table 2.6 lists a number of these mineral phases, often associated with spoil heaps and the walls of mine workings. These accumulated precipitates thus represent

Table 2.6: Summary of secondary sulfate minerals (after Langmuir, 1997)

Mineral Phase	Formula	*$-\log K_{so}$
Alunite		
amorphous	$KAl_3(SO_4)_2(OH)_6(s)$	83.4
crystalline	$KAl_3(SO_4)_2(OH)_6(c)$	85.6
Alunogen	$Al_2(SO_4)_3 \cdot 17H_2O$	7.0
Anglesite	$PbSO_4$	7.76
Anhydrite	$CaSO_4$	4.36
Aphthitalite	$NaK_3(SO_4)_2$	3.80
Barite	$BaSO_4$	9.97
Basaluminite		
amorphous	$Al_4SO_4(OH)_{10} \cdot 5H_2O(am)$	116
crystalline	$Al_4SO_4(OH)_{10} \cdot 5H_2O(c)$	117.7
Celestite	$SrSO_4$	6.62
Coquimbite	$Fe_2(SO_4)_3 \cdot 9H_2O$	3.58
Ferrosic Hydroxide "Interlayered Green Rust"	$Fe(II)_4Fe(III)_2(OH)_{12}(SO_4 \cdot 3H_2O)$	125
Gypsum	$CaSO_4 \cdot 2H_2O$	4.59
Jarosite		
Hydronium Jarosite	$(H_3O^+)Fe_3(SO_4)_2(OH)_6$	75.4
Natrojarosite	$NaFe_3(SO_4)_2(OH)_6$	89.3
Potassium Jarosite	$KFe_3(SO_4)_2(OH)_6$	93.2
Jurbanite	$AlSO_4OH \cdot 5H_2O$	17.8
Kieserite	$MgSO_4 \cdot H_2O$	0.12
Melanterite	$FeSO_4 \cdot 7H_2O$	2.21
Schwertmannite	$Fe_8O_8(OH)_6SO_4 + 22H^+ \rightleftharpoons 8Fe^{3+} + SO_4^{2-} + 14H_2O$	18
Syngenite	$K_2Ca(SO_4)_2 \cdot H_2O$	7.45
Szomolnokite	$FeSO_4 \cdot H_2O$	0.91

*Thermodynamic data have not been critically reviewed and readers are referred to Langmuir (1997) and included references. Langmuir (1997) indicates significant uncertainty in the solubility constant for amorphous Alunite and for Potassium Jarosite, while Bigham et al. (1996) indicate significant variability in composition and solubility for Schwertmannite. Other than for Schwertmannite, the solubility products cited are based on dissolution of the mineral to their constituent ions; i.e. Alunite:

$$KAl_3(SO_4)_2(OH)_6(s) \rightleftharpoons K^+ + 3Al^{3+} + 2SO_4^{2-} + 6OH^-.$$

stored or *vestigial* acidity which is released to solution only upon cessation of dewatering or upon flooding or flushing of a waste deposit.

Contact with inflowing groundwater during mine water rebound, after dewatering is stopped, will provide dilution water that flushes the voids causing these secondary phases to dissolve. This releases Al^{3+} and Fe^{3+} to solution where they contribute significantly to the acidity load in the rising waters. In addition, H^+ that has accumulated in the previously immobile void water will also contribute to the acidity. Because dissolution of many of these secondary hydroxysulfate minerals is rapid compared to hydraulic transport during mine water rebound, their constituent ions will accumulate progressively as mine waters rise, resulting in a potentially catastrophic release of contamination when the discharge emerges. The contamination will then subsequently subside over time as the workings are continually flushed by recharge water. When the vestigial acidity is flushed from the system, any zones of active pyrite oxidation will continue to produce juvenile acidity until pyrite is depleted.

2.2.6 THE IMPACT OF BIOTA ON ACIDITY AND ALKALINITY

In addition to the oxidative weathering of pyrite and the weathering of carbonate and silicate minerals, there are a number of other processes which can influence acidity and alkalinity of natural waters. Table 2.7 lists a number of these processes, and represents

Table 2.7: The influence of biomass and anaerobic processes on the proton balance of mine waters (after Schnoor and Stumm, 1985)

Biomass production and decomposition produce and consume acidity

Nutrients		*biomass*	*acidity and O_2 release*
$800\ CO_2$		$(CH_2O)_{800}$	
$6NH_4^+$		$(NH_3)_8$	
$4Ca^{2+}$	production	(H_3PO_4)	
$1Mg^{2+}$	$\longrightarrow$	(H_2SO_4)	
$2K^+$		$(CaO)_4$	$+16H^+ + 8O40_{2(g)}$
$1Al(OH)_2^+$		(MgO)	
$1Fe^{2+}$	$\longleftarrow$	$(Al(OH)_3)$	
$2NO_3^-$		(FeO)	
$1H_2PO_4^-$	decomposition	$(H_2O)_m$	
$1SO_4^{2-}$			

Aerobic respiration and fermentation reactions do not affect acidity

$O_{2(g)} + CH_2O \rightarrow CO_{2(g)} + H_2O$
$2CH_2O \rightarrow CO_{2(g)} + CH_{4(g)}$

Anaerobic respiration

$5/4CH_2O + 1/4NO_3^- + H^+ \rightarrow 5/4CO_{2(g)} + 1/2N_{2(g)} + 7/4H_2O$
$CH_2O + 2MnO_{2(s)} + 4H^+ \rightarrow 2Mn^{2+} + CO_{2(g)} + 2H_2O$
$CH_2O + 4Fe(OH)_{3(s)} + 8H^+ \rightarrow 4Fe^{2+} + CO_{2(g)} + 11H_2O$
$CH_2O + 1/2SO_4^{2-} + H^+ \rightarrow CO_{2(g)} + H_2S_{(g)} + H_2O$

Anaerobic oxidation of pyrite

$2NO_3^- + FeS_{2(s)} + H^+ \rightarrow N_{2(g)} + SO_4^{2-} + Fe^{2+}$

them as stoichiometric reactions showing uptake or release of H^+ as a sink or source for acidity. Anaerobic respiration can be particularly important in wetlands and stream sediments due to the presence of detrital organic carbon which consumes O_2 by aerobic respiration at the water-sediment interface. In underlying anaerobic sediments, microbial sulfate reduction by organic carbon can occur, consuming acidity and providing sulfide ions (S^{2-}) which can remove metal ions by forming new sulfide minerals ($FeS(s)$, $ZnS(s)$, etc). These sulfur reduction process reverse the effects of sulfide mineral oxidation that occur in oxic environments, thus providing natural attenuation of acidity and metals contamination.

Processes involving biomass growth and dieback can have a large influence on soil chemistry. This may occur for example during reclamation of derelict mine sites through composting or establishment of vegetation covers. Table 2.7 represents biomass growth as uptake of nutrient ions and their incorporation into plant tissue. The predominance of base cations as nutrients means that biomass growth extracts alkalinity from the soil. According to the definition of strong acids and bases presented earlier, loss of base cations corresponds to loss of strong base or a decrease in alkalinity; i.e. acidification of the soil. Decomposition of organic matter after dieback returns these nutrients and the associated alkalinity to the soil. Composting on cover layers thus returns alkalinity, stored in the biomass, to the soil.

Any processes that influences the charge balance of natural waters; either through proton consumption or release by redox processes or uptake and release of either cations or anions, will generally influence the acidity and pH. In mine waters, sulfate is a tracer for input of strong acid by pyrite weathering while base cations are tracers for inputs of strong bases by carbonate or silicate weathering or other attenuation processes. In summary, "processes that produce or consume the cations of strong bases or the anions of strong acids will alter the acidity, alkalinity and pH of a mine water".

In the following sections, the composition of mine water will be used together with the discharge flow rates to determine solute fluxes. These fluxes can be related to the rates of key processes that generate and attenuate contamination loads. If the mass of contaminant source minerals can be estimated, these rates can be used to estimate the contamination lifetime of a mine site, and the time to depletion for sources of natural attenuation. With such information, the basic concepts of acidity, alkalinity and pH buffering can now be extended to assess trends in discharge water quality over time.

2.3 ASSESSING CONTAMINATION LOADS

Both hydrological and water quality data are needed to carry out even very preliminary environmental assessments of mine sites. Because the critical contaminants are solutes, water flows ($L\ d^{-1}$) and ion concentrations ($mole\ L^{-1}$) are required in order to determine contaminant fluxes ($mole\ d^{-1}$).

Examples in the following sections take known discharge flow rates and combine them with chemical information, particularly water quality data, in order to assess solute fluxes, contamination loads and the rates of processes that generate and attenuation the contamination. Chapter 3 deals in more detail with the hydrology of mine workings and the controls on discharge flow rates.

2.3.1 METHODS FOR ASSESSING THE POTENTIAL FOR CONTAMINATION

Both laboratory and field methods are used to determine the extent and rates of the weathering reactions that generate and attenuate contaminant loads. The focus of this chapter is on hydrochemical methods that use solutes in existing mine water discharges as tracers for weathering processes; e.g. sulfate as a tracer for the weathering of pyrite, zinc ion as a tracer for the weathering of sphalerite, etc. There are also a number of laboratory tests that can help provide an *a priori* assessment of the environmental risk presented by mine waste or ore rock, before a site is developed for mining or for deposition of mining waste.

The US Environmental Protection Agency published a standard method of *Acid Base Accounting* (ABA, Sobek *et al.*, 1978), to provide a pre-mining assessment of pollution potential and possible site monitoring requirements. The method compares the acid-neutralising capacity, termed *Neutralisation Potential* (NP), and the acid-producing capacity, termed *Acidity Potential* (AP), in order to determine if a particular sample of material is expected to produce acidic effluents when weathered.

The NP is determined in a laboratory assay where a fixed mass of crushed rock is reacted with a known excess of concentrated hydrochloric acid while heating. The acidity remaining in the cooled, reacted fluid is titrated with strong base and compared with the initial acidity in order to determine the amount of protons consumed. The NP is generally reported in equivalents of calcite per kilogram of rock that have reacted with the hydrochloric acid $(CaCO_3(s) + 2H^+ \rightarrow Ca^{2+} + H_2O + CO_2(g))$. In other words, two moles of protons consumed corresponds to neutralisation by 1 mole of calcite. The molar equivalents of calcite is obtained by dividing the moles of protons consumed, by two, and then dividing by the molecular weight of calcite (100 g mole^{-1}). The value obtained is the "equivalents of calcite reacted" and has units of grams of calcite.

The AP is determined by chemical analysis of the rock, relating total sulfur content to equivalents of sulfuric acid (H_2SO_4) that could be produced from sulfide weathering. Each mole of pyrite corresponds to two moles of H_2SO_4 and thus 4 moles of H^+. The moles of protons calculated, the AP, is likewise expressed as equivalents of calcite per kilogram of rock. The *Net Neutralisation Potential* (NNP) is the difference between the two.

$$NNP = NP - AP \qquad g \; CaCO_3(s) \; Kg^{-1} \qquad (2.60)$$

A number of shortcomings and subsequent modifications to the procedure have been presented (Li, 1997; Kwong and Ferguson, 1997; Lawrence and Wang, 1997). For example, AP should only consider sulfide content, since sulfate minerals such as gypsum contribute to the total sulfur, but do not produce acidity when weathering. Reacting samples with concentrated HCl at high temperature overestimates NP since rock that reacts under these laboratory conditions may not be reactive at field conditions. Mineralogical information is needed when interpreting NP, since neutralising capacity can originate from a variety of minerals that react at very different rates under field conditions (see Section 2.3.3).

Additional assays have been developed. These include measuring *Net Acid Generation* (NAG) by reacting rock with hydrogen peroxide at room temperature until available mineral sulfide is chemically oxidized (Finkelman and Griffen, 1986; O'Shay *et al.*, 1990).

Humidity cell experiments are carried out on uncrushed mine rock to assess if the rock will become acid-producing (ASTM, 1996; Price, 1997) during the lifetime of a mine. The cells operate cyclically under hydraulically unsaturated conditions with forced air flow that is designed to accelerated rock weathering, but without the harsh chemical treatment used in ABA or NAG tests. A typical procedure is to subject rock samples to dry air flow for several days, followed by an equal period of exposure to water-saturated air and finally followed by flushing with water flow to mobilize solutes from the weathered material. The cycle is repeated, usually for a period of several weeks up to one year. Onset of acidic conditions is noted by the corresponding drop of pH in the effluent.

Kinetic Tests include laboratory batch or column reactors, containing mine rock in contact with aqueous solutions, where solute concentrations are followed with time. The tests attempt to extrapolate observed weathering rates for acid-producing and acid-neutralising minerals, and the associated impact on site discharge quality, to field conditions over the lifetime of a mine. The very large differences in the physical scale and complexity between laboratory tests and field sites makes such extrapolations very uncertain. Scale effects which are commonly cited as contributing to discrepancies between lab and field observations are:

1. Particle size effects where large rock aggregates at field sites contribute significant mass but little reactive surface area for weathering. Laboratory tests usually focus on smaller particle sizes which have greater specific surface area and thus weathering more quickly.
2. Temperature effects can be significant for field sites in cold weather regions. Laboratory tests at room temperature are expected to yield weathering rates that are significantly faster than those at field sites with low average temperature. Due to the activation energy effects described in Section 2.1.2, mineral weathering is faster at higher temperatures.
3. Spatial variation in mineralogy at field sites can have a large impact on contaminant generation in discharges. Material that is suspected to be reactive is usually used for the types of laboratory tests outlined above. At field sites, the extent of such material may be limited. Contaminant solutes originating from such "hot spots" may be diluted in the discharge.
4. Preferential flow through mine workings results in some contamination being retarded in immobile water, rather than contributing to contamination in the discharge.
5. The availability of O_2 may be much more restricted in the interior of mine sites, particularly below the water table where O_2 diffusion is much slower. This means that the actual extent of reacting rock may only a small fraction of that present, due to large parts of a site being anoxic and therefore unreactive.

The current state-of-the-art in the practical application of static and kinetic tests in the context of surface coal mining has been summarized by Kleinmann (2000). At present, no similar consensus exists for applications of such tests to hard rock mining environments. Nevertheless, recent research suggests that methods may soon be developed to reliably extrapolate weathering rates determined in lab tests to field conditions in the hard rock mining context. For instance, in one recent study, weathering rates determined from kinetic tests using batch (0.15 Kg rock), large column (1820 Kg) and field (9.5×10^{10} Kg) studies have been compared for a single waste rock deposit (Malmström et al., 1999).

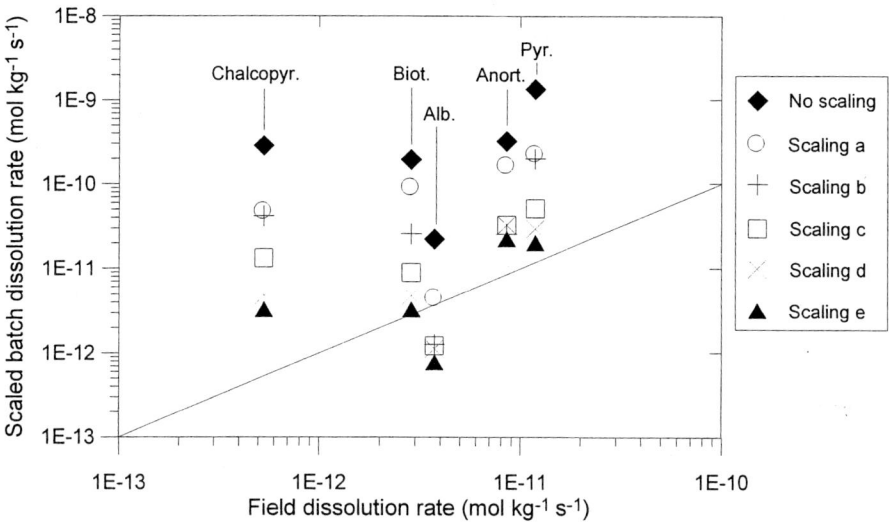

Figure 2.5: Cumulative effect of upscaling dissolution rates from kinetic tests in laboratory batch reactors (0.150 kg rock) to the field (9.5 × 10¹⁰ Kg rock) (Banwart *et al.*, 1998). The solid line denotes the ideal case of "perfect prediction", where the scaled laboratory rate equals the observed field rate. The additive impact of considering the following effects in turn is shown: (a) environmental temperature, (b) pore water pH, (c) particle size distribution, (d) reactive mineral content and (e) water flow patterns. Individual minerals are denoted as Alb (Albite), Anort (Anorthite), Biot (Biotite), Chalcopyr (Chalcopyrite) and Pyr (Pyrite).

Results showed that the scale-dependence observed for the site in question, is to a large degree predictable by quantifying the effects of temperature, reduced oxygen content, particle-size distribution, spatial variability in mineralogy and the content of immobile water at the site. Figure 2.5 illustrates the relative contribution of these effects to the differences observed between rates of weathering observed in the laboratory tests, and those determined from the solute fluxes at the field site.

Kinetic tests that avoid this scale dependence include large scale reactors constructed *in situ* at mine sites. Sulfide weathering rates are measured at field scale in order to monitor consumption of O_2 in isolated sections of tailings deposits (Blowes *et al.*, 1991) or in packed-off sections of boreholes drilled into mine rock (Harries and Ritchie, 1985). Detailed monitoring of site hydrochemistry can be used to supplement the methods outlined here. This includes gauging discharge flows and monitoring water quality. Lysimeters can be installed in the field in order to sample water flows in hydraulically unsaturated mine rock (Ericksson *et al.*, 1997). Sampling from borehole sections located below the water table can be used to monitor groundwater quality.

The following section presents a hydrochemical approach to estimating weathering rates from solute fluxes at field sites, and using these to predict the contaminating lifetime of the site. The basic data required is some knowledge of water flows draining a site and the corresponding water quality. This represents a kinetic test at field scale, and is appropriate for preliminary assessment of contaminant loads and their evolution with time, at existing or abandoned mines.

2.3.2 MINERAL WEATHERING RATES: A SCALING PROBLEM

If the average weathering rate (R, mole d^{-1}) of minerals within a mine working is known, then the solute flux could be predicted from mineral stoichiometry. For pyrite weathering, 2 moles of sulfate are produced for each mole pyrite that dissolves. The sulfate flux is therefore defined as follows.

$$F_S = 2R_{py} \qquad \text{,mole d}^{-1} \qquad (2.61)$$

This concept can be extended to other minerals and provides a crucial link between rates of weathering processes and solute fluxes at field sites. Extensive data on the weathering rates of individual minerals, determined in laboratory studies, are available in the literature (Sverdrup, 1990; White and Brantley, 1995; Nordstrom and Southam, 1997). In principle, applying literature values of weathering rates for a field site would allow the contamination intensity and loading to be calculated. Although laboratory studies may report effects of pH and other solutes, as well as microbial activity, on weathering rates, these impacts are generally difficult to predict quantitatively. This is even more difficult if they must be extrapolated to field scale.

As discussed in Section 2.3.1, there is currently no general method for reliably extrapolating laboratory test results to the complexity and physical scale of mine sites. In some cases it is possible to get order-of-magnitude estimates of contaminant loadings by transferring information about weathering rates and solute fluxes from one site to another. In order to understand how data can be transferred, it is necessary to briefly review the concepts introduced earlier regarding the kinetics of mineral dissolution.

Because dissolution reactions are proportional to the amount of mineral surface area in contact with aqueous solution, laboratory-derived rates are usually determined under hydraulically-saturated conditions with rates normalized to the physical surface area of the reacting sample. These rates (r) are reported in units such as mole m^{-2} d^{-1}. Scaling parameters are the specific area of mineral; a (m^2 g^{-1}), and the mineral mass, m (grams). Equation (2.62) shows the relationship between laboratory rates and these scaling parameters; where the subscript "py" refers by way of example to the mineral pyrite.

$$R_{py} = r_{py} \, m_{py} a_{py} \qquad \text{,mole d}^{-1} \qquad (2.62)$$

The specific surface area of an individual mineral at a field site presumably varies considerably with the mineral composition, morphology and size distribution of weathering particles in the rock mass, and therefore on the physical structure of geological material associated with a mine workings. Because of the spatial heterogeneity of field sites, this information is extremely uncertain. White and Petersen (1990) also present considerable discussion on what constitutes "reactive" surface area (Ar, m^2) and how this may relate to the physical surface area of weathering rock. There is not a simple relationship between the two parameters.

An alternative approach to this *scaling problem* is to estimate solute fluxes at the field site, and compare these with laboratory rates in order to obtain an empirical measure of A$_r$ for rock that is weathering at the site. This "calibrated" value could then be used to scale laboratory weathering rates to similar types of field sites. Table 2.8 shows calibrated

Table 2.8: Reactive surface area estimated at field scale

[1]Mineral Abundance in Field Vol. %	Tracer in Discharge	[2]Weathering Rate from Tracer Flux in Field mole kg^{-1} s^{-1}	Lab Rate pH 4 mole m^{-2} s^{-1}	Specific Reactive Surface Area at Field Scale m^2 kg^{-1}
Albite 13%	Na$^+$	3.7×10^{-12}	[3]3.1×10^{-11}	0.9
Anorthite 6%	Ca^{2+}	8.5×10^{-12}	[3]3.1×10^{-11}	4.6
Biotite 8%	Mg^{2+}	2.8×10^{-12}	[3]1.9×10^{-11}	1.8
Pyrite 0.6%	SO$_4^{2-}$	1.2×10^{-11}	[4]5×10^{-10}	4.0
Chalcopyrite 0.1%	Cu^{2+}	5.3×10^{-13}	[4]2.5×10^{-10}	2.1

[1]Biotite gneiss with mica schist reported in Table 1.
[2]From Banwart *et al.* (1998) determined for field site with mineralogy given in column 1.
[3]From Stumm and Morgan (1996, Table 13.3). Albite (Al) and anorthite (An) dissolution rates correspond to weathering of a single plagioclase with a composition 70%Al–30%An.
[4]From Strömberg and Banwart (1994).

values for A_r determined from a variety of solutes in the discharge from a field site. The range of values spans a five-fold difference, which is remarkably good given the uncertainties involved.

This approach assumes, using pyrite as an example, that sulfate ions are conserved in the discharge; i.e. not accumulated within the workings as secondary precipitates or in immobile water. The limitations posed by these assumptions need to be tested as discussed in subsequent sections. As an example here, the flux of sulfate, F_S, is treated as a direct measure of pyrite weathering at field scale.

Equation (2.63) relates reactive surface area, Ar, to the solute flux at a field site and to the laboratory weathering rate. This relationship implies that solute fluxes originating from a single mineral, e.g. pyrite, must be corrected for the amount of pyrite in the rock. The simplest approach is to assume that the fraction of reactive surface area corresponding to pyrite is similar to the mass fraction of pyrite in the rock, X_{py}.

$$A_r = \frac{F_s}{2X_{py}r_{py}} \qquad \text{,m}^2 \text{ of reactive rock} \qquad (2.63)$$

Values such as those in Table 2.8 can be used as a very crude estimate of solute fluxes in the absence of water quality data; for example when assessing possible impacts of a *planned* mining operation for sites with similar geology and hydrology. Calculations of this type thus help to answer the question "how bad will the contamination be". Better site information such as discharge rates, water quality data, geometry of workings and mineralogical information are all important pieces of information to improve estimates of (or to help interpret) contamination loads.

A common feature of these studies is that empirical values for specific surface areas derived from A_r and mass of rock at the field sites, are always 2–3 orders-of-magnitude lower than the specific surface area determined in lab studies. This apparent discrepancy is a direct result of the scaling factors listed in Section 2.3.1. The range of values for specific reactive surface area in Table 2.8 suggest that this scaling discrepancy can be similar for many minerals at a given field site (Strömberg and Banwart, 1994). Although the exact weathering rate for any individual mineral at a field site will probably not be possible to predict from laboratory rates alone, at least the *relative* rates of weathering for different minerals at a field site may similar to those determined from lab studies. Because the development of net-acidic and net-alkaline mine water discharges depends on the relative rates of pyrite, calcite and aluminosilicate mineral weathering, the corresponding laboratory data is still useful for understanding development of net-acidic and net-alkaline discharges and their temporal trends.

2.3.3 RELATING MINERAL MASS AND WEATHERING RATES TO EVOLUTION OF DISCHARGE QUALITY

Figure 2.6 compares laboratory weathering rates (mole m^{-2} s^{-1}) for calcite and pyrite with those for plagioclase, K-feldspar and biotite. Calcite dissolves approximately 3 orders-of-magnitude more quickly than pyrite, which in turn dissolves nearly 3 orders-of-magnitude more quickly than the aluminosilicates. These relative rates of dissolution suggest that if the minerals are present in similar quantities, calcite weathering will produce alkalinity at a rate that is sufficient to neutralize the acidity produced from pyrite weathering. On the other hand, if calcite is effectively depleted, then net-acidic discharges are expected.

In addition to providing a simple prediction of whether a discharge will be net-acidic or net-alkaline, this concept can be extended to changes in discharge water quality over time. Figure 2.7 illustrates a conceptual batch experiment where calcite is initially present but

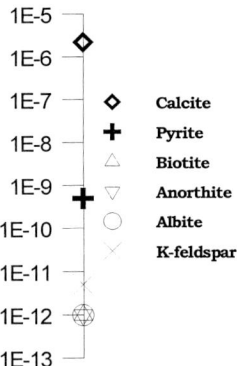

Figure 2.6: A comparison of laboratory rates for abiotic dissolution of calcite, pyrite and primary silicate minerals at pH 7. Of particular importance to mine waters is the fact that calcite dissolves much more quickly than pyrite, which in turn dissolves more quickly than the silicates. Because dissolution is a surface reaction process, weathering rates are influenced strongly by the amount of each mineral and its reactive surface in contact with waters draining mine workings. Pyrite weathering rates can increase by at least a factor of 25, particularly at low pH, due to microbial activity and are dependent on the presence of dissolved oxygen and ferric iron as oxidants; calcite weathering increases dramatically while abiotic rates of silicate weathering generally increase with decreasing pH according to a fractional-order dependence on proton activity $(H^+)^n$ (0 < m < 1) as pH decreases from circumneutral conditions (Strömberg and Banwart, 1994; and included references).

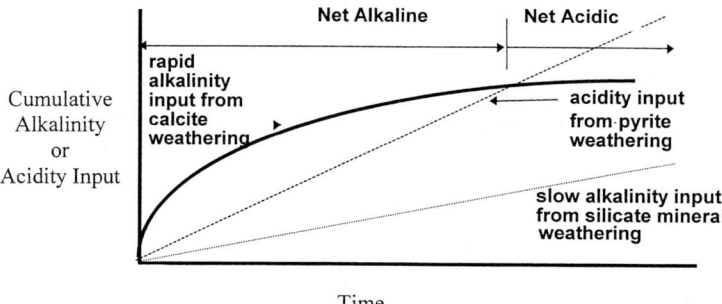

Figure 2.7: Development of net acidic conditions within a closed reaction system (no flow) due to rapid depletion of calcite, steady release of acidity due to pyrite oxidation until total acidity released equals that of alkalinity from calcite dissolution. Eventually, pyrite weathering will also cease, with subsequent alkalinity release due only to slow weathering of aluminosilicate minerals. This diagram emphasises the conservative behaviour of acidity and alkalinity; net acidic or alkaline mine waters results directly from the relative amounts of acidity and alkalinity generated within mine workings as they drain.

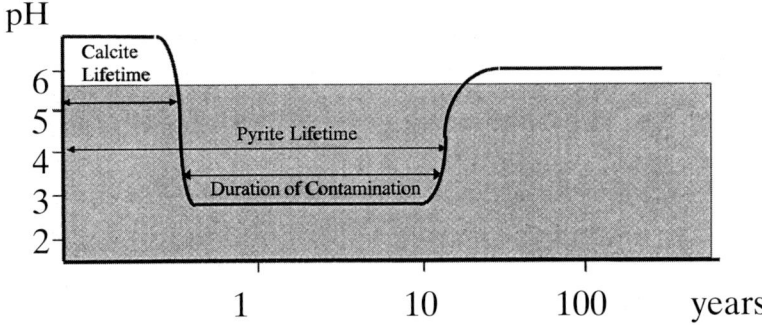

Figure 2.8: Because calcite dissolves rapidly, it will maintain a discharge in solubility equilibrium with the mineral, and thus net-alkaline, as long as it is appreciably present. When calcite becomes depleted, pH is expected to drop if pyrite is not yet depleted. Subsequent improvement in discharge quality is then expected when pyrite becomes depleted. This scenario applies to juvenile acidity produced within zones of active weathering. Hydraulic flushing times control evolution of discharge quality when vestigial acidity is the dominant source of contamination.

becomes depleted with time, leading to net-acidic mine waters. Figure 2.8 shows the conceptual development of mine water discharge pH due to depletion of first calcite (τ_c, y) and subsequently pyrite (τ_{py}, y).

Because metal ion mobility is linked so strongly to net-acidic conditions, depletion of pyrite and cessation of acidity production provides a critical time scale corresponding to the severest contamination release from a mine site. In broad terms τ_{py} corresponds to the contaminating lifetime of the site. This model for evolution of mine water discharge quality is summarized as follows.

- If $\tau_c < \tau_{py}$, the discharge will become net-acidic at time τ_c.
- If $\tau_{py} < \tau_c$, the discharge will remain net-alkaline.
- τ_{py} defines the contaminating lifetime for the highest metal contamination loads.

2.3.4 ASSESSING THE LONGEVITY OF CONTAMINATION

Critical parameters that need to be assessed for the evolution of mine water discharge quality are the times to depletion for pyrite and calcite. These values can be estimated from solute fluxes in a discharge, and estimates of the mineral mass within a mine workings. Sulfate ions in the discharge provide a natural tracer for oxidation of sulfide, while base cations and metal ions provide tracers for dissolution of their respective source minerals. If these tracers are conserved during dissolution and transport with flow from the site, then knowledge of solute concentrations and discharge rates yields the solute fluxes and provides an estimate of the mineral weathering rates at the site. Section 2.4 will illustrate how to test these assumptions about conservative behavior of tracer ions in solution.

For a discharge rate of Q (L y^{-1}) and with a sulfate concentration of [SO$_4^{2-}$] (moles L^{-1}), the sulfur flux in the discharge F$_S$ (moles y^{-1}) is given by the following equation.

$$F_S = [SO_4^{2-}]Q \qquad ,\text{moles y}^{-1} \qquad (2.64)$$

The pyrite weathering rate is then defined as ½ the sulfur flux from the site, by considering that weathering of 1 mole of pyrite yields 2 moles of sulfate in the discharge.

$$R_{Py} = \tfrac{1}{2}F_S \qquad ,\text{moles y}^{-1} \qquad (2.65)$$

If the total amount of pyrite (M$_{Py}$, moles) is known, the lifetime of the pyrite can be estimated by dividing the total moles of pyrite by the weathering rate. This measure is a zero-order estimate that assumes the rate of weathering is constant over the lifetime of the source mineral.

$$\tau_{Py} = M_{Py} / R_{Py} \qquad ,\text{years} \qquad (2.66)$$

Worked Example 2.7 carries out these calculations to illustrate the pyrite and calcite lifetimes for a metalliferous mine waste deposit. In the example, this simple mass balance approach provides an initial assessment of how long the site will produce acidity. This estimate for contaminating lifetime helps assess the longevity of juvenile acidity and guides decisions on the design lifetime for engineered remediation works for its treatment. The example also includes a comparison with results arising from a first-order estimate where weathering rate is assumed proportional to the amount of mineral present. Worked Example 2.8 illustrates the calculation of solute fluxes, pyrite and calcite mass associated with an underground mine workings and the associated lifetimes for pyrite and calcite at the site.

The time scales for release of vestigial acidity, on the other hand, are related strongly to hydraulic residence and flushing, rather than rates of weathering processes. This is because these minerals dissolve essentially instantaneously when void spaces are flushed.

Worked Example 2.7 The Lifetime for Generation and Attenuation of Contamination Loads from a Metalliferous Mine Waste Deposit

The mine waste deposit is composed of biotite gneiss with muscovite schist having a mineralogical composition as listed in Table 2.1. The discharge from the site is collected in drainage ditches with a total mean flow of $0.17 \ m^3 \ s^{-1}$. The hydrochemical composition of the discharge is that given in Worked Example 2.2. The waste rock deposit has mean depth of 20 m with an areal extent of 260 ha and an estimated porosity of 35%; the density of the rock is $2800 \ kg \ m^{-3}$ (after Banwart et al., 1998). Calculate the time scale for depletion of pyrite and calcite.

Step 1. Calculate total volume and mass of rock.

$$M = HA(1-n)\rho_s \qquad \text{where: M = mass, kg; H = height, m; A = area, } m^2;$$
$$n = \text{porosity; } \rho_s = \text{rock density, kg } m^{-3}.$$

$$M = (20) \ (2.6 \times 10^6) \ (1 - 0.35) \ (2800) = 9.46 \times 10^{10} \ Kg \ (\text{or } 9.46 \times 10^7 \ t)$$

$$V = M \div \rho_s = (9.46e1010) \div (2800) = 3.38 \times 10^7 \ m^3 \text{ of waste rock}$$

Step 2. Calculate mass of pyrite and calcite

$$M_{py} = X_{py}.V\rho_{py} \qquad \text{where: } M_{py} = \text{pyrite mass, kg; V = volume of rock,}$$
$$m^3; \ X_{py} = 0.01 \text{Vol. \% pyrite in rock;}$$
$$\rho_{py} = \text{pyrite density (5000 kg } m^{-3})$$

$$M_{py} = (0.0057) \ (3.38 \times 10^7) \ (5000) = 9.63 \times 10^8 \ kg \text{ pyrite}$$

Likewise for calcite ($\rho_{ca} = 2500 \ kg \ m^{-3}$):
$$M_{ca} = (0.001) \ (3.38 \times 10^7) \ (2500) = 8.45 \times 10^7 \ kg \text{ calcite}$$

Step. 3. Calculate total moles of pyrite and calcite

Atomic weights (A_{wt}, g mole^{-1}):	Fe	S	Ca	C	O
	55.85	32.07	40.08	12.01	16.00

Formula weights (FW) for pyrite ($FeS_2(s)$) and calcite ($CaCO_3(s)$):

$$FW_{py} = 55.85 + (2) \ (32.07) = 120 \text{ g mole}^{-1}$$
$$FW_{ca} = 40.08 + 12.01 + (3) \ (16.00) = 100 \text{ g mole}^{-1}$$

Moles of each mineral:

$$m_{py} = M_{py} \div FW_{py} = (9.63 \times 10^{11}) \div (120) = 8.03 \times 10^9 \text{ moles pyrite}$$
$$m_{ca} = M_{ca} \div FW_{ca} = (8.45 \times 10^{10}) \div (100) = 8.45 \times 10^8 \text{ moles calcite}$$

Worked Example 2.7 continued overleaf

Worked Example 2.7 continued

Step 4. Calculate Sulfate and Calcium ion fluxes in discharge.
Worked Example 2.2 showed 0.013 mole L^{-1} of SO_4^{2-} in the discharge originating from pyrite weathering. The Ca concentration is 0.185 g L^{-1}.

$[Ca^{2+}] = 0.185 \div 40.08 = 4.62 \times 10^{-3}$ moles L^{-1}

Multiply concentrations by discharge rate (170 L s^{-1}) to obtain molar fluxes.

$F_S = [SO_4^{2-}]_{py}Q = (0.013)(170) = 2.21$ mole s^{-1}
$F_{ca} = [Ca^{2+}]Q = (4.62 \times 10^{-3})(170) = 0.785$ mole s^{-1}

Calculate annual fluxes (1 year $= 3.15 \times 10^7$ s).

$F_S = (2.21)(3.15 \times 10^7) = 6.96 \times 10^7$ moles y^{-1}
$F_{ca} = (0.785)(3.15 \times 10^7) = 2.47 \times 10^7$ moles y^{-1}

Step 5. Calculate mineral lifetimes.
Weathering rates for pyrite and calcite:

$R_{py} = \frac{1}{2}F_S = 3.48 \times 10^7$ moles y^{-1}
$R_{ca} = F_{ca} = 2.47 \times 10^7$ moles y^{-1}

Mineral lifetimes calculated from mineral masses and weathering rates:

$\tau_{py} = m_{py} \div R_{py} = 8.03 \times 10^9 \div 3.48 \times 10^7 = 231$ years
$\tau_{ca} = m_{ca} \div R_{ca} = 8.45 \times 10^8 \div 2.47 \times 10^7 = 34$ years

Step 6. Review assumptions.
The calculated lifetime for calcite is similar to that of the waste rock deposit which has been in operation for 3 decades. Calculating the saturation index for calcite (see section 2.3.1) also shows the discharge to be several orders of magnitude undersaturated with respect to calcite. Because calcite dissolves rapidly this is also consistent with the absence of calcite. This is not consistent with the inherent assumption that dissolved Ca originates from calcite. However, as discussed in section 2.3.1, the saturation state of the discharge may not reflect local conditions within the deposit. On the other hand, if the discharge were assumed to be at saturation equilibrium with calcite in contact with the atmosphere, the calculated calcium ion flux is several orders of magnitude smaller than the one observed. This indicates that there must be a significant source of Calcium ions other than calcite. Table 2.1 suggests that the other Ca-bearing mineral in the rock is the anorthite component of the plagioclase.

Comparison with the first-order estimate for mineral lifetime assuming pyrite weathering rate decreases in proportion to the amount of remaining mineral:

Worked Example 2.7 continued opposite

Worked Example 2.7 continued

The assumption is stated mathematically by setting the rate as proportional to the amount of mineral present, and writing the corresponding first-order differential equation.

$$R_{py} = -\frac{dM_{py}}{dt} = k_{py}M_{py} \quad , \text{moles y}^{-1}$$

The value for the first-order decay constant k_{py} (y^{-1}) can be estimated if the values for M_{py} and R_{py} are known at a particular point in time. The decay constant is calculated by dividing the weathering rate, as estimated by the ion fluxes at the site, by the moles of mineral present. This value is the reciprocal of τ_{py} calculated previously.

$$k_{py} = \frac{R_{py}}{M_{py}} \quad , y^{-1}$$

Solving the differential equation with the initial condition that $M_{py} = M_0$ at time zero yields the following expression for decay of mineral mass with time.

$$\frac{M_{py}}{M_0} = e^{-k_{py}t}$$

$$\ln\frac{M_{py}}{M_0} = -k_{py}t$$

The value for the half-life of the mineral $t_{1/2}$ (y); i.e. the time for one-half of the mineral mass to weather, is expressed by solving this equation for t when $\frac{M_{py}}{M_0} = 0.5$.

$$t_{1/2} = \frac{\ln0.5}{-k_{py}} \quad , y$$

For the example above, $\tau_{py} = 231$ years, so $k_{py} = 1/231 = 0.0043$ y^{-1}. The half-life for the pyrite as a contamination source is 161 years. This reasoning can be extended, for example, to calculate when the mineral mass reaches 1% of its original value; $t_{0.01} = \ln0.01/-k_{py}$. The value for $t_{0.01}$ is 1071 years. The zero-order estimate of 231 years to completely deplete the pyrite corresponds to loss of 63% of the initial amount of pyrite based on this first-order estimate. The zero-order approach is therefore more optimistic in terms of the contamination lifetime. Given the uncertainties due to sparse data sets at most field sites, the difference between these two methods may well be within the margin of error.

Worked Example 2.8 The Lifetime for Generation and Attenuation of Contamination Loads from an Underground Coal Mine Workings.

An abandoned coal mine has a discharge of 2.6×10^6 L day^{-1}. The water quality data are as follows.

pH	6.5	SO_4^{2-} 460 mg L^{-1}	Ca	127 mg L^{-1}
Mg	23 mg L^{-1}	Cu	0.002 mg L^{-1}	Fe	62 mg L^{-1}
Na	130 mg L^{-1}	K	7.4 mg L^{-1}	Zn	0.005 mg L^{-1}
As	0.001 mg L^{-1}	Cr	0.001 mg L^{-1}	Ni	0.007 mg L^{-1}

The abandonment plan indicates a single worked horizon with an areal extent of 3×10^6 m^2 (300 ha) and a seam height of 1.22 m. The confining strata consist of mudstones with an assumed composition given by the UK mudstone listed in Table 1. The coal has a density of $\rho_c = 1260$ kg m^{-3}, a porosity of 3% and sulfur content of 0.3 wt %, assumed to be mainly pyrite. From the age of the workings it can be assumed that pillar and stall methods were used to extract the coal, with an extraction volume of 60% of the worked seam. Abandonment and mine water rebound is expected to cause collapse of the confining strata into the mined void. This collapsed material is presumed to be the confining mudstone with density $\rho_m = 2500$ kg m^{-3}, and with the mined void having a porosity of n = 0.40.

Step 1. Calculate total volume and mass of remaining coal.

Volume of worked seam:

$$V_{seam} = (3 \times 10^6 \text{ m}^2)(1.22 \text{ m}) = 3.66 \times 10^6 \text{ m}^3$$

Volume of remaining coal:

$$V_{coal} = (0.4)(3.66 \times 10^6) = 1.46 \times 10^6 \text{ m}^3$$

Mass of remaining coal:

$$M_{coal} = V_{coal}\, \rho_c = (1.46 \times 10^6)(1260) = 1.84 \times 10^9 \text{ Kg coal}$$

Step 2. Calculate the volume of collapsed mudstone within the workings.

Volume of removed coal:

$$V_r = V_{seam} - V_{coal} = 3.66 \times 10^6 - 1.46 \times 10^6 = 2.2 \times 10^6 \text{ m}^3$$

Volume of mudstone in workings.

$$V_m = (1-n)V_r = (0.6)(2.2 \times 10^6) = 1.32 \times 10^6 \text{ m}^3$$

Step 3. Calculate the amount of pyrite in coal and mudstone within workings.

Moles of sulfur (S) and pyrite ($FeS_2(s)$) in coal:

$$m_{Sc} = (M_{coal})(0.01)(\text{wt.\%}) \div (A_{wt})$$
$$= (1.84 \times 10^{12})(0.01)(0.3) \div (32.07) = 1.72 \times 10^8 \text{ moles S}$$

$$m_{pyc} = \tfrac{1}{2} m_{Sc} = 8.61 \times 10^7 \text{ moles pyrite}$$

Worked Example 2.8 continued opposite

Worked Example 2.8 continued

Moles of pyrite in mudstone:

m_{pym} = (V_m) (0.01)(Vol.%) ρ_{py}) = (1.32×10^6) (0.01)(2.5) (5000)
= 1.65×10^8 kg = 1.65×10^{11} g pyrite
= $(1.65 \times 10^{11}) \div (120$ g mole^{-1}) = 1.38×10^9 moles pyrite

Total moles of pyrite in workings:

m^{py} = $1.40 \times 10^9 + 8.61 \times 10^7 = 1.47 \times 10^9$ moles FeS$_2$(s)

Step 4. Calculate sulfate flux in discharge (FW$_{so4}$ = 96 g mole^{-1}) and pyrite lifetime.

$[SO_4^{2-}]$ = (0.460 g L^{-1}) $\div$ (96 g mole^{-1}) = 4.79×10^{-3} mole L^{-1}

F_s = $[SO_4^{2-}]Q$ = (4.79×10^{-3}) (2.6×10^6) = 1.25×10^4 mole day^{-1}

R_{py} = $\frac{1}{2}F_s$ = 6.23×10^3 mole day^{-1} = 2.27×10^6 mole y^{-1}

τ_{py} = $m_{py} \div R_{py}$ = $1.47 \times 10^9 \div 2.27 \times 10^6$ = 647 years

Step 5. Calculate the lifetime of calcite in the mudstone.

Moles of calcite:

m_{ca} = [(V_m) (0.01)(Vol.%) (ρ_{ca}) (1000)] $\div$ FW$_{ca}$
= [(1.32×10^6) (0.045) (2500) (1000)] $\div$ (100) = 1.49×10^9 moles CaCO$_3$(s)

Calcium ion flux in discharge:

$[Ca^{2+}]$ = (0.127 g L^{-1}) $\div$ (40.08 g mole^{-1}) = 3.17×10^{-3} mole L^{-1}

F_{ca} = R_{ca} = $[Ca^{2+}]Q$ = (3.17×10^{-3}) (2.5×10^6)
= 8.24×10^3 moles day^{-1} = 3.0×10^6 moles y^{-1}

τ_{ca} = $m_{ca} \div R_{ca}$ = (1.49×10^9) (3.0×10^6) = 497 years

Step 6. Review assumptions.
There are a number of assumptions which can be checked against additional data if it is available. For example, the discharge rate can be compared against pumping records and against regional groundwater recharge values (see Chapter 3). The assumption that all sulfur in the mudstones and coal is pyrite can be checked by mineralogical examination of samples taken from nearby spoil or outcrops. The assumption that only pyrite within the mined void and remaining coal contributes to the discharge is difficult to assess. The pyrite lifetime is probably a best case scenario for time to pyrite depletion since additional weathering may be occurring in other strata draining into the workings. A rational for including only contaminant sources within the workings is that the greater porosity and hydraulic contact with mudstone and coal surfaces within the collapsed material causes more rapid weathering of this material. An alternative scenario is that ironstone containing ferrodoloan minerals such as siderite (FeCO$_3$(s)) and ankerite ([Fe,Ca,Mg]CO$_3$(s)) could be the source of both the iron and the calcium ions. In this case, sulfate originates from pyrite weathering, but the ochre contamination could result substantially from weathering these other carbonate minerals. Detailed mineralogical examination is recommended to assess their content in the confining strata.

The source strength therefore depends strongly on the total amount of vestigial acidity that is available within the flow path, rather than how fast it dissolves. The importance of these time scales for management, and particularly remediation, decisions are discussed more extensively in Chapter 5.

Figure 2.9 shows the dynamic behavior of pH and metal ion mobility that occurs due to calcite depletion. As expected, depletion of calcite corresponds to a significant drop in pH. Metal ions that had been previously immobilized at neutral pH are mobilized and flushed with the effluent as pH drops. This behavior illustrates the important of anticipating whether mine water discharges will deteriorate as existing attenuation capacity is depleted. If deterioration is expected, the timing of the pH drop is critical since it is expected to coincide with a potentially serious, elevated loading of metals contamination.

2.4 THE GEOCHEMISTRY OF METAL IONS IN AQUATIC ENVIRONMENTS

The dramatic change in metal ion solubility with pH is due to the fact that metal ions do not exist in a single ionic form when dissolved, but form many different dissolved *species*. For example, zinc ions that are released by sphalerite weathering do not necessarily remain as *free ion* (Zn^{2+}) in solution.

$$ZnS(s) + 2O_2 \rightarrow Zn^{2+} + SO_4^{2-} \qquad (2.67)$$

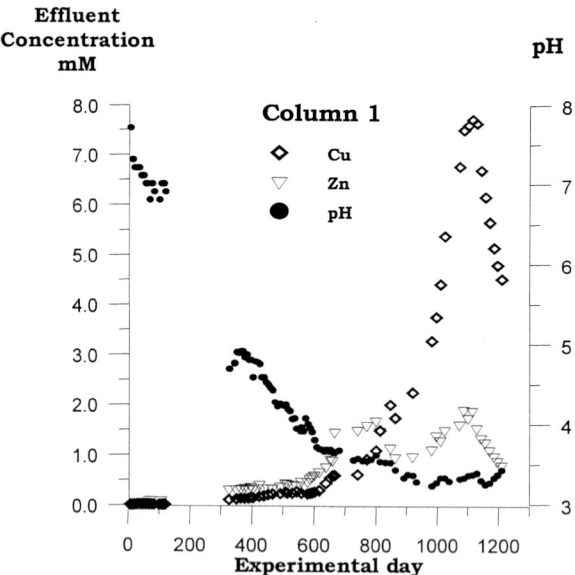

Figure 2.9: Evolution of column reactor effluent with time as calcite becomes depleted (Strömberg, 1997). There is a subsequent drop in pH and an associated release of metal ions that accumulated previously as sorbed and precipitated contaminants under circumneutral pH.

The free ion will react with other solutes to form more complex ions termed zinc "complexes". Examples of complexes include *mononuclear* hydroxide ($ZnOH^+$), sulfate ($ZnSO_4(aq)$) and carbonate complexes ($ZnCO_3(aq)$) as well as *polynuclear* complexes such as $Zn_2(OH)_6^2$ which contain more than one metal atom. It is important to note that different metal species have different molecular sizes and different charge. Because transport through cell membranes is influenced by these factors, different metal species can have profoundly different biological effects such as the extent of biouptake and toxicity impacts.

The activity of free ion may be fixed by solubility equilibrium with mineral phases such as the corresponding metal oxide or hydroxide solid phase. Applying the thermodynamic law of mass action for the solubility equilibrium of zinc hydroxide mineral illustrates that $[Zn^{2+}]$ is fixed by pH and the conditional (corrected for ionic strength and temperature effects) solubility constant K_{so}. The activity of pure phases such as water and the mineral can be approximated as unity; $[Zn(OH)_2(s)] = [H_2O] = 1$. Taking the logarithm of the resulting mass action equation and solving for zinc ion concentrations gives a linear equation relating zinc concentration and pH where pH = $-\log[H^+]$.

$$Zn(OH)_2(s) + 2H^+ \rightleftharpoons Zn^{2+} + 2H_2O \qquad ,\log K_{so} = 12.5 \text{ (I = 0M, 25°C)} \qquad (2.68)$$

$$K_{so} = \frac{[Zn^{2+}]}{[H^+]^2} \qquad\qquad\qquad (2.69)$$

$$\log K_{so} = \log[Zn^{2+}] - 2\log[H^+] \qquad\qquad\qquad (2.70)$$

$$\log[Zn^{2+}] = 12.5 - 2pH \qquad\qquad\qquad (2.71)$$

By analogy, the equilibrium concentrations of other metal ions (Cu^{2+}, Pb^{2+}, etc) can be calculated as a function of pH. Figure 2.10 shows a log concentration – pH diagram for various free metal ions in equilibrium with the corresponding hydroxide minerals.

2.4.1 HYDROLYSIS AND SOLUBILITY OF METAL IONS

Often the most important species with respect to metal solubility are hydroxide complexes formed by hydrolysis of metal ions. These reactions are written as deprotonation of water that is chemically bound to the metal. The subscript 1,1 indicates that 1 metal ion is bound to 1 hydroxide ion; a mononuclear monohydroxo complex. The thermodynamic law of mass action can be applied and the resulting equation solved for $[ZnOH^+]$. The logarithm of this expression relates the log concentration of the complex to that of the free ion and to pH.

$$Zn^{2+} + H_2O \rightleftharpoons ZnOH^+ + H^+ \qquad \log K_{1,1} = -9.0 \qquad (2.72)$$

$$K_{1,1} = \frac{[ZnOH^+][H^+]}{[Zn^{2+}]} \qquad\qquad\qquad (2.73)$$

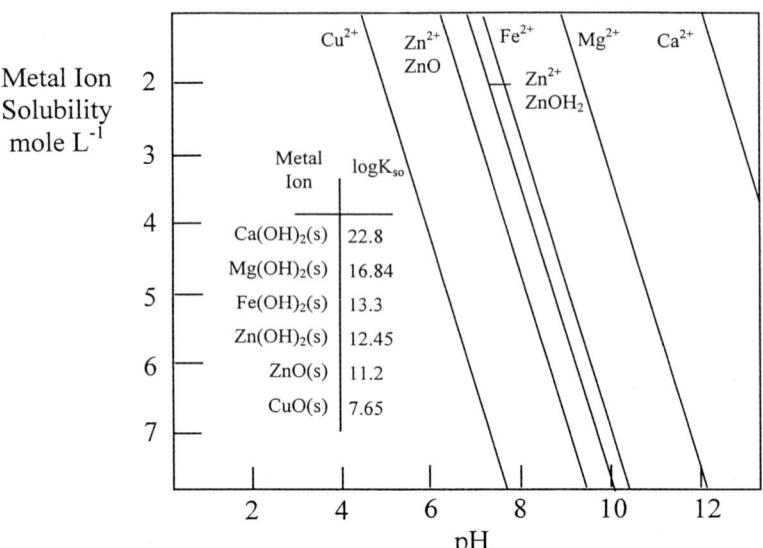

Figure 2.10: A plot of −log[M²⁺] against pH for solubility equilibrium with metal oxide or hydroxide phases. The conditional solubility constants are based on: $M(OH)_2(s) + 2H^+ = M^{2+} + 2H_2O$ ($logK_{so}$, I = OM, 25°C); Ca^{2+}, $logK_{so}$ = 22.8; Mg^{2+}, $logK_{so}$ = 16.84; Zn^{2+}, $logK_{so}$ = 12.45; Zn^{2+}, $logK_{so}$ = 11.2 (ZnO(s)); Fe^{2+}, $logK_{so}$ = 13.3; Cu^{2+}, $−logK_{so}$ = 7.65. Experimentally-determined solubility products can (and often do) vary considerably due to the non-ideal behaviour of solids, slow kinetic effects and the existence of dissolved species that have not been accounted for.

$$[ZnOH^+] = \frac{K_{1,1}[Zn^{2+}]}{[H^+]} \tag{2.74}$$

$$\log[ZnOH^+] = -9.0 + \log[Zn^{2+}] + pH \tag{2.75}$$

Deprotonation occurs sequentially, with more and more protons released as pH increases. In this way, metal ions can be viewed as multi-protic weak acids, where the relative concentrations of the different hydrolysis species are dependent on pH. Polynuclear complexes form as well.

$$Zn^{2+} + 2H_2O \rightleftharpoons Zn(OH)_2(aq) + 2H^+ \qquad logK_{1,2} = -16.9 \tag{2.76}$$

$$Zn^{2+} + 3H_2O \rightleftharpoons Zn(OH)_3^- + 3H^+ \qquad logK^{1,3} = -28.4 \tag{2.77}$$

$$Zn^{2+} + 4H_2O \rightleftharpoons Zn(OH)_4^{2-} + 4H^+ \qquad logK_{1,4} = -41.2 \tag{2.78}$$

$$2Zn^{2+} + H_2O \rightleftharpoons Zn_2(OH)_3^+ + 6H^+ \qquad logK_{2,1} = 3.4 \tag{2.79}$$

$$2Zn^{2+} + 6H_2O \rightleftharpoons Zn_2(OH)_6^{2-} + 6H^+ \qquad logK_{2,6} = -45.4 \tag{2.80}$$

If a mine water discharge is in equilibrium with $Zn(OH)_2(s)$, then the concentration of the free ion is fixed by Equation (2.71). Substituting this expression for $[Zn^{2+}]$ into equation (2.75) yields an expression for $[ZnOH^+]$ that varies only with pH. By analogy, expressions for the concentrations of the remaining hydrolysis species can be derived.

$$\log[ZnOH^+] = 3.5 - pH \tag{2.81}$$

$$\log[Zn(OH)_2(s)] = -4.4 \tag{2.82}$$

$$\log[Zn(OH)_3^-] = -15.9 + pH \tag{2.83}$$

$$\log[Zn(OH)_4^{2-}] = -28.75 + 2pH \tag{2.84}$$

$$\log[Zn_2(OH)_3^+] = 15.9 - 3pH \tag{2.85}$$

$$\log[Zn_2(OH)_6^{2-}] = -32.9 + 2pH \tag{2.86}$$

Figure 2.11 shows the concentrations of all species and their dependence on pH. The solubility of $Zn(OH)_2(s)$ is determined by the sum of concentrations of all species across the entire pH range. This analysis of zinc speciation and the resulting diagram illustrate why metal ion solubility is dependent on pH, and is dramatically higher at low pH than at neutral pH.

This pH dependence results from solubility equilibrium of the dissolved species with the mineral $Zn(OH)_2(s)$. This situation occurs in a mine water discharge if sphalerite dissolution is sufficiently rapid, compared to the hydraulic residence time of the workings, so that zinc ions accumulate in solution until the discharge becomes over-saturated with respect to $Zn(OH)_2(s)$ which will then precipitate. Another possible scenario is discharge of mine waters that are undersaturated with respect to $Zn(OH)_2(s)$, due to very low pH conditions, into a receiving stream with sufficient alkalinity to raise the pH significantly. In this case, the solubility limit for the mineral becomes much lower, and the mineral can precipitate from solution.

The *Degree of Saturation* (DS) for a mineral is determined by comparing the *Ion Activity Product* (IAP) to the conditional solubility constant for the mineral (K_{so}). A related parameter is the *Saturation Index* (SI) which is defined as log(DS). The IAP is the solubility product for the mineral based on the ionic composition of the discharge water. If DS < 1 (SI < 0) the solution is undersaturated with respect to $Zn(OH)_2(s)$. If DS > 1 (SI > 0), the solution is over-saturated. Due to the uncertainty in both IAP and Kso, a solution is operationally defined as being at saturation if SI = 0 ± 0.5.

$$IAP = \frac{[Zn^{2+}]}{[H^+]^2} \tag{2.87}$$

$$DS = \frac{IAP}{K_{SO}} \tag{2.88}$$

$$SI = \log IAP - \log K_{so} \tag{2.89}$$

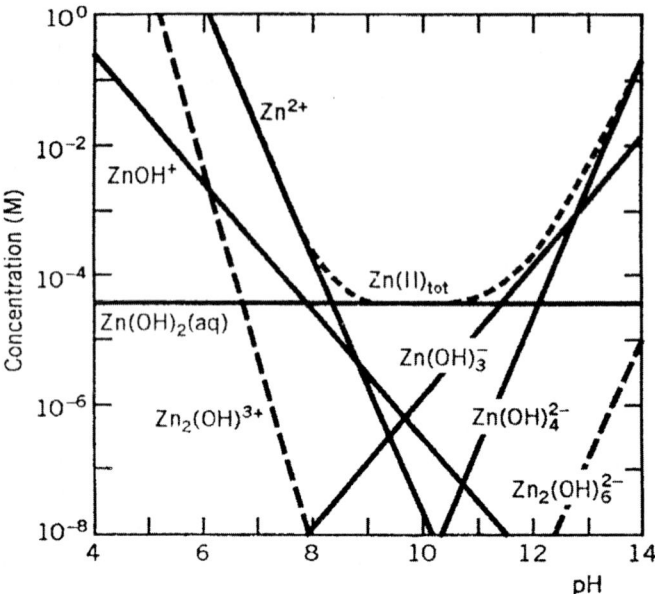

Figure 2.11: The logarithm of concentration for dominant hydrolysis species of Zn, in equilibrium with Zn(OH)$_2$(s), as a function of pH (from Stumm and Morgan, 1996). The figure emphasises that dissolved metals exist simultaneously in many different forms; each with a unique free energy, size and charge, and thus reactivity regarding mobility, bioavailability and toxicity.

The solubility of Zn(OH)$_2$(s) is used here to illustrate the necessary calculations. If the concentration of total dissolved Zn in a sampled discharge is 12 mg L^{-1}, with pH = 6.4, the IAP can be calculated. The IAP depends on the concentration of Zn^{2+}, not the total concentration. Inspection of Figure 8 shows that for pH 6.4, the concentration of Zn^{2+} is several orders of magnitude greater than ZnOH$^+$ and the other zinc species. The total dissolved Zn concentration thus corresponds closely to [Zn^{2+}]. The atomic weight for Zn is 65.39 g mole^{-1}, so the molar concentration of Zn^{2+} is $(12 \times 10^{-3}$ g L$^{-1})/65.39$ or 1.84×10^{-4} mole L^{-1}, and [H$^+$] = $10^{-6.4}$ or 3.98×10^{-7} mole L^{-1}. Note that logK$_{so}$ = 12.5.

$$IAP = \frac{(1.84 \times 10^{-4})}{(3.98 \times 10^{-7})^2} \tag{2.90}$$

$$IAP = 1.16 \times 10^9 \tag{2.91}$$

$$SI = \log(1.16 \times 10^9) - 12.5 = -3.44 \tag{2.92}$$

Because SI is significantly negative, the discharge is undersaturated with respect to Zn(OH)$_2$(s) and the mineral would not be expected to precipitate. On the other hand, if the mineral is present, it is possible that Zn^{2+} could be mobilized by dissolution.

One tentative conclusion that can be drawn from these types of calculations is that if the discharge is representative of water quality throughout the workings, it is unlikely that Zn^{2+} is accumulating locally by precipitation as $Zn(OH)_2(s)$. This reasoning can be extended to consideration of other minerals by checking the saturation index for phases such as $ZnCO_3(s)$. If such calculations support a conclusion on conservative behavior of the ion, then it becomes increasingly reliable as a tracer for the weathering of $ZnS(s)$ as a source mineral.

Table 2.6 lists a number of secondary metal precipitates that may form from mine waters, and their respective solubility constants. It is important to remember that these are conditional constants that need to be corrected for the temperature and ionic composition of the mine water being considered.

Saturation indices are prone to misinterpretation. A particular problem when interpreting the saturation state of a discharge is the possibility that mine waters with very different compositions within a mine workings are being mixed before emerging at the point of sampling. Zones of very rapid weathering ("hot spots") or low rates of flushing may result in oversaturation with respect to secondary minerals. If waters draining these zone are diluted by groundwater recharge, the mixed water may be significantly undersaturated with respect to minerals that are actually precipitating within the mine workings. Some rules for interpreting saturation indices are:

- Minerals will not precipitate from solutions that are grossly undersaturated.
- Minerals may dissolve in solutions that are undersaturated, if the mineral is present.
- The local state of saturation within a mine workings may be very different than that of the discharge.

2.4.2 ADSORPTION OF METAL IONS

Metal ions can adsorb onto both mineral and organic surfaces. If these surfaces are fixed in place, as for minerals in mine rock, then the metal ion will be immobilized. On the other hand, mineral colloids such as neogenic iron hydroxide gels that precipitate during ochre formation can remain suspended in solution. Adsorbed contaminants will be transported with the suspended particles carried by a mine water discharge. Immobilisation can subsequently occur by deposition in quiescent zones in receiving streams. Mechanical erosion of stream beds can subsequently remobilize any contaminants that are adsorbed on the sediment. By analogy, such adsorption reactions can also play a role in active treatment processes. For example, metal ions will sorb strongly to coagulant floc that forms during water treatment. Subsequent settlement of the floc allows the contaminant to be concentrated and disposed of with the dewatered sludge. For detailed discussion of the surface chemistry of metal ion adsorption, the reader is referred to summaries on modelling adsorption reactions and their relevance to groundwater chemistry (Banwart,1997) and the role of surface chemical processes in water treatment technology (Banwart,1994).

The predominant association of ochre formation with mine water pollution gives firm place to adsorption reactions as a key attenuation mechanism for metal loads to surface waters. Iron hydroxide precipitates are notorious for their sorbent properties, particularly their high specific surface area and strong sorptive interactions with metal ions. The book "Surface Complexation Modelling – Hydrous Ferric Oxide" by Dzombak and Morel

(1990) summarizes the surface chemistry of iron hydroxide precipitates and compiles and reviews the relevant binding constants for adsorption of metal ions. Some basic concepts are presented here, particularly the role of surface complexation reactions to describe the pH dependence of metal ion adsorption on ochre.

Iron hydroxide possesses a variable-charge surface; i.e. the surface charge changes with changing pH. This results from water that is chemically bound to metal atoms at the interface between the mineral phase and aqueous solution. The symbol $>$ Fe refers here to such an adsorption site on the iron hydroxide surface. A positively charged surface is associated with fully protonated water molecules bound to the mineral surface: $>FeOH^{2+}$. Analogous to hydrolysis reactions for metal ions in solution, the chemically bound water on the iron hydroxide surface can also deprotonate. This reaction is written as dissociation of a weak diprotic acid, forming uncharged and negatively charged surface complexes. The surface acidity constants are those given by Dzombak and Morel (1990).

$$>FeOH^{2+} \rightleftharpoons \, >FeOH + H^+ \qquad -logK_1 = 7.29 \ (25°C, I=0.5 \ M) \qquad (2.93)$$

$$>FeOH \rightleftharpoons \, >FeO^- + H^+ \qquad -logK_2 = 8.93 \ (25°C, I=0.5 \ M) \qquad (2.94)$$

These dissociation reactions are characterized by dissociation constants that can be used to define corresponding thermodynamic mass action expressions for the reactions.

$$K_1 = \frac{[> FeOH][H^+]}{[> FeOH_2^+]} \qquad (2.95)$$

$$K_2 = \frac{[> FeO^-][H^+]}{[> FeOH]} \qquad (2.96)$$

The corresponding mass balance for sites can be written where S_T is the total number of sites. By combining the mass balance with the mass action expressions, mathematical expressions can be derived that give the concentration of surface species as function of the adsorbing solution species; in this case, the proton. These expressions are generally termed adsorption isotherms. They describe how concentrations of adsorbed species depend on solute activities in solution.

$$ST = [>FeOH_2^+] + [> FeOH] + [> FeO^-] \qquad (2.97)$$

For the case described here, Equations (2.95) and (2.96) are solved for $[> FeOH_2^+]$ and $[> FeO^-]$, respectively. These expressions are then substituted into the mass balance.

$$[> FeOH_2^+] = \frac{[H^+][> FeOH]}{K_1} \qquad (2.98)$$

$$[> FeO^-] = \frac{[H^+][> FeOH]}{K_2} \qquad (2.99)$$

$$S_T = \frac{[H^+]}{K_1} + 1 + \frac{K_2}{[H^+]} [> FeOH] \tag{2.100}$$

Solving the mass balance for [> FeOH] yields an isotherm for this surface species as a function of proton concentration (Equation 2.101). Substituting this expression back into Equations (2.95) and (2.96) gives the corresponding isotherms for $> FeOH^{2+}$ and $> FeO^-$ (Equations 2.102 and 2.103).

$$[> FeOH] = \frac{S_T K_1 [H^+]}{[H^+]^2 + K_1[H^+] + K_1 K_2} \tag{2.101}$$

$$[> FeOH_2^+] = \frac{(S_T[H^+]^2)}{[H^+]^2 + K_1[H^+] + K_1 K_2} \tag{2.102}$$

$$[> FeO^-] = \frac{(S_T K_1 K_2)}{[H^+]^2 + K_1[H^+] + K_1 K_2} \tag{2.103}$$

The charge balance for the surface is defined as follows. Figure 2.12 shows the concentration of each species as a function of pH, and also the dependence of surface charge on pH.

$$Q = [> FeOH_2^+] - [> FeO^-] \tag{2.104}$$

In general the pH of zero net charge (pH_{znc}) corresponds to the titration equivalence point midway between the two pK values for the surface acidity constants. In the case for iron hydroxide described here, $pH_{znc} = \frac{1}{2}(7.29 + 8.93) = 8.11$. This thermodynamic model for the acid-base chemistry of mineral surfaces can be extended to a number of inorganic and in some cases organic surfaces. Figure 2.13 shows the dependence of surface charge as a function of pH for a variety of mineral surfaces. This figure emphasizes the underlying concept of variable-charge surfaces on minerals and their representation as simple acid-base reactions.

   The uncharged surface is defined as the reference state for surface complexation. As surface charge accumulates at relatively high and low pH values for proton adsorption and desorption, non-idealities occur due among other things to electrostatic interactions between charged adsorbates and the charged surface. These effects are generally small and can be neglected for cases of adsorption of solutes with a charge of the same sign as the solutes that are displaced from the surface, or when charged surface species make up less than 90% coverage of the surface; i.e. 90% of S_T. Here it is worthwhile to remember the rule of thumb that weak acids/conjugate bases increase/decrease in concentration by one order-of-magnitude for every 1 unit decrease in pH from the pK value for the dissociation constant. This suggests that the protonation described in Figure 2.12 holds for pH values in the range $6.3 < pH < 9.9$. At lower pH values, corrections for surface charge should be considered when calculating proton adsorption. These corrections are described by Stumm and Morgan (Chapter 9, 1996) but are beyond the scope of this chapter.

(a)

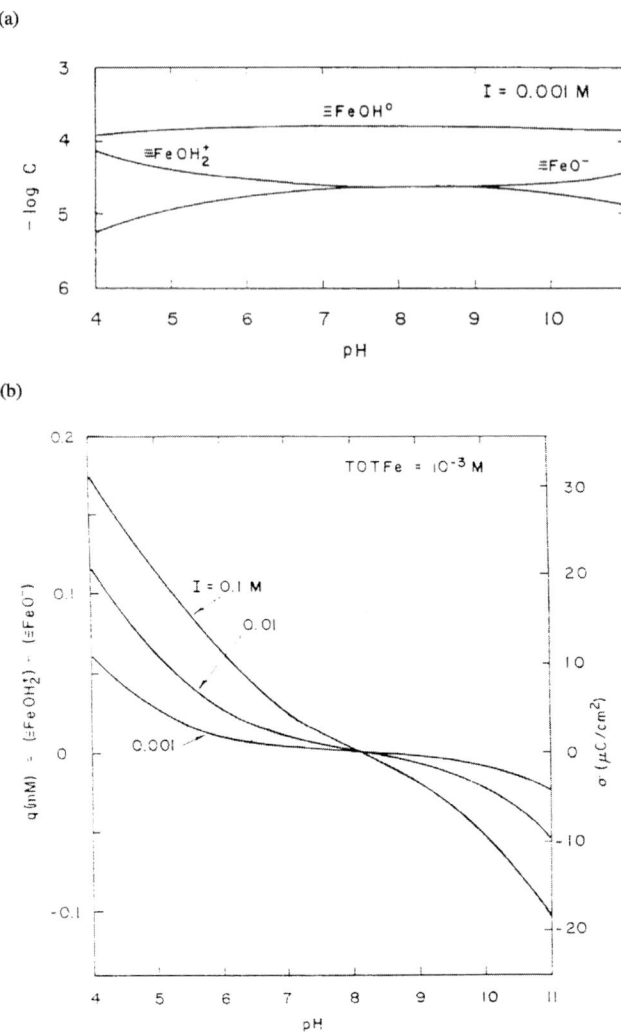

Figure 2.12: The surface chemistry of iron hydroxide (from Dzombak and Morel, 1990). (a) Surface complexes with positive, neutral and negative charge occur due to chemisorbed water acting as a weak diprotic acid. (b) The net charge on the surface varies with pH as a result of the change in concentration of the charged surface species.

Analogous to protons, metal ions can also adsorb through formation of surface complexes. Chromium adsorption is used here as an example to illustrate application of thermodynamic laws of mass action and mass balances in order to calculate metal ion adsorption on iron hydroxide. The example is drawn from Banwart (1997, Table VII.3) using data from Dzombak and Morel (1990). The adsorption of Cr^{3+} and chromate ion CrO_4^{2-} are compared and are described by the following stoichiometric reactions and

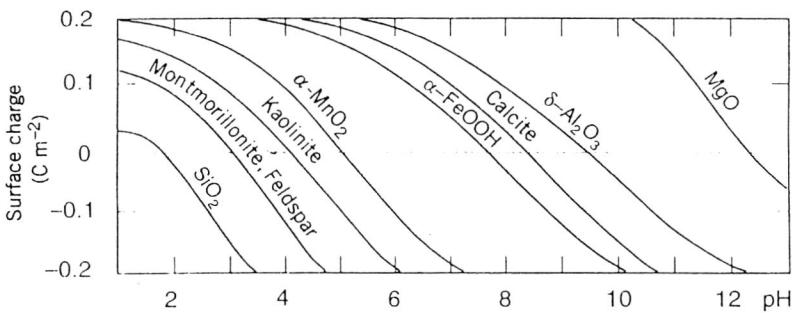

Figure 2.13: Surface charge dependence with pH for a variety of mineral surfaces (from Stumm and Morgan, 1996). These concepts can also be extended to carbonate and sulphide mineral surfaces, as well as complex oxides such as clay minerals. Note that charge is plotted with units of Coulomb m^{-2}. This can be converted to molar equivalents of charge by dividing by Faraday's constant (96,485 Coulomb per mole).

corresponding mass action expressions. The subscripts III and VI refer to the valence state (or "redox state") of Cr(III) as Cr^{3+} and Cr(VI) as CrO_4^{2-}, respectively.

$$> FeOH + CrO_4^{2-} + H^+ \rightleftharpoons > FeCrO_4^- + H_2O \qquad logK_{VI} = 10.85 \qquad (2.105)$$

$$K_{VI} = \frac{[> FeCrO_4^{2-}]}{[> FeOH][CrO_4^{2-}][H^+]} \qquad (2.106)$$

$$> FeOH_2^+ + Cr^{3+} + H_2O \rightleftharpoons > FeOCrOH^+ + 3H^+ \qquad -logK_{III} = 5.23 \qquad (2.107)$$

$$K_{III} = \frac{[> FeOCrOH^+][H^+]^3}{[> FeOH_2^+][Cr^{3+}]} \qquad (2.108)$$

For adsorption in the relevant pH ranges for both types of Cr ion (discussed below) and for small concentrations of chromium ions compared to the total number of surface site (relevant for ferruginous discharges; i.e. high ochre loading), the following simplified mass balances hold.

Adsorption of Cr^{3+}:

$$Cr_T = [Cr^{3+}] + [> FeOCrOH^+] \qquad (2.109)$$
$$S_T = [> FeOH_2^+] \qquad (2.110)$$

Adsorption of CrO_4^{2-}:

$$Cr_T = [> FeCrO_4^-] + [CrO_4^{2-}] \qquad (2.111)$$
$$S_T = [> FeOH] \qquad (2.112)$$

Combining these mass balances with the corresponding mass action expressions yields the following isotherms for adsorption of Cr^{3+} and CrO_4^{2-}, respectively, on iron hydroxide.

$$[> FeOCrOH^+] = \frac{K^{III}Cr_T}{[H^+]^3 + K_{III}S_T} \qquad (2.113)$$

$$[> FeCrO_4^-] = \frac{K_{VI}S_T Cr_T[H^+]}{1 + K_{VI} + S_T[H^+]} \qquad (2.114)$$

Figure 2.14 shows the adsorption curve for % adsorption of chromium as a function of pH. It is clear that Cr^{3+} adsorbs more strongly as pH increases. This behavior is broadly true for all cationic metal ions. This general behavior can be explained by considering that protons compete for adsorption sites with metal ions. At low pH, the high activity of protons in solution, drives reaction (2.103) to the left, effectively causing Cr^{3+} to desorb. A general rule emerges from this, that is highly relevant to mine waters. As pH drops, cationic metal ions tend to sorb less. As for the solubility of metal hydroxide and oxide minerals, decreasing pH is associated with increasing mobility of metal ions.

The opposite trend in adsorption with pH occurs for chromate anion. As pH increases, the extend of adsorption decreases. Although increasing pH can be seen as driving reaction (2.101) to the right, this behavior is best illustrated by combining this reaction with the dissociation of water, and expressing the pH dependence in terms of OH^- rather than H^+. This is done by adding H_2O to the left side of the reaction, and adding H^+ and OH^- to the right side. The H^+ and H_2O then cancel on both sides leaving the following stoichiometric reaction.

$$> FeOH + CrO_4^{2-} \rightleftharpoons > FeCrO_4^- + OH^- \qquad (2.115)$$

This reaction shows that hydroxide ions compete with chromate for adsorption sites. As pH increases, the higher activity of hydroxide ions in solution drives chromate from the surface as hydroxide is preferentially adsorbed. Thus the pH dependence of metal ion adsorption, for both cationic and anionic forms, is caused by the relative strength of the adsorption of protons and hydroxide ions, compared to the adsorbing metal ions.

Note as well that the dominant changes in adsorption for Cr^{3+} and CrO_4^{2-} occur respectively in pH ranges that are below and above the first pK value for deprotonation of the iron hydroxide surface ($pK_1 = 7.29$, Equation 2.93). This allows the mass balance to be simplified by ignoring > FeOH and > FeO$^-$ in the mass balance for surface sites used to describe adsorption of Cr^{3+}. The same reasoning allows > FeOH^{2+} and > FeO$^-$ to be neglected in the surface site mass balance for chromate adsorption.

These principles apply to the adsorption of other cations and anions. Figure 2.15 compares the calculated adsorption behavior of a range of cationic metal ions onto iron hydroxide. Table 2.9 summarizes the thermodynamic data used to build the adsorption model that is used. The model can be extended to include all surface and solution species. As more species are included, such a model rapidly becomes computationally difficult and requires specialist geochemical modelling codes (See Appelo and Postma, Chapter 10, 1993). The assumptions applied in Table 2.9 provide a simple and mathematically tractable illustration of the pH dependence of metal ion adsorption onto iron hydroxide.

(a) Adsorption of Cr^{3+}

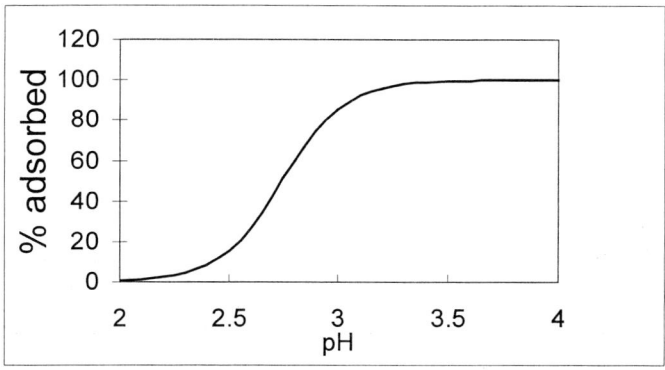

(b) Adsorption of CrO_4^{2-}.

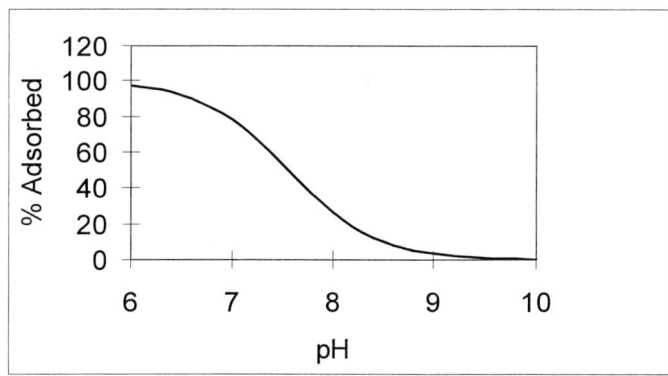

Figure 2.14: The adsorption of (a) trivalent Chromium and (b) Chromate anion onto iron hydroxide as a function of pH (from Banwart, 1997). Relatively simple mathematical expressions that describe the pH dependence of adsorption can be derived by combining the thermodynamic law of mass action for the surface complexation reactions with mass balances for dissolved and surface species.

Comparing Figures 2.14 and 2.15 shows that trivalent metal ions generally adsorb more strongly than divalent ions; i.e. that Cr^{3+} can be adsorbed at lower pH values, than is observed for the other metals. The reluctance of ions like Zn^{2+}, Cd^{2+} and Ni^{2+} to adsorb strongly in the weak acid pH region (pH 5–7), attests to the difficulty in removing these ions from mine water discharges. Difficulty in removal also arises because at neutral and alkaline pH, carbonate ions can form stable, dissolved metal complexes such as $ZnCO_3(aq)$ that compete with the iron oxyhydroxide surface for the metal ion. Formation of these complexes helps maintain the ions in solution, rather than immobilising them on the mineral surface.

Table 2.9: Development of an adsorption model for metal ions on iron hydroxide

Adsorption of metal ions is described as formation of surface complexes. The adsorption model is based on that described in Dzombak and Morel (1990), and discussed in the text. Simplifying assumptions are:

1. The total metal ion concentration is much lower than the total number of adsorption sites. Mine sites with heavy ochre precipitation, by definition, are generally expected to have metals other than Fe present only in much smaller concentrations.
2. Negatively charged surface sites are neglected since they only form in significant concentration at pH values above the second acidity constant for the surface: $pK_2 = 8.9$.
3. No solution species other than the metal ion Me^{2+} are considered. This neglects formation of hydrolysis species ($MeOH^+$, $Me(OH)_2(aq)$, etc.) or carbonate complexes ($MeCO_3(aq)$) that can occur at alkaline pH. This is due to higher activity of OH^- at high pH, and higher activity of CO_3^{2-} at high pH for waters that are open to $CO_{2(g)}$.

Stoichiometric reactions:

$$> FeOH_2^+ + Me^2 \rightleftharpoons > FeOMe^+ + 2H^+ \qquad\qquad logK_{Me}$$
$$> FeOH_2^+ \rightleftharpoons > FeOH + H^+ \qquad\qquad logK_1 = -7.29$$

Assumptions: $[> FeO^-]$, $[> FeOMe^+]$, solution species $<<[> FeOH_2^+], [> FeOH]$

Mass balances: Surface sites: $S_T = [> FeOH_2^+] + [> FeOH]$ (a)

Metal ions: $Me_T = [Me^{2+}] + [> FeOMe^+]$ (b)

Thermodynamic mass action laws:

$$K_{Me} = \frac{[> FeOMe^+][H^+]^2}{[> FeOH_2^+][Me^{2+}]} \qquad\qquad (c)$$

$$K_1 = \frac{[> FeOH][H^+]}{[> FeOH_2^+]} \qquad\qquad (d)$$

For a fixed total concentration of the adsorbing metal, the two mass balances and two mass action laws give a series of four equations with four unknowns: $[>FeOH_2^+]$, $[FeOH]$, $[FeOMe^+]$ and $[Me^{2+}]$. These can be combined into an isotherm for the adsorbed metal.

Step 1.   Solve mass action law (d) for $[> FeOH_2^+]$ and substitute into mass balance (a). This results in the following expression

$$S_T = \left(\frac{[H^+]}{K_1}[> FeOH] \right) + [> FeOH]$$

Step 2.   Solve this expression for $[> FeOH]$

$$[> FeOH] = \frac{K_1 S_T}{K_1 + [H^+]}$$

Step 3.   Solve mass balance (b) for the concentration of the dissolved metal ion.

$$[Me^{2+}] = Me_T - [> FeOMe^+]$$

Table 2.9 continued opposite

Table 2.9 continued

Step 4.   Substitute these two expressions into mass action law (c) and solve the resulting expression for the concentration of adsorbed metal ion [> FeOMe²⁺].

$$[> FeOMe^+] = \frac{K_{Me}K_1S_TMe_T}{K_{Me}K_1S_T + K_1[H^+] + [H^+]^2}$$

Dividing both sides by S_T yields an expression for the fraction of metal adsorbed as a function of pH, Me_T and S_T. The results plotted in Figure 2.15 are based on $S_T = 10^{-3}$ M and $Me_T = 10^{-5}$ M. The following thermodynamic data, corresponding to K_{Me} in mass action law (c) were obtained from the compilation of Dzombak and Morel (1990).

Me^{2+}	$logK_{Me}$
Pb^{2+}	−2.64
Zn^{2+}	−6.30
Cd^{2+}	−6.82
Hg^{2+}	0.47
Cu^{2+}	−4.40
Ni^{2+}	−6.92
Co^{2+}	−7.75

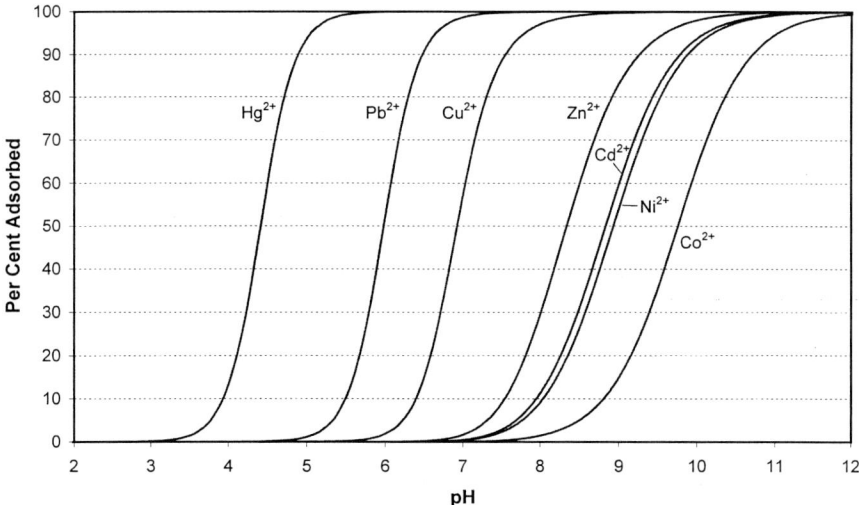

Figure 2.15: The pH dependence of adsorption for metal ions on iron hydroxide. The adsorption model that is used to calculate these curves is presented in Table 2.9.

Regardless of the complexities arising from consideration of a wider range of possible solute and adsorbed species, Figure 2.15 emphasizes the role of pH as a master variable for metal ion mobility and bioavailability for this class of attenuation reactions. It is clear that trends in adsorption indicate that increasing pH will tend to remove dissolved metals from discharges, while decreasing pH will act to keep them in solution. Furthermore, a drop in pH can cause previously adsorbed metal ions to be remobilized from iron hydroxide that has been deposited in the discharge channel or receiving stream.

CHAPTER THREE

MINE WATER HYDROLOGY

3.1 HYDROLOGY AND MINING

Hydrology is a relatively young science, which has developed gradually out of the hydraulic expedients which underpin much of civil engineering practice. Indeed, the emergence of truly scientific hydrology is such a recent development that major textbooks and learned journals have until recently devoted many pages to discussing its scientific credentials (Bras, 1990; Wilby, 1997). Until the 1980s most hydrological analysis was concerned with purely *physical* processes and practices of engineering interest, such as rainfall-runoff modelling and flow-net analysis of seepage pathways. By the start of the 21st Century, the scope of hydrology has expanded to such an extent that it now embraces relevant areas of chemistry and ecology. Sub-disciplines such as hydrogeochemistry (e.g. Appelo and Postma, 1994) and hydroecology (or ecohydrology; Baird and Wilby, 1999) are now firmly established, and account for a large proportion of the innovation in hydrological science (see, for instance, Wilby, 1997, and Wheater and Kirby, 1998). The "scientification" of hydrology has now proceeded to such an extent that Wilby (1997) could claim that hydrology provides the most logical basis for 'holistic environmental science', since water is a prominent medium in all of the earth and life sciences which deal with the natural environment. In the light of these trends, a contemporary definition of hydrology might be given as follows:

> "Hydrology is the science which deals with the nature, movement and environmental
> functions of terrestrial natural waters"

Mining, of course, has never suffered from any crises over its scientific identity: it is the quintessential example of "the scientific art of the possible" in relation to the exploitation of mineral resources. Not that mining has ever shrunk from the rigours of scientific innovation: society owes many of its most cherished technological facilities to the economic impulses of the mining sector. For instance, the roots of all modern railways and propelled shipping are to be found in the ingenuity of 19th Century mining engineers in north-east England, who needed to move coal from mines to wharves, and thence by sea to London (then the world's greatest metropolis) in order to realise the full economic value of their products (Taylor, 1858). Steam power was first harnessed in the tin and coal mines of England for drainage purposes, and was rapidly adapted for hauling run-of-mine product to the surface, and the refined product to market (Watkins, 1979). Indeed it was the development of steam pumping technology by the mining sector which first facilitated the widespread exploitation of ground water for public water supply; to this extent, modern ground water hydrology owes its very existence to the mining sector. It is therefore ironic that the hydrology of mine waters should have been so little studied over the centuries. While mainstream hydrology was pursuing its stumbling pilgrimage from engineering trade to scientific discipline during the second half of the 20th Century, mine water hydrology struggled to advance beyond a collection of pragmatic "rules of thumb", derived and cherished by a small group of mining professionals. The relative isolation of mine water hydrology from mainstream hydrological science can be ascribed to two causes:

(i) the fact that mine water hydrology is generally practised by mining engineers, mine surveyors and other mining professionals who have little or no formal hydrological training, and

(ii) the peculiarities of the hydraulic behaviour of mined systems, which often defy the application of "standard" subsurface hydrological analyses.

Nevertheless, substantial advances in mine water hydrology have been made since the early 1980s, following the First International Mine Drainage Symposium in Denver (Argall and Brawner, 1979). These advances have arisen in response to the gradual expansion in scope of mine water management problems from traditional dewatering design problems to the prediction, prevention and remediation of wider environmental impacts. This chapter presents an overview of the current state-of-the-art in mine water hydrology, emphasising techniques related to water quantities, but nevertheless taking advantage of the grounding given in the previous chapter to draw upon water quality information as appropriate.

3.2 ISSUES AND PROCESSES IN MINE WATER HYDROLOGY

The impacts of mining on the water environment were reviewed briefly in Section 1.4. In this Chapter, most of these impacts are revisited in greater detail, with an emphasis on the processes involved and the practicalities of quantifying them. Thus specific sections of this chapter cover each of the following major issues:

- Physical impacts of mineral extraction on natural hydrological systems
- Water inflows to active workings (considering typical and extreme magnitudes of such inflows, their provenance and possible approaches to controlling them)
- Dewatering techniques and their design
- Impacts of dewatering on the wider water environment, in terms of both water quantity and water quality
- Hydrological behaviour of waste rock piles, backfill and tailings impoundments
- Physical and chemical changes arising from mine abandonment and the so-called "rebound" process, in both deep and surface mines
- The longer-term hydrological behaviour of abandoned mine sites, including:
  - The physical and chemical dynamics of pit lakes
  - The hydrogeological behaviour of back-filled, former surface mine voids
  - The hydrological behaviour of flooded deep mine voids
  - Discharges from flooded deep mines

  – Migration of contaminants in aquifers[1] and streams and their environmental impacts
  – The longevity of pollution from abandoned mines
  – The hydrology of mine water wetlands (natural, volunteer and constructed)

Before proceeding to discuss these specific issues it is necessary to equip ourselves with some basic concepts (Section 3.3) and practical tools (Section 3.4).

3.3 MASS CONTINUITY AND HYDROLOGICAL FLUXES

3.3.1 MASS CONTINUITY AS THE CORE PARADIGM IN HYDROLOGY

Every mature scientific discipline has its core paradigm, which is often as not summarised in one or more mathematical laws. For instance, one can think of Newton's Laws and the Theory of Relativity in physics, the Periodic Law and Le Chatelier's Principle in chemistry, and the various Darwinian and Linnean paradigms in biology. Part of the scientific inferiority complex of hydrology has been due to its apparent lack of such a core paradigm. Reflecting on this issue, Dooge (1988) concluded that the core paradigm of hydrology is the principle of conservation of mass, or 'mass continuity', which states that matter can not be created or destroyed, but merely changed from one state to another. This is a paradigm which is by no means unique to hydrology. It may be its very simplicity which has led to it failing to attract the respect which it is due as the founding principle of virtually all hydrological analysis.

  Continuity in these terms is at the very core of virtually all hydrological investigations, from the simple water balance of a pond or a catchment, to the geochemical budget of an entire ocean (Baird, 1997). It may seem trite to emphasise the centrality of mass continuity to hydrological investigations, yet most major calamities in water resources and mine water management have arguably arisen from the temporary neglect of the concept. Its ramifications should therefore repay brief examination.

  The most fundamental application of mass continuity in hydrological systems is the "water balance", which is commonly applied over the full range of scales from a small volume of soil to an entire river catchment. The water balance for a given spatial system (e.g. a specific volume of an aquifer, a given river reach, or a particular lake) may be most simply expressed as:

  "Water entering" minus "Water leaving" = "Change in stored volume of water"

More detailed versions of this expression are the backbone of most hydrological analysis, and it is thus discussed at greater length in Section 3.3.2. below.

  Similar mass balance expressions may be written for a myriad of other hydrologically mediated mass transfers. For instance, the transport of iron in a reach of a stream receiving mine drainage could be expressed:

---

[1]An aquifer is a saturated body of rock which stores and transmits significant quantities of water.

$$(f_{\text{Fe-nat}} + f_{\text{Fe-mw}}) - f_{\text{Fe-ds}} = \Delta_{\text{Fe}} \qquad (3.1)$$

where: $f_{\text{Fe-nat}}$ = Iron entering the reach from natural sources (i.e. in upstream and lateral inflows, and by ground water discharge through the stream bed) $(M.T^{-1})$.

$f_{\text{Fe-mw}}$ = Iron entering the reach in mine water discharges $(M.T^{-1})$

$f_{\text{Fe-ds}}$ = Iron leaving the reach downstream in the dissolved and suspended loads $(M.T^{-1})$, and

Δ_{Fe} = Change in mass of iron stored in the reach $(M.T^{-1})$, which will mainly be in the form of ferric hydroxide precipitates on the streambed, with lesser quantities present as dissolved iron in 'dead zones' (areas of little or no water movement) and incorporated into biological materials.

In Section 3.3.3, we will consider how such geochemical mass balance expressions can be expanded upon to assist us in accounting for the transport and fate of pollutants in natural systems and in water treatment facilities.

3.3.2 NATURAL HYDROLOGICAL BALANCES – THE BASIC PROCESSES

In this section, some key hydrological variables will be introduced, and their significance for mined systems briefly noted. For the reader who is well-versed in surface and subsurface hydrology, this section may be largely superfluous, save as a guide to the notation used in this book. For the reader with no background in this material, this section will serve as an introductory guide to issues of hydrological analysis and modelling. However, it is beyond the scope of the book to provide thorough coverage of the major mathematical formulations used in hydrology, or the techniques by which their component equations are solved. Readers requiring such detail are directed to standard texts such as Huyakorn and Pinder (1983), de Marsily (1986), Anderson and Woessner (1992) and Chanson (1999). Here the focus is on concepts and major themes, rather than the practicalities of computer-based modelling.

Returning to the basic expression:

"Water entering" minus "Water leaving" = "Change in stored volume"

this may be written in simple mathematical terms as follows:

$$q_e - q_l = \Delta V_s \qquad (3.2)$$

where q_e is flux of water entering the system $(L^3.T^{-1})$

q_l is the flux of water leaving the system $(L^3.T^{-1})$, and

ΔV_s is the change (Δ) in the volume of water stored in the system (V_s) $(L^3.T^{-1})$

Equation (3.2) is equally applicable to surface water and ground water systems. It is now worth considering in a little more detail how each of the items in this expression relate to reality.

Water entering the system (q_e)
Water can naturally enter a given reach of a *stream* in three main ways:

- as upstream flow crossing into the reach of interest
- as tributary inflows of smaller surface water courses
- by discharge of ground water flowing up through the bed of the stream (Figure 3.1a).

Direct rainfall into the river channel is usually a negligible source of water save in the largest of rivers. In addition to these natural inputs, significant quantities of water enter many rivers artificially, in the form of sewage effluents, industrial effluents, runoff from roads and impermeable urban surfaces, and of course as mine water discharges (pumped or decanting by gravity).

Most of the water entering *aquifers* originates as rainfall landing on the overlying soils (Lerner *et al.*, 1990). In many arid and semi-arid regions, significant quantities of water enter aquifers by leakage of surface waters through stream beds (Winter *et al.*, 1998). This can only occur when the water table lies at a lower elevation than the level of water in the stream (which is termed the "stage") (Figure 3.1b), and is particularly favoured by conditions in which the water table lies below the elevation of the stream bed (Figure 3.1c). Where the latter condition persists all year round, surface water courses will be ephemeral, flowing only after intense rain storms.

As in the case of surface waters, artificial discharges to aquifers can amount to a significant proportion of the total resource. While sewage and industrial effluents are often discharged directly to ground water systems, substantial artificial discharges to ground water systems can be unintentional.

Water leaving the system (q_l)
It has already been noted that ground water may enter or leave a surface watercourse (Figure 3.1) under a range of natural hydrological conditions. Hence, subsurface leakage is a potentially important component of q_l for streams, and discharge to streams is usually a major element of q_l for a given aquifer system. When stream-aquifer systems are studied, therefore, the precise definition of q_l depends very much on the perspective of the investigator. For both aquifers and rivers, lateral outflows are usually a major component of q_l. In a stream, the flow crossing the downstream study reach boundary usually accounts for a large part of q_l. In an aquifer, lateral outflows may be to another portion of the same aquifer, to an adjoining aquifer, to springs or as subsurface discharge to the ocean. For both streams and aquifers, artificial losses to abstractions (i.e. water pumped or otherwise diverted from the system) are often quantitatively important. Bearing in mind the interconnectedness of surface- and ground waters, ground water abstractions can induce substantial flow losses in nearby streams.

A form of water leaving the system which is too frequently overlooked or underestimated is evapotranspiration, i.e. the combined losses to the atmosphere arising from open water evaporation and transpiration by plants. Evapotranspirative losses are responsible for ensuring that only a small proportion (usually between 5 and 40%) of the total rainfall on a given area actually remains available to infiltrate to aquifers and/or form surface runoff. (The fraction of the rainfall left after evapotranspirative losses is usually termed the "effective rainfall", or "hydrologically effective precipitation" (HEP)). Evaporation can be a very important process of water loss from deep mine voids, since evaporation of mine water produces humid air which is typically removed from underground workings in the ventilation stream. The quantities of water removed from a deep mine in this manner

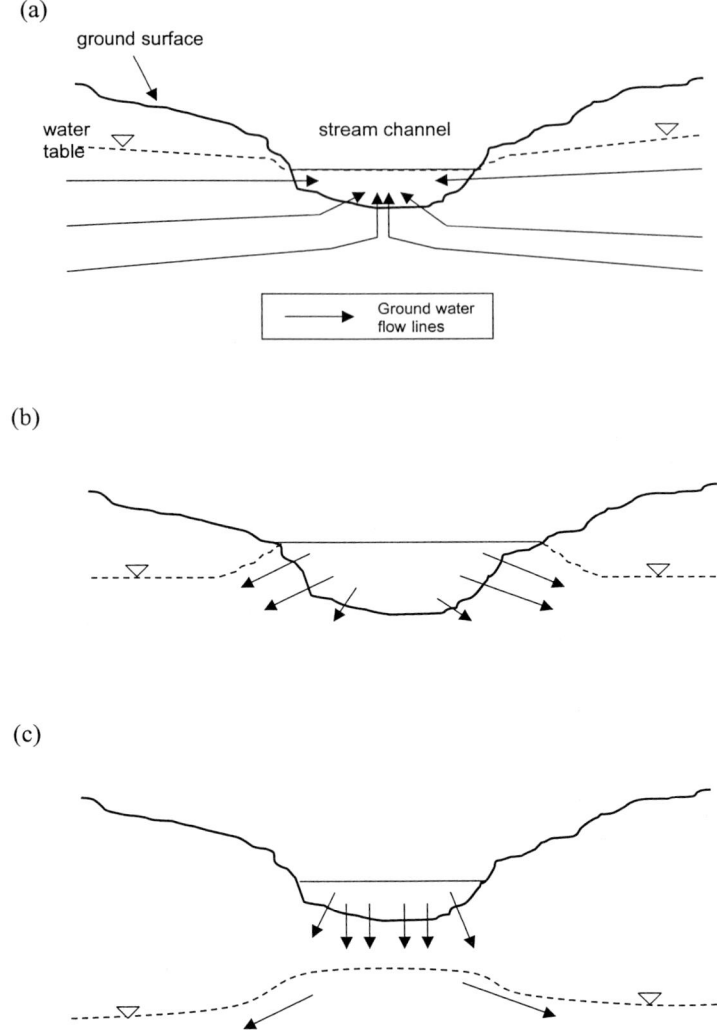

Figure 3.1: (a) Ground water discharge through a stream bed. Flow in the channel can be sustained through very long dry periods in this manner. (b) The 'bank storage' condition: ground water recharge from a river when the water table is higher than the stream bed. This typically occurs over short time periods during and shortly after storms, when surface runoff is high. After the storm runoff has passed, the stream will revert to condition (a), returning the 'bank stored' water to the channel. (c) The 'losing stream' condition: ground water recharge when water table lies below the stream bed. Flow in the channel wil decline to zero unless replenished by surface runoff.

will be non-negligible in many cases (see Section 3.4.5 below). Evaporation can also significantly affect geochemical processes in deep mines. For instance, Nordstrom *et al.* (2000) have shown that evaporative concentration of hydrogen ions and other solutes is responsible for the generation of persistently negative pH values and extreme total acidities in waters draining the Iron Mountain Mine in California. The abnormally high geothermal gradient in the vicinity of the mine is an important factor in sustaining the high evaporation rates which drive pH below zero. Under favourable circumstances, evaporation of mine waters can be economically beneficial. Thus at the Kori-Kkollo gold mine, near Oruro, Bolivia, a zero-discharge strategy is being pursued by disposal of all mine waters and processing effluents to a very large evaporation pond. Taking advantage of the extremely high evaporation rates operative at this high-altitude, low-latitude site, some 27 Ml.d^{-1} of water is disposed of in a series of lagoons and a vast tailings dam with a total surface area of around 5 km^2 (Arze-Quintanilla, 1994).

Changing storage – surface waters
The storage change term in equation 3.2, ΔV_s is almost invariably manifest as a rising or falling water level within the study system. (The only documented exception to this relates to biologically-mediated storage processes in highly compressible peat systems, in which storage changes can occur beneath a stationary water table Brown and Ingram, 1988). Water level changes in a lake or wetland are easy to relate to changes in storage: the volume of void occupied by the water as it rises to a new level obviously equals the amount of water added to storage. Less obviously, much the same reasoning is applicable to the "wedge storage" of water in a river reach as the stage rises and falls in response to the passage of a flood wave (e.g. Chow *et al.*, 1988): For a while during a storm runoff event, inflow to the reach exceeds outflow, so that the flood waters are effectively "stored" in the reach (or in extreme cases out of the channel, on the floodplain) as stage rises, albeit the period of storage in this case is usually several orders of magnitude less than that associated with a typical lake.

Changing storage – ground water systems
The relationship between ΔV_s and water level changes in ground water systems is considerably more complicated than in surface water systems. As a clear understanding of the processes of ground water level change is key to the prediction and interpretation of hydrological changes following mine abandonment, it is important to set out these processes here in some detail. In contrast to surface water systems, where water level changes can be readily related to the volume of water added or removed, void space in an aquifer typically accounts for only a small fraction of the total rock mass volume. Consequently a unit depth of rainfall[2] infiltrating a given volume of rock will invariably raise the water table by several units. The ratio of unit depth of water added per number of units rise in water is referred to as the "specific yield" (S_y) of a given rock. For instance, if 0.1 mm of infiltrating rainwater enters a rock mass with a S_y of 0.25, then the resultant rise in the water table (assuming no lateral outflows over the time-scale of interest) will be 0.1/0.25 = 0.4 mm.

Specific yield is closely related to effective porosity[3] (n_e) by the expression:

$$n_e = S_y + S_r \qquad (3.3)$$

---

[2]Rainfall and other forms of precipitation are measured in unit depths, such as mm or (historically, and to date in the USA) inches. According to this convention, the total volume of rainfall falling on a given catchment area is divided by the area of that catchment to yield a depth figure
[3]n_e = volume of inter-connected voids/total rock volume

where S_r is the "specific retention", i.e. that fraction of the pore space which remains wet after the rock is allowed to drain completely under gravity. The specific retention accounts for water held by electrostatic attraction as a pellicle around grains etc, and represents an irreducible water content. In coarse-grained rocks, S_r tends to be very small, so that S_y is closely approximated by n_e alone. In fine-grained rocks, S_r might account for most of the porosity, so that S_y is far smaller than n_e. Younger (1993) provides tables of typical values of S_y as a fraction of n_e for a range of rock types.

The use of S_y to determine storage changes applies only to unconfined aquifers, in which the upper boundary of the saturated aquifer thickness is a water table (Figure 3.2a,b). Where an aquifer is hydraulically confined (Figure 3.2c), the definition of storage is a little more complicated. Variations in the pressure within the aquifer, which are manifest in fluctuations in the piezometric surface, can certainly represent changes in storage, yet the aquifer always remains fully saturated. The pores of the rock neither drain nor fill, and yet the volume of water stored in the aquifer changes. How can this be explained? Two possible processes are commonly invoked in explanation (Freeze and Cherry, 1979):

(i)   expansion or compression of the water itself
(ii) expansion or compaction of the rock mass

Of the two, the second is far the more important in most cases (Younger, 1993b). Rock mass compressibility controls storage in the following manner: When the volume of water in storage increases, the walls of pores in the rock mass are forced slightly further apart; when water is withdrawn from storage, the pore walls can relax, moving more closely together once more. Ground water storage in accordance with these processes is commonly termed "elastic storage", and it is amenable to a fairly close analogy with the storage of air in a bicycle tire (Price, 1998): As air is pumped into a tire which is already inflated, the walls of the tire move slightly further apart and the whole structure becomes more rigid. When the tire is deflated, it always remains full of air at lower pressure, and the walls of the tire collapse in towards each other. This link between water storage and rock mass compressibility is one of the principal reasons why ground water withdrawal can lead to land subsidence.

In recognition of this fundamental difference in storage behaviour between confined and unconfined aquifers, distinctive notation is used to denote the storage parameter for the confined case: the parameter is termed "storativity" and is denoted by S. In practice, S is used in precisely the same way as S_y: for instance, 0.1 mm of water added to storage in a confined aquifer with an S value of 1×10^{-3} will give rise to $0.1/1 \times 10^{-3} = 100$ mm rise in the piezometric surface.

Depending as it does on marginal changes in rock mass status, values of S are typically many orders of magnitude smaller than the equivalent value of S_y for the same aquifer where it is unconfined. Thus S_y values commonly fall in the range 1×10^{-2} to 3×10^{-1}, whereas S values are invariably less than or equal to 1×10^{-3}, and can range as low as 1×10^{-7} (Freeze and Cherry, 1979). The dependence of S on rock compressibility, which in turn correlates quite closely with rock type, allows reasonably accurate estimates of S to be made provided that the nature and thickness of a given confined aquifer is known (Younger, 1993b). Table 3.1 presents the necessary information. One simply multiplies the

(a)

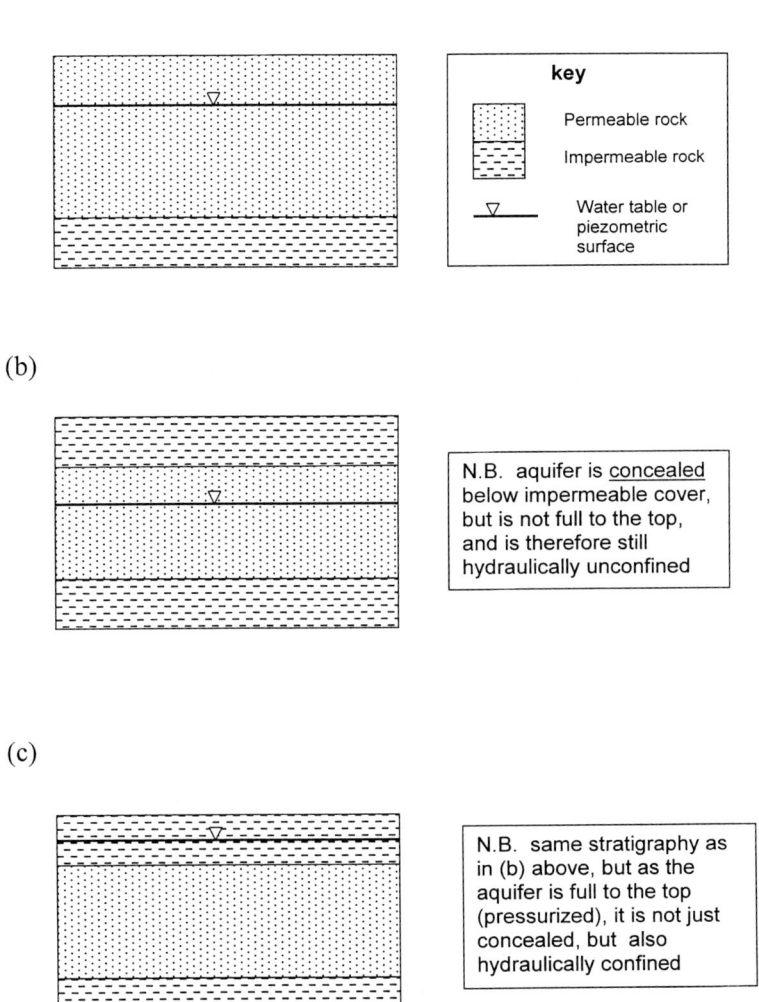

key

Permeable rock

Impermeable rock

▽   Water table or
         piezometric
         surface

(b)

N.B. aquifer is concealed
below impermeable cover,
but is not full to the top,
and is therefore still
hydraulically unconfined

(c)

N.B. same stratigraphy as
in (b) above, but as the
aquifer is full to the top
(pressurized), it is not just
concealed, but also
hydraulically confined

Figure 3.2: Sketch illustrating the occurrence of unconfined and confined aquifers. (a) simple unconfined case (b) concealed, unconfined aquifer (c) confined aquifer.

factor given in Table 3.1 by the thickness of the aquifer in order to obtain an estimate of S. This approach is extremely useful in the parameterisation of predictive models for systems lacking S values obtained directly by test-pumping analysis.

A further important consequence of the relationship between rock compressibility and water storage in confined aquifers is that deformation of the rock mass can cause marked changes in ground water levels *without* any change in the volume of water present. For instance, most confined aquifers display subtle fluctuations of piezometric surface in response to changes in surface loading due to changes in barometric pressure. Perhaps

Table 3.1: Factors for calculating the storativity of a confined aquifer using lithology and thickness data only (after Younger, 1993b). To obtain an estimate of the storativity of a confined aquifer, multiply the aquifer thickness (in metres) by the specific storage value given in the right hand column below

Lithology	Specific storage (m^{-1})
Clay	9.8×10^{-3}
Silt/fine sand	9.8×10^{-4}
Medium sand/fine gravel	9.9×10^{-5}
Coarse sand/medium gravel/highly fissured bedrock	1.05×10^{-5}
Coarse gravel/moderately fissured rock	1.6×10^{-6}
Unfissured rock	7.5×10^{-7}

more interestingly in the context of this book, loading and unloading of flooded deep mine voids has been shown to cause dramatic changes in ground water levels in shafts and boreholes open to the old workings. Figure 3.3 presents a three day digital logger record of mine water levels in the abandoned Frances Colliery, a coastal deep coal mine in Fife, Scotland. The mine worked extensively beneath the bed of the North Sea, with longwall extraction extending as far as 8 km offshore (Younger *et al.*, 1995). Pumping ceased in 1995 and water began to rise through the workings. The overall rising trend is evident, even over this three-day period, in the gradual increase in the elevations of the peaks and troughs of the water level plot. Far more striking in this plot are the major diurnal fluctuations of up to 0.4 m. These have been found to correlate perfectly (albeit with a lag time of about two hours) with tidal records for the North Sea. It is tempting at first to ascribe these fluctuations to ingress of sea water during the high tides. There are two reasons why this explanation is not credible:

(i) There are no direct hydraulic connections between the workings and the sea bed, and
(ii) As the water levels in the workings remain more than 50 m below sea level, they therefore have nowhere to drain to and are not subjected to pumping. Hence any physical input of water during a high tide would lead to a stepped increase only, not an increase followed by a decrease.

It is therefore clear that the diurnal fluctuations shown in Figure 3.3 must reflect cyclical loading and unloading of the deep flooded voids by the mass of tidal water. This is underlined by the fact that the net rise in water level between successive tidal cycles is not more than 0.03 m, which is an order of magnitude less than the amplitude of the fluctuations.

3.3.3 COMPLETE DESCRIPTIONS OF FLOW PROCESSES: THE BASIS FOR HYDROLOGICAL MODELLING

A great deal of analytical effort in mainstream hydrology has been expended on determining functional relationships between q_e, q_1 and V_s, for purposes such as reservoir spillway design and estimation of design floods. The methodologies developed for this

Frances Colliery

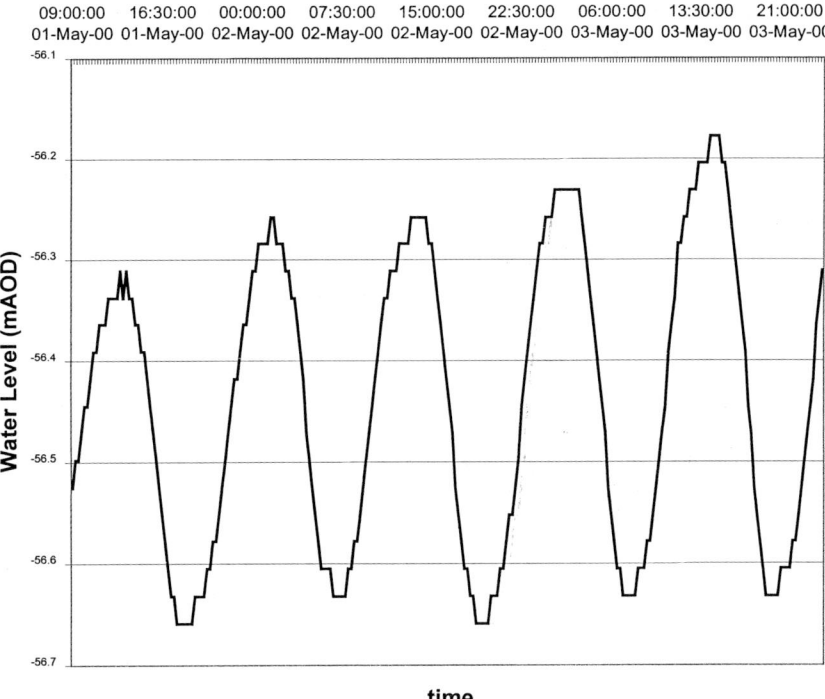

Figure 3.3: Tidal loading and unloading of a gradually flooding mine void. A three-day data-logger record of water levels in the main shaft of the abandoned Frances Colliery (Fife, Scotland) shows marked diurnal fluctuations in response to tidal cycles in the adjoining North Sea (Firth of Forth). The fluctuations can only be explained by mass loading of the flooding voids by the mass of water imposed and then removed in each tidal cycle. (N.B. 'mAOD' means 'metres above Ordnance Datum', i.e. above the official sea level datum of the UK). (Data supplied by the Coal Authority, analysed by Dr C A Nuttall).

purpose are generally termed "flow routing" (e.g. Shaw, 1994), and they have much to recommend them for applications to analogous problems in the mining sector. Nevertheless, for all but the simplest of cases, robust analyses of hydrological dynamics need to go beyond the simple storage accounting flow which is the basis of flow routing calculations, to specify the physical controls on *rates* of inflow and outflow. These controls are typically expressed using hydraulic formulae appropriate to the situation under analysis.

Describing ground water flow
For ground water flow systems, for instance, the relevant hydraulic formula is the well-known Darcy's Law, which in its simplest one-dimensional form may be written:

$$Q = K\,A\,i \qquad\qquad (3.4)$$

Where: Q is the flow of water passing through a given area of the medium (which is disposed at right angles to the hydraulic gradient $(L^3.T^{-1})$

A is the cross-sectional area through which flow occurs (L^2)

i is the hydraulic gradient (i.e. the difference in head of water which drives water across a specified length of subsurface media) (dimensionless)

K is a linear coefficient of proportionality, commonly termed the "hydraulic conductivity"[4] (L/T).

For purposes of flow modelling, equation 3.4 is usually simplified further by dividing through by the cross-sectional area A, yielding a quantity (q) known as the "specific discharge" or "darcy velocity":

$$q = Q/A = K\,i \qquad (3.5)$$

q has dimensions of L/T, and is thus an expression of the bulk velocity of flow averaged over the entire cross-sectional area A. It is the basic building block for turning the mass continuity expression (equation 3.2) into a general equation describing ground water flow. This is achieved by recognising that the left hand side of equation 3.2 can be represented by a simple differential expression. Assuming that flow occurs in a single direction x, we can express this as:

$$q_e - q_i = \frac{\delta q_x}{\delta x} \qquad (3.6)$$

where q_x is the specific discharge acting in the x-direction, and

δ is the differential operator

In reality, ground water flow occurs as three-dimensional fields, and the true ground water flux at any one point is a vector resultant from specific discharge components acting in all three spatial directions. Hence similar expressions for q_y and q_z can be written, and then the three components summed to express the total flux in and out of the system in three dimensions. Hence the LHS of equation 3.2 becomes:

$$q_e - q_i = \frac{\delta q_x}{\delta x} + \frac{\delta q_y}{\delta y} + \frac{\delta q_z}{\delta z}$$

The RHS of 3.2 can also be expressed as a differential, by recognising that the net change in water storage must be defined over some small, but finite, period of time (which we may

---

[4]Hydraulic conductivity (K) is best understood as the permeability of a rock with respect to fresh water, and it is related to intrinsic permeability (k) by the relation $K = k(\rho g/\mu)$, where ρ is the fluid density, g the acceleration due to gravity and μ the dynamic viscosity of the fluid. k is most commonly quoted in units of millidarcies (mD). For freshwater at 10°C, 1 mD $\approx 6.4 \times 10^{-4}$ m/d, or at 20°C, 1 mD $= 8.3 \times 10^{-4}$ m/d. In the remainder of this book, the terms "hydraulic conductivity" and "permeability" will be used inter-changeably)

denote as δt). To obtain an adequate differential expression we must first define the water level or potential within the ground water system. This is customarily termed the "head" of the ground water system. At any one point, head is a simple function of the water pressure and the elevation. This can be expressed as follows (Hubbert, 1940):

$$h = Z + \frac{P}{\rho g} \tag{3.7}$$

Where: h is the total head (L)
Z is the elevation of the point of measurement (L)
P is the pore water pressure ($M.T^{-2}.L^{-1}$; usual units in the kg-m-s system: $N.m^{-2}$)
ρ = the density of the water ($M.L^{-3}$; typically 1000 $kg.m^{-3}$ for fresh water)
g = acceleration due to gravity ($L.T^{-2}$; invariably 9.8 $m.s^{-2}$)

(h is also termed the *piezometric head, potentiometric head*, or simply the "ground water potential"; Freeze and Cherry, 1979).

Having thus defined head, we can denote changes in head (in time or space) in differential notation as dh. Using this convention the RHS of equation 3.2 can be rewritten as follows:

$$\Delta V_s = S_s \cdot \frac{\delta h}{\delta t} \tag{3.8}$$

Where: δt is the increment of time over which the head change (dh) occurs (T^{-1}).
S_s is the "specific storage" (L^{-1})

Specific storage is a point property, which is related to the bulk storativity discussed above by the following integral:

$$S = \int_{Z_b}^{Z_t} S_s \, \delta z \tag{3.9}$$

Where S_s can be reasonably assumed to be constant throughout the thickness of the aquifer, equation (3.9) collapses to:

$$S = S_s \cdot b \tag{3.10}$$

Where   b is the saturated thickness of the aquifer.

With these definitions in mind, for application to ground water systems equation 3.2 can be replaced by the following, generalised 3-D equation of time-variant (transient) ground water flow:

$$\frac{\delta q_x}{\delta x} + \frac{\delta q_y}{\delta y} + \frac{\delta q_z}{\delta z} = S_s \cdot \frac{\delta h}{\delta t} \tag{3.11}$$

It is possible to further expand the above expression to obtain a solution in terms of the basic physical parameters K and h. This is done by recalling that $q = K i$ (Eq. 3.5), and noting that i (the hydraulic gradient) can be written as a simple differential of head, for instance $i = \delta h/\delta x$ for flow components in the x-direction. Expanding equation 3.11 on this basis yields the following equation, which may be regarded as a complete and generalised description of subsurface flow in the saturated zone:

$$\frac{\delta\,(K_x\cdot\delta h/\delta x)}{\delta x} + \frac{\delta\,(K_y\cdot\delta h/\delta y)}{\delta y} + \frac{\delta\,(K_z\delta h/\delta z)}{\delta z} = S_s\cdot\frac{\delta h}{\delta t} \qquad (3.12)$$

where K_x, K_y, K_z are the values of K appropriate for the different directions. If K has the same value in all directions (in which case the aquifer is said to be "isotropic"), then Equation 3.12 can be factorised to:

$$K\left[\frac{\delta^2 h}{\delta x^2} + \frac{\delta^2 h}{\delta y^2} + \frac{\delta^2 h}{\delta z^2}\right] = S_s\cdot\frac{\delta h}{\delta t} \qquad (3.13)$$

In most regional aquifer systems, groundwater flow tends towards horizontality away from boundaries. The assumption of horizontal flow in extensive aquifers (the so-called "Dupuit-Forchheimer assumption") is commonly made in hydrological analysis. This allows a simplification of the analysis by integration of the problem over the vertical thickness of the aquifer, to yield a 1-D (profile) or 2-D (areal) horizontal flow model. The most common 1-D profile analyses are the well-known "test pumping analysis" methods, such as the Theis and Jacob techniques (Krusemann and de Ridder, 1990). 2-D areal flow models are commonly used in groundwater resources studies (Anderson and Woessner, 1992).

If we average K over the vertical dimension we obtain a "bulked-up K", known as the Transmissivity (T; dimensions L^2/T; usual units m^2/d). T is the integration of K between the lower and upper aquifer limits. Strictly speaking, this is defined as follows:

$$T = \int_{z_b}^{z_t} K\,\delta z \qquad (3.14)$$

where z_t is the elevation of the top of the aquifer, and z_b is the elevation of the base of the aquifer. It should be noted that while the definition of z_b is usually unequivocal (being a geological horizon in most cases), the definition of z_t differs between the unconfined case (z_t = water table elevation) and the confined case (z_t = geological horizon). This distinction is important in practice because it means that the range over which integration occurs is fixed in confined aquifers, but varies over time in unconfined aquifers. This introduces complications into the simulation process which are most satisfactorily addressed by the use of iterative algorithms.

If the aquifer is stratified, so that K variations can be approximated to multiple layers, then integrating by parts yields:

$$T = K_1 b_1 + K_2 b_2 + \ldots.\ K_n b_n \qquad (3.15)$$

where b_1, b_2 ...b_n are layer thicknesses.

In the special case that K is (or can be reasonably assumed to be) constant over the full aquifer thickness (b), then by analogy with equation 3.10:

$$T = K . b \qquad\qquad (3.16)$$

The form of equation 3.16 is quoted as the definition of T without qualification in many textbooks. In systems as vertically heterogeneous as mined ground, it is important to remember that 3.16 is a simplification, and that the full definition of transmissivity (3.14) may need to be used in practice.

Having thus defined the product of integration of K with respect to z, it remains only to integrate the ground water flow equation as a whole. Integration of equation 3.12 with respect to z has the effect of gathering all terms in z into a boundary flux term (R), which accounts for water added to, or taken from the aquifer, for instance by recharge, leakage, or pumping. We can thus write the appropriate equation for 2-D areal flow as:

$$T_x \frac{\delta^2 h}{\delta x^2} + T_y \frac{\delta^2 h}{\delta y^2} + R = S . \frac{\delta h}{\delta t} \qquad\qquad (3.17)$$

Equations 3.11 through 3.17 form the basis for orthodox physically-based ground water modelling as practised world-wide (Anderson and Woessner, 1992). With minor modifications to account for variations in hydraulic conductivity in partially wet pores (see, for instance, Parkin et al., 2000), equation 3.12 can also be used to simulate 3-D movement of water in the unsaturated zone, a task which is particularly relevant to the simulation of leachate generation and movement within waste rock piles, strip mine backfill and abandoned tailings dams (e.g. Eriksson and Destouni, 1997).

Describing surface water flows
The development of process descriptions for surface water systems follows the same general logic as that presented above for ground waters, in that the equation of mass continuity (3.2) provides the foundation, while appropriate hydraulic expressions are used to quantify the fluxes. The development of the most general form of the equations describing surface water flow combines continuity of mass with conservation of momentum. For one-dimensional flow in a stream channel, with the x-direction corresponding to the azimuth of the channel axis, continuity of mass can be written:

$$\frac{\delta Q}{\delta x} + \frac{\delta A}{\delta t} = 0 \qquad\qquad (3.18)$$

Where: Q is the flow rate of the stream ($L^3.T^{-1}$)
        A is the stream cross-sectional area (L^2)
        t is time (T).

It is perhaps worth noting here that Equation 3.18 is a somewhat simplified representation of reality which is strictly applicable only where the bed gradient and the channel cross-sectional shape remain the same (i.e. are 'conserved') downstream; for this reason 3.18 is formally termed the *conservation form* of the mass continuity equation (e.g. Chow et al., 1988).

For the same circumstances, the conservation of momentum can be expressed as follows:

$$\frac{1}{A}\frac{\delta Q}{\delta t} + \frac{1}{A}\frac{\delta (Q^2/A)}{\delta x} + g\frac{\delta y}{\delta x} - g(S_f - S_o) = 0 \quad (3.19)$$

| Local acceleration term | Acceleration downstream | Force due to water depth (pressure) | Down-stream fall of stream bed | Frictional resistance |

Where: g is the acceleration due to gravity $(L.T^{-2})$
        y = depth of water in the channel (L)
        S_o = the slope of the streambed (dimensionless) and
        S_f = the "friction slope" (dimensionless), representing down-gradient head loss
        arising from the loss of energy due to frictional resistance to flow by the streambed.

Equations 3.18 and 3.19 taken together are termed the "Saint-Venant Equations" after the engineer who first derived them in the mid 19th Century. Full and formal derivations are presented by Chow et al. (1988) amongst others. Direct solution of the Saint-Venant equations demands the use of numerical methods such as finite difference approximations. Such methods are demanding in terms of computer storage and execution times, and consequently they have only enjoyed widespread uptake since the advent of powerful digital computers (e.g. Abbott et al., 1986; Ewen et al., 2000). For certain simple cases, the Saint-Venant equations may be replaced by less detailed approximations which are amenable to analytical solutions. Kinematic wave models are a common example of this genre (Chow et al., 1988). These have the advantage that they deliver solutions rapidly, often requiring no more than manual calculations. As the analytical solutions are also exact solutions for the situations they represent, they also provide useful test data sets for verifying the performance of the inherently approximate full numerical solutions of the Saint-Venant equations.

One of the most obvious simplifications of the general case represented by the Saint-Venant equations is to assume that flows are steady, i.e. they do not change over time. In this case all terms in δt will vanish. While this assumption might at first seem excessively limiting for practical purposes, it is not so in practice. This is because flows may well be almost steady during the very brief periods of time over which stream flow estimates are typically needed (e.g. "What was the peak flow during a flood?", "What was the flow at the time the water sample was collected?"). Hence steady flow formulae enjoy much wider practical use than might at first be envisaged. A further important aspect of steady flow formulae must also be mentioned: because of the limited number of variables involved in them, these formulae have proven ideal for making comparative studies of the frictional resistance characteristics of natural and man-made channels (see, for instance, Chanson, 1999). Frictional resistance coefficients derived from them have therefore been fed back into full solutions of the Saint-Venant equations for unsteady (time-variant) flow.

While it is possible to work back from the Saint-Venant equations to obtain steady flow approximations, the history of hydraulics has been such that the most popular formulae were derived empirically, by mathematically reproducing observational data, long before the Saint-Venant equations were formulated. (The same can be said in relation to the development of ground water flow representations, since Darcy's Law was derived empirically in 1856, almost a century before equation 3.12 was first written down). A number of superficially different, empirical steady flow formulae are used in practice; the relationships between the various formulae have been documented by Chanson (1999). Possibly the best-known steady flow equation is the Chézy Formula:

$$Q = A \, C \, \sqrt{(R_H \sin\theta)} \tag{3.20}$$

where   Q is the flow rate $(L^3.T^{-1})$
A is the cross-sectional area occupied by the flow (L^2)
C is the Chézy Coefficient $(L^{0.5}.T^{-1})$
R_H is the "hydraulic radius" (L), which is defined as A/P_w
θ is the channel gradient (in degrees)
P_w is the wetted perimeter (L), which can be measured directly in small channels, or else calculated from survey data.

C is a function of R_H and bed roughness. In units of metres and seconds, typical values of C range from $30 \ m^{0.5}.s^{-1}$ (for small rough channels) to $90 \ m^{0.5}.s^{-1}$ (for large smooth channels). Chanson (1999) describes a number of techniques for calculating values of C for various channel conditions.

In most of Europe, the preferred steady flow formula is the Strickler Formula, which states:

$$Q = A \, k_s \, R_H{}^{2/3} \, \sqrt{\sin\theta} \tag{3.21}$$

Where k_s is the Strickler coefficient which has the advantage over the Chézy coefficient of being a function of the surface type *only*, and not of the hydraulic radius. In units of metres and seconds, k_s values range from $20 \ m^{1/3}.s^{-1}$ for rough surfaces (e.g. a gravel bed in a stream) to $80 \ m^{1/3}.s^{-1}$ for smooth surfaces (such as man-made concrete and steel pipes) (Chanson, 1999). For some unknown historical reason, a peculiar variant of the Strickler formula holds sway in Britain and many Commonwealth countries. This is Manning's Formula:

$$Q = (A/n) \, R_H{}^{2/3} \, \sqrt{\sin\theta} \tag{3.22}$$

in which the coefficient *n* (widely known as "Manning's n") is simply the reciprocal of the Strickler coefficient, i.e.:

$$n = 1/k_s.$$

Values of *n* range from $0.01 \ s.m^{-1/3}$ for smooth surfaces to 0.15 for very rough surfaces.

Equations 3.18 through 3.22 provide sufficient means to characterise surface flows for most practical purposes. It should be noted that they are also widely applicable for calculating flows in partially-full pipes, though pipe flow under nearly-full or full-bore conditions does not conform to these formulae, and must be analysed using specialised pipe-flow formulae. Some of these formulae will be introduced briefly in Section 3.4 below in the context of estimating flows on mines sites.

Transport of solutes – ground water systems
In describing the movement of solutes in ground water systems, three classes of process need to be considered:

- the movement of solutes with the bulk flow of the water (commonly termed "advection")
- the migration of solutes in response to other physical processes, such as mechanical mixing and molecular diffusion (commonly termed "dispersion"), which tend to result in the dilution of a volume of solutes introduced into the flow system, and
- alterations in the nature and distribution of solutes due to geochemical reactions.

The description of advection is nothing more than the description of ground water velocities. Equation 3.5 introduced the simplified version of Darcy's Law which yields the specific discharge (q), which has the same units as velocity ($L.T^{-1}$). The definition of specific discharge implies that flow takes place through the entire cross-sectional area of the aquifer, whereas in reality it occurs solely in the pores. As these occupy a relatively small volume of even the most permeable aquifers, the true interstitial velocity will be considerably higher than the flux rate given by the specific discharge. Taking this into account, the average interstitial velocity (v_i) of the groundwater is given by:

$$v_i = q / n_e \qquad\qquad (3.23)$$

where n_e is the effective porosity.

The processes responsible for dispersion in ground water flow systems are many and varied. Molecular diffusion arises from Brownian motion of solute ions, and it is operative in all ground water systems. In low permeability strata it may be the dominant solute transport process. Molecular diffusion has long been adequately described by Fick's Law, which represents the spreading of solutes along "concentration gradients" (i.e. from areas of high concentration to areas of low concentration). As Fick's Law ignores the random motions which actually give rise to diffusion it might be regarded as a "useful fiction", which mathematically reproduces observed behaviour despite lacking a strong basis in physical process. Mechanical mixing usually results in far greater solute displacements than can be explained by molecular diffusion alone. Much previous work on dispersion in ground water systems has been based on the assumption that mechanical mixing can be adequately described by a law of similar form to Fick's Law (see de Marsily, 1986, for a discussion). In this formulation, dispersion due to mechanical mixing is described by a dispersion coefficient (D; dimensions: $L^2.T^{-1}$) which is in turn calculated from the average interstitial velocity (v_i) as follows:

$$D = \alpha_L . v_i \qquad\qquad (3.24)$$

where α_L is the longitudinal dispersivity (L), which is supposedly an intrinsic property of the porous medium in question. In reality, α_L is arguably devoid of physical meaning, for it cannot be readily measured nor accurately calculated from other petrophysical parameters. The non-physical nature of the parameter is also betrayed by the fact that values obtained by solving the inverse problem (i.e. from fitting model solutions to field data) yield values which are dependent on the scale of measurement. Dispersivity, and thus the dispersion coefficient, is thus more of a "fudge factor" than an objective parameter, and should be used with caution. A suitable replacement has long been sought, though no universally-accepted alternative formulation has yet been developed (Fetter, 1999).

Geochemical reactions between ground water and the enclosing rock occur in myriad ways, as described in Chapter Two. For the purposes of tracking solute transport, all geochemical reactions can be represented by a simple source/sink term , which either adds solutes to or removes them from solution. The particularities of each reaction govern the direction, rate and magnitude of the exchange of solutes which it effects. Where reactions are rapid in comparison to the flow rate of the water, then it may be reasonable to assume that the water equilibrates with the rock at every point along its flowpath; this is the so-called "Local Equilibrium Assumption" (LEA). The LEA has been widely applied to simulating the consequences of dissolution of relatively soluble minerals such as calcite, gypsum etc, and to the simulation of rapid sorption processes, in aquifers subject to low gradients (e.g. Hongze et al., 1997; Chen et al., 1999). Where less soluble minerals or rate-limited sorption occur in fast-flow settings, the LEA is highly unlikely to be adequate. In such cases, realistic simulation of transport will require the use of expressions which represent the rates of the reactions in question; simulation undertaken using this approach is generally termed "kinetic reactive transport modelling".

The summation of all of the above processes yields the following expression for solute transport in ground water systems (given here for the simplest, one-dimensional case):

$$\frac{\delta}{\delta x}\left[\frac{D}{R_d}\cdot\frac{\delta C}{\delta x}\right] - \frac{v_i}{R_d}\cdot\frac{\delta C}{\delta x} \pm C_s.q_s = \frac{\delta C}{\delta t} \qquad (3.25)$$

Mixing term	Advection term	Source – sink term	Rate of change in concentration

Where: C is the concentration of the solute of interest ($M.L^{-3}$)

R_d is a retardation coefficient (dimensionless), describing the apparent "slowing down" of solutes relative to the bulk flow by sorption processes (in this formulation, these are assumed to be reversible, linear and instantaneous) and other reactions with similar consequences.

$C_s.q_s$ is the source-sink term ($M.T^{-1}$) representing addition or removal of water with a concentration of the solute of interest C_s and a flow rate of q_s.

Equation 3.25 is usually termed the "advection-dispersion equation", and more formal derivations of it may be found in standard ground water texts such as Freeze and Cherry (1979) and Fetter (1999). It is possible to further generalise Equation 3.25 to three dimensions, and it can also be developed further to simulate multi-species transport. This is achieved by establishing as many simultaneous versions of 3.25 as there are solutes of interest and then solving them all concurrently (see, for instance, Hongze et al., 1997).

Generally, Equation 3.25 and similar expressions derived from it are solved numerically, by a range of methods such as finite differences, finite elements or by the Method of Characteristics (e.g. Huyakorn and Pinder, 1983). However, analytical solutions of the one-dimensional version have been developed for particular boundary and initial conditions (Fetter, 1999). An example of this genre will be introduced in Section 3.11.

Transport of solutes – surface water systems
The transport of solutes in surface water systems is effected and affected by an assemblage of processes similar to those which occur in ground water systems. As in ground water systems, movement of solutes with the bulk flow of the water is the dominant process. Dispersion by mechanical mixing processes is also very commonly observed, albeit the processes by which it occurs are somewhat different. Slow exchange of waters between active flow zones and stagnant pools is also significant in many cases, and effectively results in retardation of solutes. Retardation can also occur by sorption to bed materials. On the other hand, accelerated transport of some reactive chemical species can occur if they are sorbed on to rapidly travelling sediment grains or colloids. Sources and sinks of solutes occur in the form of water added to or removed from the channel. The summation of all of these processes is described by an equation which is exactly analogous to 3.25, save that v_i (the average interstitial ground water velocity) is replaced by V (the velocity of water flowing in an open channel), which is in turn obtained from solutions of equations 3.18 through 3.22 by the simple expression:

$$V = Q / A \qquad\qquad (3.26)$$

The version of equation 3.25 which is applied to transport in streams is usually termed the "convection-diffusion equation". Like its ground water counterpart, numerical solutions are usually necessary, though a limited range of analytical solutions are also available. For further information on the application, formulation and solution of the convection-diffusion equation in surface water quality modelling, the reader is referred to the text by James (1993), which includes computer code listings and numerous case studies.

Of particular relevance to mine water systems are the OTIS and OTEQ modelling codes of the US Geological Survey (Runkel, 1998; Walton-Day *et al.*, 2000), which are available for free down-loading from their web-site (*http://co.water.usgs.gov/otis*). OTIS (*O*ne-dimensional *T*ransport with *I*nflow and *S*torage) numerically solves the convection-diffusion equation, and represents reactions by simple source-sink accounting. OTEQ uses the same transport algorithm as OTIS, but represents reactions by equilibrium sorption and/or precipitation/dissolution reactions, calculated in accordance with the well-known speciation code MINTEQA2. The principal applications of OTIS and OTEQ to date have been to upland streams receiving polluted mine drainage (Walton-Day *et al.*, 2000). As the codes are readily available as public domain software, they should enjoy widespread applications in mining regions.

Whatever the flow or quality problem faced, however, no amount of modelling will advance our understanding unless we have data to constrain our conceptualisation and predictive modelling. It is to the acquisition of the relevant data that we now turn.

3.4 HYDROMETRY IN MINING ENVIRONMENTS

3.4.1 DEFINITION OF HYDROMETRY

Hydrometry is the measurement of hydrological parameters such as water levels, water velocities etc. As time goes on, the in-situ measurement of water quality and the collection of water samples are also coming to be recognised as hydrometric activities.

Hydrometry is the very backbone of hydrological research and water resources management. The advent of new information and communications technologies within the last few decades have transformed hydrometry into a field of rapid technical development Readers seeking acquaintance with the basics of the subject should consult standard textbooks (e.g. Strangeways, 2000). Exhaustive coverage of modern hydrometric practices is given in the authoritative manual of Herschy (1995). Clearly a complete exposé of modern hydrometry is neither necessary nor desirable in this book; rather this section offers comments on how best to undertake hydrometric work in the mining environment, focusing on the quantification of parameters which are peculiar to mines (such as water loss in forced ventilation streams and underground measurements of head in overlying strata).

3.4.2 HYDROMETEOROLOGICAL PARAMETERS

For all hydrological investigations, it is necessary to be able to characterise the *hydrologically effective precipitation* (HEP) falling in and around the mine site. HEP is defined as the difference between total rainfall in a given period and the losses to satisfy any soil moisture deficit (which is essentially a measure of crop water use prior to the onset of rainfall) and evapotranspiration. To evaluate HEP, it is therefore necessary to have a record of changes in soil moisture content (usually obtained by continual application and upgrading of moisture budgeting models) and those meteorological variables which govern the rate of evapotranspiration (i.e. solar and net radiation, wind speeds and directions, air temperature etc).

In many countries, the data required to calculate HEP are gathered by public authorities, and the mine water investigator will be able to purchase time-series of quality-controlled HEP values. However, publicly available HEP time series are generally given as spatially-averaged values for a region, and are rarely readily available at any greater temporal resolution than weekly. While weekly or even monthly averages may suffice for analyses of subsurface flow, studies of storm runoff for surface mine drainage design will generally require much finer resolution.

For detailed mine water studies, consideration should be given to the installation of rain gauges and automatic weather stations. Figure 3.4 shows the two principal types of rain gauge. Storage gauges (Figure 3.4a) are simply containers of standardised proportions in which the accumulation of rainfall is recorded by an observer at regular intervals (daily, weekly or at most monthly). Recording rain gauges automatically log the incoming rainfall over specified periods. Historically the logging was done by using a pen tracing on a paper chart wrapped around a rotating cylinder, and a number of different devices were used to automatically measure incoming rainfall (see Shaw, 1994). Nowadays the most common type of recording rain gauge is the tipping-bucket gauge (Figure 3.4b), with the recording of data being made by means of a solid state digital logger. Each time one of the small buckets tips (which typically occurs every time 0.15 mm of rainfall has entered the gauge), a magnetic rod passes a static electromagnet, and an impulse is recorded by the logger. Logging of individual tips allows very high resolution rainfall records to be obtained. A modern automatic weather station (AWS) is shown in Figure 3.5, which identifies the various sensors attached to the instrument, covering the full range of parameters necessary to calculate potential evapotranspiration. Most AWS will also have a tipping-bucket rain gauge attached to them so that HEP can be determined for that particular site. In recent years, the progressive success of miniaturisation in electronics has served to reduce both the size and cost of AWS substantially, so that a complete AWS system may now be purchased for less than $5000 (year 2001 prices). With cost a less formidable barrier than hitherto, AWS installations are gradually spreading in the industry. The key considerations in the siting and optimal operation of AWS and rain gauges are given extensive coverage elsewhere (see, for instance, Strangeways, 2000). It is therefore only necessary to mention here that where AWS are installed on mine sites they need to be sited carefully so that net radiation measurements are made above representative ground surfaces. In essence, this means avoiding the manicured lawns in front of the office block in favour of areas of stripped ground, overburden, or bare spoil which are more representative of the mine site as a whole. However, to avoid clogging of rain-gauges and coating of radiation sensors, it may also be necessary to compromise a little on instrument siting to avoid the dustier areas of the site.

Given accurate values of HEP it is possible to begin to develop a water-budget for a mine site, since it is HEP which is apportioned between surface runoff and groundwater recharge, the two principal forms of water movement around and into mine workings. The details of calculations necessary to partition HEP into surface runoff and groundwater recharge are complex, and have been reviewed in great detail by Lerner $et\ al.$ (1990). More recently, a very useful, simple algorithm has been devised by Calver (1997), which takes a monthly time-series of HEP and transforms it into an equivalent recharge time-series. The basic algorithm may be written:

$$R_n = \sum_{i=1}^{m} I_{n-i+1} * u_i \qquad (3.27)$$

where R_n is the recharge in month n (L)

I_n is the infiltration in month n (L); this equals the difference between HEP and surface runoff in the month.

u_i are "calculational weights" (dimensionless), which describe what fraction of I_n actually arrives at the water table in each of m successive months (starting with month n).

In the original work of Calver (1997), the weights u_i were defined such that:

$$\sum_{i=1}^{m} u_i = 1$$

However, Younger (1998e) argued that the algorithm should be modified such that:

a

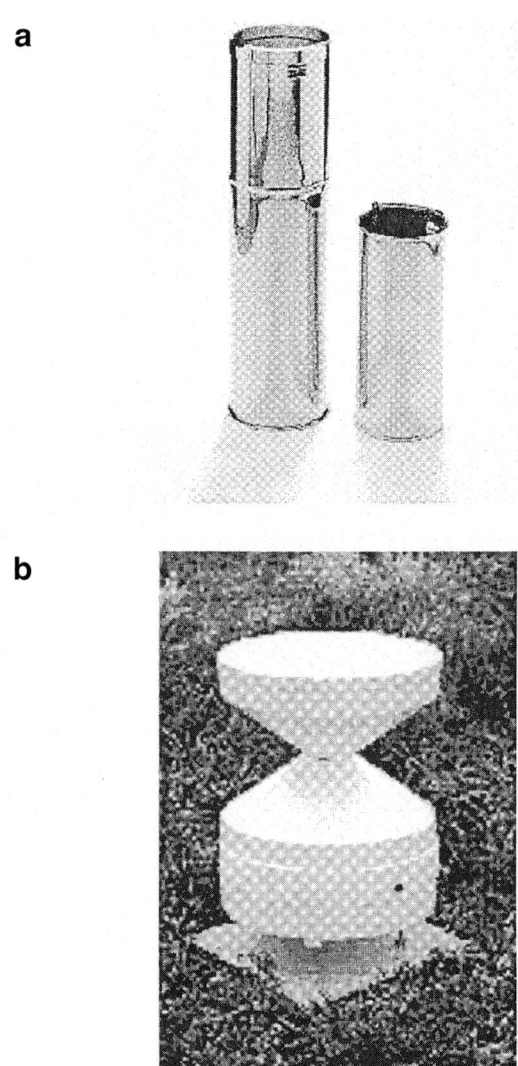

b

Figure 3.4: The two main types of raingauge: (a) storage gauge, manually read daily, weekly or (at least) monthly (b) automatic tipping-bucket raingauge, which contains a device which moves every time a certain amount of rain has entered the gauge (usually 0.3–1 mm, depending on setting), and records this movement on a digital logger. (Courtesy of Environmental Measurements Ltd).

$$\sum_{i=1}^{m} u_i = BFI$$

where BFI is the "base flow index", i.e. the ratio of base flow to the total runoff in the principal river draining the study catchment over the period of record. This summation

Figure 3.5: A modern, modular automatic weather station (AWS). Incoming radiation is measured by the short mast on the top of the instrument, and net radiation by a transparent sensor in the arm sticking out to the left. Wind speed and direction are measured by the vane and anemometer attached to the top right of the frame. Air temperature and humidity are measured by devices mounted in the louvered white box on the right of the AWS. The other white box (non-louvered) attached centrally to the frame houses the data logger, from which Geoff Parkin is preparing to download data using a portable computer. (Photo: P. Younger).

makes sense in situations where almost all base flow in local streams is ground water, which is very often the case but should be checked locally.

It should be noted that the accuracy of recharge calculations obtained using monthly data is usually significantly less than that obtained with models operating at finer time-steps (Lerner *et al.*, 1990). However, the algorithm described above does at least provide a simple, robust starting-point for developing a general mine water-budget using data which are usually publicly available at little cost.

3.4.3 FLOW GAUGING

Numerous textbooks and manuals provide detailed guidance on standard flow gauging techniques (e.g. Gordon *et al.*, 1992; Herschy, 1995). Here we provide a commentary on the adaptation of these standard techniques to the quantification of flows in and around active and abandoned mines.

Estimating flows from pump run-time data
In active mines it is common for all of the mine water to be captured by dewatering pumps and thus removed from the mine. In such circumstances, the most common flow "gauging" technique (really a crude estimation) is to log pump run-times and multiply these by known pump capacities (i.e. water delivery rates) to obtain time-series of water removed from the mine (or a sub-district of it). While run-times can be accurately logged, estimation of pumped quantities in this manner is rendered inaccurate by deviations of actual pump performance from design values. These can be caused by fluctuations in pumping head, perturbations of the power supply (particularly where local generators are used), snarling of pump impellers by suspended solids, and gradual abrasion/corrosion of pump components (a problem so marked where strongly acidic minewaters are pumped that it is at times necessary to drip lime into the pumping sump to extend pump life). Notwithstanding these *caveats*, the method enjoys widespread application in the mining industry. In practice, data gathered in this manner should be quality-checked by perform-ing spot measurements of true pump yields at various times over the life of the investigation. Such spot measurements can be made using the methods outlined below.

Measurements of flows in pipes
The most simple way to measure a flow in a pipe is to measure the flow leaving it. This may be done with ease where the pipe ends. Alternatively, depending very much on local conditions it may be possible to temporarily breach a pipe range at a junction between two pipe lengths and make a quick measurement before re-joining the two pipes once more. Wherever access is gained to water flowing from the pipe, then flow can be accurately determined using the "time-volume" method, more popularly known as the "bucket-and-stopwatch technique". As the latter name suggests, this consists merely of recording the time taken for a "bucket" (or other suitably shaped/sized container of accurately-known volume) to fill with water flowing from the pipe. The timing of filling is usually repeated three times for each measurement to be sure an accurate figure is obtained. The bucket-and-stopwatch technique is highly accurate as long as it takes more than ten seconds for the bucket to fill. Where the bucket fills more rapidly than this more accurate results will be obtained using one of the following methods.

An effective variant of the bucket-and-stopwatch technique, which is applicable where flows are so high that a bucket would fill in less than ten seconds, has been proposed by Wegelin (1996). In this method, a bucket with a circular hole in it is filled with the water, and the water overflows from the hole. The level to which the water settles in the bucket (measured as the height of the water above the top of the hole) is recorded and the flow determined from a calibration graph.

Where neither of the above two techniques is feasible, but the end of the pipe is accessible, then for a partially full pipe it is possible to estimate the flow using the Chézy, Strickler or Manning formulae (Equations 3.20–3.22). To do this, it is necessary to note the diameter of the pipe, the angle at which it is inclined (θ), and the maximum depth (d) of the water flowing in the pipe. For circular pipes, the cross-sectional area of the flow (A) can be estimated from:

$$A = D^2/4(\alpha - \sin \alpha \cos \alpha) \tag{3.28}$$

where: D is the diameter of the pipe (L), and
α is an angle given by $\alpha = \cos^{-1}(1-2\ (d/D))$.

The wetted perimeter (P_w) can be estimated from:

$$P_w = \alpha \cdot D \tag{3.29}$$

which is used to calculate the hydraulic radius R_H (from $R_H = A/P_w$). Thus equipped, equations 3.20 to 3.22 can be applied directly to estimate the flow in the pipe.

Automatic logging of flows in pipes can be achieved using "dopplers", i.e. devices which are emerged in the water and measure the water velocity from the doppler shift in transmitted and reflected ultrasound impulses. These are not very useful in small diameter pipes in which the doppler device is large enough to disturb the flow patterns. Where pipes are flowing full-bore, conventional in-line flow meters (essentially the same as domestic water meters) can be used, though these are often beset with problems of corrosion and/or mineral encrustation in mine water applications. A recently-developed alternative is the use of non-invasive electromagnetic flow meters for full-bore pipe flows, which are simply clamped onto outside of the pipe.

Where such devices are not available, and the flow cannot be diverted through a weir or flume (see below) simple estimates of full-bore flow rates can be estimated from observations of the curvature of the water jet leaving the end of the pipe. For the case of a horizontal pipe, the curvature of the jet is characterised very simply by the lateral distance "X" from the end of the pipe to the point at which the upper surface of the jet of water has fallen by a height "Y" below the level of the top of the pipe (Figure 3.6). For a given value of Y and specified pipe diameter, there is generally a simple linear relationship between flow rate (Q) and X, i.e.:

$$Q = m X + c \tag{3.30}$$

For instance, if Y is chosen always to be equal to 15 cm, and with X in cm and Q in litres per minute, then for 100 mm (nominal 4") internal diameter pipes, m = 31 and c = 43.

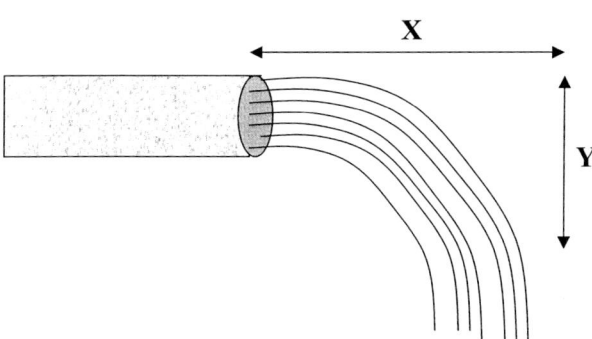

Figure 3.6: Sketch of jet curvature from a horizontal pipe discharging at full-bore. For discussion of the meaning and use of X and Y, see the text, especially Equation 3.30.

For 160 mm (nominal 6") internal diameter pipes, m = 74.5 and c = –43. Appropriate values for other pipe diameters and Y values, and corrections for partially-full flow conditions, may be calculated from information presented by Driscoll (1986), who also describes a similar method for estimating observed discharge from vertically-oriented full-bore pipes.

Measurements of open-channel flow
Open channel flow measurements are necessary in many settings in and around mine sites, including underground in deep mines, with two main purposes:

(i) to quantify mine water flows and
(ii) to demonstrate (or refute) dewatering impacts on flows of natural watercourses.

For the latter purpose, standard hydrological measurement techniques will suffice. Detailed accounts of these techniques are widely available (e.g. Gordon *et al.*, 1992; Shaw, 1994; Herschy, 1995). For single, one-off measurements of flow (usually termed "spot gauging") the velocity-area method is the most accurate and hence most widely-used technique. When the point at which channel flow is to be measured has been identified, the total cross-sectional area of the stream at this point is sub-divided into a number of small sub-areas (usually at least ten across the channel width). In each sub-area, the mean velocity is determined using a flow meter suspended at three fifths of the total water depth (at this point the actual velocity should equal the mean throughout the depth profile). Flow meters are most commonly impeller devices (the revolutions of which are automatically counted and converted into an equivalent velocity). Heat-pulse flow meters, which measure the downstream movement of pulses of heat introduced into the water, are increasingly popular as they have no external moving parts and are less prone to clogging and abrasion than impellers. In cases where a flow meter is unavailable, floats may be used to obtain rough estimates of the mean velocity. An orange fruit makes a good float due to its density and visibility, and velocities tracked using a floating orange should be multiplied by 0.8 to estimate the mean velocity in the channel. Once velocity measurements are available, the mean velocity ($L.T^{-1}$) in each sub-area is multiplied by the corresponding area (L^2) to yield a flow value ($L^3.T^{-1}$) for the sub-area. The flow values for all sub-areas are summed to give the flow rate of the full stream cross-section at the measurement point. Discharge values accurate to within ± 10% should be readily obtainable by this method. However, because of the physical difficulties of sub-dividing the width of narrow channels, such velocity-area measurements may have errors in excess of ± 40% for channels narrower than about 4 m. Where channels approach such diminutive dimensions, and for all sections where long-term records are required, consideration should be given to the temporary or permanent installation of a "control structure" to facilitate flow measurements.

Control structures are weirs or flumes which constrict the flow in the channel such that the flow conforms to a hydraulic condition known as "critical flow". Under critical flow conditions, the relationship between flow rate and water depth is linear under both rising and falling flow conditions. (For other hydraulic conditions, the relationship may differ between the rising and falling conditions). Hence a control structure which is properly designed, constructed and maintained should facilitate the determination of flow simply by measurement of the water level (usually termed "stage") immediately upstream. For small weirs and flumes of precise dimensions, stage measurements alone will suffice. For larger structures in major channels, it will often be necessary to "calibrate" the structure by making occasional velocity-area flow measurements downstream. In mine water studies, the majority of appropriate control structures are likely to be thin-plate weirs or small flumes. The selection of the appropriate structure depends on the magnitude of the flow to be gauged and maintenance considerations. For small flows, v-notch weirs (Figure 3.7) are the most common choice. They can accurately measure flows from just above zero to several $Ml.d^{-1}$, depending on the depth of the weir. Great depths will not always be easy to arrange on specific sites. Even where they can be accommodated, v-notch weirs are highly prone to clogging by sediment or ochre in mine water applications, particularly where flows are small. For relatively small flows, a good alternative to a v-notch weir is an H-flume (Figure 3.8). These flumes have jaws which are open to the base of the structure, and are therefore far less prone to clogging than v-notch weirs. The 0.18 m H-flume shown in Figure 3.8 is capable of carrying flows up to 6.5 $l.s^{-1}$; construction details and ratings for H-flumes suitable for measuring greater flows are given by Ackers *et al.* (1978).

For large flows, v-notch weirs and h-flumes will often be inappropriate, due to the great depths to which they must be built. For flows in excess of about 20 $l.s^{-1}$, it may be best to use a rectangular thin-plate weir (Figure 3.9). Table 3.2 lists some stage-discharge values for H-flumes and weirs of v-notch and rectangular shape, which are useful not only in estimating flows from water level measurements, but also provide a useful guide for selecting and designing a structure appropriate for particular circumstances.

Where none of the foregoing methods can be applied for whatever reason, dilution gauging may be used to measure open-channel flows. In mainstream hydrology, dilution

Figure 3.7: A small v-notch weir tank providing flow measurement for clean, treated water leaving a coal processing plant at a South African colliery. (Photo: P. Younger).

Figure 3.8: An H-flume in use to measure flows of ferruginous mine water at Colpitts Grange mine water discharge, Slaley, Northumberland (UK) (Photo: P. Younger). Although 'self-cleansing' in relation to inert sediments due to the high velocities in the 'jaws', such H-flumes need to be brushed out occasionally when used to measure ferruginous waters, as accretion of ochre from fast-flowing waters (by means of surface-catalyzed oxidation of ferrous iron) can gradually clog them.

Figure 3.9: A rectangular weir recording flow existing a major treatment wetland (Woolley Colliery, West Yorkshire, UK) (Photo: P. Younger). Note the accumulation of weed in the weir orifice – a common problem affecting the accuracy of weir measurement, and requiring regular inspection and clearing.

Table 3.2: Selected rating equations for H-flumes and v-notch and rectangular weirs, yielding flows (Q) from measurements of head (h)

Type of flow gauging structure	Simplified rating equation	Comments
H-flume	$Q = (0.014\, h^2 - 0.56\, h + 9.3)/60$	Head measured above floor of flume. Yield Q in $l.s^{-1}$ if h is in mm.
V-notch weir (90°)	$Q = 0.29\, h - 32.5$	Head measured above the apex of the 'V'. Yield Q in $l.s^{-1}$ if h is in mm.
Rectangular thin-plate weir	$Q = L \cdot [1.83\,(1{-}h) \cdot h^{1.5}]$	Head measured above the crest of the weir. L is the width of the weir in m. The formula yields Q in $m^3.s^{-1}$ if h is in m.

gauging is principally used in steep, rocky, mountain streams, in which as much as 40% of the total flow will locally occur within the bed rather than in the open channel. In mine settings, dilution gauging can be used to measure flows in channels to which access is limited for safety reasons, in old adits with rocky floors, and in steep, rough channels in open pits. Dilution gauging is undertaken by introducing a chemical tracer into a water course (either as a single "slug", or at a constant rate over a period of a few hours or days) and measuring the changes in tracer concentration at one or more monitoring points downstream. Suitable tracers can be as simple as common table salt, in which case the conductivity (specific electrical conductance) of the water can be easily monitored at downstream monitoring points using portable meters. Where a single slug of tracer is added to a stream, the flow rate at a given observation point is given by:

$$Q = \frac{V_t \cdot C_t}{\Delta t \sum_{i=1}^{n} C_i - C_o} \tag{3.31}$$

where C_t is the concentration of the tracer in the injected "slug" of water ($M.L^{-3}$)
        C_i is the measured concentration of the tracer at the observation point at each of n time-steps during the test ($M.L^{-3}$)
        C_o is the background concentration of the tracer in the water course (ie prior to the test)
        V_t is the volume of the slug of water containing the tracer (L^3)
        Q is the flow rate of the water course ($L^3.T^{-1}$).
        Δt is the length of the time-steps between successive measurements (T)

For accurate measurements, Δt should be kept as short as possible. For all but the largest and most slow-flowing watercourses, a Δt value of one minute is advisable.

Where tracer is injected at a constant flow rate over a sustained period, the downstream concentration of tracer will gradually rise to a constant concentration. In this case, the calculation of the total flow simplifies to:

$$Q = \frac{C_t - C_s}{C_s - C_o} \cdot q \qquad\qquad (3.32)$$

where C_s is the constant concentration attained in the stream after injection has been underway for a long period (M.L^{-3})

q is the rate of injection (L^3.T^{-1})

and all other symbols have their previous meanings.

Applications of tracer tests to mine water problems are described in detail by Trexler (1979), Aldous and Smart (1988), Walton-Day *et al.* (2000) and Wolkersdorfer (2002), amongst others.

Measuring underground feeders and drippers
Here we consider flows which are never described in standard hydrological texts: the various modes of water ingress to underground workings.

Feeders are relatively large flows, resembling underground springs, which can cause considerable nuisance for miners. As with surface springs, the simplest and most accurate approach to measuring feeder flow rates is the bucket-and-stopwatch technique. In many cases, this is most readily achieved by diverting the feeder through a section of pipe (Figure 3.10). Where the feeder is sufficiently voluminous, one or other of the pipe flow or open-channel flow measurement techniques may be applicable.

"Drippers" is the term used to describe water ingress in the form of numerous, isolated drip-feeds through the ceiling, which causes discomfort for the miner akin to the experience of walking in heavy rainfall. It is often beneficial both to operations and to hydrological measurement to capture the drippers in a tarpaulin or corrugated sheeting suspended below the ceiling, which has the effect of gathering the drippers from a wide area of roof to a single point, where they may be measured using a bucket-and-stopwatch (Figure 3.11) or even using a tipping-bucket rain gauge. An example of the use of tarpaulins for this purpose in the Bunker Hill Pb-Zn Mine, Idaho, is given by Lachmar (1994).

3.4.4 MEASURING GROUND WATER HEADS

Standard head measurements using vertical boreholes and old mine shafts
The usual technique for determining ground water heads (also known as "piezometric heads") is deceptively simple: the depth from the top of a borehole to the water level is measured, and the absolute elevation of the water relative to some fixed datum is calculated by subtracting the depth to water from the surveyed level of the top of the borehole. Such measurements yield the total head in the borehole at the point of measurement. Most commonly ground water heads are expressed relative to a national datum which will be mean sea level as measured at a specified location. However in mining settings, to avoid dealing in negative numbers, surveyors often express all values relative to the "mine datum", which will be locally defined. For instance the national mine datum in the British coal mining industry was previously set such that zero occurs at an elevation equivalent to 10,000 feet (3048 m) below mean sea level. In the world's deepest mines (South African

Figure 3.10: Measuring a feeder undergound, by diverting it through a pipe whence it is readily meaasured by bucket-and-stopwatch. John Evans is making the measurements on the Upper Whin Sill Dib, Frazer's Grove Fluorspar Mine, Weardale, UK. (For details of the mine and the use of the data collected, see Younger, 2000b). (Photo: A. Doyle).

Figure 3.11: Collecting numerous small drippers associated with an old ore hopper using a tarpaulin, facilitating bucket-and-stopwatch measurement. (The tarpaulin is improvised from 'brattice cloth', i.e. canvas normally used for sealing unwanted ventilation routes). John Evans and Paul Younger are making the measurement on the 285E Level of Frazer's Grove Fluorspar Mine, Weardale, UK. (For details of the mine and the use of the data collected, see Younger, 2000b). (Photo: A. Doyle).

gold mines), where workings are advancing to depths of as much as 5 km at the time of writing, a correspondingly deeper mine datum is needed.

In order to correctly interpret borehole measurements of head it is necessary to understand something about the borehole construction (Figure 3.12). If the borehole pierces the entire saturated zone of a given aquifer or flooded body of mine workings, then the total head will represent average head for that position, which will correspond to the position of the water table (in unconfined settings) or piezometric surface (in confined settings; see Figure 3.2). Such a fully-penetrating borehole can be termed an "observation borehole" (Figure 3.12a). Where a borehole is open to the aquifer layer only over a very narrow interval, that borehole is termed a "piezometer" (Figure 3.12b). The head measured in a piezometer will be the local value of head at the depth corresponding to the open interval. The head at a specific depth may be considerably higher (position Z1 in Figure 3.12b) or lower (position Z2 in Figure 3.12c) than the water table/piezometric surface. Where piezometric head increases with depth, ground water flow is directed upwards and will likely discharge to the surface or an adjoining aquifer nearby. Indeed, if the piezometer is tightly sealed above the open interval, and the head in this sampled interval exceeds the ground level at the top of the piezometer (e.g. position Z3 in Figure 3.12d), then water can be made to flow from the well without pumping. By contrast, where piezometric head decreases with depth below the water table (Z2 in Figure 3.12c), the observations are being made in a zone of active recharge, in which the ground water is moving downwards.

For most purposes, the depth to water in a borehole will be measured using a tool known as a "dipper". This is simply a graduated tape containing electrical wires, arranged such that the circuit between the two wire ends is completed when the end of the tape touches water (see Brassington, 1998, for further details). A number of problems can beset the use of such dippers in mining settings. For instance, where water levels lie far below ground level due to dewatering, it may be difficult to obtain a dipper with sufficient tape to reach the water. In such cases, improvisation may be necessary, such as suspending a float on a graduated cable which can be raised whenever a measurement is required until the float leaves the water surface and its weight is sensed on the cable. Even where spools of 1000 m of dipper tape can be obtained, considerable stretching of the tape occurs under self-loading at great extensions, reducing the accuracy from the norm of ± 1 cm to as much as ± 1 m. Furthermore, where measurements are made in old mine shafts rather than purpose-drilled boreholes, it is likely that old shaft fittings and/or floating debris will be present, preventing the tape reaching the water and in some cases snaring it and preventing its retrieval. The installation of a solid "dip tube" (typically a 25 mm or 50 mm pipe of galvanized steel, PVC or HDPE) extending from the shaft collar to a short distance below the water surface may overcome this problem.

Head measurement using dippers yields discontinuous records of head, which will be insufficient for many detailed investigations. Where frequent head measurements are required, it is common practice to install automatic pressure transmitters in boreholes, which will digitally record heads at pre-programmed intervals in solid-state loggers. Obviously, pressure transmitter cables are prone to similar problems to those listed above for deeper tapes where water levels are deep and/or shafts contain obstructions. In addition to simple cable stretching, most reliable pressure transmitters/transducers must be vented to the atmosphere. This requires the use of cable with built-in venting ducts, which is expensive for deep applications. Even where budget is not limiting, the cable is typically

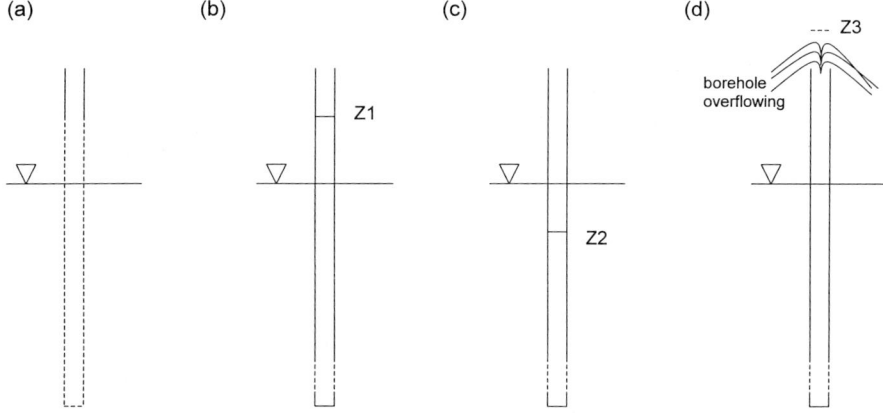

Figure 3.12: Sketch illustrating the nature of boreholes fitted as either an "observation borehole" (a) or "piezometers" (b–d). In all sketches, solid lines are where the borehole is tightly sealed against the rock and will not allow water to enter, whereas dashed lines are 'screened' intervals, where the boreholes are open to the aquifer. The observation borehole (a), being open over the entire pierced interval of the saturated zone, records the mean head, which approximates to the water table. The piezometers record the head at the specific depth at which they are open. See the text for further explanation.

supplied from the factory in maximum lengths not exceeding 100 m. Where greater depths must be attained, successive lengths of cable need to be spliced, which is difficult to achieve and results in joints of dubious long-term integrity. Furthermore, stretching of these cables not only induces errors in nominal logged depths, but can also attenuate the venting ducts to the point of uselessness. To complicate matters even further, in abandoned coal mines, it is also often required by safety authorities that "intrinsically safe" electronic logging equipment be used, which is considerably more expensive than conventional equipment (though no more accurate). Recent developments in pressure-logging may provide solutions to these problems. For instance, the "Orphimedes" logger (manufactured by OTT Hydrometrie GmbH, Germany) uses the resistance to release of an air bubble from a small nozzle to estimate water depth above the logger.

Measurement of head in inclined drifts which cannot be entered safely
Many deep mines are accessed by inclined drifts rather than vertical shafts. After mine abandonment, it may be desirable to monitor water level rises in such inclined drifts, yet they may be inaccessible in the absence of forced ventilation and maintenance. This can make it difficult to measure water levels using standard equipment. For instance, dippers cannot be lowered from surface along low-angle drift floors. Even if a pressure transmitter had been left in the inclined drift before loss of access, it would be difficult to use for two reasons:

- excessive lengths of expensive vented cable would be needed to reach the water table in all but the shallowest of inclined drifts, and
- as the head of water above a transmitter exceeds the calibrated range of the particular transmitter (they are usually factory-calibrated to 10 m or 100 m ranges), they could not reliably be hauled to higher positions.

To overcome these problems, a home-made device may be used. This consists of a bundle of insulated wires connected to a series of independent float switches. The float switches are laid out along the floor of the inclined drift at regular intervals (determined by trigonometry according to the angle of inclination) so that float switches closer to the surface are successively closed as the water level rises. In the only application of this technique to date, in a 1-in-4 drift at Frazer's Grove Mine (County Durham, UK; see Johnson and Younger, 2000), float switches were positioned at 4 m spacings along the drift floor to obtain a metre-by-metre record of water level rise.

Measuring head in overlying or adjacent flooded workings
Inclined and horizontal underground boreholes (Figure 3.13a) are generally drilled ahead of development headings and active faces where there is any chance that the workings are approaching an accumulation of water in old workings. Where these boreholes encounter significant quantities of water, they may be allowed to flow in order to relieve water pressures in the old workings in advance of re-mining (e.g. pillar winning). However, decisions with regard to such drainage may well require a knowledge of the head in the flooded old workings. As a dipper cannot be used in such circumstances, it is necessary to measure the head by means of a pressure gauge, which can be fitted to the near end of the borehole. When a measurement is required, the borehole must be temporarily sealed (usually with a flanged blanking plate) and the pressure allowed a few minutes or hours to stabilise as necessary before taking a reading. This pressure reading gives one component of the total head in the old workings. The other component (equation 3.7) is the elevation head, which corresponds simply to the elevation of the point at which the gauge is fixed to the borehole. The total head of water is therefore the sum of the elevation head and the pressure head. Pressure readings are usually given in units of kilonewtons per square metre ($kN.m^{-2}$), also termed kilopascals (kPa), which can be converted into an equivalent head of water in metres simply by multiplying by 0.1. (In older mining engineering literature, water gauge pressures are usually quoted in units of pounds per square inch (psi), for which the conversion to an equivalent head of water in metres is achieved by multiplying by 0.7). Hence if a borehole in contact with flooded workings has a pressure gauge installed at 2305 m above mine datum, and this gauge records a steady pressure head of 250 kPa, the total head of water in the old workings equals $2305 + (250 \times 0.1) = 2330$ m above mine datum. Worked example 3.1 illustrates the practical application of this approach in mine safety assessments.

Mine water head measurements using this approach are not restricted only to subsurface boreholes. Figure 3.13b illustrates how an access tube left in an adit plug at the time of mine abandonment can be used to obtain head measurements in the flooded workings. Such measurements can have considerable value in the assessment of management options for abandoned drift mines.

3.4.5 MEASURING VENTILATION WATER LOSSES

The potential importance in mine water-budgets of subsurface evaporation due to forced ventilation was noted in Section 3.3.2. Where mines are very deep they are invariably also hot, reflecting the geothermal gradient. Since warm air will hold significantly more moisture than cold air, the humidity of the air in deep workings is often high. Warm, humid mines typically require very high rates of forced ventilation to maintain working conditions within tolerable limits. Consequently, considerable quantities of water can leave a deep mine in vapour form, in the upcast ventilation exhaust air stream. These quantities can amount to a significant proportion of the total mine water-budget. For example, ventilation water losses of around 130 m^3/d are quoted by Plumptre (1959) for each of two coal mines in Kent, UK; this rate corresponded to about 10% of the quantity of water pumped from those mines. More recently Grmela and Tylcer (1997) calculated ventilation water loss rates of 103 m^3/d and 216 m^3/d for two coal mines in the Czech Republic (corresponding to 3% to 18% of the pumped quantities respectively).

Determination of the quantities of water leaving a mine in the ventilation exhaust stream can be achieved to a first approximation by manipulation of a few simple descriptors of the moisture content of air. The mass of water present in a given volume of air is termed the "absolute humidity" (H_a), and is usually expressed in units of g.m^{-3}. The capacity of air to hold water increases with temperature, and the absolute humidity at which a given volume of air at a specified temperature will begin to release moisture from the gaseous phase (by condensation to form droplets of liquid water) is termed the "saturation humidity" (H_s). In the range of temperatures relevant to most deep mine settings (i.e. 0 to 30°C) the dependency of saturation humidity (H_s, in g.m^{-3}) on air temperature (t, in °C) is well-described ($R^2 = 0.999$) by the following function:

$$H_s = 5 \cdot e^{0.06t} \tag{3.33}$$

Using Equation 3.33, it is simple to calculate the maximum humidity of a cubic metre of air at a given temperature. Thus given the temperature of the ventilation exhaust stream from a mine, we can already obtain an upper-bound estimate of the associated rate of water loss by multiplying the exhaust rate (m^3 of air per second) by H_s.

If a more accurate value than the upper-bound estimate is required, then further steps must be taken. For a complete analysis of the water losses by subsurface evaporation, it is necessary to identify the net gain in humidity on passage of the air through the workings. This is achieved by determining the *absolute humidity* (see below) of the downcast ventilation current, and subtracting this from the absolute humidity of the upcast ventilation current. (For weekly or monthly average values, this simple subtraction will suffice, but where daily or hourly losses are required, it will be necessary to take the subsurface residence time of the ventilation air into account).

The absolute humidity (H_a) of a given volume of air at a known temperature can be calculated from its H_s value by application of the concept of "relative humidity" (H_r). The relative humidity of a body of air at a specified temperature is the percentage ratio of its absolute humidity to the saturation humidity for the same temperature. For instance, the saturation humidity of a body of air at 12°C is given by Equation 3.33 as 10.3 g.m^{-3}. If the

(a)

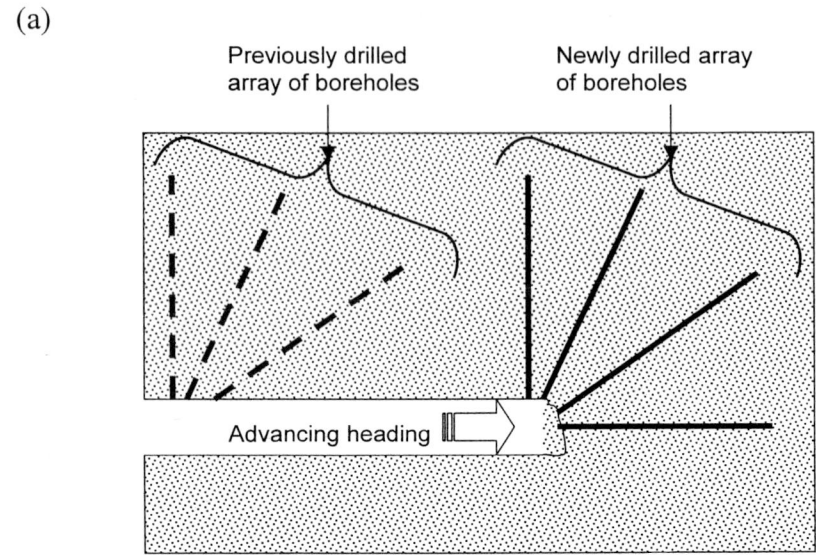

(b)

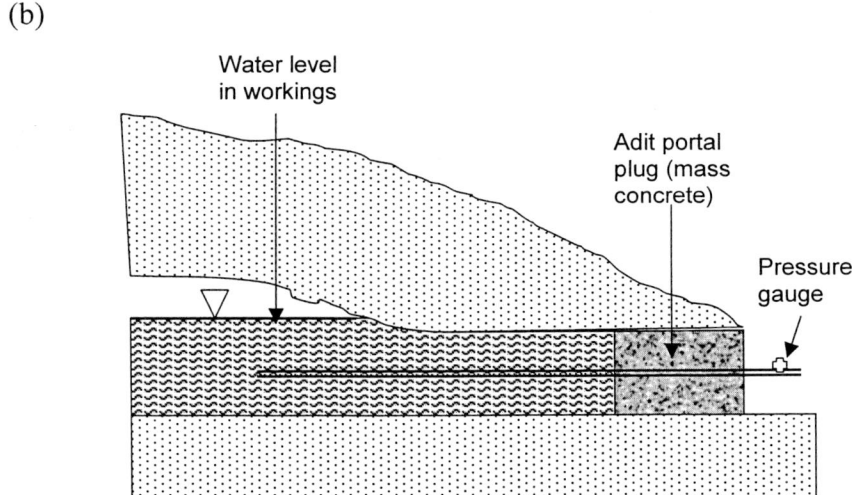

Figure 3.13: (a) Boreholes drilled ahead of an advancing underground heading as a precautionary measure in situations where accumulations of water may be present (e.g. in possible old workings). The boreholes are drilled at set distances in the progress of the heading, with the numbers, inclinations and frequencies of boreholes being prescribed in local mining law. (b) An open pipe installed through the plug in an abandoned drift, facilitating measurement of the head of water in the workings (and therefore the head being held back by the plug) by means of pressure measurement. If left open, this pipe can be used to drain the workings, but if the head is to be measured, the flow must be closed off for a while and the pressure measured until a steady value is obtained.

Worked Example 3.1 Use of underground head measurements in mine safety assessment

The practical use of such measurements in mine safety assessments is well illustrated by the following example from an undergound coal mine in northern England. The former Kelloe Colliery, in the Durham Coalfield, UK, worked a rich sequence of 8 coal seams, with extraction commencing in 1836. By 1983, after almost 150 years of working, reserves were restricted to a block of coal in the lowest seam, known as the Busty Seam. This block of coal had been previously left as a pillar to support overlying workings in the Tilley Seam (27 m above the Busty Seam) and the Harvey Seam (40 m above the Busty) which had been active in the 1960s. In outline, the layout was thus:

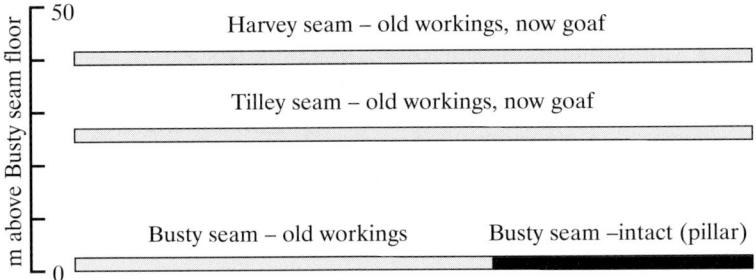

Human access to the Tilley and Harvey seams was no longer possible by 1983. However, the records from the 1960s reported little or no water in the Tilley and Harvey seams, so it was not anticipated that there would be any risk of flooding from the overlying workings if the untouched block of coal in the Busty was now removed. Nevertheless, in line with government regulations, an inclined borehole (38° from vertical) was drilled upwards from the edge of the unworked block of coal in the Busty to intersect the overlying Tilley workings. On 19th April 1983, the borehole reached an area of goaf in the Tilley Seam, and to the surprise of the mine management, water immediately began to flow from the hole. The hole was reamed and fitted with a pressure gauge, which returned a reading of 68 psi. This value was so high that is was suspected that the gauge was faulty. A second, new gauge subsequently returned the same reading. This equates to $0.7 \times 68 = 48$ m head of water above the pressure gauge, which meant that the Tilley and Harvey seams were both completely flooded, with $48-40 = 8$ m of head above the base of the Harvey Seam.

The hydrogeological situation was now seen to be thus:

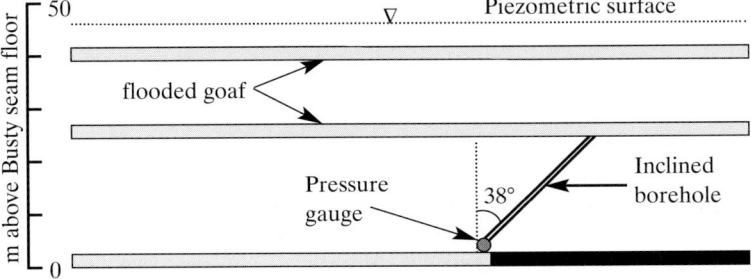

Thus a major accumulation of water was lying only 27 m above the Busty workings, whereas UK law prohibits working closer than 105 m to such a known body of water. With its last reserve of coal unworkable, then without resort to a major advance dewatering operation (which the finances of the mine could no longer stand) Kelloe Colliery had to close.

body of air in question has a relative humidity of 60%, then its absolute humidity (H_a) is given by:

$$H_a = H_s (H_r/100) \qquad (3.34)$$

which, for the example here gives $H_a = (60/100) \times 10.3 = 6.18$ g.m^{-3}.

Fortunately, relative humidity is amenable to rapid measurement using simple, portable, readily-improvisable equipment. Little more is required than a pair of identical glass thermometers. One thermometer has its bulb suspended dry in the air column, to measure air temperature (t). The second thermometer (the "wet-bulb thermometer") has its bulb wrapped in muslin connected by a wick to a small reservoir of distilled water. The heat loss associated with the evaporation of the distilled water chills the wet bulb, so that the "wet-bulb temperature" (t_w) which it records is lower than the ambient air temperature. From simultaneous dry and wet-bulb temperature values it is possible to calculate the relative humidity, as follows:

(i) Calculate the saturation vapour pressure (E_s, in millibar) for both the air temperature and the wet-bulb temperature, according to:

$$E_s = 6.1 \times 10^{(7.5T/(237.7 + T))} \qquad (3.35)$$

where T represents t or t_w as appropriate. For convenience, call the wet-bulb saturation vapour pressure E_{swb} and the saturation vapour pressure at air temperature E_{sa}

(ii) Calculate the "actual mixing ratio" (W) as follows:

$$W = \frac{(t - t_w) - 2500(E_{swb}/P)}{(-(t - t_w) - 2500)} \qquad (3.36)$$

where P = atmospheric pressure at the ground surface (varies with altitude, but is 1013 mb at sea level).

(iii) Calculate the "saturation mixing ratio" (W_s) from

$$W_s = E_{sa}/P \qquad (3.37)$$

(iv) Finally calculate the relative humidity (H_r, as a %) as follows:

$$H_r = 100 (W/W_s) \qquad (3.38)$$

Worked example 3.2 illustrates the practical application of this suite of equations to quantifying the equivalent flux rate of water removed as vapour from a deep mine in the upcast ventilation stream.

Worked Example 3.2 Water loss in the ventilation exhaust of a deep mine

A 1200 m-deep base metal mine located close to sea level in western Europe maintains a ventilation circuit of 185 $m^3.s^{-1}$. Dry and wet-bulb thermometer readings were taken in the downcast and upcast shafts, yielding values for t of 10°C (downcast) and 29°C (upcast), and corresponding values of 5°C and 24°C for t_w. Assuming these values are representative of a period longer than the residence time of the air in the workings, it is possible to calculate the rate at which water is leaving the mine in the ventilation stream. From equation 3.33, the saturation humidities at the downcast and upcast air temperatures are found to be 9.1 and 28.5 $g.m^{-3}$ respectively. Actual saturation vapour pressures at the two shafts are given by Equation 3.35 using the t values, ie downcast $E_{sa} = 6.1 \times 10((7.5 \times 10)/237.7 + 10) = 6.1 \times 100.3 = 12$ mb, and

upcast $E_{sa} = 6.1 \times 100.8 = 38.5$ mb.

In the same way, the wet-bulb saturation vapour pressures are calculated from the t_w values, obtaining downcast $E_{swb} = 6.1 \times 100.15 = 8.7$ mb and upcast $E_{swb} = 6.1 \times 100.67 = 28.5$ mb.

Equations 3.36, 3.37 and 3.38 are now used to yield H_r values for the downcast and upcast shaft ventilation streams as follows:

Downcast: $W = (5–2500(8.7/1013))/(–5–2500) = –16.5/–2505 = 6.6 \times 10^{-3}$
$\qquad\qquad W_s = 12/1013 = 1.2 \times 10^{-2}$
$\qquad\qquad$ Hence $H_r = (0.0066/0.012) \times 100 = 56\%$

Upcast: $W = (5–2500(28.5/1013))/(–5–2500) = –65.3/–2505 = 0.026$
$\qquad\qquad Ws = 38.5/1013 = 0.038$
$\qquad\qquad$ Hence Hr $= (0.026/0.038) \times 100 = 68\%$

Consequently the absolute humidity in the downcast air current is $(56/100) \times 9.1$ $g.m^{-3}$, i.e. 5.1 $g.m^{-3}$. In the upcast air current the absolute humidity is $(68/100) \times 28.5 = 19.4$ $g.m^{-3}$.

This means that there is a net gain of $19.4–5.1 = 14.3$ $g.m^{-3}$ of moisture in the ventilation stream during its passage through the workings. At a ventilation rate of 185 $m^3.s^{-1}$, this means that the ventilation stream is removing $185 \times 14.3 = 2645.5$ $g.s^{-1}$ of water from the mine. This is equivalent to 2.6×10^{-3} tonnes per second, which equals 2.6×10^{-3} $m^3.s^{-1}$ liquid water equivalent, or about 230 $m^3.d^{-1}$. For this particular mine, this was equivalent to around 10% of the total amount of water pumped from the mine during the period of measurement.

3.5 PHYSICAL IMPACTS OF MINERAL EXTRACTION ON NATURAL HYDRO-LOGICAL SYSTEMS

3.5.1 EXCAVATION AND FRACTURING

In this section we consider the effects of mineral extraction *per se* on rivers, wetlands, and aquifers. The focus is on the hydrological effects of the excavation of voids, and of the consequent propagation of fractures beyond those voids. The hydrological impacts arising from dewatering activities, though sometimes difficult to differentiate in practice, are conceptually different; discussion of these is therefore reserved to Section 3.8 below.

In most cases, the hydrological impacts of mining *per se* tend to be far more localised in scale, and far smaller in magnitude, than those due to dewatering. This is probably because, as Kesserû (1995) has pointed out, the miner and the water resources manager share a common interest in avoiding the ingress of fresh water into a mine void: the water manager's loss of resource is the miner's increase in nuisance. With both parties suitably motivated, efforts are generally made wherever feasible to minimise the direct interaction of mine voids and water resources.

Nevertheless, when excavation of a deep mine or surface mine results in the removal of rock from the ground it inevitably disturbs the local geomechanical stress fields. Where the enclosing rock is extremely competent and the excavations relatively modest, the degree of disturbance may be immeasurably small, or may take so long to develop that it can effectively be ignored over time-scales of human interest. This is the case, for instance, with many apparently stable bord-and-pillar arrangements. Although a number of "rules of thumb" for the design of stable bord-and-pillar systems enjoy wide currency in the mining industry, most of these are sufficiently misleading (Taylor *et al.*, 2000) that the sizing of meta-stable bords and pillars is usually determined by local experimentation. Once a suitable local formula has been identified, however, it frequently proves possible to mine for decades without causing fracturing of the roof strata. For instance, Younger (2000e) reported the development of a 8×8 m bord-and-pillar system in a UK salt mine which was so stable that no roof fracturing occurred over four decades, resulting in zero ingress of water from overlying saturated sands.

Where strata are weak, or pillars are too slender, or a caving method of mining is used (see Section 1.3.2), closure of mine voids will be reflected by the development of *strain* in the surrounding rocks, i.e. deformation as the rock mass adjusts to the altered stress regime. Rock masses most commonly adjust to excavation by brittle failure, which is characterised by extensional fracturing and caving, i.e. the breaking of a hitherto dense mass into a jumble of smaller fragments which together occupy a larger volume than the unmined rock mass. Only at great depths (and thus temperatures and pressures) will most common rock types deform by ductile failure, i.e. by irreversible (plastic) deformation of a flowing rather than fracturing nature. Nevertheless, certain types of rock (such as rock salt and certain mudstones) may display ductile deformation at relatively shallow depths. A thorough exposition of the mechanics of brittle and ductile failure is beyond the scope of this work; readers seeking such details are directed to the excellent work of Brady and Brown (1993). Here it is sufficient to note that the response of most enclosing strata to the excavation of deep mine galleries or open pits will be by brittle failure, principally by fracturing. In some cases these fractures will propagate for considerable distances vertically and laterally, causing surface subsidence. Differential subsidence at the ground surface often leads to alterations in surface drainage patterns. Even where the mining-induced fractures do not daylight, they can cause substantial changes in the permeability and porosity of underground strata. The consequences of these sorts of changes will now be considered.

3.5.2 IMPACTS OF EXCAVATION ON SURFACE WATER SYSTEMS

Subsurface mining

Impacts on surface water courses are at their most extreme where a channel is directly intersected by one of the following types of mining-related features:

(i)   Deep mine stopes which have been driven inadvertently to the surface underneath a stream (Figure 1.34).
(ii)   Extensional fractures caused by subsidence of roof strata above a collapsed mine void. Figure 3.14 shows one of several large open fractures in a dolerite sill which have developed above zones where three underlying coal seams were subjected to total extraction. The family of fractures to which the illustrated example belongs intercept the entire surface drainage of a South African mountain plateau, so that the underlying deep mine system now receives the entire effective rainfall falling on the plateau.
(iii)   Subsidence hollows caused by upward migration of abandoned mine voids. Figure 3.15 shows an example from the KwaZulu-Natal coalfield in South Africa, in which shallow bord-and-pillar coal workings have formed a crown hole in a small ephemeral stream valley. When the stream flows in wet weather, the entire flow is diverted underground at this point.

Although such hollows, fractures and even open stopes will often become so choked with silt and vegetation over the years that they are no longer conspicuous, they will often continue to divert surface runoff into deep mine workings for many decades. For instance, Trexler (1979) described a surface subsidence hollow formed above block-caving stopes in the Bunker Hill Pb-Zn mine in Idaho, which was still accepting the entire flow of three streams some thirty years after the stoping had ended.

In some cases, mine workings have propagated up into the beds of lakes, wetlands or *in extremis* the ocean, with drastic consequences. For instance, 27 miners were drowned in 1837 in Workington Colliery (Cumbria, UK) when extraction of a pillar in under-sea workings caused fracture propagation through 50 m of cover to the seabed. The sudden inrush of sea water to the mine caused a visible whirlpool on the surface of the sea, and the mine was entirely flooded within minutes (Duckham and Duckham, 1973). In 1908 three miners were killed by a deluge of liquefied peat which flooded Roachburn Colliery (also in Cumbria), when bord-and-pillar workings advancing at a depth of more than 80m below ground level came unexpectedly into contact with an unusually thick sequence of unconsolidated Quaternary deposits overlying the coal seams. The sudden imposition of a very high hydraulic gradient prompted liquefaction of a peat mire, which rushed into the workings as a slurry, leaving behind on the surface a crater measuring 15 m in depth and 40m in its longest dimension. A similar incident occurred at Knockshinnoch Castle Colliery (Ayrshire, Scotland) in 1952, though in that case quick thinking and intrepid action on the part of some of the older miners resulted in the safe evacuation of all personnel via long-neglected old workings (Duckham and Duckham, 1973). Such experiences have provided vivid lessons to mining engineers, who have subsequently devised techniques for safe working of deep mines beneath major bodies of water (rivers, aquifers and the ocean; e.g. Orchard, 1975; Atkins and Whittaker, 1985; see Section 3.6 below). These techniques facilitated safe exploitation of undersea coal reserves in Britain and Canada during most of the 20th Century.

Even where mine voids do not prompt such major inrushes, fractures propagated from the workings can cause sufficient under-drainage of previously perennial streams that the latter may dry out in summer months (e.g. Grapes and Connelly, 1998). Where mining has been extensive, dry valleys may be as common a geomorphic form as they are in natural

Figure 3.14: An open fracture caused by extensional deformation of a dolerite sill overlying superimposed zones of total extraction in three coal seams (KwaZulu-Natal Coalfield, South Africa). A stone dropped down this crack kept falling beyond the limit of audibility. Surface watercourses encountering such cracks are intercepted to form recharge to the underlying mine workings. William Pulles of PHD Inc. daringly gives the scale. (Photo: P. Younger).

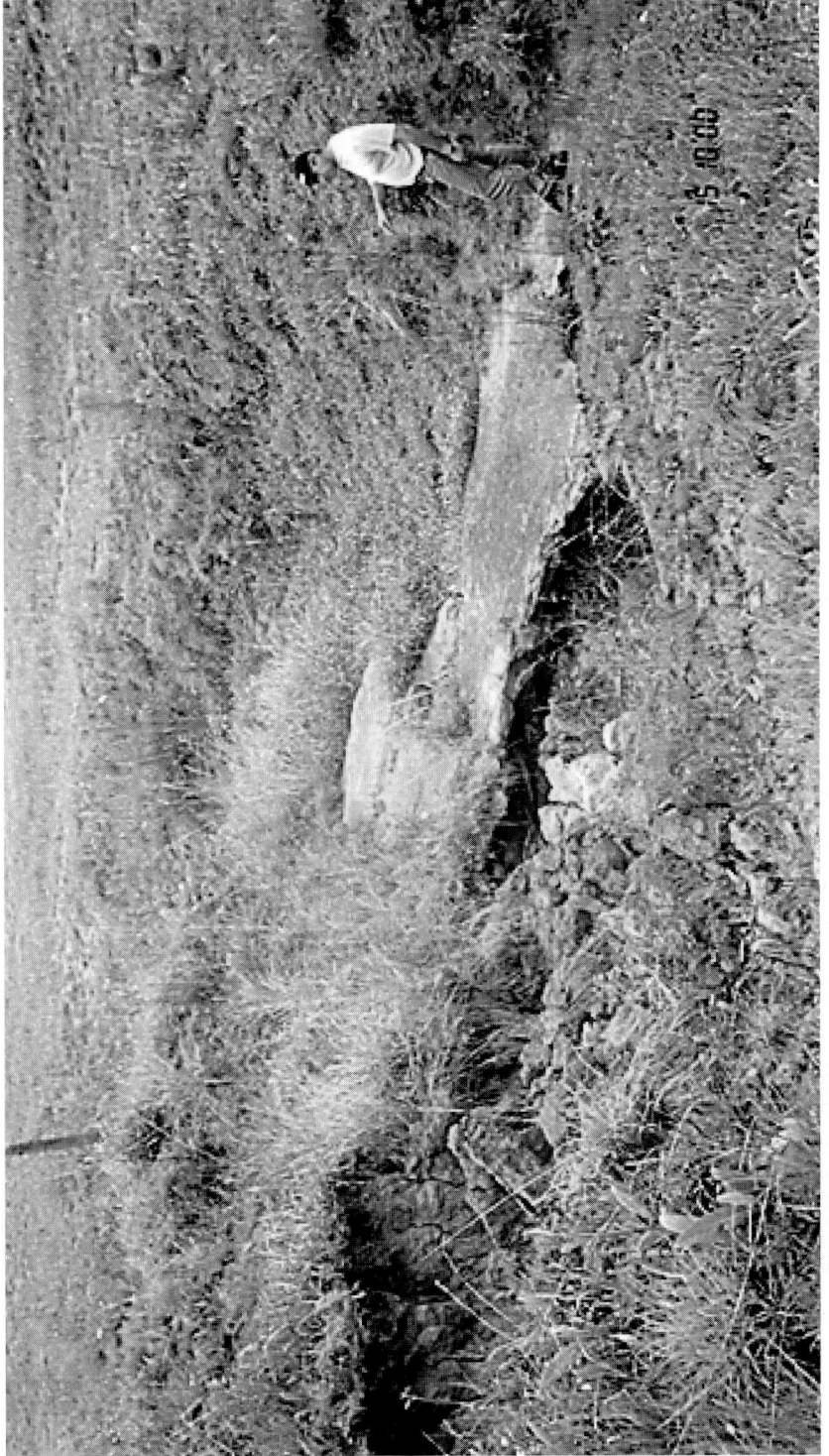

Figure 3.15: A 'crown hole' formed by void migration above shallow bord-and-pillar coal workings in the KwaZulu-Natal Coalfield, S. Africa. Rob Hattingh of PHD Inc observes the feature. This crown hole lies in the axis of the valley of an ephemeral stream, and the angle of view is looking downstream in this valley. During the rainy season, the entire flow of the stream disappears undergound at this point: note the presence of ordinary streambed sediment on the upstream (near) side of the crown hole, but the absence of any evidence of recent flow (i.e. the presence of dense grass in the valley floor) on the downstream (far) side. (Photo: P. Younger).

karst terrains. The hydrology of entire river catchments has been found to change in response to the enhanced subsurface drainage caused by deep mining. For instance, Puente and Atkins (1987) found that mining in West Virginia led to an increase in infiltration to the deep ground water system from only 9% of total rainfall pre-mining to about 26% after mining. However, this enhanced infiltration did not manifest itself as increased base-flow in local streams as would be expected, for mine dewatering beyond the limits of the studied mined catchment led to export of the extra infiltrated water to adjoining catchments.

Even where open fractures do not propagate into the shallow subsurface and hence cannot lead to increased infiltration of surface waters, subsidence can still exert powerful influences on topography and therefore on drainage. For instance, Figure 3.16 shows a small lake which has a hedge line passing right through its mid-line. This hedge line betrays the origins of the lake, which was farmland until the 1970s, when subsidence above superimposed longwall workings in two coal seams led to the formation of a closed depression, which was fed by breached land drains to form a lake. Small lakes of this type are relatively common in areas of extensive longwall mining. Save for the loss of valuable agricultural land, these subsidence lakes can actually be an ecological asset, since they represent a local expansion of habitat for migratory waterfowl.

More subtle impacts of mine subsidence on surface morphology and hydrology have been reported by Sidle *et al.* (2000), who examined temporal changes in the distribution of consecutive pools and cascades of a mountain stream underlain by actively-subsiding longwall workings. In the two years following the onset of subsidence due to longwall mining, it was found that cascades and pools in the mountain stream both lengthened, and that the channel width had narrowed in places. By the end of the third year, with no further measurable subsidence occurring, the fluvial system had already adjusted such that the dimensions of the channel and its pools and cascades were no longer distinguishable from those which had been measured before longwall mining. Although short-lived, such changes in habitat could negatively affect the local diversity of benthic macro-invertebrates, potentially compounding stresses arising from water quality problems.

Surface mining
The impacts of open pit extraction on surface water systems are often less subtle than those due to deep mining. Two major issues commonly arise in this regard:

(i)   the interception of natural streams by open pits
(ii) the loss of natural runoff source areas
(iii) the hydrological behaviour of back-filled surface mines.

The third of these is considered in Section 3.9 below. The other two will be discussed briefly below.

Where streams cross the site of an open pit, the original channel will be obliterated by mining. It is good practice to divert such streams around the perimeters of the open pit during mining, for at least two reasons:

• if the streams are not diverted, they will add to the pit dewatering burden. As dewater-ing costs can account for a large proportion of total operating costs (Section 3.6), it makes sound economic sense to prevent surface waters flowing into open pit mines.

Figure 3.16: A perennial pond occupying a subsidence trough (cf. Figure 3.17) in the centre of a fully-extracted longwall panel worked by Brenkley Colliery, Newcastle Upon Tyne. Note the pre-existing hedgerow passing right through the centre of the lake, betraying its previous identity as well-drained arable farmland. (Photo: P. Younger).

Furthermore, as surface waters pass through open pit workings they invariably become contaminated, at least with silt and other suspended solids (which can be deposited in downstream watercourses, smothering the benthos and thereby causing ecological damage), if not with ecotoxic metals. Under most environmental regulatory regimes, it will be necessary to treat such waters to comply with rigorous emission standards.

• if streams are allowed to fall into a pit only to be pumped out again, the natural continuity of the original stream will be utterly lost, with potentially grave consequences for the natural migration of pelagic and benthic fauna and flora.

The diversion of streams inevitably involves trauma for the aquatic life-forms inhabiting the affected reaches. The diversion of streams through buried pipe lines or culverts is the least ecologically appropriate solution, and should be avoided wherever possible. Where surface diversion is achieved, the preservation of continuity will allow colonisation of the new reach of the stream by downstream migration. Hence, in all but the most extreme cases, substantial recovery of channel biology should be possible within one to two years after channel diversion is complete. Considerable care must be taken in re-routing substantial streams, however, so that the bed gradients, channel depths and flow velocities remain roughly comparable to those which obtained before diversion. If this is not the case, the channel will react to the new hydraulic conditions by accelerated erosion or deposition (as appropriate) until a new dynamic equilibrium is attained. Sound guidance on the ecologically-sensitive restoration of channels which were previously degraded by human activities (e.g. canalised, robbed of floodplains etc) is provided by Hey (1997).

Even where care is taken to site surface mines so that they avoid stream axes, accidents can still happen. For instance, in 1988 a major slope failure induced the River Aire (Yorkshire, UK) to flow into the the nearby St Aidan's Extension Opencast Coal Site. Soon almost the entire flow of the Aire was entering the mine, briefly forming a spectacular cascade down the highwall (Hughes and Clarke, 2001). The problem was so difficult and costly to remediate that coal production was not resumed for nearly ten years (Wilson and Brown, 1997; Hughes and Clarke, 2001).

More subtly, large-scale surface mining can remove from the landscape geomorphological features which formerly played important roles as source areas for surface runoff. No artificial dewatering regime is likely to emulate the temporal behaviour of such natural features. Consequently, where surface mining removes such features, significant alterations in the runoff behaviour of downstream watercourses can be confidently expected. Similarly, once restoration of surface mines is complete, back-fill is likely to behave very differently to natural soils in terms of runoff generation (Section 3.9).

3.5.3 IMPACTS OF EXCAVATION ON GROUND WATER SYSTEMS

Subsurface mining

The general effects of subsurface mining on overlying strata were discussed in Section 1.3.2. In particular, Figure 1.15 summarised orthodox views on the patterns of deformation above typical longwall panels in coal-bearing strata of Carboniferous age (NCB, 1975). Figure 3.17 is a re-labelled version of Figure 1.15, in which the hydraulic properties pertaining to the various zones of extensional and compressional deformation are identified

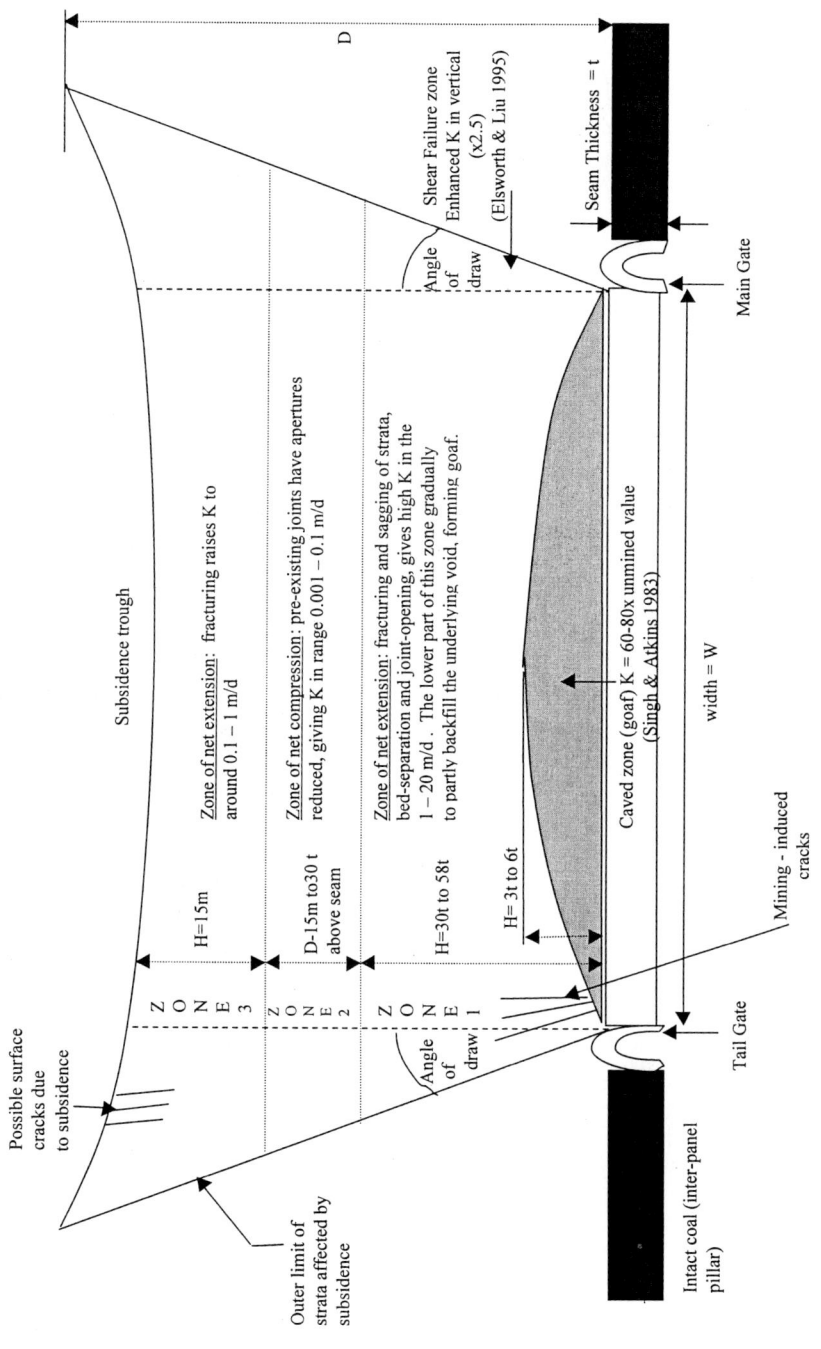

Figure 3.17: Schematic cross-section summarizing the changes in permeability above a fully-extracted longwall panel. Compare this with the summary of zones of deformation shown in Figure 1.15. (Adapted from Younger and Adams, 1999, and Adams, 2000).

in accordance with the findings of Singh and Atkins (1983). To simplify and summarise the hydrogeological consequences of these processes, three key points should be noted here:

(i)   The fanning out of the zone of subsidence in a manner defined by traces inclined 35° from the vertical above the edge of the void has been widely observed in practice (NCB, 1975; Smith and Colls, 1996). The 35° angle is commonly termed the "angle of draw" (Zipper et al., 1997). While the precise value of the angle of draw will vary as a function of rock mass strength (Brady and Brown, 1993), detailed analyses of 157 longwall panels (at depths varying from 24 to 833 m) at 50 collieries in the British coalfields revealed most angles of draw to fall in the range 20° to 40°, about an average of 35° (NCB, 1975)[1]. In applying the concept of the angle of draw to hydrological analyses, Zipper et al. (1997) introduced a very similar "angle of influence", with the intention of including all influences of deep mining on water levels in suprajacent water wells, irrespective of whether the effects are directly attributable to subsidence or to some other process (e.g. lateral drainage into subsided areas). The inference is that the "angle of influence" may well be greater than the geotechnical "angle of draw"; however, values for both angles reported by Zipper et al. (1997) are generally similar, with values in excess of 40° only rarely being reported, and these only in cases where steep surface topography complicates strata deformation patterns.

(ii)   Stress fields which develop above an unsupported, recently created void cause fracturing and subsidence of the overlying strata which extends over a vertical distance calculated to be half as high as the void is wide (e.g. a 200 m-wide longwall panel can be expected to affect around 100 m of overlying strata). There are at least three noteworthy caveats in this regard:
        – Closely adjoining panels may act in consort to disturb ever higher portions of ground.
        – In strata which differ in lithology and structure from the Carboniferous Coal Measures of the UK, the height of disturbance may far exceed the void width, and vice versa.
        – These observations cannot be easily transposed to the failure of narrow voids, for the lack of extreme stresses associated with the cantilever effects of long beams may result in different modes of failure, such as chimney caving (Brady and Brown, 1993).

(iii)   The changes in permeability in each of the three zones of deformation which develop above the extracted seams are distinct, and may be summarised as follows (in ascending order above the void):
        Zone 1: This is the first zone above the unsupported void, and is typically one-third as high as the void is wide (eg over a 200 m-wide panel, Zone 1 would be

---

[1]The average angle of draw of 35° is enshrined in assessing claims of property damage arising from subsidence above state-owned coal mines (NCB, 1975). As any affected property owner will attest, the government did not adopt the 35° angle as a charitable policy; therefore there is every reason to accept that subsidence effects do indeed fan out in the manner described by the angle of draw

about 65m thick). Zone 1 is characterised by collapse and fracturing of the strata (by a combination of sagging, extensional fracturing and bed separation), which increases the permeability of the strata (when compared to the pre-mining permeability) by factors as high as 60 to 80 (Whittaker *et al.*, 1979; Singh and Atkins, 1983; Fawcett *et al.*, 1984). Field tests suggest that, in typical Carboniferous coal-bearing strata, hydraulic conductivities in the range 1 to 20 m.d^{-1} are common in this zone (Whittaker *et al.*, 1979; Minett *et al.*, 1986; Aljoe and Hawkins, 1994).

Zone 2: This is usually about 25–30% as thick as the underlying Zone 1 (eg over a 65m thick Zone 1, Zone 2 may be 15–20 m thick). In Zone 2, the net effect of the subsidence-related stress fields is compression, so that the permeability of the strata remains at or below pre-mining values (i.e. it may even decrease). For instance, in Carboniferous coal-bearing strata, hydraulic conductivities (K) in the range 10^{-3} to 10^{-1} m.d^{-1} are likely in Zone 2 (Minett *et al.*, 1986; Aljoe and Hawkins, 1994). Zone 2 can thus function as a valuable "low-permeability barrier" above workings. For this reason, the position of Zone 2 in relation to overlying aquifers or flooded old workings critically controls whether subsidence will induce greater flows of water from above into the mine.

Zone 3: This is usually similar in thickness to Zone 2. Zone 3 is an extensional zone, so that permeability again increases (though not by so much as in Zone 1).

Where Zone 1 intersects an aquifer, the tendency will be for it to drain down into the underlying workings, so that any drop in water level in the aquifer will be persistent. This effect has been observed in Illinois by Booth *et al.* (1998), who noted that it is particularly marked in low transmissivity aquifers, where lateral flows and renewed recharge cannot keep pace with under-drainage. In the UK, more than 10 m of persistent regional drawdown was induced in a high-transmissivity dolomite aquifer by coal mining below it (Younger and Adams, 1999). The persistence of drawdown in this case can be ascribed to the location of the mine close to the edge of the dolomite outcrop; even though transmissivity was high, lateral inflow was not possible due to the impersistence of the aquifer. The drawdown was eventually reversed after the mine was abandoned and allowed to flood, when water migrating from the flooded workings began to recharge the dolomite aquifer. This restoration of water levels was accompanied by contamination of the dolomite aquifer by mine waters high in sulphate (Younger and Adams, 1999).

Where Zone 2 lies between the workings and any aquifer horizon, physical disturbance of the aquifer horizon may occur *without* direct drainage into the underlying workings (Singh and Atkins, 1983; Aston and Whittaker, 1985). In this case, Zone 3 deformation of the aquifer may increase its permeability by as much as an order of magnitude (Booth and Spande, 1992), and may also result in less extreme increases in porosity. These changes may result in temporary, but locally pronounced, declines in the water table (e.g. Booth and Spande, 1992). In most cases, the water table has been observed to return to its original position within months of the cessation of measurable subsidence (Booth and Spande, 1992; Booth *et al.*, 1998). While the increase in aquifer transmissivity arising from Zone 3 deformation may serve to improve well yields (Booth *et al.*, 1998), it may also induce less desirable changes in water quality, as rising permeabilities also affect aquitards, mobilising poor quality waters which were previously virtually immobile within them. It should be emphasised that this deterioration in water quality is due not to mine water migration, but to changes in the circulation of natural waters induced by mine subsidence (Booth and Bertsch, 1999).

In a study of mining impacts on springs and shallow wells above deep coal mine workings in Appalachia, Zipper *et al.* (1997) observed a number of instances in which the increase of permeability in aquifers overlying mine workings led to a lowering of the water table. This caused decreases in flow rates in some springs, and a number of cases in which wells were left "high and dry". The most shallow wells are obviously the most vulnerable, and are also the most likely to be the sole source of water supply for isolated residences; hence the potential importance of this impact can go far beyond the economic cost of replacing the well. The data collected by Zipper *et al.* (1997) indicate that shallow wells are at greatest risk where:

- deep mining occurs within about 130 m of the base of a well
- the mine uses a high extraction technique (i.e. longwall, or pillar-retreat mining)
- the geological succession is naturally fractured, and
- the strata between the well and the coal seam include little mudstone.

Where traditional room-and-pillar extraction techniques were used, mine roof collapse was minimised, and subsidence fractures were inferred to rarely propagate more than 30 m above the worked seam. Hence the risk to wells from room-and-pillar workings is considerably less than the risk from longwall or pillar-retreat workings at similar depth (Zipper *et al.*, 1997).

Surface mining
Surface mining involves the wholesale excavation and manipulation of previously intact rock. The hydrological behaviour of surface mine waste rock and backfill are explored in Section 3.9 below. There are two further hydrological influences of surface mining which repay examination:

(i)   The loss of unsaturated zone storage due to the removal of intact rock
(ii) The impacts of open pit excavations on the enclosing rock mass.

The first of these influences remains somewhat contentious and poorly characterised. The excavation of an open pit above the water table inevitably removes a large volume of rock which previously hosted unsaturated zone drainage pathways. Indeed, the removal of unsaturated zone water storage which an open-pit working represents has been one of the key elements of controversy surrounding a number of surface limestone mines in south-west England (Hobbs and Gunn, 1998).

The impacts of surface mining on the enclosing rock mass can be regarded as being less severe than in the case of deep mining, simply because surface mines are excavated *beside* (rather than beneath) *in situ* strata. Nevertheless, the process of surface mining can induce considerable increases in permeability immediately around the void. For instance, a compilation of data from open-pit limestone mines throughout the UK (Figure 3.18) clearly shows that permeabilities measured in boreholes *within* the open-pits are significantly higher than those measured in boreholes in the undisturbed rock beyond the limits of surface mining activities. This is probably ascribable to two main processes:

(i) Increases in permeability due to the common practice of sub-floor blasting, which is usually carried out to a depth of around 2 m to facilitate easy grading of the pit floor.

(ii) Stress release in response to bench extraction, which commonly induces extensional fracturing in a circumferential belt of strata immediately surrounding an open pit (see Figure 1.25). Such mining-induced fractures pose an ever-present slope failure hazard where they intersect the surfaces of benches in many quarries (Figure 3.19; see also Hughes and Clarke, 2001).

It is a matter of common observation that such fractures are rarely developed more than a few tens of metres into the virgin rock beyond pit high walls. Although the zone of extensional fracturing behind pit high walls is of such limited lateral extent, it can be expected to be reflected in a "halo" of increased permeability around a surface mine. Field and modelling results presented by Hawkins (1994) suggest that hydraulic conductivities in the fracture zone beyond the high-wall of a surface coal mine in West Virginia are as much as two orders of magnitude greater than the undisturbed strata ($K \approx 4$ m/d in the fracture "halo", versus 4×10^{-2} m/d in the undisturbed strata). The presence of such zones of higher permeability may contribute to the development of turbulent ground water flow conditions in the immediate vicinity of dewatered open pits excavated below the regional water table (Dudgeon, 1985a, 1985b, 1997). This has important consequences for the assessment of dewatering impacts (see Section 3.7).

3.6 WATER INFLOWS TO ACTIVE WORKINGS

3.6.1 SEEPAGES, DRIPPERS, FEEDERS, INRUSHES AND DIRECT PRECIPITATION

As outlined in Section 3.4.3, specific terminology has evolved to describe the various modes of water ingress to mine workings. There are five major modes of ingress. The first four below are listed in order of increasing magnitude.

Seepages are areas of diffuse ingress, usually through the floors of mine voids, or else along specific geological contacts. (This term is used in both deep and surface mines). Ingress by seepage at any one point within a seepage zone may be imperceptibly slow, even though the overall quantity of water coming from the seepage zone as a whole may be very large in total. Evaporation of seepages leads to the accumulation of crusts of evaporite minerals (Figure 3.20). Depending on the chemistry of the initial seepage water, the minerals forming these crusts may include those listed in Table 2.6. If the site of the seepage zone is later inundated, such crusts usually dissolve very rapidly, and if they contained the minerals listed in Table 2.6, the resultant drainage is likely to be acidic and metal-laden (Younger, 2000a,b).

Drippers: Diffuse inputs of water in the form of numerous, isolated drip-feeds entering deep mine workings through the ceiling, which causes discomfort for the miner akin to the experience of walking in heavy rainfall. (Clearly this term is specific to deep mines). Where the dripper water is sufficiently mineralised, it may form 'forests' of stalactites upon entering the workings (Figure 3.21); in some cases, these 'forests' can grow to be so dense that they impede ventilation, and must be regularly cleared.

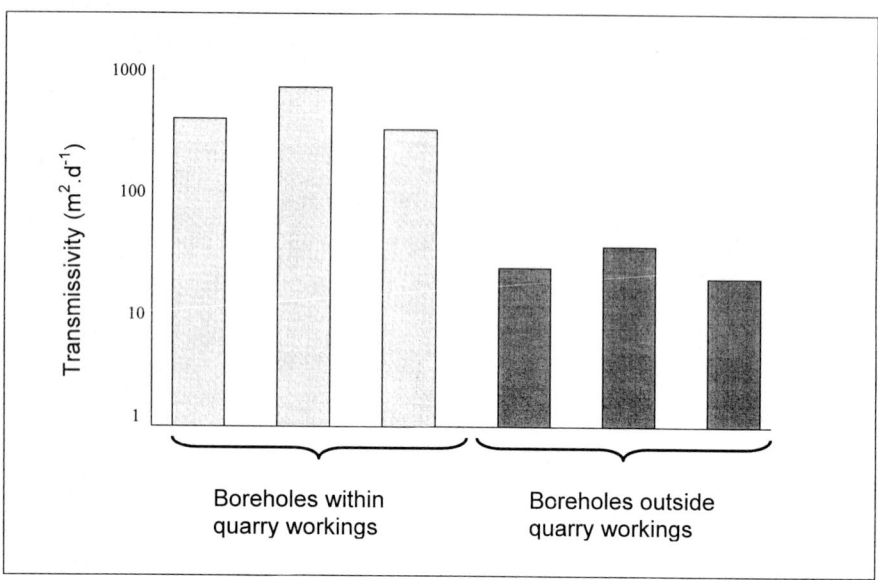

Figure 3.18: Evidence of an increase in transmissivity due to extensional fracturing induced by blasting and stress relief, in a major limestone quarry in the UK. All boreholes lie within a few hundred metres of each other and access the same lithostratigraphic unit, which displays no noticeable difference in karstification between the quarry workings and the surrounding area. (Data provided by G.H. Karami).

Feeders: Relatively vigorous, point inflows (Figure 3.10), resembling underground springs (J.C., 1708), which can cause considerable nuisance for miners, and which typically account for the bulk of the total flow entering a mine. (The term 'feeder' is also used in surface mines). Often feeders will commence flowing very suddenly during the course of mining. Typically, the rate of flow of the feeder increases rapidly at first as flow pathways into the workings are scoured, and then gradually declines as the driving head (in the aquifer giving rise to the feeder) gradually declines in response to drainage by the feeder. This time variant behaviour of feeders has been known to cause alarm, if not panic, for it is not always clear that the final inflow rate will prove to be manageable; in extreme cases, the emergence of a new feeder is the herald of a more substantial inrush. However, in many cases, careful monitoring of the situation (coupled with nerves of steel) will allow mine managers to cope with most new feeders without resort to rash measures. For instance, substantial feeders from overlying Permian limestones into a coal mine at Wistow (N Yorkshire, UK) in the early 1990s totalled as much as 65 Ml.d^{-1} soon after their first emergence. Feeder flows were monitored carefully, and were all found to decline over a few months, yielding steady flows of no more than a few hundred cubic metres per day.

Inrushes: Inrushes are dangerously large feeders which arise suddenly during mining. Generally short-lived, they most commonly arise due to the inadvertent connection ("holing") of workings into a water body (be this adjoining flooded workings, a major aquifer, a stream, lake or wetland). Such inrushes have been the cause of many deaths amongst miners over the centuries. The inrush of sea water at Workington in 1837 has

Figure 3.19: Major collapse blocking a former haulage road in a limestone quarry, caused by failure of massive limestone along the plane of a large extensional fracture in the highwall, which is thought to have been formed (or at least expanded) by blasting and/or stress relief associated with the working of the pit. (Photo: P. Younger).

Figure 3.20 (*see overleaf*): Evaporitic crusts formed by evaporation of mine waters in underground workings. (a) Cementing track ballast at the edge of a major haulage roadway at 860 m depth in the Mina de Raposos gold mine, Minas Gerais (Brasil) Field of view approximately 0.5 m (Photo: P. Younger). (b) Looking into a 100-year old, 2 m high roadway (bord) in the High Main Coal Seam, Morrison North Pit, County Durham (UK). (These workings were accessed from open-pit workings of Chapman's Well Opencast Coal Site). While the walls of the bord are black (intact coal), the floor is light in colour (pale yellow), due to the presence of a thick evaporitic crust of acid-generating salts. Where these salts had dissolved in puddles found in adjoining bords, the water was found to have up to 500 mg.l^{-1} Fe and a pH of 3.5. (Photo: A. Witcomb).

Figure 3.21: A dense forest of stalactities and stalagmites of halite, formed from a line of drippers across a major roadway in Westoe Colliery (Durham, UK). Although the roadway was driven in clastic sediments of Carboniferous (Stephanian) age, the incoming dripper water had acquired high concentrations of Na and Cl from dissolution of sedimentary halite in the overlying Permian strata. Upon entering the mine, the ventilation current evaporates the water (the direction of the current is reflected in the shapes of the stalactites). These halites speleothems grew so rapidly and densely that they actually hindered the ventilation circuit and therefore had to be 'harvested' regularly. (Photo: A. Doyle).

already been mentioned. Even earlier than that, in 1815, miners in Heaton Colliery (Newcastle Upon Tyne) broke into abandoned, flooded workings of the nearby Jesmond Colliery. Altogether, this inrush cost 90 lives, with most being killed by starvation and asphyxiation, after being trapped above the water in isolated workings (Doyle, 1997). A very similar incident in 1877 in South Wales, when workings of Tynewydd Colliery holed into old flooded workings, causing an inrush which immediately claimed 4 lives. In this case, however, ten other miners survived for up to 7 days below the water table, in air pockets trapped in isolated up-dip workings. One of these men was sadly killed at the moment of rescue, when the release of compressed air sucked him into the hole dug from above by the rescue brigade. The remaining nine men survived, although following the rapid decompression at the time of rescue, all suffered from the first documented cases of the bends in medical history (Llewellyn, 1992). Nearly a century later, in March 1973, an advancing longwall face at Lofthouse Colliery (Yorkshire) was suddenly flooded upon unexpectedly holing into an ancient, flooded shaft. Prolonged rescue attempts were pursued in the hope that at least some of the seven trapped miners might have made it to possible air pockets identified from the mine plans. This sadly proved not to be the case (Calder, 1973). Cases such as these have prompted the development of ever more stringent legislation stipulating precautions which must be taken against inrushes (see Worked Example 3.1 for an example of some of these precautions in action).

The most infamous inrush to date fortunately failed to claim any lives, but it very nearly caused loss of what was then (1968) the world's richest gold mine: West Driefontein, South Africa (Cousens and Garrett, 1969; Cartwright, 1969). The ore body at West Driefontein occurs as stratiform layers in Precambrian metamorphic rocks underlying a regionally karstified dolomite aquifer. In 1968, the normal pumping rate in the mine was about 68 Ml.d^{-1}. However, as a precaution against unforeseen inrushes, the total capacity of pumps installed in the mine was 286 Ml.d^{-1}. Nevertheless, no single feeder in the orefield had ever exceeded 27 Ml.d^{-1}, and none had ever persisted at such a high rate. West Driefontein mine boasted a well-planned system of water-tight doors in major roadways to ensure that a total void volume of 4500 Ml could be used to store waters in the event of an inrush. On the morning of October 26th 1968, a stope progressing upwards from the second highest active level in the mine (at about 850 m below the surface, and more than 100 m below the base of the dolomite aquifer) encountered a new feeder associated with a fault (which was later deduced to continue upwards into the overlying dolomite aquifer) at an astonishing flow rate of about 370 Ml.d^{-1}. This feeder exceeded the maximum installed pumping capacity by a sufficiently large margin that the mine would have been flooded within about 30 days if no action was taken. The ultimately successful 23-day struggle which followed has now entered the global annals of mining history (e.g. Saul, 1970). Extra pumps were installed, taking the total pumping capacity to 363 Ml.d^{-1}. These helped to slow the rate of flooding, while high-pressure concrete plugs were constructed to isolate the area of the mine immediately around the source of the feeder. Even after the mine itself had thus been saved, further exploitation of the reserves was only assured when regional dewatering of the overlying dolomite was undertaken (Forth, 1994), albeit this had undesirable consequences of its own (see Section 3.8).

The foregoing classification of modes of mine water ingress is not the only one that can be applied. For instance, Fernández-Rubio et al. (1998) favour the use of a classification system of Russian origin, which classifies water inflows solely on the basis of volumetric flux rate (Table 3.3). A study of 2600 coal faces in the former Soviet Union yielded the percentage attributable to each of the four classes given in the final column of Table 3.3. It is not clear how these four classes relate to the four types of ingress introduced above, since details of the physical nature of the various inflows are not presented (Fernández-Rubio et al., 1998). However, one would intuitively expect the four classes of increasing flow to correspond to seepages, drippers, feeders and inrushes respectively. If this is indeed the case, Table 3.3 provides some guidance of the proportions of the various forms of inflow likely to be encountered during mining, which has implications for the planning of dewatering systems (see Section 3.7 below).

Table 3.3: A flow-rate based classification of mine water inflow features derived from studies in Russian coal mines (after Fernández-Rubio *et al.*, 1998)

Category of inflow	Flow rate range		Percentage in category
	$Ml.d^{-1}$	$m^3.hr^{-1}$	
Small inflows	<0.12	<5	82
Medium inflows	0.12–0.24	5–10	10
Heavy inflows	0.25–0.36	11–15	4
Very heavy inflows	>0.36	>15	4

In the investigation of relatively large surface mines, *direct precipitation* into mine voids may require consideration[2]. In some of the larger open-pit operations, this is one of the most important hydrological problems (NCB, 1982), demanding careful planning to ensure the dewatering system can cope for design storms of a prescribed annual probability of exceedance (see Section 3.7).

The sum total of all of the waters entering a mine from all of the feeders, drippers and seepages etc accounts for most of the mine "water make", to the quantification of which we now turn.

3.6.2 WATER MAKE AND MINE WATER BUDGETS

Many hydrological questions concerning mining will revolve around quantification of a quantity which mining professionals commonly term the "water make". The "water make" of a given body of mine workings may be defined as the total volume of water entering those workings over a specific period of time, and which must therefore be removed by any active or passive dewatering system. The terminology of "water make" has spawned related verbs, used in phrases such as: "the mine makes ten cubic metres an hour from the 260 horizon"; "most of the water in the mine makes from the Permian"; "this one fracture makes 70% of the total dewatering burden".

In most cases, the water make corresponds closely to inflows of natural waters, though it is important to note that in some particularly dry mines the total amount of water pumped from the workings may include a significant component of "process water" (i.e. water deliberately brought into the workings for dust suppression, cooling of drills and shearers etc). The magnitude of the water make of a given body of active mine workings has direct economic significance, since it governs the costs of dewatering. Hence quantification of the water make is often an important task in overall mine planning. In particular, failure to anticipate the scale of future water makes can lead to costly errors in the sizing of major pipe ranges. In extreme cases, water makes greatly in excess of those anticipated can threaten the viability of a mine if shafts and adits become so crowded with multiple, parallel pipe ranges that haulage of run-of-mine product is hindered.

---

[2]While not logically associated with deep mines, snow has been seen to fall many tens of metres below ground in deep mine shafts, thanks to unusual ventilation arrangements in winter time!

In most cases, the water make cannot be directly quantified, since water enters the mine as numerous drippers, feeders and seepages etc, few of which will be readily accessible/measurable throughout the life of the mine. Rather, quantification of make usually involves determination of the overall mine water-budget (Grmela and Tylcer, 1997), which is a particular expression of the basic principle of mass continuity (equation 3.2). A flow-chart for a generic mine water-budget is shown in Figure 3.22. Some of the budget elements shown in Figure 3.22 are specific to deep mines (e.g. all pathways related to the ventilation system), and others (surface evaporation from sumps, direct precipitation into the void) are restricted to surface mines. The quantities of water removed from the mine in the form of moisture in run-of-mine product will usually be non-negligible in both surface and deep mines; they can be estimated from records of masses of material entering and leaving mineral beneficiation plants (Grmela and Tylcer, 1997).

Examples of real mine water-budgets are given in Figures 3.23 and 3.24 for a surface mine and a deep mine respectively. These budgets were developed by the operators of the mines concerned for the following reasons:

1. To monitor the rate at which the water make was changing over time, as mineral extraction proceeded, to provide a basis for planning future water "standage" (i.e. storage volumes in mine sumps etc).
2. To determine seasonal variations in water make, and the long-term annual average, providing a basis for planning water environmental water management (i.e. arrangements for treatment and discharge to natural watercourses).

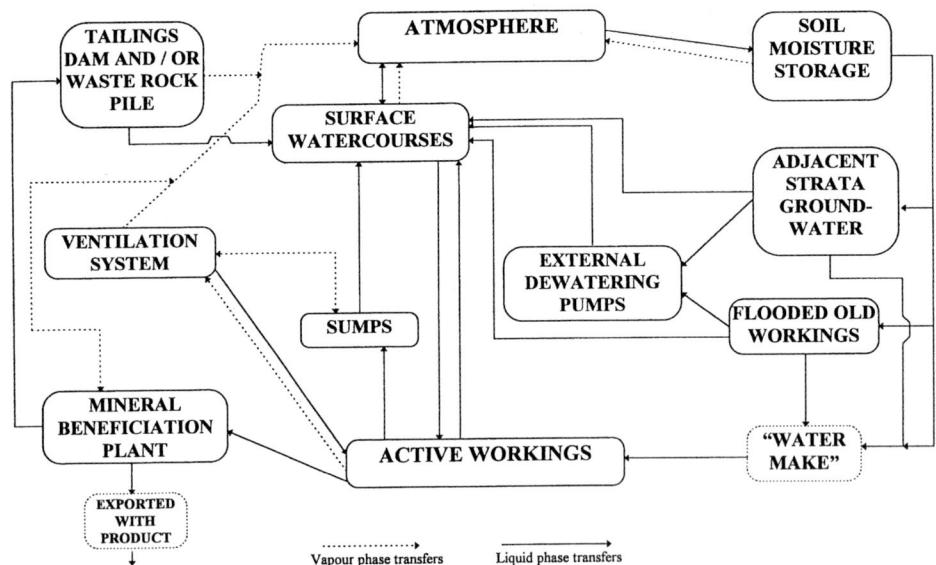

Figure 3.22: A generic mine water budget, identifying the principal compartments and fluxes to be considered when constructing a budget for a specific case.

3. To discriminate between the different rates of water ingress at different horizons in the mine, for safety planning procedures.
4. To provide a starting point for calculating the rate of mine flooding after abandonment, and the likely long-term discharge rate from the mine after it is completely flooded.

One obvious critique which can be made of Figures 3.23 and 3.24 is that they contain no error terms. This is a common omission, particularly in cases where one or more terms in the budget are determined by differences between directly-measured components. In such cases, the overall error is implicitly (and unreasonably) added into the terms estimated by difference. This should be avoided: it is good practice to explicitly identify the terms which are most approximate, and to estimate the magnitudes of errors wherever feasible.

One motivation for determining mine water budgets is to clarify the ways in which the water make changes during the working life of the mine, since this information can influence the selection of the most appropriate mineral extraction method (Orchard, 1975). For instance, it may be found that the rate of increase in water make per tonne of rock mined is higher in the case of long-wall extraction with 200 m-wide panels than in the case of short-wall extraction with 30 m-wide panels. Depending on the scale of the total water make, such considerations can have a significant influence on the cost–benefit analysis of a preferred extraction strategy. Of course, temporal variations in water make are not restricted to long-term increases as the mined area expands; seasonal fluctuations are also important, particularly in shallow workings. Where data are sufficient, seasonal variations in water make can be correlated with rainfall records to obtain annual probabilities of exceedance (NCB, 1982), thus providing a basis for planning facilities to store flood waters in the mine and/or for specifying peak pumping rates for dewatering equipment (see Section 3.7).

3.6.3 DEEP MINE WATER MAKES: SOME GENERAL OBSERVATIONS

Problems caused by water entering mine workings are amongst the most prominent issues discussed in the very earliest of mining engineering treatises, For instance, Georgius Agricola, writing in 1556 about the metal mines of eastern Germany, noted that excessive ingress of water was (along with failure of reserves and bad air) one of the three main reasons for the abandonment of mines:

> "... sometimes the miners can neither divert this water into the tunnels, since the tunnels cannot be driven so far into the mountains, or they cannot draw it out with machines because the shafts are too deep; or if they could draw it out with machines, they do not use them, the reason undoubtedly being that the expenditure is greater than the profits of a moderately poor vein ..."
>
> (Hoover and Hoover, 1950)

Accordingly, he devoted much of Book VI of his monumental work 'De Re Metallica' to the description of contemporary mine dewatering pumps and their use. Despite the wide uptake of such pumps, one and a half centuries later water remained one of the most serious limitations on the expansion of production in the Great Northern Coalfield of England. Thus in "The Compleat Collier" (J.C., 1708) we find the following complaint:

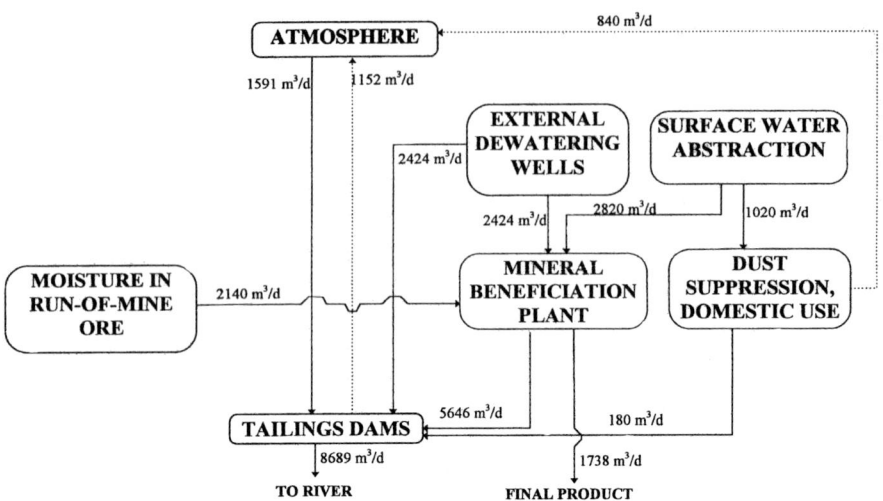

Figure 3.23: Example of an average water budget for a very large surface mine in a humid tropical area of Brasil. Note that in practice it is always advisable to quote the errors in each of the boxes and transfers, but these are omitted here for clarity. Note the substantial quantities of water leaving the mine with the final product, as well as the major losses to evaporation from the tailings dams (these would not be nearly so high in other climatic zones).

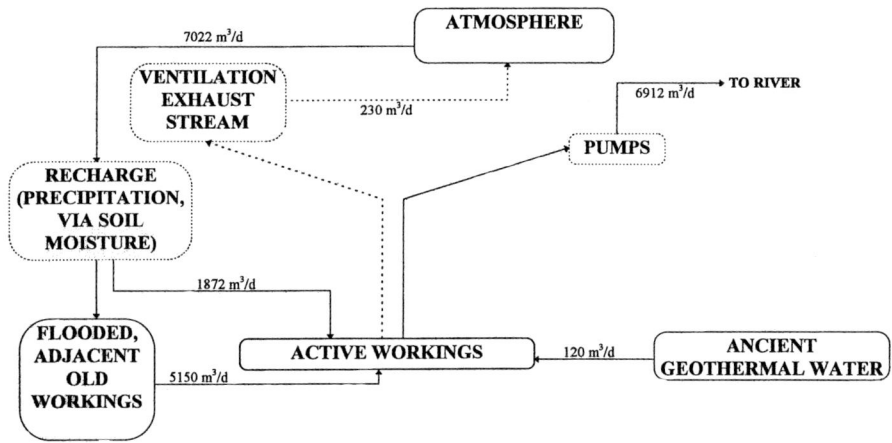

Figure 3.24: Example of an average water budget for a deep mine: in this case a 1000 m deep tin mine in warm granite country rock. (South Crofty, Cornwall, UK). Note that in practice it is always advisable to quote the errors in each of the boxes and transfers, though these are omitted here for clarity of illustration. The high ventilation rate needed to maintain bearable working conditions in this warm mine resulted in a very substantial loss of water from the workings by evaporation. The component of ancient geothermal water entering the deepest workings is a classic example of an unequivocal 'head-dependent inflow', and is one of the sources of such water discussed in relation to the later flooding of this mine (see Figure 3.39b).

"... it is very rarely found, that a Pit of 40, 50 or 60 fathoms is sunk, without going
through several sorts of Feeders; indeed, were it not for Water, a collery [*sic*] in these
Parts might be termed a *Golden Mine* to Purpose, for *Dry Colleries* [sic] would save
several Thousands *per Ann.* which is expended in drawing Water hereabouts ..."

With engineering interest in mine water control having such an ancient pedigree, it is
perhaps surprising that little systematic work was undertaken on the provenance of mine
waters until relatively recently. Indeed, the only explanation of mine water provenance
offered by the author of 'The Compleat Collier' is that:

" ... all which [mine] Water we suppose to come from the sea, and so being fed by
that inexhaustible fountain, we call it by the name of a *Feeder*, and that it may rise to
the top of any mountain we are subject to believe no great Matter of Wonder,
because we are so often, by the Curious and Learned, told, That the Sea, this
Fountain Head, is higher than the Earth ..." (J.C., 1708)

(One presumes that the author of this passage had Genesis 1 (vv 6–7) in mind when he
considered mine water provenance, since that verse of scripture ascribes the blue sky to
the presence of a vault holding back primordial waters from the earth below). It was not
until the mid 20th Century that systematic, scientific investigations of the origins of deep
mine waters were commenced, focusing on experiences in some of the world's longest- and
most thoroughly-worked coal sequences. In the USA, the US Bureau of Mines was responsible for many of the earliest insights (e.g. Ahmad, 1974). In Europe, much pioneering
work was carried out by the National Coal Board in the UK, most notably by Saul (1936,
1948, 1949, 1959, 1970). These works of Saul profoundly influenced the thinking of several
generations of mining engineers world-wide, right down to the present day. Four major
conclusions of Saul's work are most pertinent here:

1. In previously-unmined Carboniferous Coal Measures, the main sources of water to a
   new deep mine will be the major sandstones in the sequence; the future water make of
   a new deep mine can be expected to amount to between 20% and 40% of the mean annual
   rainfall landing on the outcrops of the sandstones in the local sequence (Saul, 1948).
2. Water makes its final entry into mine workings in a highly localised manner, appearing
   either as feeders or drippers. Inclined boreholes driven upwards from headings into
   virgin ground, or backwards into goaf, can sometimes successfully gather numerous
   drippers into a single flow, which will gradually diminish in rate as the goaf compacts
   under its own weight, losing some of its permeability in the process (Saul, 1970).
3. It is most useful to regard mined ground not so much a porous medium as a network
   of interconnected "breaks" (i.e. vertical water-bearing fractures, mostly corresponding
   to dip-parallel faults, and "laterals", i.e. beds of sandstone or worked seams); the discrete nature of most inflows to mines noted above is explicable in terms of the eventual
   interception of these discrete features by mine headings and panels (Saul, 1948, 1949).
4. In the absence of adjoining shallow workings or a steeply-dipping permeable sandstone
   in the sequence, new mine voids deeper than about 140 m (or more than 140 m below
   the sea bed, or the base of overlying aquifers as appropriate) are unlikely to encounter
   major feeders (Saul, 1948), save where faults provide short-circuiting connections to
   higher horizons (Saul, 1970).

Given the remarkable comparability of Carboniferous and older coal-bearing sequences world-wide (at least in terms of basic lithostratigraphy; Ager, 1993) there seems no reason to doubt that some of these insights will be applicable to some degree elsewhere in the world. Indeed, international similarity is suggested by the data given in Table 3.4, which is a compilation of areally-normalised water makes for deep mines around the world. The values have been normalised simply by dividing the reported water make by the area underlain by mine workings. Intuitively, one would not expect very comparable values to emerge from such a simplistic normalisation, which does not account for factors such as mean annual rainfall, mining technique or overburden thickness and lithology. Yet the values for coalfields in Table 3.4 show a remarkably low range of values, suggesting that it is the generally low permeability of coalfield overburden (rather than climatic variables) which exerts most influence on the average water makes of major coalfields. Indeed, even where the Coal Measures underlie major water bodies, such as an aquifer or the sea bed (i.e. the Durham and Dysart-Leven Coalfields in Table 3.4), such that the availability of water above the workings will never be limited by scarcity of rainfall, normal inflow rates remain in the same narrow range. The persistent narrowness of this range suggests that Table 3.4 may be used to obtain a first approximation of future dewatering requirements in the planning of deep coal mines.

All of the coalfields listed in Table 3.4 have shallow structural dips, so that the outcrop of a given seam will often be many kilometres from the active workings. By contrast, most of the metalliferous orebodies covered by Table 3.4 (and certainly those with the more prolific water makes) occur as steeply-inclined structures which often outcrop close to the

Table 3.4: Areally-normalised water makes for various deep-mining fields around the world[3]

Mining district name (country)	Normalised water make (ml.d^{-1}.km^{-2})	Source reference from which calculation made
(a) Coalfields		
Durham Coalfield (England)	0.3	Younger (1993a)
Dysart-Leven Coalfield (Scotland)	0.6	Younger *et al.* (1995)
Jharia Coalfield (India)	0.3	Gupta & Singh (1994)
Lower Silesian Coalfield (Czech Republic)	0.2	Grmela & Tylcer (1997)
Nottinghamshire Coalfield (England)	0.2	Dumpleton & Glover (1995)
Polkemmet, West Lothian (Scotland)	0.1	Sadler and Rees (1998)
Ruhr Coalfield (Germany)	0.2	Coldewey & Semrau (1994)
Upper Silesian Coalfield (Poland)	0.4	Rózkowski & Rózkowski (1994)
(b) Metalliferous ore fields		
Cleveland Ironstone Field	0.1	Younger (2000c)
West Cornwall Tin Mines (England)	4 to 15	Environment Agency records
North Pennine Orefield (England)	4 to 8	Nuttall (2000)
Troya Pb–Zn Mine (Basque Country)	80	Iribar *et al.* (2000)
Rzrebionka Zn–Pb Mine (Poland)	3 to 5	Kowalczyk *et al.* (2000)
Witwatersrand Goldfield (South Africa)	60 to 90	Cook (1982)

----

[3]It should be noted that no North American examples are included due to an absence of sufficiently complete published datasets

areas of underground working, thus favouring greater water ingress. This perhaps explains why most metalliferous fields have water makes one to two orders of magnitude greater than those of the coalfields. (The one clear exception (the Cleveland Ironstone Field) is actually a flat-lying, stratiform , sedimentary orebody with substantial thicknesses of mudstone in the overburden (Younger, 2000c), which was worked by bord-and-pillar methods. As such it is very hydrogeologically similar to most coalfields, with which it shares a modest water make value).

3.6.4 MINING BENEATH WATER BODIES

In many parts of the world, highly valuable mineral deposits underlie major water bodies, i.e. rivers, lakes, wetlands, aquifers and the ocean. There has long been incentive, therefore, to develop safe methods of deep mining below such water bodies. Most of these have been based on empirical studies of the relationship between thickness and nature of overburden and the rates of water ingress to workings. One of the earliest such empirical studies was undertaken by Clarke (1962), who sought to derive a quantitative rule relating the magnitude of "normal" inflow rates (i.e. excluding extreme inrushes) to coal mines in Carboniferous strata with two principal, logical controls:

(i)   overburden thickness, and
(ii) distance to outcrop, which is a surrogate measure of the ease of direct recharge to mined voids.

Although Clarke (1962) invoked Darcy's Law (equation 3.4) in developing his quantitative rule, he did serious violence to the true meaning of that formula in the process. Nevertheless, this has not hindered the widespread adoption of his findings, not least his coining of the term "apparent hydraulic gradient". Amongst disciples of Clarke (1962), "apparent hydraulic gradient" is taken to signify "the head of water measured in the workings divided [by] the minimum distance to the point of replenishment by a free body of water" (Orchard, 1975). The "point of replenishment" equates to the outcrop (subaerial or submarine) of a water-bearing bed in the sequence above a worked seam, or to its subcrop "against unconformable water-bearing strata" (Orchard, 1975). The "apparent hydraulic gradient" thus defined is more a measure of structural dip than of the trend of any potentiometric surface. According to Clarke (1962), where the "apparent hydraulic gradient" exceeds about 1:5, then coal extraction will induce permanent feeders. Where this ratio is 1:7 or more, workings will generally remain dry.

It is only in the last three decades that attempts have been made to improve upon these empirical "rules of thumb" by means of process-based geotechnical research. Building upon the success of the NCB's "Subsidence Engineer's Handbook", Orchard (1975) sought a geomechanical basis for "safe working depth" criteria for coal mines advancing below major bodies of water. In terms of the zones of stratal deformation shown on Figures 1.15 and 3.17, the strategy for working beneath bodies of water is basically to ensure that Zone 2 lies considerably below the base of the aquifer or sea-bed. Specifically, Orchard (1975) and Singh and Atkins (1983) recommended that mining should be limited such that the tensile strain at the base of an overlying water body does not exceed 10 mm/m. More recently, Aston and Whittaker (1985) demonstrated that for most undersea workings increased water ingress does not occur even where the tensile strain at the sea bed reaches 14 mm/m. Kesserü (1995) has also argued that a more rational approach to the minimisation of the risk of massive inflow would be to determine rock stress patterns and calculate the degree of safety they imply in relation to the available driving head in the water body. Nevertheless, the conservative 10 mm/m tensile strain criterion has served well as the basis for UK mining regulations, in which typical Coal Measures rock properties are used to translate the recommendation into a rule which stipulates that no working should be undertaken within 105 m of the base of the overlying water body (Orchard, 1975). Workings adhering to such regulations have been documented as remaining virtually dry in the UK (Saul, 1970; Orchard, 1975; Aston and Whittaker, 1985), Canada, Australia, Japan, Chile (Singh and Atkins, 1983), Russia, China, Hungary and the USA (Kesserü, 1995). Indeed, on the basis of detailed statistical evaluations of maximum and long-term water yields of several hundred undersea longwall faces, Aston and Whittaker (1985) have concluded that wet conditions are most usually due to the presence of major faults which bring aquifers into contact with the roof strata of workings. Further examples of otherwise "safe" tensile strains inducing excessive water yields in the vicinity of faults are given by Singh (1986).

Where extraction is observed to induce increased inflows of water from above, it is common practice to attempt to draw Zone 2 closer to the mined void by narrowing the width of voids left unsupported. One option is to revert to the system of room-and-pillar, in which large pillars of intact mineral are left in place to support the roof indefinitely. However, this may result in mineral recovery rates which are too low to be economic. Consequently, in longwall mining situations, it is more common to reduce the width of panels from the typical 100–200 m to "shortwall" widths of 50 m or even less (see Worked Example 3.3). There is a two-fold rationale for a switch to "shortwall":

(i) the tensile strain induced by subsidence is lessened, hence inducing less fracturing at the base of overlying water bodies, and
(ii) the rate of retreat of the working face is sufficiently rapid that feeders will be "left behind" in the goaf, rather than encountered on the face (Orchard, 1975).

It is now accepted that new workings can proceed beneath flooded old workings as long as the principles outlined above are observed (e.g. Orchard, 1975; Singh and Atkins, 1983; Cain et al., 1994; Reddish et al., 1994). Nevertheless, where flooded old workings lie up-dip (i.e. along the same geological horizon) from a new mine, problems of water ingress can be much more difficult to avoid. Indeed, from the early 18th Century onwards, numerous instances have been recorded in which the cessation of pumping in abandoned workings has led to the inundation of newer workings down-dip (e.g. J.C., 1708; Taylor, 1858; Saul, 1936; Coldewey and Semrau, 1994). These experiences prompted the development of regional, external dewatering schemes in many areas (see Section 3.7), commencing as early as 1858 (Taylor, 1858; Saul, 1936).

3.6.5 SURFACE MINE WATER MAKES: A QUESTION OF STABILITY

Because surface mines inevitably intersect all surrounding strata as they are sunk deeper, they often have higher water makes per tonne of run-of-mine product than deep mines in

---

Worked Example 3.3 Switching from longwall to shortwall mining to limit water ingress

Wistow Colliery is a modern deep mine in the Selby Coalfield (Yorkshire, UK), which works a single seam of coal (the Barnsley Seam) beneath a prolific dolomite aquifer. During the early months of operation of the mine, extraction using standard 200 m-wide longwall faces induced substantial inflows from the overlying aquifer. Peak flow rates were sometimes as high as 65 $Ml.d^{-1}$, though feeders always declined over a few months to steady flows of only a few hundred cubic metres per day. After the width of faces was reduced to 60 m ("shortwall"), no feeder greater than 13 $Ml.d^{-1}$ was ever encountered, and all feeders have also declined dramatically in flow rate, so that the total residual water ingress from the overlying limestones five years after mining commenced at Wistow was very low, at around 0.7 $Ml.d^{-1}$.

Recourse to shortwall extraction in an attempt to reduce water ingress has also been described from undersea workings by Saul (1970), though in the case he describes, absolute water yields were not greatly reduced following the switch to shortwall; however, the rapid movement of the shortwall faces was of benefit to face workers as feeders were "left behind" in the goaf. Aston and Whittaker (1985) subsequently explained that the apparent failure of the shortwall strategy to reduce water inflow rates in the case described by Saul (1970) could be ascribed to the occurrence of a major fault which has since been discovered adjacent to the workings. For these inflows to have been avoided, the total tensile strain would have had to be minimised much closer to the workings than would have been predicted from the thickness of cover to the base of the overlying aquifer.

---

the same sequence. Where a well-designed deep mine can be worked below substantial water bodies without inducing them to enter the underground workings, it is not possible for a surface mine to sink through a water-bearing horizon without draining it (albeit for reasons of convenience the drainage may be deliberately undertaken outside the void using wells; see Section 3.7). Nevertheless, drowning hazards are far less marked in active surface mines than in deep mines, owing to the generally large volumes of potential storage for flood waters, the lack of a roof to prevent egress of workers, and the relative ease with which pumps can be deployed from above. Hence it is generally possible to adopt a more relaxed attitude to problems of water ingress in surface mines than in deep mines. This is not to say that elevated water makes in surface mines are not a problem. On the contrary, the following problems are commonly associated with the presence of excessive quantities of water in surface mines:

(i)   The discomfort of miners working in wet areas
(ii)   The expense and inconvenience associated with gathering water to principal drainage sumps
(iii) The costs of draining sumps by pumping
(iv)   The difficulties of operating automated extraction equipment where the floor has been softened by water (Saul, 1970)
(v)   The need to use more expensive waterproof explosives (Ngah et al., 1984)
(vi)   The extra costs of transporting loose rock when it is wet, and therefore heavy, slippery and prone to freezing (Norton, 1983)
(vii) The destabilisation of excavation slopes and floors by water with high pore pressures (Cook, 1982).

The last of these problems is by far the most important in practice (Cook, 1982), and merits some further discussion here.

It is a fundamental property of virtually all rocks and soils that they are weaker when wet. Saturated rocks and soils tend to settle to far more gentle slopes than they will adopt when dry. Hence a previously stable slope may fail if the water table is allowed to rise within it. Similarly, a pit floor that safely bore heavy loads when dry can become unstable if pore water pressures exceed critical thresholds. Criteria for predicting when these conditions are likely to arise are well established in the soil and rock mechanics literature.

With regard to slope stability, competent rocks with a high shear strength (> 10 MPa) may be regarded as unconditionally stable for most engineering purposes, and can be engineered with stable slope angles as high as 90°. For all other rocks and soils, the stable angle of repose decreases as both slope height and pore water pressure increase. Table 3.5 summarises the relationships for slopes of heights typically found in surface mines. At its most marked, the effect of saturation is to reduce the stable slope angle by about 40%. The practical consequence of this is that an open pit in relatively weak material (< 0.8 MPa) which is worked *without* active dewatering of the wall rock will require such gentle slopes to be left that it may jeopardise the profitability of the mine (see Worked Example 3.4).

Excess water pressures can give rise to two main problems in surface mine floors (Figure 3.25):

(i) Fluidisation, in which the force exerted by upwardly flowing ground waters forces soil grains apart, so that the floor loses strength and allows imposed loads to founder into it. This phenomenon is also known as "quick conditions" (as in 'quicksand'), "boiling", or (where the fluidisation is localised to particular spots) "piping". In soil mechanics terms, fluidisation occurs when the pore water pressure value is close to the vertical total stress value (due to the weight of the soil), so that the vertical effective stress approaches zero (Preene *et al.*, 2000). In situations where the vertical total stress *just* exceeds the pore water pressure, the surface will remain stable until a further load is imposed (e.g. a miner walking across the surface, or a machine driving across it). Spontaneous loss of strength will then ensue, with costly (and sometimes fatal) consequences. The prediction of these alarming conditions is actually straightforward (Capper and Cassie, 1976) and can be undertaken easily as part of mine water safety assessments. It is first worth noting that, under the head differentials feasible in surface mining scenarios, quick conditions are highly unlikely to develop where the pit floor comprises moderately strong rocks or soils (shear strength > 0.5 MPa). However, where pits are floored with sands, silts or muds, fluidisation will occur when the vertical hydraulic gradient (i in Equation 3.4) exceeds a value known as the "critical hydraulic gradient" (i_c), which is defined by (Capper and Cassie, 1976):

$$i_c = \frac{G_s - 1}{1 + n_e} \tag{3.39}$$

Where G_s is the specific gravity of the soil, and
            n_e is the effective porosity of the soil (see equation 3.3).

Table 3.5: Stable slope angles under dry and water-saturated conditions for various earth materials, for a range of slope heights relevant to surface mining. (Adapted from information given by Cook, 1982)

Material	Shear strength (MPa)	Slope height (m)	Stable slope angle (degrees) under *dry conditions*	Stable slope angle (degrees) under *saturated conditions*
Competent rock with occasional major joints/fault planes	0.1–0.8	10 50 100	90 78 63	90 65 54
Heavily fractured rock or rock fill material	2–8	10 50 100	90 80 58	70 53 40
Soil or clay	0.02–0.5	10 50 100	90 48 35	85 32 22

Worked Example 3.4 The case for open-pit dewatering to minimise working of sterile country rock

A surface mine sinking is planned in a heavily-weathered hydrothermal ore body in central Brasil. Pervasive metasomatic chloritisation has left the ore body and the country rock very weak (mean shear strength 0.4 MPa). The water table at the site is 2 m below ground level. Pay dirt is known to be present to a depth of at least 100 m, in a belt some 100 m wide.

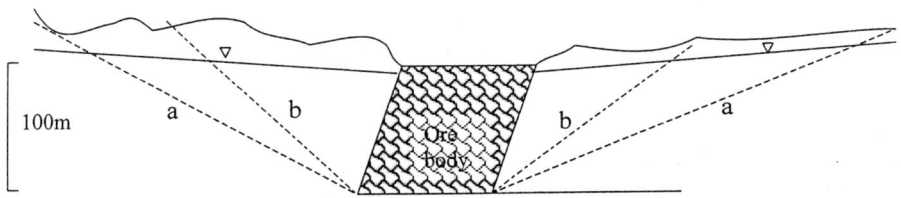

The superficially cheapest way to dewater the open pit would be to simply "sump it", i.e. pump water out of the bottom of the excavation as it deepens. If the entire ore body is to be exploited, then the surrounding rock will remain largely saturated, and by the time they are 100 m tall, the stable slope angle for the high walls will be on the order of 22° (see Table 3.4). The pit profile marked "a" on the diagram above shows how much wide the pit would need to be to safely access the full width of the pay zone to a depth of 100 m if no dewatering of the country rock is implemented. By contrast, if the water table in the country rock is lowered to the same elevation as the sole of the pit, profile "b" could be used, representing a saving of around 50% on moving sterile (i.e. non-paying) rock outside of the ore body. When entered into the overall economic planning of the mine, dewatering to obtain profile "b" was far and away more cost-effective than merely sumping this particular sinking.

For typical values of these parameters (e.g. $G_s = 2.65$, $n_e = 0.1$), i_c is around 1.5.

(ii) 'Floor heave', in which low permeability beds are forced to bulge upwards under the force of unrelieved artesian head in underlying aquifer layers (Figure 3.25b). Here the constraint on the degree of deflection of the floor of the excavation is the strength of the confining layer. Where this layer has a shear strength in excess of 10 MPa, no heaving is likely; otherwise it cannot be ruled out without careful geotechnical testing and modelling. In practice this is rarely undertaken, since the cure is relatively simple: to locally depressurise the underlying aquifer using dewatering wells, such that the piezometric level in the underlying aquifer is depressed below the base elevation of the confining bed (Figure 3.25b). (In reality, this is a fail-safe option, and it will rarely be necessary to lower the piezometric level quite so far).

3.7 DEWATERING TECHNIQUES AND THEIR DESIGN

3.7.1 THE NECESSITY OF DEWATERING

Virtually all deep and surface mines which work below the water table will require some form of dewatering. The only real exceptions to this rule are:

- shallow (usually ≤ 10 m) surface workings for sand and gravel, which can be worked wet using a specially designed drag-line
- temporary, deep tunnelling works, in which it may prove more cost-effective to exclude water by maintaining a very high air pressure within the void in preference to external dewatering

The water makes which can be expected in deep and surface mines have been discussed in the preceding section. Rational design of any dewatering system clearly demands that such makes be catered for. In deep mining operations, the sizing of mine access infrastructure (shafts, declines etc) may be dependent on a knowledge of the likely space requirements for rising mains etc. Consequently, considerable effort in prior hydrogeological characterisation of the target strata may be necessary, including:

- determining pre-mining piezometry
- estimating or measuring hydraulic parameters (T, S)
- modelling the changes in water levels likely to be achieved by various options for mine sinking and dewatering, and the likely pumping rates which will be needed to obtain these changes

In surface mines, although prior characterisation is often helpful in making economic projections, the relative ease of access means that most contingencies can usually be handled without making drastic alterations to mine design.

Where very permeable, saturated strata are to be mined, it is very important to bear in mind that a transient phase of heavy pumping will generally be needed during the early stages of the operation, until ground water levels have been lowered sufficiently that mine access

(a)

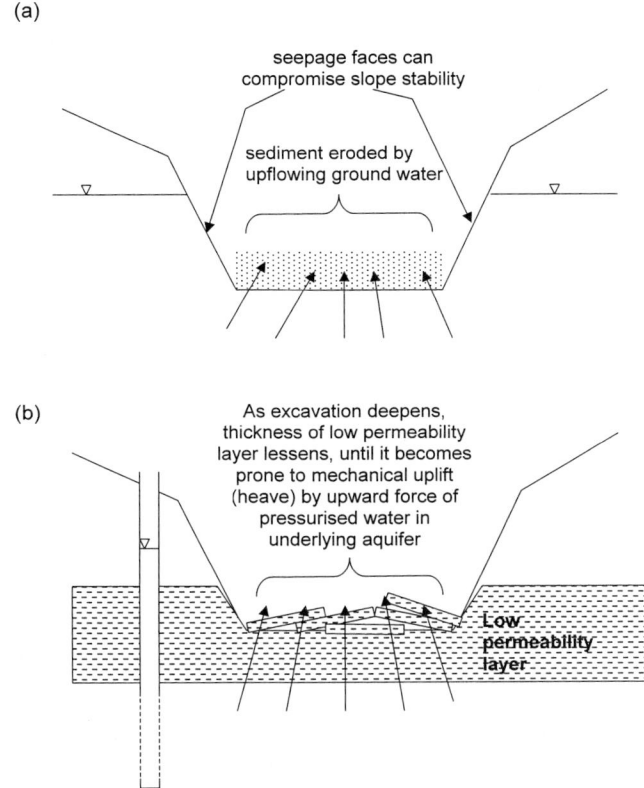

(b)

Figure 3.25: Physical stability problems due to ground water in the vicinity of an open pit (a) quick conditions in base and slope stability on walls (b) floor heave.

is feasible. The pumping rate during this early stage is effectively removing water from ground water storage, and will often be several times greater than the rate which will be required to maintain steady drawdowns during the remainder of the life of the mine. Typical ratios of early dewatering rate to long-term dewatering rate are in the range 2:1 to 10:1.

While the variety of geological and hydrological conditions around the world ensure a vast range of possible dewatering strategies, three general approaches are pre-eminent (Table 3.6). We will now consider each of these in turn.

3.7.2 SUMP DEWATERING

Principles of application
Sump dewatering involves the creation of basins in the floors of workings, to which all water is diverted, and from which it is all pumped out of the mine ((Table 3.6; Figure 3.26a). Sump dewatering (also termed "passive protection" by Fernández-Rubio, 1979) is the most widespread dewatering method in both surface (Figure 3.27) and deep mines (Figure 3.28) world-wide Indeed, even where other dewatering methods are employed to

Table 3.6: The three principal methods of mine dewatering

Type of Dewatering	Range of appropriate conditions	Outline of Process
Sump Dewatering	(i) If the mine is above the water table, or (ii) If the mine is below the water table, but the strata are of low permeability, or (iii) The mine is an open pit operation in chemically inert rocks, or (iv) The mine is small and/or isolated.	Roads and ditches within the mine are routed and graded to deliver all water to one or more centralised sumps, whence it is pumped from the mine.
Adit Dewatering	If the maximum depth of working is above the minimum geographical elevation to which an adit can economically be constructed.	A drainage adit with a gradient of >1:500 is driven from a portal in a valley beneath the area to be mined, and all mine drainage is routed to the adit via roadways, shafts, pipe-work etc.
External	(i) If the mine is an underground mine surrounded by highly permeable aquifers and/or large volumes of flooded old workings, or (ii) If the mine works pyritic or otherwise highly reactive strata which may cause a deterioration in water quality if water is allowed to enter the workings, or (iii) The mine is very deep, but external dewatering wells can intercept the water make at much shallower depth.	Boreholes and/or shafts in the aquifers or old workings outside the mine site are used to pump water, either to prevent this water entering the mine by gravity flow, or to lower the water table below the areas of active mining.

handle the bulk of the water make, there will nearly always be some need to support these with some localised sump dewatering.

In deep excavations, it may often be necessary to have a staged series of pumping sumps (or 'lodges') installed at various elevations above the base of the workings, with separate batteries of pumps lifting the water from one stage to another. Similarly, in widespread workings, it may be necessary to have more than one major basal sump to ensure adequate drainage of the entire mine.

The principal advantage of sump dewatering is the flexibility it offers in pumping arrangements: throughout the life of a mine, pumping stations can be relocated as needs arise, and as areas of working are abandoned. The principal disadvantage of sump dewatering is that water flowing to the sumps can continue to cause a considerable nuisance as it passes through the mine, which might only be minimised by considerable investment in ditching and pipe-work. There are a number of circumstances in which it can be desirable to keep the void as dry as possible. For instance:

(i)   to avoid contamination of the water make with suspended solids or contaminants leached from the crushed ore/gangue, which might necessitate expensive treatment of the mine water prior to its disposal to a natural watercourse

(a)

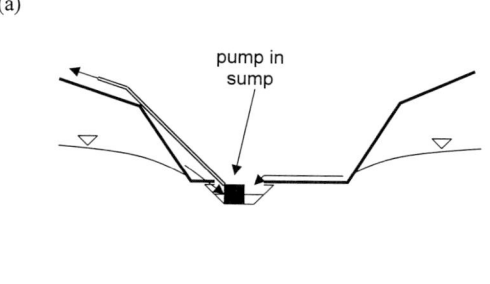

(b)

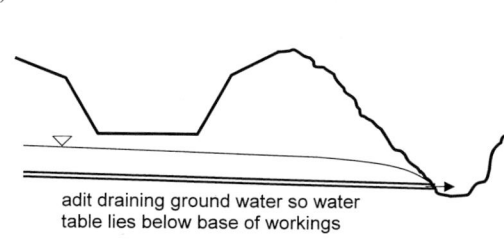

(c)

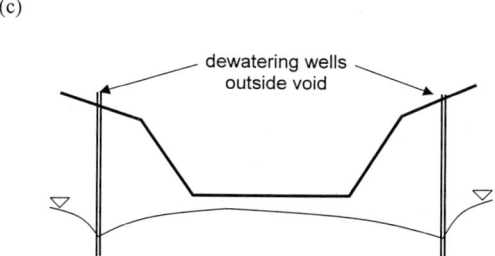

Figure 3.26: The three main types of mine dewatering (a) sump dewatering (b) adit dewatering (c) external dewatering.

(ii)   to avoid slurrying of roadways and destabilisation of unsupported faces

(iii)   to minimise problems related to freezing of water during the winter

In such circumstances, sump dewatering may not be the best option, as it necessarily entails tolerating water within the working void.

Although access restrictions may impose some upper limit on the amount of water which can be handled by sump dewatering in a deep mine, there is no real upper limit on the volumes of water that can be handled by sump dewatering in a surface mine, and total sump-pumped water makes on the order of 20–100 Ml.d^{-1} are by no means uncommon where surface mines work in permeable strata.

Design of sump dewatering systems

Theoretical guidelines for the design of sump dewatering systems have been developed to the extent that computer-aided design is now feasible (Desouza and Penner, 1993). However,

Figure 3.27: A major sump in a surface mine: a limestone quarry in NW England.

Figure 3.28: A small underground dewatering sump: the main pumping lodge on the 295 Level,
Frazer's Grove Fluorspar Mine, Weardale (UK). John Evans checks the end of a rising main feeding
into the sump, whence the large Flygt™ pump lifts the water to the next sump in the system. For
details of the overall dewatering system in this mine see Younger (2000b). (Photo: A. Doyle).

design in practice is almost always organic, responding opportunistically to changes in the
locations and magnitudes of feeders. The principle attributes of a sump dewatering system
which require design effort include:

– Sizing of the sump(s) to allow "standage" (spare storage volume) to cope with likely
  changes in water make and / or the periods necessary to replace failed pumps etc. In
  most cases the sump itself will simply be an open excavation, albeit more elaborate
  designs have been proposed for use on cramped sites (Preene *et al.*, 2000).
– The sizing and layout of drainage channels leading to the sump. These should be sized
  in accordance with standard hydraulic procedures, using formulae such as Equations
  3.20–3.22 to design adequate wetted perimeters and bed gradients for the flows they are
  intended to carry. Bed gradients towards the sump should never be set less than 1:500.
– The pumps used to dewater the sump(s), which must be able to:
  • handle the necessary quantity of water
  • lift this quantity of water to a sufficient elevation that it can reach the intended
    discharge point (be this beyond the mine void or just another stage sump)
  • cope with turbid water
  • run fully submerged, or partially submerged (i.e. drawing both air and water, in what
    is known as "running on snore", due to the noise pumps make under these conditions)

The Flygt™ pump shown in Figure 3.28 is a typical 50 Hz, aluminium submersible sump pump, commonly used in both deep and surface mine dewatering systems. The particular model shown is capable of lifting as much as 3 Ml.d^{-1} of water to a height of 50 m. Pumps capable of lifting much greater quantities through greater heights are widely available, though the economics of selecting the right pump for the given job can only be assessed on a case-by-case basis. A good starting point in all cases is technical advice from pump manufacturers. Web-sites of some of the leading mine dewatering pump suppliers include:

http://www.mono-pumps.com
http://www.flygt.com
http://www.grundfos.com

(While we can accept no responsibility for any products supplied by these companies, we can certainly say that we have successfully used their equipment on numerous occasions).

3.7.3 ADIT DEWATERING

Principles of application
Adit dewatering is achieved by the use of long drainage adits which provide an artificial, highly-permeable hydraulic connection between local mined strata and some more distant, lower, base-level of gravity drainage (Table 3.6; Figure 3.26b). Drainage adits of several hundred metres extension have been commonly used in European orefields for millennia. For instance, Davies (1935) documents Roman drainage adits in Spain, Greece and Slovakia, some of which reached lengths as great as 2 km. Many ancient adits continue to under-drain mined ground many centuries after their construction. For instance, in Germany, 11th Century drainage adits (constructed by formally-constituted drainage cooperatives, called *Wassergemeinschaften*) are still flowing today (Coldewey and Semrau, 1994). With the advent of inexpensive gunpowder in the 17th Century, ever-longer adits could be constructed, most of which continue to flow at high rates, representing major, permanent modifications of the hydrology of their host catchments. For instance, the County Adit, which was initiated in 1748 to under-drain the rich tin and copper deposits of west Cornwall (UK), had attained a total drivage length of some 64 km by 1880 (Buckley, 1992), and today transmits around 60% of the effective precipitation falling on the overlying surface catchment, leaving only 40% to form direct surface runoff (Knight Piésold and Partners, 1995). An adit of similar age in the Ruhr Coalfield of Germany, known as the *Schlebuser Erbstollen*, was more than 12 km long by the 18th Century and still drains freely (Coldewey and Semrau, 1994). By the 19th Century, drainage adit construction was reaching new heights of excellence. In the North Pennine Pb–Zn Orefield (UK), for instance, two major drainage adits achieved total drawdowns in their respective catchments of 180 m (the Blackett Level, Allendale) and 200 m (The Nent Force Level, Nent Valley) (Younger, 1998a).

   Adit construction has, of course, continued to the present day. One very recent European example is afforded by a new gold mine in northern Spain (Río Narcea, Asturias). The gold deposit is a "skarn" (metasomatised limestone adjoining a granitic intrusion) in which extraction commenced by open-pit methods in January 1997. To obtain permission to mine in this environmentally sensitive area, the company committed itself to a policy of "zero discharge" from the mine void (see Section 3.8.2). Compliance with this policy necessitated the diversion of surface runoff around the void, and the lowering of the water table such that the open pits will never intersect it even at their maximum working depth. To achieve the necessary drawdown, a 1 km adit was driven below the pits, and three inclined boreholes were drilled at the forehead of the adit. These boreholes flowed by gravity, and in little more than a month they achieved a steady drawdown of more than 50 m below the sole of the pits (Younger, 1998a).

Adit dewatering can readily secure access to ore reserves down to adit level. However, where payable rock persists to even greater depths, there will always be an incentive to mine below adit level. This requires active pumping (by sump or boreholes) to dewater the deeper strata. In this stage of mine development, the original adit can often be neglected, at best functioning as a convenient saving in pumping head for waters extracted from deeper workings accessed by shafts. However, lack of regular man-access through the old adits often results in a lack of maintenance. Although roof falls very rarely result in a total loss of water-passage, they can raise the head sufficiently in the old adit that shaft inset sumps are drowned-out, and continued use of the adit becomes impractical. In such cases, renovation of the adit may be less cost-effective than re-arranging pipe ranges so that it is possible to pump directly to the shaft collar. Hence, the old adit becomes neglected, and often forgotten, until the mine is abandoned and the workings left to flood. At this stage in the mine life-cycle, the old adit can take out a new lease on life, as the principal gravity discharge route from the mine workings (see Section 3.11.2 below). Examples of this concatenation of circumstances are reported by Younger (1998a) and Johnson and Younger (2000).

Design of adit dewatering systems
Drainage adits are probably the most difficult dewatering technology to design effectively, at least in terms of the hydrological response they will induce. While an efficient adit will undoubtedly cause significant drawdown in the surrounding rock mass, the extent of the drawdown and the period of time necessary to accomplish maximum drawdown after completion of the adit are both difficult to predict with confidence. A few simple predictive models are available for this problem, in the form of exact analytical solutions for cases in which adits are driven through homogenous saturated strata (Freeze and Cherry, 1979). Solutions of relevance to adit design are available for the following two cases:

(i) Calculation of 'steady-state' inflow rates to deep adits located *within* the saturated zone, below a static water table (e.g. Lei, 1999). This case is implicitly applicable only to cases in which the adit intercepts so little ground water that no significant drawdown is caused, and as such is of limited direct interest in mining (as opposed to linear tunnelling) applications, where the dewatering of ground above the adit is an important aim.
(ii) The transient hydraulic response of saturated ground to adits which intercept sufficient water that they cause drawdown of the water table (Goodman *et al.*, 1965; Freeze and Cherry, 1979). This is the case of most interest in the context of mining.

For the first of these cases, the exact analytical solution to the steady-state ground water flow equation yields the rate of inflow to the adit per unit length, which can be adapted from the work of Lei (1999) to read as follows:

$$q_i = \frac{2\pi\, K(h_w - h_a)}{\ln[D/R + \sqrt{(D/R)^2 - 1}]} \qquad (3.40)$$

Where q_i is the rate of inflow per unit length of adit $(L^3.T^{-1})$
        h_w is the water table head (L)
        h_a is the head at the perimeter of the adit (L)
        D is the depth of the centreline of the adit below the water table (L)
        R is the radius of the adit (L)

For the second case, q_i is a function of time, and hence should be written $q_i(t)$, which is the inflow per unit length of adit at a given time t after commencement of drawdown. This is quantified using the following expression (after Goodman *et al.*, 1965, and Freeze and Cherry, 1979):

$$q_i(t) = \sqrt{(c \cdot K \cdot (h_w - h_a)^3 \cdot S_y \cdot t)} \qquad (3.41)$$

Where c is a coefficient with a value in the range 1.33 to 2 (with higher values being favoured in practice; Goodman *et al.*, 1965), and Sy is the specific yield (see Equation 3.3). While equations 3.40 and 3.41 provide interesting illustrations of the interplay of the factors which determine adit water makes, we concur with Freeze and Cherry (1979) that "they may be suitable for order-of-magnitude design-inflow estimates, but [they] should be used with a healthy dose of skepticism". Certainly, the limiting assumptions upon which these analytical solutions are based preclude their application to heterogeneous rock masses. In such cases (which are the majority in practice) numerical models may be used. A thorough discussion of the application of numerical methods to adit dewatering design is beyond the scope of this book, but the interested reader will find suitable models described in detail by Fawcett *et al.* (1984) and Adams and Younger (2001).

3.7.4 EXTERNAL DEWATERING

Principles of application
External dewatering is accomplished by pumping boreholes (or, in some cases, old mine shafts) outside of the current zone of extraction (Table 3.6; Figure 3.26c). The drawdown caused by these boreholes intercepts water which would otherwise have flowed into the active mine-workings (Figure 3.26). By dropping the water table below the floor of the working area, much of the need for sump dewatering will be removed. Even surface waters arising from direct precipitation into surface mine voids will tend to infiltrate and migrate to the external dewatering pumps, save where the floor of the void is naturally lowly permeable, or has become lowly permeable due to "blinding" (i.e. accumulation of fine-grained rock debris). In the latter case, local ponding will occur, requiring some sump dewatering.

There is a range of circumstances in which external dewatering may prove to be appropriate. Some of the more common circumstances are as follows:

(i)   Where lateral inflows to a mine from adjoining aquifers or flooded workings cannot readily be controlled by other means (such as sump dewatering).

(ii)   Where environmental or rock stability considerations make it imperative to minimise the amount of water coming into contact with the mine workings. This can be the case, for instance, where the mine works highly soluble rock salt, or where the ore body is radioactive, acid-generating or polluted in some way (e.g. through previous *in situ* acid leaching; see Section 1.3.6), so that water entering the mine would become contaminated and thus require expensive treatment prior to disposal or re-use.

(iii)   Where sinking of a deep void intersects very shallow, but highly prolific, water-bearing strata. If large quantities of water from the shallow strata can be intercepted by external dewatering using "interceptor wells", then it will not be necessary to lift this water all the way from the mine sump by expensive pumping.

(iv)   Where a surface mine works in a cold region, problems associated with freezing of large quantities of water can be avoided by minimising the ingress ot water to the void by means of external dewatering.

(v)   Dewatering of strata overlying and adjoining deep mines can facilitate maximisation of extraction, since it will remove the need for implementing the usual precautions against inrushes.

(vi)   The stability of slopes within surface mines will be increased if they are pre-drained before excavation (see Section 3.6.5). This not only enhances working safety, but also allows steeper boundary walls to be left, thus maximising the mineral take (Cook, 1982; Ngah *et al.*, 1984).

The literature contains a variety of terms which refer to external dewatering as defined in Table 3.6. For instance, "active protection" (Fernández-Rubio, 1979) and "peripheral dewatering" are virtually synonymous with external dewatering as the term is used here. "Peripheral dewatering" has been applied in particular to the context of US surface mines. The term "advance dewatering" is applied to external dewatering which is implemented prior to the commencement of excavation, a strategy which remains very popular in the UK opencast coal sector (e.g. Norton, 1983; Minett *et al.*, 1986). Where it is practised at a very large scale, external dewatering has been termed "regional dewatering". This is achieved where a number of shafts/boreholes distributed throughout a mining district are used to control water movement in and around many inter-connected mines. The earliest regional dewatering scheme to be implemented appears to have been that which operated in the South Staffordshire Coalfield, UK, from 1873 to 1920 (Saul, 1959). Pumping at rates of up to 65 Ml/d from old workings in each of four "pounds" (i.e. ponds in modern terminology; see page 231) was maintained to facilitate active mining at depth. Subsequently, similar schemes have been implemented in several UK coalfields, including Fife (Younger *et al.*, 1995a); Durham (Harrison *et al.*, 1989; Sherwood and Younger, 1994; Younger, 1993; Younger, 1998b); South Yorkshire (Saul, 1936); Derbyshire (Peters, 1978); Nottinghamshire (Peters, 1978; Awberry, 1988; Lemon, 1991) and Leicestershire (Smith, 1996). As an example, Figure 3.29 illustrates the layout of the regional dewatering scheme in the Durham Coalfield. In Germany, the deep mines of the Ruhr coalfield are dewatered regionally by a network of 28 pumping shafts, which has gradually expanded in area as deep mines have closed since 1920 (Coldewey and Semrau, 1994). Similar systems have also been used for many decades in Hungary (Kesserü, 1997) and Poland (Rogoz, 1994; Rózkowski, 1997).

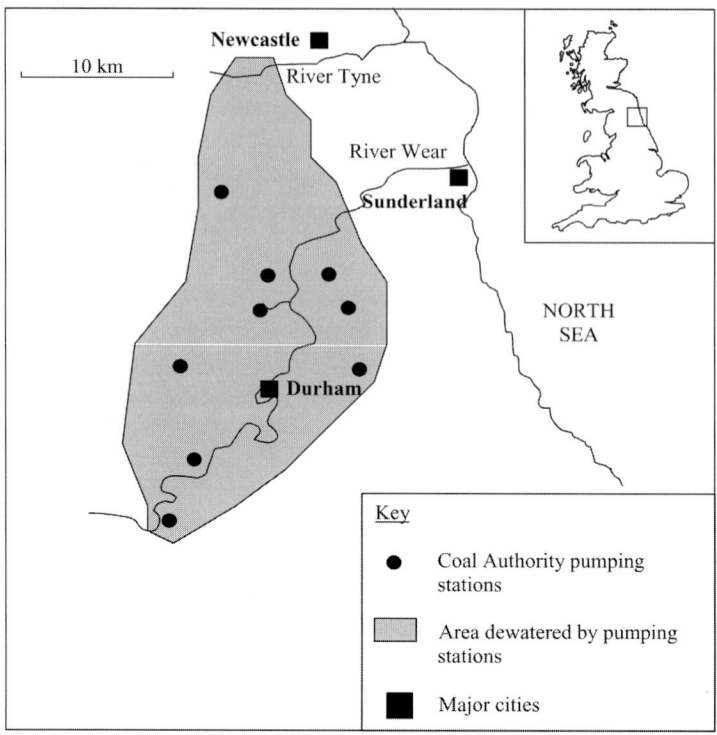

Figure 3.29: Regional-scale dewatering of part of the Great Northern Coalfield, County Durham (UK), by means of nine pumping shafts (former mine shafts fitted with electric submersible pumps).

Design of external dewatering systems
At their simplest, external dewatering systems can be designed using simple calculations of drawdown around one or more relatively shallow wells. These are usually made using inverse applications of standard Theis and Jacob test-pumping analysis methods, which are widely described in standard hydrogeology texts (e.g. Freeze and Cherry, 1979; Fetter, 1994; and Kruseman and de Ridder, 1990). Examples of the applications of these calculations to shallow excavations are presented by Preene *et al.* (2000). There are two main issues to be determined in the design process:

(i) The establishment of the necessary drawdown. Dewatering involves lowering the water table to a specific depth (which will be defined by the mining geologists and economists) and then retaining it at that depth. In other words, a lot of water must be removed from storage until the drawdown is sufficient. In most cases, this will mean pumping more heavily during the early phases of dewatering than in the later stages. Whatever pumping rate is used, it is necessary to calculate whether this will yield sufficient drawdown on an acceptable time scale. This involves calculating the drawdown at specific times from the start of pumping, using robust estimates of the transmissivity (T; see Equation 3.14) and storativity (S; see Section 3.3.2) of the strata.

The relationship between the rate of pumping in a well and the drawdown at a particular distance from the well at a specified time was defined by Theis (1935), and his formulae continue to provide the basis of dewatering calculations for many simple applications. In the form most useful to us, the Theis (1935) equation may be written:

$$s = \frac{Q}{4 \pi T} W(u) \qquad (3.42)$$

Where s is the drawdown at a specified distance (r) from the pumping well and at a given time (t) since the start of pumping, Q ($L^3.T^{-1}$) is the rate at which the well is pumped and W(u) is the "well function of u", where u = $r^2S/4T$. Values of W(u) for a wide range of values of u are tabulated in standard ground water texts (e.g. Freeze and Cherry, 1979, Fetter, 1994), but are also easily calculated (with sufficient precision for our purposes) from the following expression:

$$W(u) = -0.5772 - \ln(u) + u - (u^2/2\cdot2!) + (u^3/3\cdot3!) - (u^4/4\cdot4!) + (u^5/5\cdot5!) - (u^6/6\cdot6!) \quad (3.43)$$

(The astute reader will realise that this is a truncation of an infinite series, but later terms are negligibly small). Only one more piece of information is needed to apply equation 3.42 in practice: the 'principle of superposition', which states that the drawdown at a given point due to the pumping of more than one well nearby is simply the sum of the individual drawdowns which each of the wells would produce were it pumping alone. The application of this approach to a real external dewatering design is given in Worked Example 3.5.

(ii) The long-term pumping rate necessary to maintain drawdown at the target level. In many cases, it will not be technically feasible to deepen a mine at a rate greater than the rate of drawdown which can be achieved by external dewatering. In such cases, the long-term pumping rate can be calculated readily from Equation 3.42 by simply setting t equal to the anticipated end of mining at the deepest level. In other cases, it will be necessary to achieve target drawdown rapidly (by pumping at a higher rate) and then reduce the pumping rate to the minimum needed to hold water levels steady at the target level. If the pumping rate needed to achieve the target drawdown by the specified date is calculated using Equation 3.42, then it is certain that the long-term pumping rate will be less than this. Precise quantification requires knowledge of the 'radius of influence' (r_i) of the wells, i.e. the distance from the well to the point on the water table at which drawdown is zero. This can be determined using observation wells disposed at various distances from each pumping well, or by estimation using the Jacob distance-drawdown method (see Driscoll, 1986, and Kruseman and de Ridder, 1990 for full details). If r_i is known, the long-term pumping rate is readily calculated using the following adaptation of the well-known Thiem Formula (cf Fetter, 1994):

$$Q = \frac{T \cdot 2\pi \cdot S_w}{\ln (r_i/r_w)} \qquad (3.44)$$

Worked Example 3.5 Simple design calculations for an external dewatering system

A small surface mine is planned to win placer diamonds from the uppermost five metres of a 24 m-thick body of saturated alluvial sediment in a remote location in central Africa. Test pumping of the sediment has revealed a transmissivity (T) of 120 $m^2 \cdot d^{-1}$ and a storativity (S) of 0.1. In this remote location, equipment for wet working cannot easily be obtained, so a plan is hatched to dewater the sediment to a depth of up to 10 metres below the initial water table level and develop a rectangular box-cut 125 m long by 32 m wide. Because the sediment body is relatively thin, it is not possible to externally dewater the pit using a single well. Four wells are proposed by the mine manager, to be drilled at the extremities of the pit. The question is, what pumping rate will be required from the four wells in order to dewater the centre point of the pit to a depth of ten metres within two weeks?

Parameterising Equation 3.40 for this problem gives t = 14 days, s = 10 m total. From the principle of superposition, this amounts to a contribution of drawdown from each of the four wells of 10/4 = 2.5 m.

Also, for all of the four wells, r is given by half of the diagonal distance across the box-cut, i.e. by Pythagoras r = $(\sqrt{(125)^2 + (32)^2})/2$ = 64 m.

With these values, u = $((64)^2 \cdot 0.1)/(4 \cdot 120 \cdot 14)$ = 0.06, giving W(u) = 2.29

Then, for

$$s = \frac{Q}{4 \pi T} \quad W(u) \text{ we have } 2.5 = \frac{Q \cdot 2.29}{4 \cdot 3.14159 \cdot 120}$$

So Q = 1,646 $m^3 \cdot d^{-1}$ *from each well*, totalling 6,584 $m^3 \cdot d^{-1}$ altogether. This is a lot of water to demand from four shallow wells, and excessive well losses/local drying of the wells would be likely. Accordingly, the mine manager should be advised that a larger number of wells should be used, lessening the load on each individual well

---

Where   T is the transmissivity of the aquifer ($L^2 \cdot T^{-1}$) (see Equation 3.14)
   s_w is the steady drawdown in the pumping well (L)
   r_w is the radius of the pumping well (L)

Where r_i is unknown, it is sometimes assumed for the sake of argument that it will be on the order of 1000 m, which is reasonably conservative in unconfined aquifers, but may be a considerable under-estimate where the aquifer remains confined at some distance from the mine. Failing even this simplification, all that can be assumed is that the long term pumping rate is likely to fall in the range of 60% to 100% of the rate that was required to achieve the initial rapid drawdown.

As the scale of the dewatering system increases, so often does the difficulty of design. Where a new mine is developed in otherwise virgin strata, conventional ground water modelling (Anderson and Woessner, 1992) will be applicable for dewatering design.

However, where previously-mined ground must be dewatered, flows induced by renewed pumping may well be turbulent, thus violating the laminar flow assumption upon which conventional ground water flow modelling is based. In such cases, alternative hydraulic formulations may be necessary. For instance, in the design of advance dewatering systems for very large opencast coal mines, Minett *et al.* (1986) derived and applied non-standard hydraulic analysis techniques, which were formulated to cope with the peculiarities of ground water storage in flooded deep-mine workings. At the very largest scales, most regional dewatering systems (such as those of the Ruhr and Durham coalfields in Europe; Younger, 1993a; Coldewey and Semrau, 1994) were developed experientially, in an *ad hoc* manner, rather than by purposeful design. It is only at the present time that methods of analysis for such large systems are being developed (e.g. Sherwood and Younger, 1997; Adams and Younger, 2001), principally in the context of the optimisation of pumping systems for long-term environmental management purposes. This is a topic to which we will return in Sections 3.10 and 3.11 below.

3.7.5 GEOTECHNICAL ALTERNATIVES/AIDS TO DEWATERING

Where mines are surrounded by highly permeable aquifers or flooded old workings, it is often advantageous if the inflow of water to the active workings can be minimised by the installation of some form of barrier. The installation of barriers to prevent unnecessary water ingress to workings has been practised for many centuries, and was the key to the opening up of many of the world's most successful deep mines below highly permeable river gravels and limestone aquifers. Thus the author of the 'Compleat Collier' (J.C., 1708) tells us:

> "... For framing back our Shaft Feeders, we make use of Wood, but chiefly *Firr* [sic], because .. we think it swells with the Water lying against it ... we make use of Sheep-Skins with the Wool on ... which, being well Wedged in between the frame and such rough Mettle, &c ... we find to perfect our Design of stopping the Water ..."

Although mining geotechnics has advanced considerably since the days of fir timber and sheep-skins, the principles of barrier design remain the same: to place in the way of any unwanted water a physical obstruction which combines low permeability with sufficient structural strength to resist the pressure which will build up behind it as the level of impounded water rises.

The installation of watertight linings in shafts is now a well-developed field of endeavour, greatly facilitated by the availability of ground freezing and borehole dewatering techniques. Perhaps more important in terms of controlling the overall water make are barriers in roadways and other workings which are specifically designed to prevent (or at least minimise) lateral inflows from aquifers or flooded workings. Such barriers are of two main types:

1. *In situ* rock barriers, which are created by refraining from mining within specified areas, and

2. Artificial barriers, also known as "plugs" or "dams", which are typically mass-concrete
    structures of various lengths, supplemented where necessary by pressure grouting of
    the surrounding rock mass.

The design of both types of barrier is a specialist mining engineering operation, the details
of which are beyond the scope of this text. Interested readers are referred to the recent
synthesis of Fuenkajorn and Daemen (1996). Some of the key challenges which need to
be overcome in barrier design include:

*   the presence of undetected permeable fractures or beds within the strata, which might
    provide unexpected water flowpaths
*   the possibility that errors in surveying in the past might lead to under-estimation of the
    necessary width of a barrier to be left between present and former workings
*   the likelihood that future workings at deeper levels might cause fracturing of the
    barrier. This can be accommodated by widening the barrier by as much as a factor of
    two, or (if future deeper workings can be controlled), leaving a barrier in the deeper
    workings up to three-and-a-half times the width of the uppermost barrier.

Earliest recommendations for the effective width of such barriers suggested they should
be at least 30 m wide. However, as experiences mounted, the recommended minimum
width gradually rose, through 40 m in the 1930s–40s (e.g. Saul, 1936, 1949) to 50 m by the
1970s (Miller and Thompson, 1974). Nevertheless, width alone remains no guarantee of
success: in the early 1990s, for instance, a 60 m barrier of intact ground between two
adjacent sets of coal workings in the No 2 Rhondda Seam in South Wales was
comprehensively breached by rising mine waters (Younger, 1994). Many more instances
could be cited of water finding a way past apparently impregnable barriers, and barriers
of up to 140 m have been constructed in desperation where roof and floor strata are
especially permeable. In extreme cases, pressure grouting of roof and floor strata may also
be necessary. Some authors have suggested that barrier design must be based on rock
strength criteria (e.g. Kesserû, 1995). However, as Saul (1949) remarked, "there is no case
on record of the bursting of a barrier, failure being always by leakage. Computation of
shearing strength [sic] will show that the actual strength to resist bursting is normally given
by a width of barrier much less than that required to resist leakage".

Barriers have not only proved effective in long-term applications, but have also been
installed in emergency situations to prevent inundation of workings. In the case of the
West Driefontein inrush in 1968 (summarised in Section 3.6.1; and see Cousens and Garrett,
1969; Cartwright, 1969) high-pressure concrete plugs were constructed within three weeks
to isolate the area of the mine immediately surrounding the source of the inrush from the
overlying dolomite aquifer. Guidelines on plug construction have now been developed on
the basis of worldwide mining experience (e.g. Kesserü, 1994; Fernández-Rubio et al.,
1987; Fuenkajorn and Daemen, 1996). A potentially less expensive option to control
extreme inflows is by topical pressure-grouting of the particular fissure system from which
an extreme feeder arises. Guidelines on drilling and grouting procedures, including
recommendations for pre-emptive grouting of fissures ahead of development headings,
have been developed by van Schalkwyk and Bellamy (1994) on the basis of extensive
experiences in the Witwatersrand Goldfield of South Africa.

3.8 ENVIRONMENTAL IMPACTS OF DEWATERING AND THEIR MITIGATION

3.8.1 DEWATERING IMPACTS

Water quantity impacts
Dewatering of sub-water table mines inevitably causes drawdown of the water table in the surrounding rock. The consequences of such drawdown are conceptually the same as those caused by pumping ground water for other purposes (e.g. for public water supply), and can therefore be analysed in an entirely analogous manner. A substantial literature exists on the impacts of public supply abstraction wells on ground and surface water resources. None of the more recent analyses have changed the basic principles as enunciated by Theis (1940), who clearly showed that any abstraction from ground water will eventually be matched by some combination of the following three responses:

(i)   a decrease in the volume of ground water in natural storage
(ii)   an increase in the rate of ground water recharge
(iii) a decrease in the rate of natural ground water discharge

The peculiarity of mine dewatering systems lies in the deliberate maximisation of (i) above. Decreases in ground water storage are manifest in a lowering of the water table. Lowering of the water table *per se* has a number of socio-environmental consequences, not all of which are bad. On the negative side, lowering of the water table can leave pre-existing abstraction wells high and dry. It can also lead to desiccation of ponds that previously occupied enclosed hollows (Figure 3.30). Where these were of ecological value, the impact can be grave. On the other hand, the drying out of previously waterlogged land can render it more useful for agriculture, albeit at the cost of relocating a source of surface runoff.

Lowering of the water table in areas where surface runoff was previously impounding above saturated ground can induce further recharge to the subsurface (although this may be at the expense of wetland habitats). Water can also be induced to enter the subsurface from streams and rivers (Figure 3.30). For instance, the Reocín Zinc Mine in northern Spain has a total water make of about 1 $m^3.s^{-1}$, nearly all of which enters the mine from a nearby river (the Río Saja) (Fernández-Rubio, 1982). In a study of the Gays River Mine (Nova Scotia, Canada), Aston and Lamb (1993) found that induced recharge from the nearby river was one of the two major sources of inflow to this deep mine (the other being direct infiltration above the dewatered area). Similarly, dewatering of an open-pit limestone mine in Poland induced varying amounts of inflow from the Vistula River, which on hydrochemical grounds is seen to amount to as much as 80% of the total water make (Motyka and Postawa, 2000). In extreme cases, dewatering can lead to streams drying up altogether for part of their course and/or for part of the year. (Unsurprisingly in view of the legal sensitivities, few examples of this genre are reported in the open literature, though unpublished examples abound).

A decrease in natural discharge from aquifers is probably the most common impact of mine dewatering. The drying-up of springs in limestone aquifers subject to quarry dewatering is widely reported (e.g. Hobbs and Gunn, 1998). More widespread, but more subtle, are reductions in the rates of natural ground water discharge to perennial streams. In

(a)

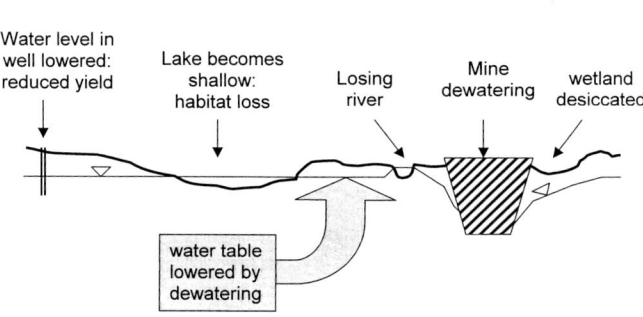

(b)

Figure 3.30: A summary of some of the major possible effects of water table lowering by mine dewatering. In any particular case, the actual effects may range from zero to one or two of those shown; it would be the most unfortunate of mine managers who had to contend with all these impacts at one site!

many cases, such reductions will be masked by the disposal of dewatering pump effluents into the same streams (see Section 3.8.2 below).

The disposal of dewatering effluents can be a considerable benefit to surface water catchments. For instance, it can provide welcome dilution for other polluted waters (especially sewage effluents; e.g. Banks *et al.*, 1996). In one catchment in northern England, pumped mine water discharges served to sustain river flows during dry periods, thus saving the regional water authority the expense of augmenting flows by pumped transfers of water from another reservoir-supported catchment (Younger and Harbourne, 1995). Wetlands can also be supported by pumped discharges from mines. In one particularly interesting case in South Africa, a perennial pumped discharge of some 100 Ml.d⁻¹ from the deep workings of the Grootvlei Gold Mine provides most of the water entering the Blesbokspruit wetland (Wood and Reddy, 1998), which was designated an ecosystem of the highest international conservation importance (under the terms of the Ramsar convention) in October 1986. The Blesbokspruit wetland lies at 1600 m above sea level, and covers a total area of 1858 ha. As it ultimately contributes to the flow of the Vaal River, which is a major water resource in the region, the Blesbokspruit wetland has considerable economic importance. However, it is for its biodiversity that it is most highly prized. Historically, the site now occupied by the wetland was a narrow meandering non-perennial stream with an associated "wet meadow" wetland. Gradual development of the area by the gold mining industry from around 1930 onwards wrought profound changes to the drainage. Roads and embankments crossing the stream impounded upstream areas, which slowly became colonized by reeds. By the time of its designation as a Ramsar site, the Blesbokspruit had been transformed into a permanent wetland which is known and valued for the variety and abundance of bird species utilising it, including a number of endangered species such as the Yellow-billed Duck (*Anas undulata*), the Spur-winged Goose (*Plectropterus gambensis*), Greater Flamingo (*Phoenicopterus ruber*), Lesser Flamingo (*Phoenicopterus minor*) and Goliath Heron (*Ardea goliath*). The permanent inundation of the Blesbokspruit wetland due to the perennial discharges of mine water is the key to its regional importance as a reliable habitat even when other, smaller wetlands dry out in droughts.

Water quality impacts
A positive water quality impact of dewatering has already been noted, in the form of pro-vision of dilution for more polluted waters. Nevertheless, where dewatering effluents are of very poor quality, the impacts on receiving watercourses can be detrimental. For instance, gold mine effluents are known to cause arsenic pollution of surface and groundwaters in the Ashanti region of Ghana and in the Mazowe district of Zimbabwe (Smedley *et al.*, 1996). Acidic effluents from active mines continue to pollute rivers in India, China and other Asian countries. The pumping of saline mine waters is a common cause of fresh water degradation in many major European coalfields (e.g. Lemon *et al.*, 1991). In the Upper Silesian Coalfield of Poland, the saline mine waters are also highly radioactive, and their disposal into the local rivers has led to alarming levels of environmental exposure to radionuclides (e.g. Lebecka *et al.*, 1994).

The discharge of acidic, pumped mine water into the Blesbokspruit wetland in South Africa has led to some controversy, even though the wetland owes its very existence to the mine water and other effluents (sewage, industrial) from mining settlements (Wood and Reddy, 1998).

In most cases of pollution by water actively pumped from deep mines in developed countries, treatment systems are used to mitigate the possible environmental damage (see Section 3.8.2).

Geotechnical impacts
The lowering of the water table caused by large-scale dewatering of mines has sometimes led to land subsidence. Most commonly this occurs in karst terrains where large voids may exist in the saturated zone, such that lowering of the water table withdraws buoyant support from roof spans which are unstable in the drained state. In such cases, the result of dewatering is collapse of the newly-drained voids, giving rise to void migration and the development of crown holes (roughly circular subsidence craters) at the ground surface. The most classic case of this type occurred in South Africa in the 1970s in the aftermath of the major inrush in 1968 at West Driefontein Gold Mine (see Section 3.6.1 and 3.7.5 above). To ensure the safety of future mining, a major project was initiated to entirely dewater the dolomite aquifer overlying the mine. This was probably the largest external dewatering scheme ever undertaken anywhere in the world, involving peak dewatering

rates of 340 Ml.d^{-1} in 1970 during the initial establishment of drawdown in the dolomite aquifer. The dewatering rate then gradually declined to a steady rate of about 75 Ml.d^{-1} by 1976. This dewatering resulted in the development of several hundred sinkholes at the ground surface, ranging from small depressions a few metres in diameter and depth to vast craters many tens of metres in diameter and depth (Forth, 1994). One of the latter variety suddenly engulfed an entire mineral beneficiation plant, with tragic consequences for the people working there.

Where mines dewater poorly consolidated strata containing soft silts and clays, land subsidence may also occur. However, in this case the settlement is usually more evenly spread over the area affected by drawdown, so that a general lowering of the ground surface is the result rather than catastrophic crown hole collapses. Domenico and Schwartz (1990) provide useful guidelines for predicting the likelihood, potential magnitude and timing of this type of subsidence as a function of the anticipated total drawdown. One important conclusion arising from the analyses they present is that a considerable lag time exists between drawdown and settlement, since the fine-grained beds primarily responsible for overall compaction are slow to drain. It is entirely feasible for dewatering to have been discontinued before settlement becomes apparent.

Unnerving as such occurrences undoubtedly are, it is necessary to remember that such extreme geotechnical impacts are relatively rare in mining world-wide, and the vast majority of mine dewatering systems will operate without ever experiencing such problems. Where mines are being developed in karst terrains or in thick sequences of unconsolidated sediment, it is prudent to assess the potential for subsidence due to large-scale dewatering.

3.8.2 MITIGATION

Assessment of problems
Before implementing a mitigation strategy, it will usually be necessary to implement a ground water/surface water monitoring system in the vicinity of the mine (e.g. Streetly, 1998). Ideally, this should be in place for several years before the commencement of dewatering. In some cases, it may prove necessary to construct numerical models to interpret and predict impacts, both to assess how reasonable claimed impacts are and/or to assist in designing adequate mitigation measures (e.g. Dottridge and Keeble, 1998). Full instructions on numerical modelling of ground water systems are given by Anderson and Woessner (1992).

Mitigating flow problems: compensation water
While dewatering removes water from areas intended for mining, it also necessitates disposal of the pumped water. In most cases, this will be into the nearest convenient surface water course. If this water course happens to have lost natural flow upstream due to the dewatering operation, then the disposal of pumped effluent will have the beneficial effect of providing "compensation flow" to make up for the water lost to the mine. It is important to note that relatively constant pumping effluent discharges to streams will result in the streams having very different hydrological regimes to those which would have obtained under natural conditions. In many cases, this will be of little consequence, but there are certain circumstances where the replacement of a formerly "flashy" runoff regime by a steady flow may be ecologically detrimental. While it is not possible to provide generic guidance for such cases, it may be necessary in some circumstances to provide storage

reservoirs for dewatering effluents, which can be released to the natural watercourse in a manner more compatible with local ecological requirements.

In some cases, the watercourses affected by drawdown may be inconveniently located with respect to the active workings, rendering direct compensation in this manner infeasible. In such cases, the environmental cost-benefit balance may favour the development of other sources of water external to the mine curtilage to support flows in impacted streams. Again, such interventions can only be designed on a case-by-case basis.

Mitigation of effects on ground water levels
The lowering of ground water levels around a mine subject to dewatering can impact other water users by lowering the water levels in their wells, thus incurring increased pumping costs or even drying the well out entirely in some cases. In practice, it can be virtually impossible to prove beyond all doubt whether an observed lowering of water level is attributable to dewatering, some other local pumping activity or even climate change (e.g. Dudgeon, 1997). Rather than waste money on fruitless litigation, most mining companies prefer to compensate claimants by supplying alternative sources of water, deepening wells or other activities (e.g. Wardrop *et al.*, 2001).

Ecological consequences of a lowering of the water table can be less easy to mitigate. In some cases, it may be necessary to install impermeable cut-off walls, re-inject water and/or create localised impoundments in appropriate parts of the mine void in order to ensure the water table in a sensitive adjoining area is sustained at a suitable level (e.g. Cliff and Smart, 1998). Successful examples of all three strategies have been recently described by Wardrop *et al.* (2001).

Mitigating water quality problems
The management of water quality in active mines is subject to a hierarchy of concern, ranging from limestone mines which may yield water of essentially potable quality to metal mines using xenobiotic chemicals (such as cyanide for gold beneficiation) in their processing operations. Obviously the amount of effort necessary to mitigate potential impacts varies dramatically between these two extremes. In the most extreme cases, it is increasingly common for mines to develop and implement a "zero discharge policy". In the most benign cases, dewatering effluents may be fed directly into public water supply reservoirs. In most intermediate cases, some form of water treatment will be necessary prior to disposal of the dewatering effluent.

Zero discharge policies effectively require that:

(i)   All external sources of water (surface and ground waters) be prevented from entering the mine by means of elaborate watercourse diversions and/or external dewatering
(ii) All water arising within the mine site (by direct rainfall into the void, or as process water deliberately introduced to facilitate working) is recycled and/or disposed of by means of evaporation.

It is immediately apparent that such approaches are likely to be expensive in terms of design, construction and operation. For instance, the relatively inexpensive option of diverting a stream by simply passing it through a culvert beneath overburden heaps is increasingly unlikely to be acceptable to environmental regulators, since it involves permanently darkening a reach of river which would previously have been receiving daylight. Alternative options will usually require much greater access to land and much greater construction costs. Similarly, if all water arising within the site in excess of the recyclable quantity has to be disposed of by evaporation, then large areas will need to be acquired and set aside for evaporation ponds. For instance, at the Inti Raymi gold mine (Bolivia), a water make of up to 27 Ml.d^{-1} is disposed of solely by evaporation in a series of lagoons and a vast tailings dam with a surface area of around 5 km^2 (Arze-Quintanilla, 1994). This approach is feasible at Inti Raymi because it lies at high altitude (3700 m) in a tropical zone of intense evaporation. At higher latitudes, evaporation may never be sufficient to cope with the amounts of water involved.

Because of the high capital and operating costs associated with them, zero discharge policies are only really feasible for high-revenue operations involving potentially hazardous minewaters and process waters. As such, they have a particular niche in the gold mining sector, in which ores are frequently associated with arsenopyrite and other pollutant-generating sulphide minerals, and in which the mineral processing activities can involve the use of cyanide (in heap-leach pads) or mercury compounds. Zero discharge gold mines are now operating in all continents; examples include the CVRD Mine in Caeté (Minas Gerais, Brasil); the Río Narcea mine (Asturias, northern Spain); and the Lake Cowal Mine (New South Wales, Australia; B.A. Dudgeon, 1997).

In the majority of cases, where less lucrative ores are mined, a zero discharge approach will not be economically feasible, and problematic dewatering effluents will have to be handled using water treatment technology. Two of the most common problems, in nearly all kinds of mine, are turbidity (due to entrainment of silt) and oil pollution (from vehicles and haulage machinery). Both problems are easily solved: sediments are easily trapped using settlement basins, which are typically designed to offer a minimum of 3 hours retention time for a flood predicted to have an annual probability of exceedance of about 0.01. Motor oils are readily retained in three-chamber oil separators. Where the mine effluent contains other contaminants, it may be necessary to apply the active or passive treatment techniques described in Chapters 4 and 5. For instance, the controversy over the Blesbokspruit wetland in South Africa has led to the installation of a high-density sludge treatment system (see Chapter Four) to treat the Grootvlei mine water prior to discharge into the wetland (Wood and Reddy, 1998). Using such treatment methods, dewatering effluents from active mines in most industrialised nations are generally treated to sufficiently high standards that river pollution is avoided. An exception to this relates to extremely saline waters pumped from some Polish and English coal mines (Lemon, 1991; Lebecka et al., 1994), treatment of which would require the use of desalination technology, which has to date been deemed prohibitively expensive in these areas.

3.9 HYDROLOGICAL BEHAVIOUR OF WASTE ROCK PILES, BACKFILL AND TAILINGS DAMS

3.9.1 WASTE ROCK PILES (SPOIL HEAPS) AND SURFACE MINE BACKFILL

Issues in waste rock reclamation
The reclamation of abandoned waste rock piles, including the back-filling and restoration of surface mines, is one of the most important environmental management activities

associated with mining. As was noted in Section 1.3.7, 70% of all material mined world-wide is waste rock, and of this nearly all is excavated by surface mines As surface mining has expanded dramatically since the middle of the 20th Century, so have problems of mine waste management become ever more pressing. Research and development in relation to environmental aspects of these problems have focused on two main areas:

- the requirements for lasting re-vegetation of spoil, and
- the prevention or mitigation of the release of polluted leachates from spoil materials

Where the focus has been on establishment of vegetation on the final surface of compacted, re-contoured spoil, most research has concerned the establishment of viable soil cover, be this previously-stripped top soil which has been spread over the waste rock, or else soil formed out of the soil-forming materials (SFMs) identified within the waste rock itself. Most work has been undertaken on SFMs within coal-bearing sequences of Carboniferous age, especially the low sulphur mudstones. A useful review of soil-related work on coal mine spoils has been published by Rimmer and Younger (1997). However, the successful establishment and maintenance of plant-bearing soil presupposes the physical stability of the underlying ground, and numerous experiences have shown that settlement and intense erosion by piping processes can continue to affect the interior of a waste rock pile for years or decades after initial restoration (Groenewold and Rehm, 1982). Thus it has been rightly claimed that "for reclamation to be successful, the entire landscape must be considered, not merely the soil zone. Any approach less than this will lead, at best, to temporary solutions" (Groenewold and Rehm, 1982). Notwithstanding this exhortation, holistic geomorphological and hydrological studies of the physical evolution of old waste rock piles and backfill are still relatively rare.

Far more common are studies of the geochemistry of waste rock, focusing on the potential for acidity generation and its mitigation. Because they are generally in far more intimate contact with the atmosphere than *in situ* rock, waste rock piles are frequently far more vigorous generators of acidic water than flooded underground workings. In response to problems of acid drainage, a number of major national research programmes have been established during the last two decades, most notably in Australia (Greenhill, 2000), Canada (Tremblay, 2000), Sweden (Höglund, 2000) and the USA (Shea, 2000). The resulting hydrogeochemical literature on waste rock piles and surface mine backfill is now substantial, but as the major findings of this literature have already been discussed in Chapter Two there is no need to repeat them here. As in the case of re-vegetation research, however, studies of the physical hydrology of mine wastes are still relatively rare.

Nevertheless, some major advances in our understanding of the hydrology of mine spoil have been made in recent years. Three key insights have emerged from this work:

(i)   Spoil materials frequently host subsurface flow systems of their own, functioning effectively as "perched aquifers" above the underlying bedrock

(ii) The heterogeneous nature of spoil gives rise to highly preferential subsurface flowpaths

(iii) The physical instability of spoil is reflected in relatively rapid changes in surface and subsurface hydrological pathways

Water tables in spoil and surface runoff from spoil
The existence of discrete water table aquifers within spoil heaps is commonly observed. Figure 3.31 shows the classic evidence for a water table within spoil, in the form of a linear seepage face along the front of the heap, resulting in vegetation die-back on the otherwise re-vegetated spoil. Water table contours constructed from borehole records within a spoil heap are shown in Figure 3.32a, along with a hydrograph for one of the boreholes (Figure 3.32b). Two aspects of ground water behaviour within the heap are apparent in Figure 3.32: seasonal changes in flow direction, and rapid fluctuations in water table elevation in response to rainfall, even in summer periods when recharge to natural aquifers in this region is usually minimal. The former reflects the influence of preferential flowpaths which become active at different water table elevations, while the latter is most marked in spoil heaps which have little or no soil cover, so that the attenuation of rainfall by evapotrans-piration is minimal.

The presence of a perched water table within spoil is of considerable significance for pollution management purposes, since it provides the basis for perennial discharges of water from the spoil. Where no water table develops, discharge will only occur by surface runoff after storms (e.g. Bayless and Olyphant, 1993). The likelihood of a given body of spoil developing its own water table system depends on the predominant lithology of the spoil. For instance in England, the predominant waste rock lithology arising from coal mining varies systematically from sandstone in the north (Northumberland Coalfield), forming permeable spoils, to mudstone in the south (East Midlands Coalfield), forming low-permeability spoils. This systematic variation is reflected strongly in spoil heap hydrology, with most of the larger (> 1 ha) spoil heaps in the north hosting perched water tables, while even the largest spoil heaps in the south (> 50 ha) predominantly give rise to surface runoff.

Spoils which are rich in pyrite commonly give rise to acidic drainage. Fortunately, acid-generating spoil heaps do not account for the majority of cases. For instance, available data suggest that less than 40% of Carboniferous coal mine spoils in Europe will ever be acid generating. Where the spoil is acid generating, temporal patterns of pollutant release are largely governed by the hydrological behaviour of the spoil (Figure 3.33). Where a perched water table within the spoil mediates most of the drainage, and surface runoff is restricted to stormy periods, the tendency is for acidity concentrations to be highest in dry spells, when there is little rain water available to dilute the steadily discharging ground water (Figure 3.33a). Although a rise in the water table within the spoil during recharge events will normally be accompanied by a flushing of further acid salts into solution, the resultant acidic water will not normally reach the discharge point from the heap for some days or weeks, normally during a dry period. By contrast, where surface runoff is the dominant (or sole) drainage mechanism, peak acidity concentrations usually coincide with (or occur slightly before) peak flows (Figure 3.33b). This is because evaporation from the spoil surface between successive rain storms leads to the accumulation of surficial crusts of white and yellow iron and aluminium sulphate and hydroxy-sulphate salts (see Table 2.6), which then rapidly dissolve to release acidity when the rains arrive (Bayless and Olyphant, 1993). The coincidence of peak flows and peak contaminant concentrations is a "nightmare scenario" as it results in extreme loadings (flow x concentration) which are difficult to treat effectively.

Figure 3.31: Panoramic view of a reclaimed coal mine spoil heap in South Africa, with the presence of a perched ground water system within the spoil clearly betrayed by the horizontal line denoting the top of the seepage face (unvegetated zone) on the heap. (Photo: P. Younger).

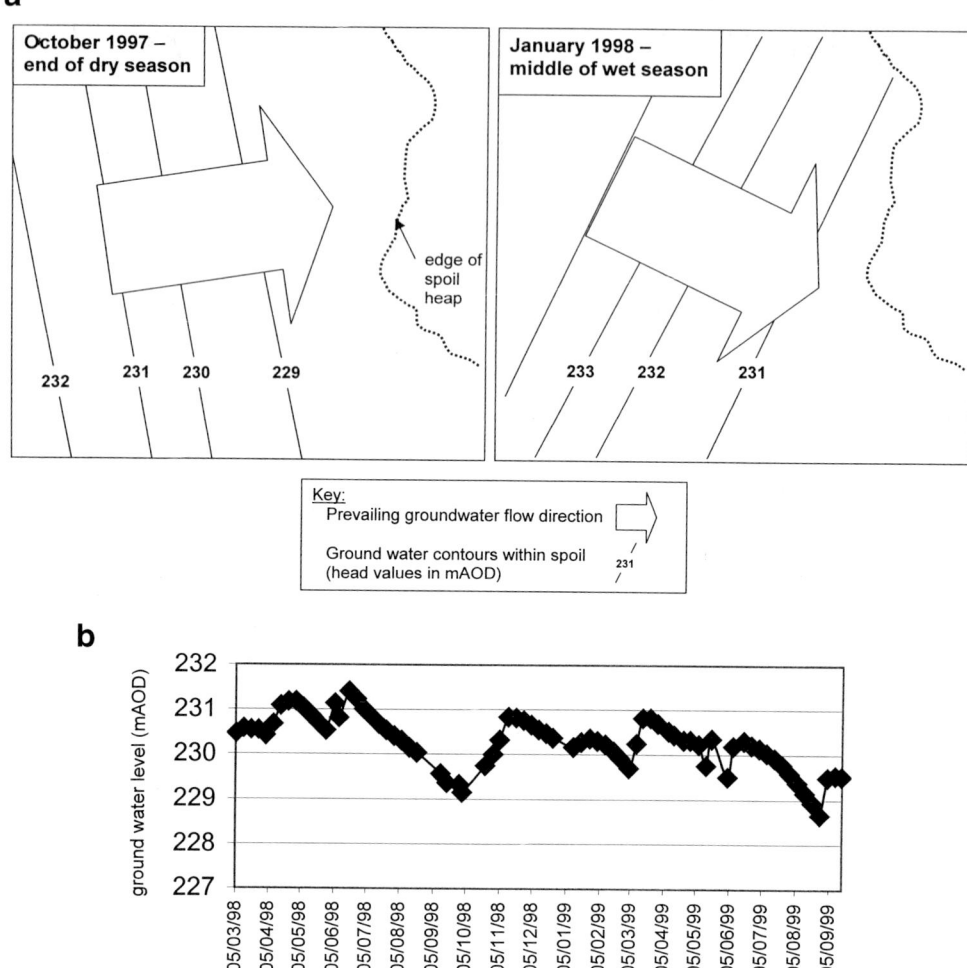

Figure 3.32: Aspects of the behaviour of an aquifer system 'perched' within a waste rock pile (spoil heap), Morrison Busty Colliery, Co Durham (UK). (a) Water table contours in two different seasons, showing a change in the prevailing ground water flow direction as the head rises in response to high infiltration rates in winter. (b) A hydrograph for one of the boreholes in the spoil heap, showing peaks in the water level occurring at all times of year. This is in distinction to natural aquifers in the same region, which tend to peak only one time per year (in late winter), and reflects the relative ease in which rainfall can infiltrate bare areas of spoil all year round.

Heterogeneity and preferential flow

The heterogeneity of spoil is a direct consequence of the processes of waste rock disposal, which were outlined in Section 1.3.7 (see also Figure 1.30). In the broadest terms, the heterogeneity in sediment fabrics resulting from end-tipping and dragline spreading is reflected in widely ranging values of saturated hydraulic conductivity, as measured by

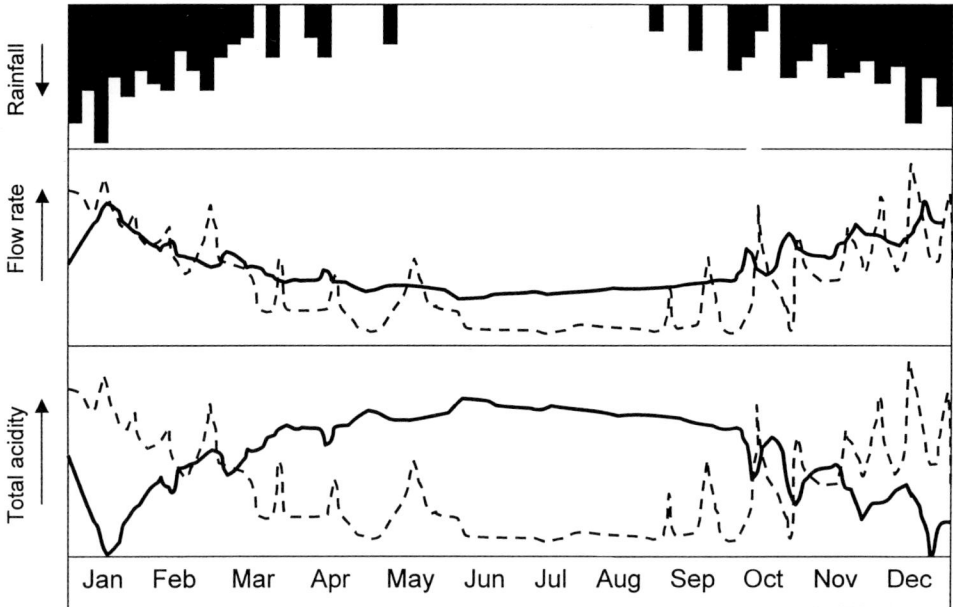

Figure 3.33: Schematic diagram of an annual cycle of rainfall, flow rates and acidity release for two spoil heaps, illustrating the contrasting acidity release regimes for waste rock piles dominated by surface runoff (dashed lines) and those in which much of the drainage is via a ground water system within the pile (solid lines). Peak acidity tends to coincide with peak flow for the surface runoff dominated system (dashed lines), whereas peak acidity coincides with minimum flow in the ground water mediated spoil drainage system (solid lines).

borehole slug/bail or pumping tests within spoil. Table 3.7 summarises hydraulic conductivity data for spoil materials from studies around the world. It is clear from these data that spoil hydraulic conductivities have been found to vary over eight orders of magnitude. This is a far wider range than is encountered in the unmined parent rocks (e.g. National Research Council, 1981; Fetter, 1994).

At first glance, such a wide range of hydraulic conductivities might seem like a basis for despair in the analysis of flow through spoil heaps. However, the extreme values tend to cluster according to the history of spoil emplacement (Groenewold and Rehm, 1982): higher values of hydraulic conductivity (> 0.1 m.d^{-1}) tend to correspond to rubbly zones (i.e. areas which were "valleys" in the spoil mass at the time of tipping), whereas lower values ($< 1 \times 10^{-2}$ m.d^{-1}) are more typical of what were originally "ridges" within the spoil, which were naturally subject to much compaction as draglines and other heavy machinery repeatedly tracked across them. Thus heterogeneity of hydraulic conductivity in spoil heaps is often rather structured (Newman et al., 1997), which results in subsurface flow within them occurring predominantly by preferential pathways.

It is crucial to note that the predominant pathways differ markedly between the unsaturated and saturated zones of the spoil. Since unsaturated hydraulic conductivity is a function of moisture content (e.g. Freeze and Cherry, 1979) the hydraulic conductivities

Table 3.7: Hydraulic conductivities (K) of mine spoils from various countries. Data sources: National Research Council (1981), Groenewold and Rehm (1982), Diodato and Parizek (1994), Aljoe (1994a), Newman et al. (1997) and Skarzynska and Michalski (1999)

Country	Origin of spoil	Reported K values (m.d^{-1})
Czech Republic	Deep coal mine spoils	1×10^{-2} to 90
Germany	Deep coal mine spoils	1×10^{-3} to 90
Kentucky, USA	Deep coal mine spoil	2
Montana, USA	Gold mine	3 to 90
North Dakota, USA	Surface coal mines	4×10^{-4} to 4
Pennsylvania, USA	Surface coal mine backfill	1×10^{-3} to 10
Poland	Deep coal mine spoils	1×10^{-3} to 90
UK	Deep coal mine spoils	1×10^{-6} to 10

of the moisture-retaining fine-grained layers will often be higher than those of the well-drained coarse layers above the water table. This means that *the fine-grained spoil layers are the preferential flowpaths in the unsaturated zones of spoil heaps* (e.g. Diodato and Parizek, 1994; Newman et al., 1997). For this reason, detailed numerical simulations of unsaturated flow through waste rock piles have often had to incorporate preferential flow pathways to accurately reproduce observed drainage and solute transport behaviour (e.g. Strömberg and Banwart, 1994; Eriksson and Destouni, 1997; Erikkson et al., 1997). In a comparison of model simulations of lab and field tracer experiments with saturated and unsaturated waste rock, Eriksson et al. (1997) found that preferential flow pathways accounted for between 55% and 70% of the total water content of the spoil, with the remaining water moving only slowly. A further practical conclusion of potentially wide application arose from detailed modelling of flow in waste rock at a gold mine in Montana (Wilson et al., 2000): preferential flow through the fine-grained layers in the unsaturated zone is predicted to occur whenever the recharge rate to the spoil (expressed in the dimensions L.T^{-1}) is numerically less than the saturated hydraulic conductivity (L.T^{-1}) of the fine-grained spoil.

By contrast, below the water table, the saturated hydraulic conductivity of the coarse, rubble zones of the former spoil "valleys" is usually several orders of magnitude greater than that of the fine-grained layers, so that *the coarse layers are the preferential flowpaths in the saturated zones of spoil heaps* (e.g. Aljoe, 1994a; Aljoe and Hawkins, 1994). This latter observation explains why the hydraulic conductivities needed to satisfactorily model flow beneath the water table in spoil heaps tend to be towards the higher end of the overall range. For instance, in simulating the effects of large bodies of surface mine backfill on regional ground water flow, Schwartz and Crowe (1985) found it appropriate to assign K values in the range 9×10^{-2} to 90 m.d^{-1} to the backfill. When simulating flow within a single spoil heap, it has sometimes been found necessary to locally assign hydraulic conductivities as high as 1000 m.d^{-1} in order to satisfactorily reproduce observed flows and water levels (Hawkins, 1994). In such settings, Aljoe (1994a) noted that the spoil aquifer operates like a "pseudokarst" system, with rapid transmission of water in a few erosional soil pipes (the formation of which is described by Groenewold and Rehm, 1982), draining water from the less permeable spoil mass which encloses them. As test pumping wells nearly always intercept the spoil mass rather than erosional pipes, borehole values for K will usually not be high enough to account for the flow rates occurring in the soil pipes. Insistence on using only field-measured values of K can lead to underestimates of spoil drainage rates by as much as 80% (Aljoe, 1994a).

The contrasting modes of preferential flow above and below the water table have profound consequences for pollutant release and transport, since much of the most reactive pyrite is found in the fine-grained spoil, while coarse-grained fractions are often poor in buffering minerals. Hence acid generation is favoured above the water table, and its rapid transport with only limited buffering is favoured below the water table.

Instability of spoil

It was pointed out in Section 1.3.7 and Figure 1.3.1 that settlement of backfill is the norm for a decade or so after infilling of a surface mine void. The process of settlement is due to closer packing of the grains in response to self-loading by the sediment body. Closer grain packing means tighter pore necks and therefore a decrease in hydraulic conductivity. Disintegration of mudstone clasts in the presence of fresh water also leads to the release of fine-grained sediment, which can block pores and therefore also decrease the overall hydraulic conductivity. The net result of these processes has been investigated by several researchers. For instance, Groenewold and Rehm (1982) undertook borehole perme-ability tests in a suite of boreholes drilled into freshly re-contoured surface coal mine spoils in North Dakota, and then repeated the measurements one year later after settlement had occurred. In nearly all of the boreholes, hydraulic conductivity dropped by an order of magnitude over the year. From analysis of data obtained in various European countries, Skarzynska and Michalski (1999) reported that decreases of two orders of magnitude are by no means uncommon in coal mine spoils. On the other hand, where rubbly zones impart a higher initial hydraulic conductivity to the spoil, water can flow so rapidly that it entrains silt and even sand grains, leading to the development of erosional pipes. These are the very features which impart "pseudokarstic" hydraulic properties to some spoils (e.g. Aljoe, 1994a; Hawkins, 1994). When such pipes collapse, they form surface depressions which serve to intercept surface runoff on top of the spoil heap, thus diverting ever more of the effective rainfall into the subsurface (Groenewold and Rehm, 1982).

Where spoils are sulphur-rich, pyrite weathering releases lots of iron into infiltrating waters. The dissolved iron commonly re-precipitates along the interface between nearly-saturated fine-grained spoil and still-aerated coarse grained spoil, forming layers of ferric hydroxide which are commonly termed "hard pans". Besides providing a sink for iron and other metals within the waste rock pile, these hard pans frequently clog the pores, reduc-ing permeability. Eventually, the hard pans become so lowly permeable that they begin to function as aquitard horizons, impeding further unsaturated flow downwards, and promoting locally saturated flow conditions above them, leading to lateral flow of perched groundwater towards the margins of the waste rock pile, where the positions of the hard pans are frequently marked by spring lines and seepage faces.

The disintegration of mudstone clasts on the surface of the spoil yields large quantities of mud in a matter of months, imparting a very low permeability to the spoil surface (Bell, 1996). Where hydraulic conductivities of surface materials drop to 10^{-4} m.d^{-1} or less, infiltration is likely to be minimal, and the hydrological response of the spoil heap will be dominated by surface runoff. Given the steepness and lack of cementation of many spoil heap flanks, this runoff can rapidly lead to surface erosion, producing a characteristic "badlands" meso-scale topography. Such runoff and erosion can result in high flows and excessive sediment loadings in receiving watercourses, with their attendant problems of flooding and/or siltation.

Remedial action
The mitigation of drainage and water pollution problems arising from mine wastes requires either:

- preventative action (such as capping the spoil to minimise the ingress of those principal agents of pyrite oxidation, water and oxygen; see Section 3.12 below), or
- treatment of the polluted spoil drainage (using the techniques described in Chapters Four and Five)

As full details of such actions are discussed elsewhere in this book, no further elaboration is needed here.

3.9.2 TAILINGS DAMS

Water is an integral part of a tailings dam
The construction of tailings dams is described in outline in Section 1.3.7, where it is noted that they were originally introduced for purposes of water pollution control. Although the original motivation for the introduction of tailings dams appears to have been the prevention of excessive siltation of navigational watercourses which were needed for shipment of mine produce, they are now most commonly used to prevent the release of silt (generated by mineral processing) which can cause ecological damage by blanketing streambeds.

To ensure the attainment of these objectives, it is necessary to consider carefully the management of water in tailings dams (Vick, 1983). There are three main considerations:

(i)   the efficient settling of fine tailings from suspension
(ii)   the control of seepage to ensure dam stability
(iii) the prevention of flood erosion

The settlement of fine tailings from suspension is a multi-variable problem. Important factors to be considered in a site-specific design include the granulometry and relative density of the tailings particles (see Vick, 1983, for guidance) and the velocity of the water in which the tailings are suspended. As in all hydraulic settings, the latter can be minimised by maximising the volume through which flow takes place. In practical terms this would generally mean increasing the depth of standing water in a tailings dam. While this may well be feasible in a dam constructed using the downstream method (Figure 1.33), structural stability considerations usually place severe limitations on the depth of open water permissible in a dam constructed using the centre-line or upstream methods (Vick, 1983). An alternative approach to maximisation of tailings settlement is to configure the dam such that there is no surface decant and the only route for drainage to take is via seepage through the bed of the impoundment. In this case, the mass of the dam will act like a slow sand filter. While this approach is virtually guaranteed to remove all of the tailings from the water, it has important implications for dam stability (due to the high pore water pressures involved), and can also lead to a considerable increase in the dissolved load of the seepage waters, since they continually come into contact with the entire contents of the dam. For these reasons seepage-only dams are little used in practice.

Where fine tailings remain in suspension despite all reasonable steps having been taken to control flow velocities, it may be necessary to resort to the use of coagulants and flocculants (see Chapter 4) to achieve compliance with effluent standards.

The observations made in Section 3.6.5 concerning the effects of pore water pressures on slope stability in surface mines are equally applicable to tailings dams. Consideration of Table 3.5 suggests that tailings dams constructed without regard for the control of pore water pressures within them may need to have slopes as gentle as 40° to be unconditionally stable. This implies very wide land-takes for dam construction. Hence it is usually cost-effective to incorporate seepage control measures into tailings dams during construction. In practice, these measures are much easier to incorporate into downstream dams than into upstream dams, thus providing another reason for preferring downstream construction wherever possible. Figure 3.34 illustrates the principles involved, by contrasting the water table profiles in dams with and without seepage control drainage layers. Where no drainage layer is present (Figure 3.34a) the water table daylights as a seepage face high on the downstream face of the dam. This has the double disadvantage of decreasing the shear strength of the dam material (thus rendering it more prone to slope failure) and giving rise to runoff which can erode the embankment. Where a drainage layer is incorporated (Figure 3.34b), most of the dam remains unsaturated (and therefore of higher shear strength), and all of the seepage leaves the dam via a toe drain, where it can be readily collected in channels and conducted away from the embankment to prevent erosion. The design of drainage layers for seepage control is beyond the scope of this book; the interested reader is referred to Vick (1983) and Klohn (1979).

Flood erosion is a problem common to all types of surface water impoundments. Consequently, a vast literature exists which describes the design and construction of spillways, bypass channels and other relevant features (see, for instance, Chanson, 1999). Of particular importance in the case of tailings dams is the prevention of overtopping, since overland flow across unconsolidated tailings comprising the dam face can be expected to cause rapid down-cutting, leading to drainage of liquefied sediments from the tailings impoundment. Overtopping is a particular risk in valley dams which have the potential to receive runoff from the surrounding land. Normally, strenuous efforts are made to divert external surface waters away from the dam, using channels which are suitably designed with erosion-resistant armour layers or liners. However, as Smith and Connell (1979) have noted, "since a single storm rarely causes major damage, the problem is generally considered to be one of maintenance; however, if maintenance is neglected, the cumulative effects of intermittent erosion can produce a failure". Hence it is essential that flood diversion channels are regularly inspected and repaired as necessary.

Tailings dams as pollutant sources
Although constructed for the best of reasons, tailings dams have too often ended up functioning as sources of aquatic pollution in their own right, in two principal ways:

(i)   through inadequate operation leading to carry-over of fine tailings and/or dissolved pollutants into receiving watercourses (e.g. Ellis and Robertson, 1999), and

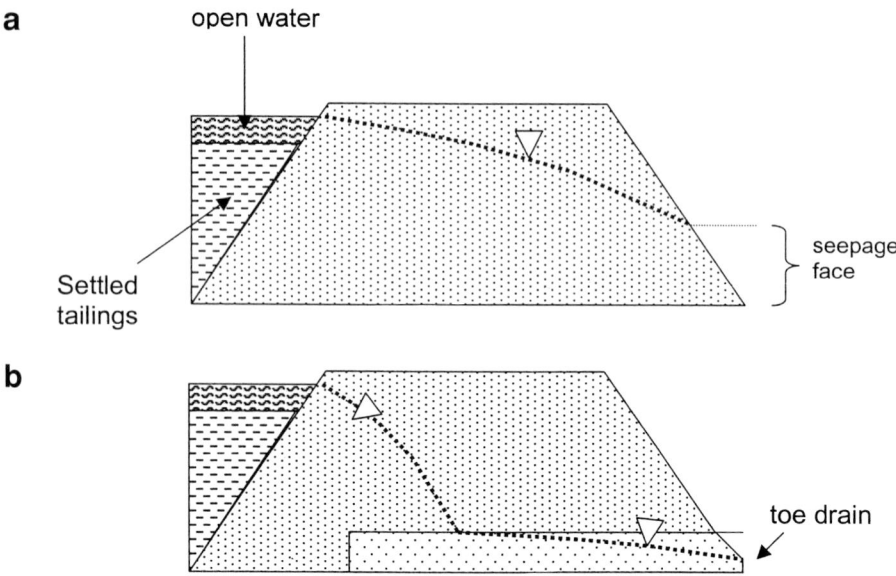

Figure 3.34: Schematic cross-sections illustrating the contrast in water table profiles in tailings dams (a) without and (b) with a seepage control drainage layer (incorporated at the time of construction). The high seepage face in case (a) represents a potential raising of slope failure risk when compared with (b).

(ii)   through dam failure leading to the catastrophic release of mudflows of liquefied mud (e.g. Blight, 1997), which frequently contains high concentrations of toxic metals (e.g. UNEP, 1996). (Major examples of this genre are listed in Table 1.1).

While it is certainly possible to reduce loadings of dissolved metals in tailings dam decant liquors by means of active chemical treatment (see Chapter Four), it is more sensible and cost-effective to minimise the oxidation of sulphides within the tailings in the first place. Traditionally, tailings were often deposited sub-aerially, and then later flooded (Vick, 1983). This approach allows sulphides in the tailings to oxidise until flooding occurs, at which point the oxidation products (i.e. the compounds listed in Table 2.6) rapidly dissolve, releasing toxic metals into solution. This problem can be prevented by ensuring that tailings are kept submerged throughout the disposal process, and subaqueously deposited into the dam (e.g. Ellis and Robertson, 1999). Such underwater tailings disposal has been mandatory in Norway since 1968, and accordingly a wealth of experience exists from which its efficacy can be gauged (Arnesen *et al.*, 1997). Analysis of data from three long-established Norwegian tailings dams confirms that underwater emplacement of tailings leads to far less contamination of receiving watercourses than subaerial emplacement. In particular, copper concentrations in the dam effluents are invariably low (always < 1 mg.l^{-1}, and usually < 100 μg.l^{-1}). Nevertheless, environmentally significant quantities of zinc (≤ 2.5 mg.l^{-1}) are still released, along with sufficient sulphate to demonstrate that sulphide oxidation is still occurring to some degree (Arnesen *et al.*, 1997). Although not suggested by Arnesen *et al.* (1997), the release of zinc from flooded tailings can probably be explained by subaqueous oxidation of sphalerite (ZnS) by ferric iron, which is commonly present in tailings:

$$ZnS + 3Fe^{3+} \rightarrow Zn^{2+} + 2Fe^{2+} + S^{o} \tag{3.45}$$

The elemental sulphur (S^{o}) released by this reaction will subsequently oxidise to sulphate when the water is aerated. Reaction (3.45) has been invoked to explain consumption of ferric iron in a Zn-rich orebody (Younger, 2000b). Details of the reaction mechanisms and kinetics have been thoroughly investigated by Boon *et al.* (1998).

As was noted in Section 1.3.7, catastrophic failures of tailings dams are almost never due to inadequate initial design. Rather, they tend to arise from a failure to fully implement the design throughout the period of construction. Thus encouragement to follow the existing excellent design guidelines for tailings dams (Vick, 1983) will not on its own ensure that the reader will never be confronted with a tailings dam failure problem. Rather, the avoidance of dam failures is a managerial issue, which requires the implementation of strict QA/QC in relation to tailings dam operation and maintenance. At a minimum this demands that the systems of regular independent inspections which are long-established in north America and northwest Europe be implemented world-wide.

After-care of redundant tailings dams
The decommissioning of tailings dams after the cessation of mineral processing is one of the most challenging issues for long-term management of former abandoned mine sites. For instance, while the benefits of underwater disposal of tailings are clear (Ellis and Robertson, 1999), the long-term maintenance of flooded conditions is no trivial task. While the cessation of mineral processing usually results in a drastic reduction in the water supply to the dam, seepage and evapotranspiration continue indefinitely, so that the tailings dam is often left in a drained, dried-out state. Under such conditions, the sulphides in the tailings will oxidise, and elevated acidity and metals loadings may be released to the surrounding environment in the form of leachate and/or storm runoff. Possible preventative measures include the continuous 'impermeabilisation' of the tailings during disposal, by repeated grouting of the uppermost surface of the tailings (e.g. Zou and Huang, 1999). Where tailings have already been deposited without grouting, it is possible to retroactively incorporate cement by means of reworking. Indeed layers of cemented tailings may themselves be used to reduce the overall permeability of waste rock piles as a means of minimising acid drainage problems (Poulin *et al.*, 1996). However, despite being shown to be feasible in laboratory experiments, the high costs of thoroughly grouting problematic tailings are likely to preclude the widespread adoption of the approach at full-scale. Furthermore, the vast majority of abandoned tailings dams were created without the application of any such strategy. In most cases, therefore, work will be required post-abandonment to prevent sulphide oxidation and/or leaching. The principal methods for achieving this are identical to those for waste rock piles and backfill, and as such are discussed in Section 3.12 below.

3.10 MINE ABANDONMENT AND REBOUND PROCESSES

3.10.1 UNDERSTANDING MINE WATER REBOUND: PONDS AND HEAD-DEPENDENT INFLOWS

The rebound phenomenon

> " ...Slyly at first, almost secretly, the water gathered itself, then impetuously broke through; a dark treacly, chemical-smelling slow-moving mass that lashed itself into a fury the nearer it got to the shaft, like some caged beast seeking release ... 'Man it stank, and it licked at our heels like a live thing' [recalled one of the miners who escaped ahead of the water]. As the cage lifted they heard it follow with a snarl, and a blow that collapsed the shaft entry ..."

(Sid Chaplin "In Blackberry Time", © Bloodaxe Books 1987, used with permission).

This graphic account of the flooding of underground coal workings following the suspension of dewatering in an adjoining mine encapsulates some of the key issues relating to the hydrology of recently-abandoned voids, such as:

- the movement of water in open roadways by turbulent flow
- the erosion of the mine voids by the flowing water
- the deterioration in water quality which often accompanies the process of mine flooding.

In this section, we will consider all of these issues and more besides, paving the way for a discussion of the long-term behaviour of already-flooded mine voids in Section 3.11.

When mine dewatering ceases, the water make of the mine does not. Hence the mine voids gradually fill with water, in a flooding process which has come to be termed "rebound"[1]. Rebound will continue until such time as the water levels in the mine voids arrive at one or more "decant points". As the mine water decants (i.e. overflows), into an adjoining aquifer and/or into a surface water body, water levels in the mine voids stabilise, and will thereafter tend to remain within a narrow range (responding to seasonal fluctuations in water make).

Predicting the details of rebound in a given situation is a complex task (see Section 3.10.2 below). Before it can even be considered, it is necessary to develop a conceptual understanding of processes of water movement and the associated hydrochemical changes during the process of rebound. First, it is important to note that there are critical differences between the mode of hydrogeology encountered in working mines (described in Sections 3.6.3 through 3.6.5) and the hydrogeological behaviour of mined systems during rebound. These differences are largely attributable to one fact: in working mines, the huge voids comprising the mine itself are kept dry by strategic pumping, whereas during rebound, the mine voids themselves become the principal conduits for water movement.

---

[1]The term is used in a variety of phrases such as "groundwater rebound" (Norton, 1983; Robins, 1990; Sherwood and Younger, 1994), "water table rebound" (Henton, 1979, 1981), and "mine water rebound". Less popular synonyms include "mine water recovery", "water level recovery", "mine water resurgence", "flooding of voids", "flooding-up", "drowning out" etc.

The hydrogeologist who wishes to predict mine water rebound must account for flows *within* the mined voids, as well as flows in the strata beyond the voids. During rebound, when hydraulic head gradients are usually high, flow through the large, open mine voids, can be confidently predicted to be turbulent. Using the analysis of the hydrogeological behaviour of karst conduits developed by Smith *et al.* (1976), it is clear that flow in most open mine voids is likely to be turbulent whenever flow velocities exceed about one millimetre per second. This has at least two important consequences:

(a) Turbulence favours erosion of the mine voids. This can have negative consequences, such as the collapse of previously stable roadways. On the other hand, erosion by rapidly flowing mine waters above the water table has been observed to prevent the blockage of open roadways which would otherwise have been sealed by floor heave processes (P Aldous, personal communication, 1995). Erosion typically results in the mine waters carrying a high suspended solids load. An example of this phenomenon is reported from Scotland by Younger and LaPierre (2000), in which rebound in one colliery led to distinct peaks in suspended solids in the water drained from two adjoining mines, as the water decanted into each of them in turn from the rebounding mine

(b) The occurrence of turbulent flow violates the assumption of laminar flow upon which all major groundwater flow modelling packages are based (e.g. Anderson and Woessner, 1992). Turbulent flows cannot be simulated realistically using such packages, necessitating the use of alternative model formulations (see Section 3.10.2 below).

The concept of ponds

During mining, the contrast in permeability between intact strata and near-void strata affected by subsidence is a matter of major interest. The contrast in that case is generally about three to four orders of magnitude in terms of hydraulic conductivity (e.g. compare Zone 2 with Zone 1 in Figure 3.17). The contrast between near-void strata and the voids themselves spans no fewer orders of magnitude, albeit open voids are so "permeable" they fall outwith the bounds of Darcian classification (Younger and Adams, 1999). If we go on to contrast intact strata with open mine voids, the contrast in terms of hydraulic conductivity is in excess of seven orders of magnitude.

Recognition of this marked contrast is nothing new. Indeed, it has long been implicit in the actions of surveyors and engineers in the coal industry, particularly in the manner in which they conceptualise volumes of interconnected workings as 'ponds'[2] separated by barriers of unworked coal (Minett, 1987). Inherent in the definition of a pond is the concept that the mine-workings within any one pond are extensively inter-connected (often on multiple levels, if mining was undertaken on more than one horizon) so that water rising within any one pond will display a common level throughout that pond. At certain elevations, adjoining ponds may be connected via discrete decant features. Typical decant features include:

---

[2]The term 'pond' can be replaced by regional variants: e.g. a 'mine pool' in the USA, a 'basin' in Scotland and a 'pound' in Staffordshire, England.

- roadways inter-connecting areas of otherwise discrete workings
- areas in which two adjoining goafed panels coalesce
- old exploration boreholes
- permeable geological features (e.g. the margins of a basaltic dyke, a limestone bed, or an open fault plane).

Systems of ponds can be defined at a variety of resolutions. At the coarsest scale, regional ponds can be identified in major coalfields. For instance, Figure 3.35 shows the distribution of five major ponds in a single Scottish coalfield. Each of the major ponds encompasses the workings of numerous collieries, but each corresponds more-or-less to the "take" of the last major colliery to be worked in its vicinity. This compartmentalisation of coalfields has its origins in the system of private ownership of mineral rights and/or mining leases, which dominated mining economics until the middle of the 20th Century. Further examples of ponds, defined over scales ranging from a few km^2 to a few hundred m^2, are documented from numerous mining districts, for instance in the UK (e.g. Minett, 1987; Robins and Younger, 1996; Younger and Adams, 1999), the USA (e.g. Aljoe, 1994b), Germany (e.g. Coldewey and Semrau (1994) and Poland (e.g. Rogoz, 1994).

  Implicit in the recognition of ponds is an expectation that the process of rebound will occur independently in two (or more) adjoining ponds until such time as the water level in one or more of the ponds reaches an inter-connecting decant feature (Figure 3.36). Inter-pond transfers of water will then occur until the difference in head between the two ponds either side of each decant point is minimised. If the decant is "unrestricted" (i.e. offers very little resistance to inter-pond flow, e.g. a large-diameter roadway), then the ultimate head difference between the adjoining ponds will tend to zero. Where the decant is "throttled" (i.e. *does* resist flow to some degree, which might be the case if the feature in question is a body of goaf, a narrow borehole, or a natural geological feature), the ultimate head difference between the two ponds may amount to a difference of several metres. Differentiation between unrestricted and throttled decants is possible where a record of seasonal water levels in the pond with higher water level is available. A pond with an unrestricted decant will transfer newly-recharged water to the adjoining pond so rapidly that water level fluctuations will be minimal ($\ll 0.5$ m), whereas seasonal water level fluctuations in a pond with a throttled overflow may be quite marked (1–20 m).

Controls on the rate of rebound
The rate of water level rise in a given pond is a function of two factors:

(i)   The distribution of available water storage volume within the pond and
(ii) The rate of water inflow to the pond.

In most cases, the available water storage volume will equate to the sum of the volume of mine voids and the porosity of the enclosing rock mass (provided the latter was actually drained during mining). For surface mines, the volume of mine void space is easy to define. For deep mines, it is somewhat more complicated, due to the instability of deep mine voids. This can lead to void migration, as roof collapse propagates open space into the overlying strata. As implied in Figure 3.17, for typical longwall panel widths in the range 100 m to 250 m, void space can be expected to be enhanced over an elevation of 40–100 m

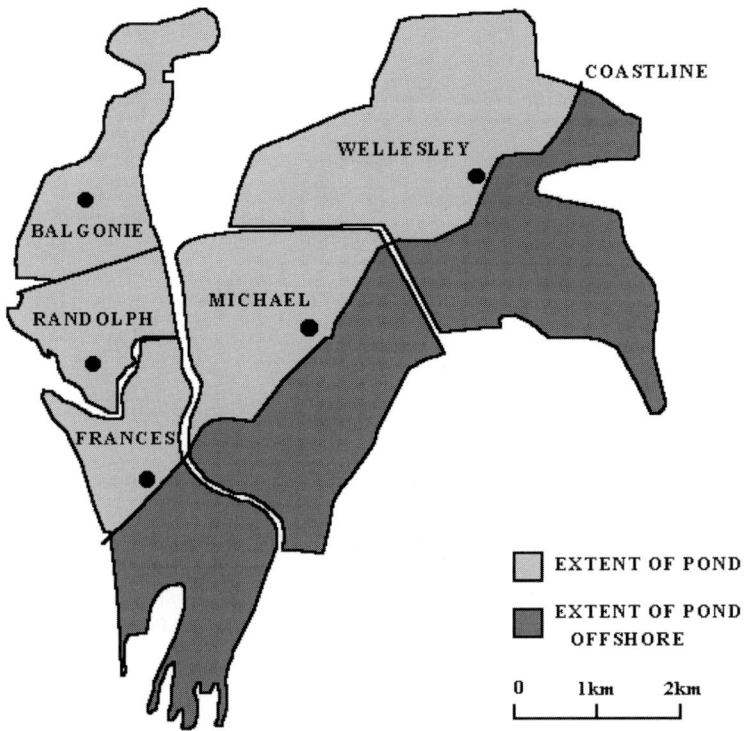

Figure 3.35: A simplified map of the Dysert-Leven Coalfield, Fife, Scotland, showing the system of five ponds identified on the basis of mine plans and water level data. For further details of this system, see Younger _et al._ (1995) and Sherwood (1997).

above each worked seam. Strata lying even higher than this will tend to have naturally low void space. In multi-layered systems of workings, this results in 'staircase' rebound curves, in which water levels rise relatively slowly during the flooding of worked horizons and associated disturbed roof strata, but relatively rapidly as inter-seam intervals of intact strata are traversed. This can be appreciated by application of the storage change calculations introduced in Section 3.3.2, using values of around 0.05 for the worked seams and disturbed strata, but values as low as 0.001 for the inter-seam intact strata. Figure 3.37 gives a good example of a real 'staircase' rebound plot from a UK coalfield, on which flattened segments of the curve correspond to the periods during which the specified seams and their disturbed roof strata were flooding.

The overall rate of water inflow to a given pond can be conceived as the sum of two categories of inflow:

(i)   head-dependent inflows, and
(ii)   head-independent inflows

Head-dependent inflows account for most of the water make in the majority of sub-water table mines. They represent inflows from adjoining aquifers, be these above, beside or

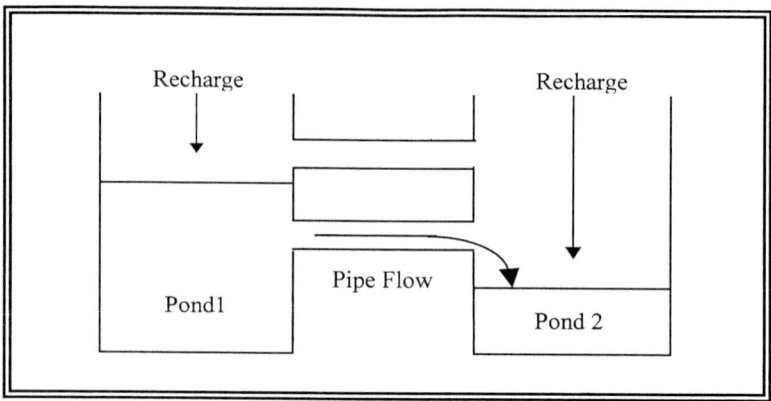

Figure 3.36: Sketch cross-section illustrating the concept of mine water rebound in two adjoining ponds which are connected only by two discrete decant routes (e.g. roadways, or areas where goafed panels coalesce). Until the water level in at least one of the ponds reaches the lowermost decant route, the rebound process will occur independently in each of the ponds. Thereafter, water will flow from one to the other (as shown) eventually resulting in an equalization of heads between the two, and subsequent common evolution of rebound in both ponds. (After Sherwood, 1997; see also Sherwood and Younger, 1997).

below the mined voids. The rates of inflow from these aquifers depend on the degree to which the hydraulic head within them exceeds the head (elevation plus atmospheric pressure) in the mine voids. The head difference between a given aquifer and an adjoining mine void will clearly be at a maximum when the mine is fully dewatered to its entire depth. As rebound progresses, the head differences between the aquifers and the mine voids will progressively decrease, leading to both a gradual reduction in the rate of inflow *and* a gradual deceleration of water level rise. Where head-dependent inflows predominate, these processes will result in the rebound curve being exponential in form (case (i) in Figure 3.38).

Head-independent inflows enter mine voids from remote sources via long recharge pathways. This means that the rate of inflow to the mine voids is independent of the water level within them, depending only on the availability of water in the remote source. The principal sources of head-independent inflows to mine voids are:

- rainfall directly into the void (surface mines) or infiltrating into strata immediately overlying the void (deep mines)
- surface water courses which cascade into the void (surface mines) or infiltrate into strata above the void (deep mines)
- drainage from aquifers which are perched high above the worked strata, such that the aquifer base lies far above the highest level which water will ever reach in the mine void.

Where head-independent inflows predominate in the water make of a rebounding mine system, the rebound is likely to be approximately linear (case (ii) in Figure 3.38).

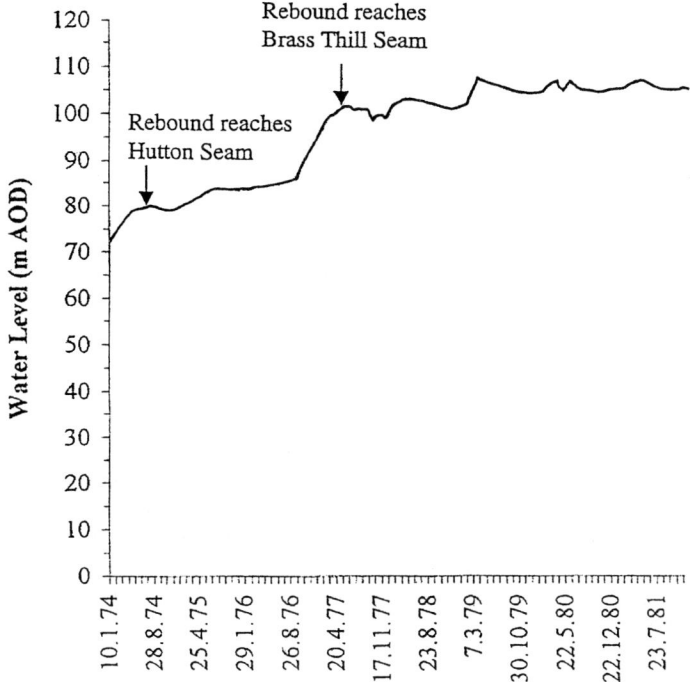

Figure 3.37: Mine water rebound record for the Ladysmith Shaft (NZ 194255) in southwest County Durham (UK) in the late 1970s, illustrating the stepped nature of the recovery curve, reflecting the presence of worked seams and disturbed roof strata at discrete depth intervals. Where the water table is rising through intact, inter-seam strata, the recovery is steep and rigid. Wherever a seam and the overlying disturbed strata are transversed by the rising water table, the recovery is more gentle and 'noisy'. (After Younger and Adams, 1999).

In most systems the true rebound curve will follow a trajectory lying somewhere between the two extreme responses shown in Figure 3.38 (modified from time to time with steps corresponding to changes in available storage, as in Figure 3.37). Two examples of real rebound curves are given in Figure 3.39. An example in which virtually the entire water make was head-independent is shown in Figure 3.39a. In this case, the bulk of the water was entering the workings by slow infiltration from sandstone aquifers present in overlying mountainous terrain, perched well above the local base level of drainage to which the water levels finally recovered (Younger and Adams, 1999). As the recovery could not affect the heads in these aquifers above the valley floor, the rebound curve is virtually linear. Figure 3.39b is an example of rebound in a system which received most of its water make from a head-dependent source, namely extensive adjoining flooded workings (Adams, 2000). As the head increased within the mine voids, so the inflow decreased, resulting in a gently exponential rebound curve. The only deviation from the exponential recovery occurred in the very late stage of rebound, when water levels entered a zone with no workings. The lack of void space for flooding led to rapid recovery of water levels.

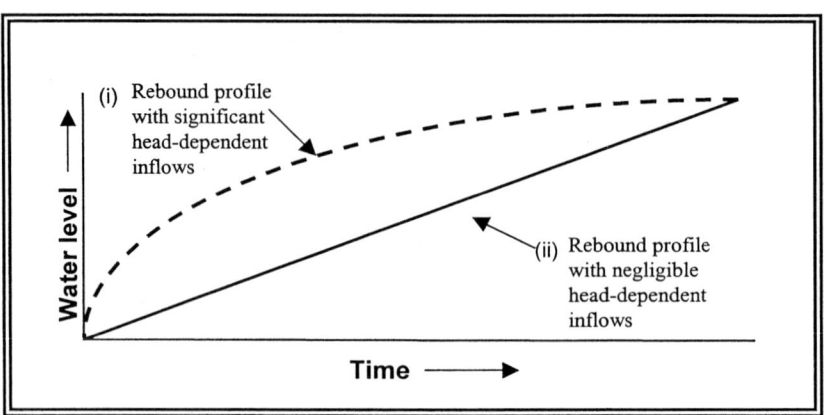

Figure 3.38: Schematic mine water rebound curves which would be anticipated for the two extreme possibilities of (i) a mine system which receives a large proportion of its total inflow from head-dependent sources (i.e. inflow from adjoining aquifers which initially have a higher head than the mine system), and (ii) a mine system in which virtually all of the inflow originates as infiltration from the land surface above the mined area, or by decant from an aquifer which is perched above the mined system and drains into it at a constant rate. (Adapted from Sherwood, 1997).

Hydrochemical changes accompanying rebound: deterioration and stratification
It has already been noted in Sections 1.4 and 2.2.5 that rebound commonly results in a marked deterioration in the quality of mine waters. This is due to the sudden dissolution of the various minerals (Table 2.6) which comprise the 'vestigial acidity' in the workings. As a general guide, this problem is only likely to give rise to significant contamination where the mined strata contain more than 1% by weight total sulphur. In such cases, dissolved concentrations of sulphate, iron and/or other problematic metals can be expected to increase by as much as an order of magnitude during rebound (Younger, 2000a). The phenomenon is as likely to occur in back-filled strip mines (e.g. Marsden *et al.*, 1997; Younger, 2001) or open pits in pyritic orebodies (e.g. Geller *et al.*, 1998) as it is in the better-documented case of deep mines (e.g. Younger, 1998b). This deterioration in water quality has important environmental consequences when the abandoned workings finally overspill to the surface environment (see Section 3.11 below).

There is an important nuance in the hydrochemical changes which accompany rebound, which has sometimes misled engineers managing mine closure programmes: stratification of water quality. Two examples of the phenomenon are shown in Figure 3.40, in the form of hydrochemical profiles obtained by discrete-depth sampling of the water columns in abandoned mine shafts prior to overspill (Younger, 2000d). It is immediately apparent in both cases that the better quality water is found at the top of the water column, albeit step changes in the different water quality parameters do not always coincide. Further examples of stratification are presented by Ladwig *et al.* (1984), from anthracite mines in Pennsylvania, and Johnson and Younger (2000) from a fluorspar mine in northern England.

The stratification of water quality during rebound implies that mechanical mixing of the water column in the mine is minimal. This in turn implies, for instance, that there are few lateral inflows and outflows at depth. As long as flow remains sluggish and laminar, there

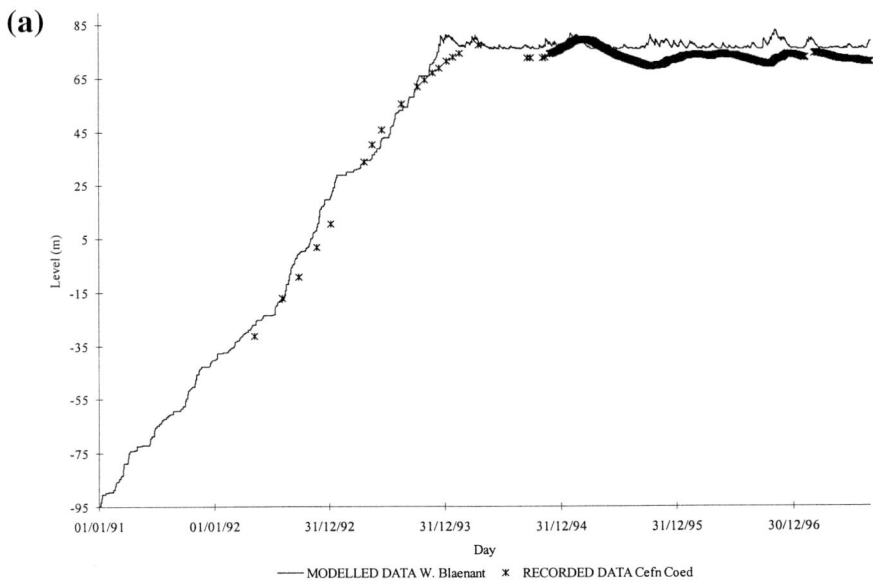

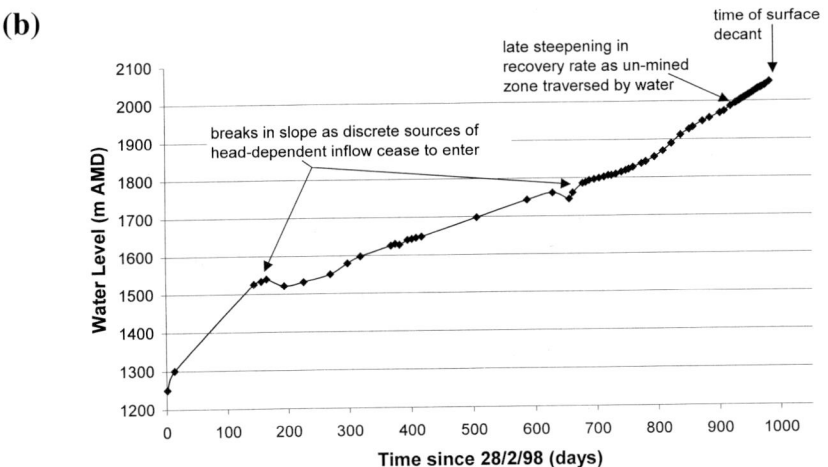

Figure 3.39: Examples of real rebounding deep mine systems with (a) negligible head-dependent inflow and (b) major head-dependent inflow. Case (a) is the Blaenant-Ynysarwed coal mine system in South Wales (UK), where the rebound is almost perfectly linear, and was readily modelled using the simple void-filling model GRAM (after Younger and Adams, 1999). Case (b) is the South Crofty tin mine in Cornwall (UK), into which some 60% of the inflow was ultimately found to be head-dependent. Note the curved nature of the early part of the curve, as head-dependent inflows at great depth were rapidly terminated. The steepening of the final portion of the rebound curve corresponded to a period when the water table was rising through a previously-unknown zone of unworked, low porosity strata with very low storage capacity. (It had previously been considered that this shallow zone had been worked in the early 19th Century, and that the lack of mine plans for this zone meant they had simply been lost). (N.B. 'AMD' in the y-axis label means 'above mine datum', which for this mine was 2000 m below sea level). (Adapted after Adams, 2000).

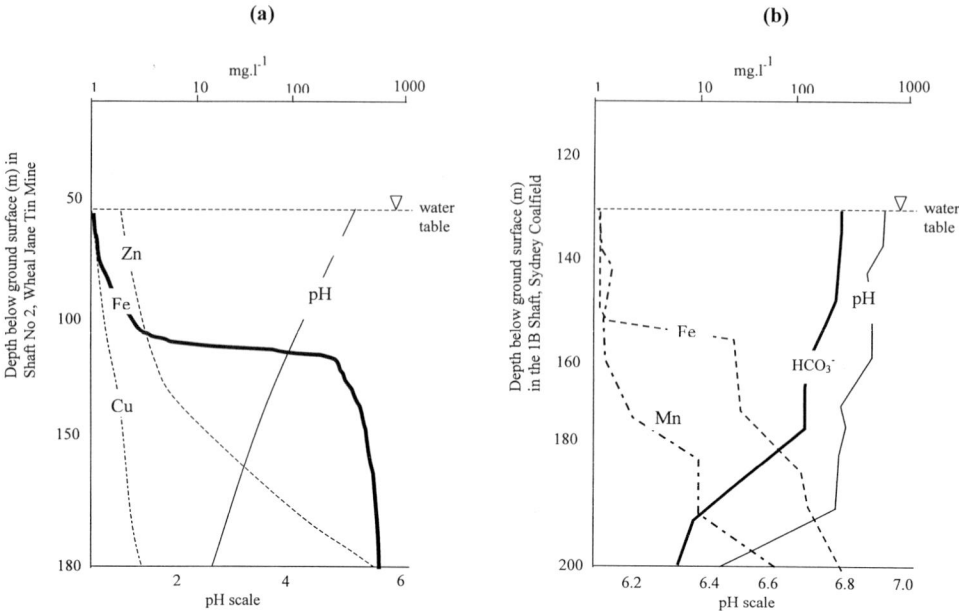

Figure 3.40: Stratification in rebounding mine waters, illustrated by variation of selected hydrochemical parameters over depth during the flooding of (a) Wheal Jane tin mine, Cornwall, UK (November 1991), and (b) The 1B Shaft of the Sydney Coalfield, Cape Breton, Canada (August 1997). (After Younger and LaPierre, 2000).

is little to disturb the stratification. However, when discharge from a mined system commences, turbulent flow in the vicinity of shafts and open roadways can cause substantial mixing of the mine water body, leading to a breakdown in the stratification. Hence the quality of a mine water discharge from a formerly stratified system is more likely to resemble a mixture of all depth intervals rather than the better quality water previously found at the top of the water column. For instance, when the Wheal Jane mine first overflowed, the water was considerably poorer in quality than the uppermost waters marked on Figure 3.40a; it in fact resembled the median of the values shown on the Figure.

This has two practical implications:

1. Mine water discharges may well be considerably poorer in quality than would have been inferred by sampling the uppermost waters alone. Hence depth-sampling is highly advisable in studies of mine water rebound.

2. Where stratification is identified, it would be imprudent to assume that future discharges at the surface will resemble the uppermost waters in the rebounding water column. While the very first waters to emerge may have this benign quality, these will usually be quickly followed by waters of a more mixed nature, with an overall quality approaching that of the mean concentrations found over the full depth of the mine water body.

Other impacts of the rebound process
These have been detailed at some length in Section 1.4, so are only reviewed briefly here. Besides the physical erosion of mine workings by fast-flowing water, the process of rebound can also lead to geotechnical problems by the simple wetting of weak strata, which are then prone to sloughing and slaking (e.g. Smith and Colls, 1996). Less commonly, the increase in pore water pressure caused by rebound has been shown to promote renewed seismicity along long-dormant extensional faults (e.g. Donnelly, 2000). Re-hydration of expansive clays during rebound is thought to explain rises in ground surface level of as much as 25 cm during mine water rebound in the Netherlands (e.g. Bekendam and Pottgens, 1995). As noted in Section 1.4, gas pockets trapped by rising mine waters have formed both refuges for miners awaiting rescue, and may also give rise to pneumatic fracturing of rocks in very deep mines, where the head of water against the gas pocket can exceed 900 m (e.g. Younger, 1999; see also Figure 1.35). Dense gases which effectively 'float' on the water table can be released from mine voids ahead of a rising water table, sometimes with tragic consequences (e.g. Burrell and Friel, 1996).

3.10.2 PREDICTIVE MODELLING OF MINE WATER REBOUND

Rebound modelling resources
A thorough guide to the modelling of mine water rebound has recently been published by the Environment Agency (England and Wales). The bibliographic reference for the guide is:

Younger, P.L., and Adams, R., 1999, Predicting mine water rebound. Environment Agency R&D Technical Report W179. (ISBN: 1 85705 050 9). Environment Agency, Bristol. 108pp.

Copies of the guide may be obtained inexpensively by contacting:

Environment Agency R&D Dissemination Centre,
c/o WRc, Frankland Rd., Swindon,
Wiltshire SN5 8YF, UK.
e-mail: publications@wrcplc.co.uk

A summary of the flow modelling methodologies presented in that guide has been published by Adams and Younger (2001), while Younger (2000a) has expanded upon the recommendations in relation to prediction of mine water quality changes. In view of the ready availability of this documentation, only a brief overview of rebound modelling methods is given here.

Why predict mine water rebound?
Before undertaking any major modelling project, it is essential to be clear about the reasons for undertaking the work. For modelling is not an end itself; rather it is one of the more important tools we can use to test the consistency between the available data and our concepts of system behaviour. There are a number of reasons why predictive modelling of mine water rebound can be desirable, foremost amongst which are (Younger and Adams, 1999):

(i)   To assess the hydrological/environmental risks and time-scales associated with mine water rebound, as a basis for deciding whether to allow rebound to continue unabated, or to implement a preventative solution (such as a pump-and-treat scheme).
(ii) To independently assess predictions of time-scales and risks associated with rebound which are made by mining companies during the formal closure application process.
(iii) To address related issues of public concern such as the risk of elevated rates of mine gas emission, or assessments of subsidence risk.

In all projects with which we have been involved to date, the costs of hydrogeologically investigating a possible rebound scenario have been on the order of 1–2% of the capital costs of implementing a remediation scheme. When the operating costs of such schemes are taken into account, the provision of a sound hydrogeological basis for system design becomes vanishingly inexpensive. Nevertheless, it usually remains important to limit the expense of the modelling method applied in proportion to the importance of the problem at hand. Even where a problem is potentially of major significance, it is desirable to have 'first pass' modelling techniques available for rapid appraisals, which can be used to decide whether a more detailed modelling exercise is justified, and to provide some initial conditions estimates for more detailed models if they are applied. Consequently in the following sections we work through a series of modelling approaches of steadily increasing sophistication and cost, from manual calculations to realistic 3-D models of complex systems. Before considering mathematical approaches, however, it is essential to consider the formulation of conceptual models, which are the *sine qua non* of all other modelling work.

Constructing the conceptual model
The conceptual model is both the basis for any mathematical work *and* the focus for interpretation of all results. Indeed it can easily be argued that "conceptual modelling" is the only kind of modelling there is, since mathematical modelling is merely concerned with using calculations to test the credibility and implications of the conceptual model. In practice, the conceptual model for a given system is normally defined before any mathematical simulations are implemented, and is continually updated in an iterative process as mathematical results become available.
    The term "conceptual model" has a formal definition in hydrogeological modelling circles, one version of which is as follows:

> "A conceptual model is an assemblage of simplifying assumptions about a complex, real system, which achieves a valid representation of that system, including all major features, whilst avoiding unnecessary detail"

As in all hydrogeological systems (see Anderson and Woessner, 1992, for instance) the principal assumptions which need to be assembled before mine water rebound can be simulated include the following:

(a) Assumptions concerning system geometry. These fall into two categories:
    • internal geometry, such as any 'ponds' or zones of distinct permeability associated with specific types of mine workings etc. These features are usually inferred from mine plans

- the external boundaries of the system. These boundaries may be simply the limits of the mine workings or geological features such as outcrops of adjoining aquifers, major faults etc. In setting up a model for a mined system with a major head-dependent water make, it may be possible to specify the boundaries of the simulation domain to coincide with the outer limit of workings, but the boundary conditions themselves will need to be represented as head-dependent flux boundaries, rather than zero-flow boundaries. Only where a major water-yielding bed lies above the final surface overflow point from the workings will it be appropriate to apply a specified-head boundary. The general lack of cases in which specified-head boundaries may be used validly in modelling mine water systems poses a problem for the application of standard groundwater modelling packages to such systems. This is because all of the common numerical methods for solving the transient groundwater flow equations have a requirement that head be specified upon at least one point (and often more than one point, depending on grid shape) on the outer boundary.

(b) Assumptions concerning initial and final conditions. The "initial conditions" at the commencement of mine water rebound are defined by the water level in the workings at the time the pumps were switched off, and the final dewatering rate before the withdrawal of pumps began. Where workings were already partly flooded before the final cessation of dewatering, a reliable value for the water level before the start of rebound may well be available. Where a mine was actively worked until a few weeks before the cessation of dewatering, then it may only be possible to estimate the initial water level by reference to the state of operations at the end of extraction. In many cases, this will mean that the effective "water level" in the workings corresponds to the elevation of the deepest part of the dewatered workings during the final phases of extraction. (Since many mines close precisely because their final reserves are concentrated in the deepest parts of the take (which are also the most expensive to work), the last workings are often also the deepest). Defining an initial water level in this way is fraught with uncertainties, because the details of local hydraulic gradients in strata around the deepest workings are often not known. It is to be expected that the latest workings will have been provoking transient hydraulic responses in the surrounding strata. These transient phenomena are complex, and are almost never characterised in day-to-day mining operations. Hence the initial rate of fill of the deepest workings is difficult to predict in all but the most simple of mining situations. In practice, we have often found it best to derive an appropriate "initial water level" by trial-and-error during rebound simulations.

   The final dewatering rate before commencement of rebound is an important figure, as it indicates the total water make of the workings up to that point. This water make will include a component of head-dependent inflow, the estimate of which is a necessary part of rebound analysis. Estimation is aided if an upper-bound can be fixed by knowledge of the patterns of dewatering in the final months before closure. In an ideal situation, a record of dewatering rates (at, say, weekly resolution) for the last few years of working will be available. In most cases, the mine manager will have only an approximate idea of the quantities of water pumped. Care should be exercised in interpreting such figures, for in most cases the total dewatering rate of the mine will apparently be declining for several weeks (if not months) before final abandonment, as salvage teams gradually withdraw local pumps from remote districts of the mine underground, and mine water begins to fill localised pockets of storage in the mine instead of making its way to the main pumping lodge.

By "final conditions" we mean the equilibrium water level and discharge rate of the mine system after rebound is complete. The equilibrium water level in the system post-rebound will be defined by the position and elevation of decant points. After studying several hundred mine water discharges in Europe and eastern North America, we have gained the impression that the overwhelming majority of decant points to surface water are man-made features. The most common features forming surface decants are old adits ("drifts"). Probably second in frequency are overflowing shafts, with old exploration/ventilation boreholes accounting for a significant minority. Mine water discharge through un-mined outcrop areas is extremely rare; indeed, we can think of no unequivocal example. That is to say, while discharges are very commonly associated with seam/lode outcrops, the actual point of emergence is almost invariably via an old mine entrance on that outcrop. In undertaking rebound analyses, therefore, the prediction of surface decant points is normally pursued as follows:
- determine the outcrop patterns of the major worked horizons (using published geological maps if available)
- scrutinise archival records to ascertain the locations of old mine entrances (adits, shafts etc) on the outcrop area
- determine which of the old mine entrances lies in the lowest topographic position on the outcrop; this will be the most probable decant point, and should be subjected to field survey, including trial-pitting to determine its condition if possible.
- determine the distribution of other old mine entrances which lie higher than the most probable decant point, for if roof falls etc prevent the most probable decant point from flowing, then higher decant points may become active.

In the case of decant to an overlying aquifer, which may occur via old shafts/boreholes or subsidence-related fractures, the final water level attained in the mine system will depend on the permeability of the decant route. Upward flow into an aquifer requires not only decant points, but a driving head, which will largely be topographically controlled. While this general point is easy to make, *a priori* prediction of the development of suitable head conditions is often far from simple. It is also difficult to predict the equilibrium discharge rate from a mined system after rebound is complete. While it is possible to calculate this directly using the more advanced mathematical models (see Adams and Younger, 2001), preliminary estimates are fraught with uncertainties. In certain cases, it is relatively easy to identify head-dependent water sources which will cease to flow after rebound, such as sea water entering undersea workings and deep geothermal waters. In the most extensively mined regions, it may well be that the entire effective rainfall in the area is intercepted by the mined system, so that the post-rebound water make is very similar to that during mining. However, in many cases it will be very difficult to arrive at an estimate for the post-rebound discharge rate *a priori*.

The simplest mathematical models of rebound: manual and graphical techniques
Once having established the preliminary conceptual model, simple rebound predictions may commence. Three approaches are currently used:

(i)   Predicting rebound on a "void filling" basis. At its simplest, rebound prediction can be made by determining the volume of coal extracted in a given mine, pond, or mining district, then comparing this volume with a long-term average recharge rate. This yields an estimate of the time which will elapse before the workings are full to overflowing. However, this approach suffers from a number of shortcomings, such as the neglect of head-dependent inflows, an overly simplistic conception of the nature of old mine voids (assuming them to be standing open indefinitely, rather than changing by void migration etc). Nevertheless, the apparent logic of void-filling calculations has led to their widespread adoption. In practice, they can be applied by constructing a "hypsometric curve", which is basically a cumulative frequency graph of mine void volume against height as digitised from mine plans. Having obtained this hypsometric curve, the future evolution of water levels in the workings is predicted by assuming that the shape of the rebound curve will follow the shape of the hypsometric curve. At best, void-filling calculations may provide a useful preliminary estimate of possible rates of rebound, which can be valuable when deciding whether a particular rebound problem is a short-term emergency or a long-term phenomenon for which there is sufficient time to develop a cost-effective solution. At worst, void-filling calculations can be misleading, leading to serious misjudgements over the scale and timing of an appropriate response to possible rebound problems. For this reason, it is suggested that they are never used as the sole means of estimating the rate of rebound in a given area.

(ii)   Predicting rebound using a "specific yield" approach. It was noted above that difficulties arise in applying the void-filling approach because of the overly-simplistic concept of the underground voids which it employs. These difficulties relate primarily to the fact that goaf has a finite porosity, and a porosity which is almost certain to be closer to that of a gravel (i.e. 0.3 or less) than to that of an open void (~ 1). The responses of some deep-mined systems to seasonal changes in recharge, and / or to fluctuations in pumping, illustrates that goaf panels do have specific yield properties analogous to those encountered in sedimentary aquifers. For instance Minett et al. (1986) have described the hydrological analysis of a pumping system installed to dewater previously-flooded old coal workings in Northumberland, in advance of opencast reworking. In their analysis, it soon became apparent that the old workings (and associated fractured roof-strata) were dewatering in a manner which suggested they had a specific yield on the order of 2 to 5%. Other studies have yielded estimates of specific yield for mined Coal Measures of a similar magnitude (Sherwood, 1997). This suggests that, instead of performing void-filling calculations which unreasonably assume that all voids remain open, it may be more appropriate to model rebound such that goaf panels and associated roof-strata are modelled as layers (40 m to 100 m thick) with specific yields on the order of 2 to 5%, with intervening intact strata being assigned low values (10^{-3} or less) reflecting the high specific retention of virgin Coal Measures lithologies, in which pore necks are very small. If this assumption is made, then rebound calculations for multiple-seam ponds will produce stepped rebound curves similar to that shown in Figure 3.37.

(iii)   Fitting an exponential curve to observed data. Where head-dependent inflows are significant, rebound curves can be expected to follow an exponential pattern (Figure 3.38a). It is possible to fit an exponential curve to rebound data by simple regression, and then to use extrapolated parts of the curve to predict the later stage of rebound. The larger the system under study, the more likely is this approach to be successful. However, as with any extrapolation method, it must be used with caution. In particular, the method implicitly assumes that the processes responsible for the decline in inflow rate will operate similarly throughout the period of rebound, whereas in situations where deep mine voids are overlain by unworked strata, the change in specific yield may result in an abrupt steepening of the curve in the later stages of the rebound process.

Semi-distributed 'pond' modelling
The next level of sophistication in rebound modelling beyond manual calculations is computer-based, semi-distributed pond modelling. By "semi-distributed" we mean a model which represents some of the regional variation in mined system geometry and properties insofar as these affect the general course of events during rebound, but which is suffi-ciently coarse that local details cannot be resolved. Such a model is most appropriately applied to relatively large systems, extending over areas of several tens to many hundreds of square kilometres (Sherwood, 1997). The basic methodology is to simulate the water budgets of each of a series of inter-connected ponds, including inter-pond flows via mutual decant features. The basic concept of such models is illustrated by Figure 3.36, which correctly suggests that the water level in each pond will rise largely independently of all others until the invert of a mutual decant feature (such as a roadway, or an area of coalesced goaf) is reached, after which inter-pond transfers will lead to equalisation of the heads in both ponds, so that future rises in water level will occur in tandem. Semi-distri-buted pond models of this type were first described by Younger and Sherwood (1993), and were subsequently developed considerably by Rogoz (1994), Sherwood (1997), Gatzweiler *et al.* (1997), Sherwood and Younger (1997) and Banks (2001). The water balance for a pond is calculated over short time-steps (usually $\leq$ 1 day), using an expression such as:

$$R - O - V_p = \Delta S \tag{3.46}$$

Where: R = recharge during time-step (Δt) (= recharge rate x pond area (A) x Δt) (L^3).
O = the sum of outflows during the time-step Δt, comprising inter-pond decant flows (positive for flow leaving the pond, negative for flow entering it) and surface outflows (L^3)
V_p = volume of water (if any) pumped from the pond during the time-step (L^3)
ΔS = change in volume of water stored in the pond during the time-step (L^3)

The change in water level (ΔL) in the pond over the time-step is related to ΔS by:

$$\Delta L = \Delta S / (AS_y) \tag{3.47}$$

where A is pond area and S_y is the specific yield of the pond at the relevant horizon.

Given that the level at the start of the time-step (L_{t-1}) is always known, then the water level at the end of each time-step (L_t) is given by:

$$L_t = L_{t-1} + \Delta L \tag{3.48}$$

Inter-pond decant flows can be modelled in a number of ways, some of which are entirely arbitrary. Given that most inter-pond transfers occur via turbulent flow in large, open voids, it is perhaps most physically realistic to represent flow between ponds using formulae which describe turbulent flow in pipes. In the GRAM model for instance (Sherwood, 1997), the Bernoulli equation is used to model decant flows. Neglecting head losses due to pipe entry and velocity head, which are very small compared to frictional head losses in mined systems (Sherwood, 1997), and arranging the Bernoulli equations so that the velocity of flow in the pipe (V) is the subject, the overall expression for decant flow converges on the well-known Darcy-Weisbach formula:

$$V = \sqrt{\frac{2\,g\,\Delta H}{\left(1.5 = \dfrac{\lambda\,L}{D}\right)}} \qquad (3.49)$$

Where: V is the velocity of flow in the pipe ($L.T^{-1}$),
ΔH is the head difference between the ponds (L),
g is acceleration due to gravity ($L.T^{-2}$),
D is the pipe diameter (L),
L is the pipe length (L) and
λ is a non-dimensional coefficient which is a function of the roughness and diameter of the pipe.

Evaluation of λ can be achieved by various means, of which the Prandtl-Nikuradse Equation was found to be the most efficient in sensitivity analyses (Sherwood, 1997). For rough turbulent flow, the Prandtl-Nikuradse Equation can be written:

$$\frac{1}{\sqrt{\lambda}} = 2 \log \frac{3.7\,D}{k} \qquad (3.50)$$

Having calculated V by means of Equation 3.49, it remains only to multiply by the cross-sectional area of the decant feature in order to yield the flow rate.

The above mathematical formulation is implemented in the modelling package "GRAM" (*Groundwater Rebound in Abandoned Mine-workings*). GRAM also includes refinements allowing for layering of specific yield within each pond in relation to areas of extensive working, some basic solute transport capabilities, and Monte Carlo simulation facilities. Further details of the algorithm are given by Sherwood and Younger (1997) and a full listing of the code with details of development and testing are provided by Sherwood (1997). GRAM is the most widely-used code of its type, and it has now been applied successfully to the simulation of several abandoned mine systems (see Younger and Adams, 1999; Burke and Younger, 2000). For example, Figure 3.39a shows a comparison of measured head data with GRAM predictions for the Blaenant-Ynysarwed system in the South Wales Coalfield. The correspondence between predicted and observed results, which was obtained without fitting, illustrates, *inter alia*, that the void volume estimates and recharge rate time-series derived for this system are highly credible.

It should be noted that GRAM and similar pond models assume the that the rock mass enclosing the mine workings is essentially impermeable. This implicitly assumes that the overall water make of the mined ground is independent of head (save for short-term head-dependent decant flows between ponds). For the largest coalfields, where the entire effective rainfall is captured by mined ground, this assumption can be reasonable (e.g. Younger, 1993; Burke and Younger, 2000). Even where this assumption is not valid application of GRAM can still be useful, as an inability to reproduce observed rebound data can provide compelling evidence that a large proportion of the water make must be from head-dependent sources. Where this proves to be the case, a modified pond model can be used which allows for some representation of the head-dependence of inflows (Banks, 2001). In some cases it may be worthwhile applying an even more detailed model (Adams and Younger, 2001), which allows full representation of the hydrogeological characteristics of the country rock as well as the mined voids. Such a model is described below.

Physically-based distributed modelling
When simulating rebound at much smaller spatial and temporal scales than are appropriate for GRAM and similar pond models, it is usually necessary to take the hydrogeological characteristics of the enclosing rock mass into account. This can be achieved readily using standard porous medium groundwater flow models, which solve Equation 3.12 (and variants of it) using numerical methods (as described by Anderson and Woessner, 1992, amongst others). However, as flow in open mine voids is unlikely to conform to the laminar conditions upon which Equation 3.12 is based, an alternative approach is necessary if this is also to be realistically simulated. While a number of alternative approaches are feasible, perhaps the most credible (and certainly the most successful to date) has been the use of pipe network models to represent the mine voids. Thus a total modelling package for simulating ground water flow into, through and around mined voids comprises two coupled elements (see Adams and Younger, 2001):

- a 3-D porous medium model to represent the enclosing rock mass and goaf etc (preferably one which allows explicit simulation of unsaturated zone flow), and
- a 3-D pipe network model to represent the major mine roadways, shafts and other open voids in the mined ground.

An integrated model with these characteristics can be described as "physically-based and distributed" on the grounds that the governing equations are based on the true physics of the situation, and details of spatial behaviour can be resolved at very finest of scales (to the nearest metre, if necessary) so that point values of state variables distributed across a wide area can be generated. Such models have the major advantage over the semi-distributed pond models that they can be used to predict the precise locations of surface decants, rather than requiring these to be specified *a priori* (Adams, 2000).

Pipe network models are commonly used in the water industry to simulate water distribution systems and sewerage networks. A number of alternative mathematical representations are employed, depending on the specific application and the preference of the modeller. Probably the most computationally-efficient and robust pipe network modelling approach in widespread use is the one which has been selected by the US EPA for their own network analysis code "EPANET". In this approach, flow in each length of pipe (between successive pipe junctions, or "nodes") is calculated using the Hazen-Williams equation:

$$Q_{ji} = 0.28 \ C_{ji} \ D_{ji}^{2.63} \ [(h_i - h_j)/L_{ji}]^{0.54} \tag{3.51}$$

where Q_{ji} is the flow from node i to node j ($L^3 \cdot T^{-1}$)
C_{ji} is the "Hazen–Williams flow coefficient" for pipe ji (T^{-1})
D_{ji} is the pipe diameter (L) and
h_j and h_i are the respective heads at nodes j and i (L)

In water supply applications, Equation (3.51) is the most popular choice where flow is turbulent but fluid viscosity is constant, which is deduced to be the case also in most mine water rebound situations (see Section 3.10.1 above).

Obviously, a large system of pipe elements meeting at numerous nodes define a complex 3-D network for which simultaneous solutions of potentials, and therefore flows, in all pipe elements must be sought. Such solutions require a numerical algorithm, which must be iterative in order to accommodate the non-linearities in the equations. One of the most efficient numerical techniques for this purpose is the Gradient Algorithm (Todini and Pilati 1989), which has been extensively tested in comparison with other methods (Salgado *et al.*, 1989), and has been successfully used for many years in EPANET, and more recently in the mine water rebound code VSS-NET (Younger and Adams, 1999). (Full details of the Gradient Algorithm and its application are beyond the scope of this book, but can be obtained from the sources just cited).

The key to successful application of this approach to the simulation of mine water rebound is the coupling of the pipe network model to the porous medium model. This has been achieved by Younger and Adams (1999) and Adams and Younger (2001) in the production of the VSS-NET code, essentially by calculating the exchange of water between each pipe element and its contiguous porous medium element for every time-step using an expression based on Darcy's Law (Equation 3.4), as follows:

$$J_p^{n+1} = \beta_p \ k_r^n \ (z_p^n - \psi^n) \tag{3.52}$$

Where J_p^{n+1} is the exchange flux at the current time-step between pipe and aquifer ($L^2 \cdot T^{-1}$)
β_p is the 'conductance' . which represents the resistance to inflow exerted by the permeability of the tunnel lining ($L \cdot T^{-1}$)
k_r^n (dimensionless) is the relative conductivity (< 1.0 if the column is unsaturated)
z_p^n, (L) the head in the pipe, and
ψ^n (L) the head in the porous medium

(The superscripts n indicate the value at the previous time-step). After the exchange flows are calculated, the pipe network model is run to calculate a new set of z_p values over the current time-step (n+1). The porous medium model then iterates for the heads in the aquifer at the current time-step including any water which has flowed from the mine voids into the surrounding rock.

A physically-based mine water rebound model with the above characteristics, called VSS-NET, has been developed (Younger and Adams, 1999) and successfully applied to a range of mining settings, including:

- large volumes of inter-connected, multi-seam coal mine workings (e.g. Parkin and Adams, 1998), and
- a multi-vein hydrothermal tin ore body worked by longhole open stoping (Adams and Younger, 2000).

VSS-NET is a member of the major SHETRAN family of catchment modelling codes (Ewen *et al.*, 2000) and as such can be integrated with detailed simulations of surface water catchments receiving mine water discharges. (Further details of VSS-NET and its practical use in mine closure management are beyond the scope of this book, but may be found in Younger and Adams (1999) and other works listed above).

3.11 ABANDONED MINE HYDROLOGY AFTER REBOUND IS COMPLETE

3.11.1 ABANDONED SURFACE MINES: PIT LAKES

The hydrological behaviour of surface mine backfill has already been discussed (see Section 3.9 above). In cases where there is insufficient waste rock or other material (e.g. municipal waste) to backfill a surface mine void, the void will persist indefinitely after mining ceases. The long-tern hydrological behaviour of the void will then depend on the elevation of its floor in relation to the local water table. There are two main scenarios:

(i)   Where the water table lies below the base of the excavation, then unless the mine worked virtually impermeable strata, the void is likely to function as a focus for ground water recharge. This is because surface runoff arising from direct precipitation into the void (or surface watercourses which enter the void) will collect in the floor of the void, whence it can leave naturally only by infiltration (forming ground water recharge), evaporation or by overspill from the lowest point of the void perimeter. As erosion leads to a gradual build-up of fine-grained silt in the base of the void a low-permeability layer can develop (a process sometimes termed "blinding"). This layer may well locally perch water within the base of the void, so that infiltration only occurs through the side walls of the void when the water level rises above the rim of the silt layer (Figure 3.41a). In other cases, where the mine worked hard, permeable rocks, previous sub-floor blasting will have ensured that the floor is one of the most permeable margins of the void, and water will infiltrate through it so readily that ponding never occurs.

(ii)   Where surface mines have operated below the water table, the cessation of dewatering will lead to a gradual flooding of the void until such time as the water level within the void equals that of the surrounding aquifer (Figure 3.41b). When this occurs, a "pit lake" is formed (e.g. Figure 1.20). Pit lakes are very often formed deliberately as a planned part of the after-use of a surface mine void, for where the water quality is appropriate they can be used as wildlife habitats, fisheries, water sports venues or other forms of amenity. The configuration of most large surface mine voids is such that the all water entering the pit lake (by lateral ground water inflow or precipitation/surface runoff into the void) must leave by lateral ground water outflow (Figure 3.42a). Lakes behaving in this manner are sometimes known

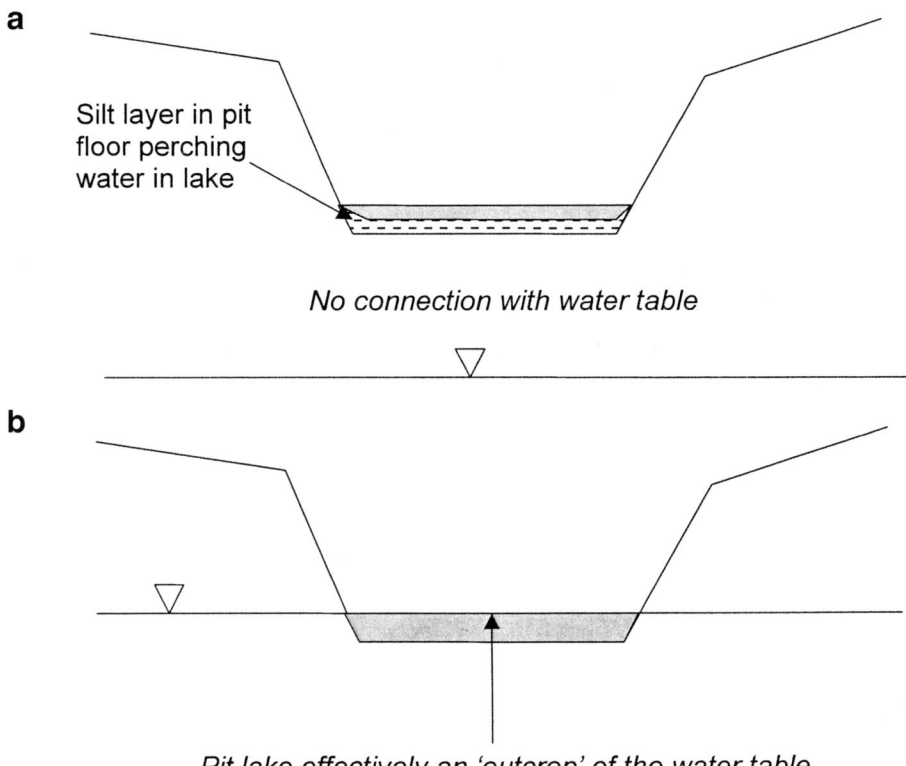

Figure 3.41: Sketch cross-sections showing two contrasting origins of pit lakes: (a) Abandoned surface mine void above the water table, with water perched by 'blinding' of pit flow with silt. Pit lakes of this type can readily be drained by pumping and will often dry up entirely in the dry season. (b) Pit lake formed where the floor of the void lies below the water table. Such pit lakes are usually perennial (may be intermittent where the water table rises and falls a lot).

as 'through flow' pit lakes. However, where surface mines have been developed by excavation from outcrop, the pit lake may drain by means of a surface decant through the low wall (Figure 3.42b).

Of these two scenarios, the second has received far the most attention by the engineering community. Three issues commonly arise in relation to the hydrology of pit lakes:

(a) the rate at which the void will flood to establish the pit lake in the first place
(b) the effects of pit lakes on the regional hydrology
(c) the water quality of pit lakes, particularly in areas where the mined strata contain sulphide minerals.

The rate of flooding of surface mine voids has been addressed by a number of workers. In concept the prediction of pit lake formation is straightforward, and can be addressed by

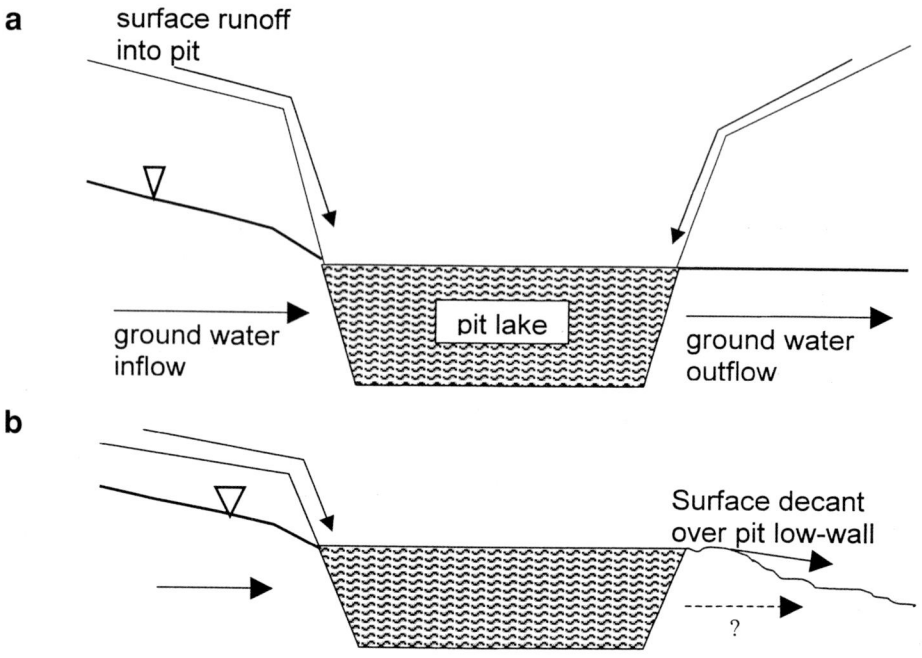

a surface runoff into pit

ground water inflow

pit lake

ground water outflow

b

Surface decant over pit low-wall

?

Figure 3.42: Drainage from a pit lake by (a) lateral ground water outflow, and (b) surface decant.

assuming that the rate of inflow from the surrounding aquifer will steadily decrease as the pit fills. Immediately after the cessation of dewatering, the rate of inflow to the mine area will correspond to the final dewatering rate. By the time rebound is complete, ground-water inflow will equal the sum of evaporative losses plus any ground water outflow (the magnitude of which will depend on the regional hydraulic gradient and the transmissivity of the country rock). In practice, complications in prediction arise from the localised nature of many components of the total water make. Where inflows occur as discrete feeders rather than as tall, laterally continuous seepage faces, the reduction in inflow rate during flooding will occur in a more step-wise fashion than would be anticipated where the surface mine had been excavated in utterly homogeneous strata. Notwithstanding such diffi-culties, analytical models have been developed which faithfully reproduce observed pit lake filling rates, and which may be applicable in the prediction of filling rates for future pit lakes (e.g. Shevenell, 2000a). For systems of complex geometry and heterogeneous aquifer properties, the most realistic predictions may demand the application of fully distributed, physically based, numerical flow models (e.g. Schwartz and Crowe, 1985).

The very fact that pit lakes represent the replacement of what was previously solid ground by expanses of open water is a cause for concern in areas where water resources are scarce, for open water evaporation rates greatly exceed the actual evapotranspiration that occurs from undisturbed, soil covered ground. Effectively, pit lakes represent net sinks for ground water, in much the same way as a water well does. Besides representing a loss of water from the ground water system, evaporation also serves to increase the concentrations of all solutes (pollutants included, of course) in pit lakes. For this reason alone it is often an important process to characterise as part of an overall evaluation of pit lake chemistry (Shevenell, 2000b).

In addition to the evaporative "pumping" effect of pit lakes, their very presence as zones of effectively 'infinite permeability' within an aquifer tends to cause a permanent shift in regional piezometric patterns so that flow lines tend to converge on them (e.g. Hobbs and Gunn, 1998). As noted by Morgan-Jones *et al.* (1984), if the pre-mining water table profile is compared with the long-term profile once a pit lake is established, it will be found that the latter is steeper than the pre-mining water table immediately up-gradient from the pit lake, and gentler down-gradient of it (Figure 3.43). Bearing in mind Darcy's Law (Equation 3.4), the steeper water table feeding into the pit lake means that much more water feeds into that zone of the aquifer (which is also now a zone of water loss to evaporation) than was the case before mining. Hence pit lakes tend to assume a regional hydrological importance out of all proportion to the area of the aquifer which they actually occupy.

The water quality of pit lakes is an area of very active research, particularly in water-scarce mining regions of Nevada and other arid and semi-arid western States of the USA (e.g. Miller *et al.*, 1996; Castro and Moore, 2000; Shevenell, 2000b). Before considering some of the water quality problems associated with pit lakes it is worth noting the comment of Bowell *et al.* (1998) that it is "a common misconception of pit lakes that they are always acidic or of poor quality". For instance, in a study of ten pit lakes in Nevada, Shevenell (2000c) found that six had persistently circum-neutral pH. In those few lakes which are prone to acidify, there is evidence that low pH and elevated concentrations of ecotoxic metals preferentially occur during the short period of initial pit lake filling and briefly thereafter (e.g. Rytuba *et al.*, 2000; Shevenell, 2000c). Only those pit lakes in which the terminal water level lies in contact with a rock unit rich in pyritic sulphur are likely to remain acidic in the long-term (Bowell *et al.*, 1998).

The phenomenon of evapoconcentration in pit lakes has already been mentioned, and has been the subject of detailed investigations by Shevenell (2000b,c), who found that this process is less important in pit lakes than in nearby natural lakes. The reason for this lies in the contrast in 'relative depth' values between pit lakes and natural lakes. Relative depth (D_R) is defined as a percentage as follows (Castro and Moore, 2000):

$$D_R = 100 \cdot (z_m / d) \tag{3.53}$$

where z_m is the maximum depth of the lake (L) and
        d is some standardised diameter (L). (For instance, for circular lakes, d is standardised in relation to the area by calculating it as $d = 2 \cdot \sqrt{(A/\pi)}$, where A is the surface area).

Most natural lakes have D_R values of less than 2%, and very few yield values in excess of 5%. By contrast, mine pit lakes commonly have D_R values in the range 10 to 40% (Castro and Moore, 2000). In terms of evapoconcentration, high D_R values mean that annual evaporative losses will account for a comparatively small proportion of the total lake volume (usually < 10%; Shevenell, 2000c).

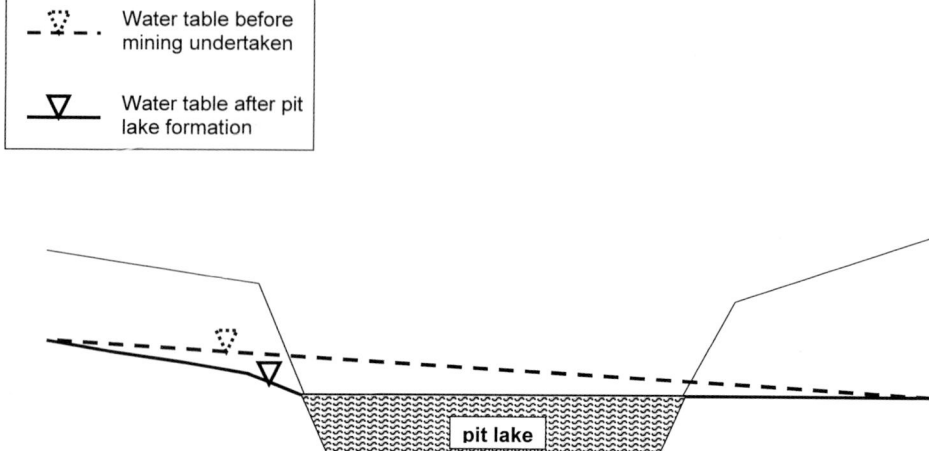

Figure 3.43: Schematic cross-section showing the contrast in water table profile around a pit lake compared with the pre-mining situation.

The extremely high values of D_R have another important consequence for pit lake lake dynamics, in relation to density stratification processes. In many natural lakes the water column seasonally stratifies such that an upper layer of warmer, less dense water overlies a lower, colder, denser layer with a relatively sharp interface (known as the pycnocline). Seasonal overturn of such stratified lakes occurs when the ambient air temperature abruptly changes. While some of the shallower pit lakes display similar behaviour (e.g. Atkin and Schrand, 2000), the high D_R values of most pit lakes promote the development of three-layer systems in which the third, deepest layer is never involved in seasonal overturn. While the two upper layers may seasonally overturn, the dense, cold bottom layer, containing the most mineralised water in the lake, will remain stagnant. Lakes behaving in this manner are termed "meromictic", and their scarcity in nature makes pit lakes particularly interesting to limnologists (e.g. Muggli *et al.*, 2000). A clear understanding of the likely dynamics of a given lake is an important pre-requisite for the development of a rational management plan for acidic pit lakes.

Where pit lakes are acidic, they can generate plumes of polluted ground water which migrate down-gradient into the surrounding aquifer. (Alternatively, if pit lakes decant at the surface, they can cause stream pollution). Consequently, the more acidic pit lakes are increasingly the focus of remedial action. Geller *et al.* (1998) have collated a large body of information on possible management options for such lakes, focusing in particular on the relatively shallow acid pit lakes associated with abandoned lignite workings in Europe. A brief review of options for the deep, hard-rock pit lakes of the western USA has been drawn up by Castro and Moore (2000). Most suggestions for long-term remediation involve one or more of the following actions:

- treatment of the runoff-generating areas which feed surface waters into the pit lakes, to minimise their acid-generating potential (e.g. by clay capping, lime treatment etc)

- releasing hydrated lime or other neutralising agents into the pit lake
- introduction of carbon-rich materials into pit lakes to promote the activity of sulphate-reducing bacteria, which can serve to raise the pH and precipitate metals in accordance with the principles outlined in Chapters Two, Four and Five.

3.11.2 DEEP MINE VOIDS

It was noted in Section 3.10.1 that large, open voids in deep mines can often be expected to host rapid, turbulent flows during the process of rebound, when the driving heads across them are frequently very large. After rebound is complete, the most rapid flows are likely to be restricted to voids above the equilibrium 'water table', in which open channel flow conditions still obtain. In such voids, the hydraulic gradient approximates the gradient of the void floor. Hence where the structural dip is substantial, or where water flows through vertical or steeply inclined tunnels or shafts, flow in supra-water table mine voids is likely to remain rapid and turbulent long after rebound is complete. For instance, in a series of tracer experiments in the Forest of Dean coalfield, England, Aldous and Smart (1988) found that roadways above the regional water table transmit water at velocities of up to 16 km.d^{-1}.

Below the post-rebound water table, the driving hydraulic gradient is that of the water table itself, rather than the structural grain of the voids. As permeable voids tend to result in relatively flat hydraulic gradients ($\ll 1 \times 10^{-3}$), sub-water table velocities tend to be far more gentle than those in the higher, unflooded voids. The observations of Aldous and Smart (1988) revealed that roadways below the water table displayed velocities less than 0.5 km.d^{-1}. Such velocities, while only a thirtieth of the supra-water table velocities in the same system, still imply turbulent flow. Where water table gradients are particularly gentle, flow may well prove to be laminar. For instance, Aljoe (1994b) used a series of bore-holes to investigate a partially flooded system of bord-and-pillar workings in Pennsylvania. Although the mine roadways were conducting a flow totalling several tens of l.s^{-1}, the hydraulic gradient within the flooded workings was immeasurably low. Tracer tests yielded flow velocity estimates of < 20 m.d^{-1} within the workings, and some of the lower values (~ 3 m.d^{-1}) implied that the flow regime within the voids was either laminar or transitional.

It is clear that there is a continuum of hydraulic response types in abandoned mine voids post-rebound, ranging from 'flashy' systems dominated by turbulent flow in supra-water table voids, through hybrid systems in which rapid storm flow inputs are superimposed on steady baseflows from ponds with gentle hydraulic gradients, to deeply-flooded mine voids hosting slow, laminar flows, which respond in a very muted way to short term changes in recharge. Various examples of these different types of system are given by Younger (1998a).

It is worth noting that, even where a mine has been abandoned for many years, unexpected perturbations in the hydraulics of the flooded mine voids can prompt a renewed phase of turbulent flow, with problems of erosion similar to those which are more often associated with the initial rebound period. For instance, around midnight on the 24th of June 1998, water suddenly began to surcharge the main storm sewer in the small coastal town of Spittal (Northumberland,UK), quickly flooding 19 homes with ochreous, sediment-laden water, which continued to flow as a torrent for 17 hours before suddenly slowing to a trickle. Subsequent investigations showed that a substantial head of mine water had built up behind a roof fall in old coal workings. When the head exceeded some critical limit, collapse and/or rapid erosion of the roof fall debris gave rise to the turbulent discharge and the consequent flooding (Younger, 2000d). Similarly, in the village of Leadhills in Scotland's Southern Uplands, post-abandonment changes in mine stability led to a serious threat of flooding. Prior to August 1991, an old drainage adit known as the Gripps Level drained some 23 Ml.d^{-1} of water from abandoned lead mines. A roof-fall in the Gripps Level at that time halted the discharge, and led to the impoundment of a vast volume of mine water in the old workings. A head of 25 m built up behind the roof fall, and eventually water began to emerge from old air shafts along the Level. A large tension crack opened up in the hillside, and this also began to yield water, raising the alarming prospect of a possible catastrophic slope failure (Schmolke, 1998).

3.11.3 THE ENVIRONMENTAL IMPACTS OF ABANDONED MINE DISCHARGES

Impacts on aquifers

Where the hydraulic head in a pit lake or flooded deep mine workings exceeds that in an adjoining or overlying aquifer, and hydraulic connection exists between the mine void and the aquifer, then a plume of mine water can be expected to enter and migrate through the aquifer. Remarkably few cases of this phenomenon have been reported to date. One example given by Younger and Adams (1999) relates how part of a public supply dolomite aquifer was rendered unusable as a drinking water source for the foreseeable future following rebound in underlying coal workings. Although any acidity in the inflowing mine waters was rapidly buffered by dissolution of the dolomite, sulphate persisted in the aquifer at concentrations as high as 500 mg.l^{-1}, which is twice the permissible limit for potable waters and very hard to remove by conventional water treatment processes.

In many cases, minor aquifers (e.g. alluvial sand and gravel) become pervasively contaminated by mine waters after rebound, but the main perceived impact is not the loss of resource in these aquifers themselves, but the impacts of the poor quality ground water on the surface watercourses into which they drain.

Impacts on surface water systems

As with pumped mine waters, it is only right to acknowledge that not all post-rebound mine waters cause degradation of receiving watercourses. For instance, drainage from abandoned mines in many parts of the world supports the development of natural fluvial and lacustrine ecosystems, which would desiccate without the mine water inputs. A good example is provided by the Blaydon Hazard Shaft (Gateshead, UK), which discharges up to 2700 m^3.d^{-1} of water with only minimal iron contamination (< 2 mg.l^{-1} Fe) into Shibdon Pond, which is a legally-designated 'Site of Special Scientific Interest' on account of its lake and wetland habitats. As the mine water maintains a temperature of 10°C all year round, it serves to maintain unfrozen conditions in a large area of the pond in even the coldest of winters, thus providing an invaluable refuge for wildfowl. Indeed, besides receiving much of its flow from mine water, Shibdon Pond owes its very existence to mining, as it has developed in a closed depression formed by mining subsidence.

Polluting discharges from abandoned mines naturally receive far more attention than inconspicuous discharges of good quality. Probably the most common pollution problem associated with abandoned mine discharges is the vivid red/orange staining of stream beds by 'ochre' (i.e. ferric hydroxides and oxyhydroxides; Section 2.1.1). Although widely considered as merely a question of aesthetics (i.e. human perceptions of the alarming appearance of ochre-stained streambeds), iron actually displays both direct and indirect toxicity in aquatic ecosystems (Vuori, 1995). This is all the more surprising in view of the fact that iron is an essential trace element for humans and other vertebrates. Although antioxidants within mammalian bodies usually counteract acute iron toxicity, when present at high concentrations dissolved iron can attack the tissues of higher animals by peroxidation of lipids. This releases hydroxyl free radicals which can go on to attack proteins, particularly those which form the intestinal linings, cardiovascular tissues and liver material (Tenenbein 1998). This is probably why high iron concentrations cause irritation of gill epithelia in fish. Iron can also act as a chronic toxin, causing 'haemochromatosis', the symptoms of which can include cirrhosis of the liver, anorexia, oligura, diarrhoea, hypothermia, diphasic shock, vascular congestion, metabolic acidosis, and, ultimately, death (Cornell University, 1999). While the precise levels at which iron becomes acutely toxic to mammals and fish are not yet well known, anecdotal evidence suggests that waters with a few tens of mg.l^{-1} Fe can be successfully traversed by salmonid fish, whereas waters with hundreds of mg.l^{-1} Fe are apparently impassable. Furthermore, experimental work in South Wales has shown that salmonid redds are far less likely to successfully hatch in stream reaches affected by iron pollution than in unpolluted reaches of the same streams (Edwards and Maidens, 1995).

With regard to invertebrates, iron is known to cause 'oxidative stress at cellular level' (Totaro *et al.*, 1992), although again, at what concentrations these stresses become lethal remains uncertain. In any case, dissolved iron concentrations in excess of 0.5 mg.l^{-1} can be expected to decimate aquatic invertebrate populations simply by disruption of their food supply. This arises because ochre coatings on streambeds (which begin to become significant at about 0.5 mg.l^{-1} Fe) effectively smother the sediment surface, preventing the penetration of light to benthic algae. Suspended iron hydroxide particles in the water column further exacerbate the interception of light. The net effect is that benthic photosynthesis is prevented, thus halting most primary production. With no algal debris on which to graze, invertebrate primary consumers cannot survive. Pathway analysis of chemical and ecological data has identified this "ochre smothering" effect as the dominant cause of impoverishment of benthic macroinvertebrate numbers and diversity in iron-polluted streams, be they acidic or alkaline in nature (Jarvis and Younger, 1997). It is also possible that the clogging of streambed gravel pores by ochre precludes oxygen circulation, which may explain the poor spawning success of redds in ferruginous river reaches (Edwards and Maidens, 1995). Thus with spawning success jeopardised and food scarce, higher fish are likely to be absent even if ambient iron concentrations are not directly toxic to them.

Where mine waters result in the receiving watercourse having a low pH and/or elevated concentrations of ecotoxic metals (such as cadmium, zinc, copper, lead etc), serious ecosystem degradation is common, even in the absence of ochre staining. For instance, Figure 3.44 shows the ecological impacts of a pumped mine water discharge which contained very little iron (there was no ochre staining in the receiving watercourse), but up to 25 mg.l^{-1} Zn. The pie charts show the numbers and relative abundances of various genera of benthic macro-invertebrates revealed by three-minute kick-sampling in the stream at

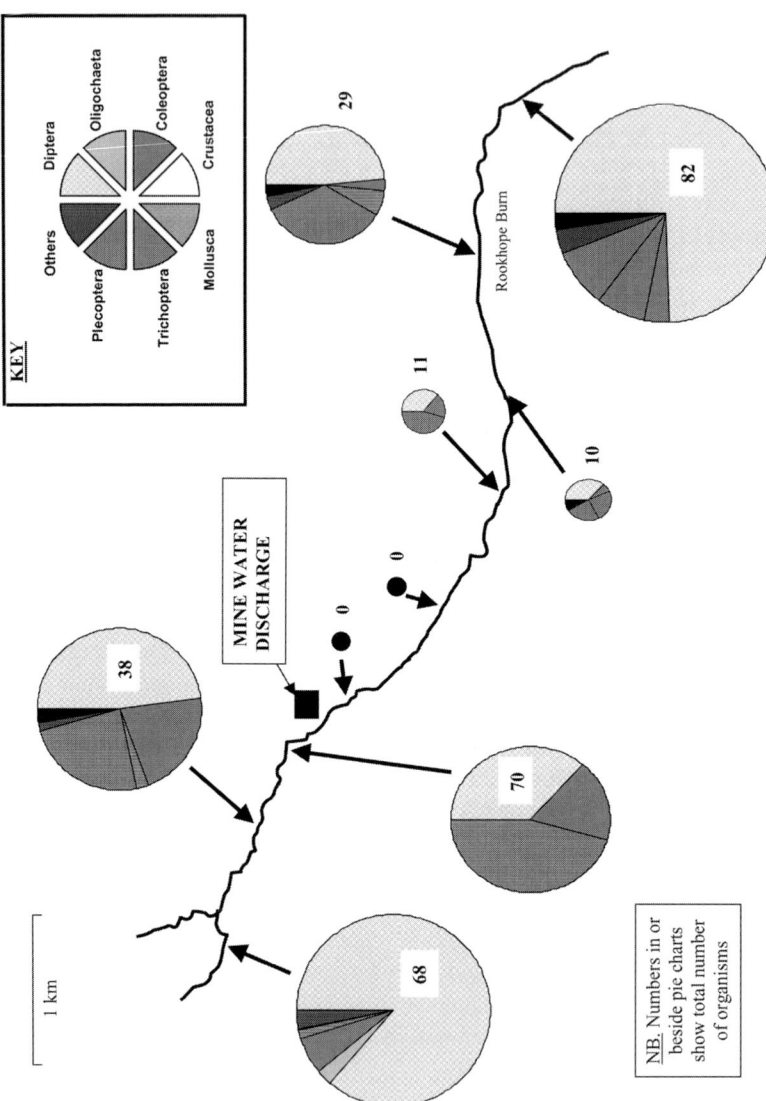

Figure 3.44: Sketch map of the Rookhope Burn, a stream in County Durham (UK) strongly impacted by mine water discharges. The pie-charts showing the impact of a zinc-rich discharge from a mine on the numbers of diversity of benthic invertebrates in the receiving watercourse. (Figure courtesy of A. Jarvis).

the points shown. The impact of the mine water discharge is immediately apparent, with zero invertebrates being captured in the sites immediately downstream of the mine, and only slow recovery thereafter, albeit the 'recovered' population is dominated by pollution-tolerant genera such as *Diptera*. Examples of this type can readily be multiplied (e.g. Scullion and Edwards, 1980; Jarvis and Younger, 1997); however, further details on the aquatic ecotoxicity of mine waters are beyond the scope of this volume, and the interested reader is encouraged to consult the sources cited and the excellent reviews of Kelly (1988, 1999).

Even where the metallic contaminants are removed from mine waters by means of conventional treatment (Chapters Four and Five), the increase in total dissolved solids which a mine water imparts to a receiving watercourse can remain a major problem. This is especially the case in semi-arid countries, such as South Africa, the western USA and parts of South America, where the loss of water resources to salinisation introduced by mine waters is generally considered a greater overall problem than ecological problems due to toxic metals. As will be seen in Chapters Four and Five, this is a major driver for ongoing developments in mine water treatment technology.

3.11.4 THE LONGEVITY OF POLLUTION FROM ABANDONED MINES

After an abandoned mine decants to the ground surface or an adjoining aquifer, the quality of the water flowing from the workings usually improves over time, until some long-term 'asymptotic' level of residual contamination is reached (Figure 3.45). The initial improvement in quality, which may take years or even as long as four decades to reach completion (Wood *et al.*, 1999; Demchak *et al.*, 2000), has come to be termed 'the first flush' (Younger, 1997a). The word 'flush' is appropriate, since the water quality improvement corresponds to the gradual flushing of the voids, such that the highly contaminated waters which typically occupy the voids immediately after the culmination of rebound (see Sections 2.2.5 and 3.10.1) are progressively displaced by fresh, less contaminated recharge. After completion of the first flush, the long-term 'asymptotic' quality is determined by the rate of ongoing pollutant release in and near the zone of water table fluctuation (Younger, 1997a, 1998b). The balance of factors which determine the nature (and temporal changes in) the long-term asymptotic water quality have already been discussed in Sections 2.3.3 and 2.3.4, and illustrated by calculations in Worked Examples 2.6 and 2.7. Hence no further discussion is warranted here; rather it is apposite now to consider those predominantly hydrological factors which influence the form and duration of the first flush.

Although a number of authors have attempted to identify the processes responsible for the first flush, most early explanations focused on unlikely hydrochemical explanations. For instance, Frost (1979) assumed that the observed decline in iron concentrations in water pumped from recently-flooded mine voids could be attributed to the exhaustion of a mineral source. Supposing that the transport of pollutants from the mineral surface was a diffusion-layer limited step, he applied a first-order kinetics formulation that yielded an exponential decay in concentrations, as long as it was assumed that the total volume of water in the flooded workings remained constant (which is patently not the case during (and immediately after) rebound). Similarly, Jones *et al.* (1994) argued that the inhibition of diffusion-limited dissolution of calcite by high ionic strength and high sulphate concentrations in the mine water might explain why, 22 years after commencement surface discharge, a mine water in Pennsylvania was far more alkaline than it had initially been.

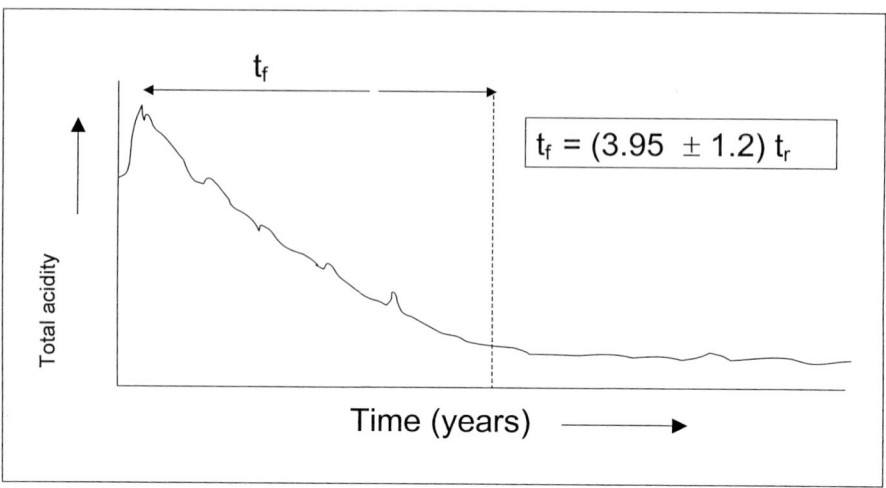

Figure 3.45: Schematic graph illustrating the 'first flush' phenomenon and its duration. t_f is the duration of the first flush and t_r is the 'rebound time', i.e. the time it took the workings to flood up to surface decant.

Crucially, however, they failed to explain why ionic strength and sulphate concentration should have declined over time.

Glover (1983) seems to have been the first to appreciate that the first flush is readily explicable in terms of hydraulic flushing, when he drew upon extensive personal experience of mine waters in the UK to suggest that the iron content of an abandoned mine discharge will typically halve in each subsequent period of time equal to that taken for the abandoned workings to fill with water after the pumps were withdrawn (i.e. an exponential decline). As discussed at length in Section 3.10.2, the time which workings take to flood is a function of the rate of water inflow and the volume of voids. These are precisely the same hydrological factors which will broadly govern the rate at which flooded mine workings can be flushed by fresh recharge (Younger, 1997a). Scrutiny of more recently available records largely vindicates the suggestion of Glover (1983), such that Younger (2000a) has proposed that:

(i)   the first flush is generally exponential in form, and
(ii) the duration of the first flush (t_f) can be related to the duration of the rebound process (t_r) as follows (Figure 3.45):

$$t_f \approx 4t_r \tag{3.54}$$

A natural corollary of the deduction that the first flush phenomenon is fundamentally hydrological in nature is that it ought to be amenable to simulation using physically-based hydraulic models. If the flushing of voids were a simple process, then satisfactory simulations might be obtained using simple "piston flow" or "well-mixed reactor" models, such as (e.g. Brusseau, 1996):

$$t_f = V_o / Q \tag{3.55}$$

where V_o is the total volume of voids containing the mine water (L^3) and
       Q is the rate of discharge from the mine ($L^3.T^{-1}$), which at steady state equals the recharge rate to the mine system

In practice, Younger (2000a) found that flushing is not completed anything like as rapidly as Equation 3.55 would predict. Clearly, the duration of flushing is a manifestation of more complex hydraulic processes. The tortuosity and heterogeneous permeability of the old mine workings clearly introduce significant delays into the flushing process. Such delays are normally described by the term 'dispersion' (Fetter, 1999), as embodied in the advection-dispersion equation (i.e. Equation 3.25 and its variants). For the particular case of the displacement of polluted water from flooded mine workings immediately following rebound, Equation 3.25 can be re-arranged and solved analytically as follows (Younger, 2000a):

$$C(t) = 0.5 \, C_o \, [erfc(\{L - v_a \, t_w\}/\{2(Dt_w)^{0.5}\}] + C_a \tag{3.56}$$

Where $C(t)$ = concentration at the decant point from the flooded mine system at time t (i.e. the elapsed time since the commencement of decant) ($M.L^{-3}$)
       $C_o = C_p - C_a$
       C_p = peak iron concentration at the end of the flooding process ($M.L^{-3}$)
       C_a = asymptotic iron concentration at the end of the main flushing period ($M.L^{-3}$)
       v_a = average groundwater flow velocity within the mine system ($L.T^{-1}$)
       t_w = "working time", the difference between the total length of the main flushing period (found by application of Equation 3.55 where records permit, or else on a trial-and-error basis) and the total number of days since overflow commenced
       D = longitudinal dispersion coefficient ($L^2.T^{-1}$). (If molecular diffusion is neglected, then $D \approx \alpha_L \cdot v_a$, where αL is the longitudinal dispersivity (L))
       erfc is the complementary error function (values for which are widely tabulated; e.g. Fetter, 1999)

Equation 3.56 is generally valid as long as advective transport is more vigorous than dispersive transport. This condition is generally measured by the Peclet Number ($P_e = v_a.L /D$); as long as P_e exceeds 10, the solution should be valid (Fetter, 1999). As Younger (2000a) has noted, however, *a priori* definition of the various terms in this equation is not straightforward, and it is likely to prove more useful as a means of indexing apparent velocities and dispersion parameters for flooded mine systems *a posteriori*, rather than as a predictive modelling tool.

3.12 HYDROLOGICAL INTERVENTIONS IN MINE WATER REMEDIATION

3.12.1 INTRODUCTION

As previously noted (Section 3.5.1), miners and water resources managers have a common interest in minimising water ingress to workings (see also Kesserû, 1995). The same common interest does not, of course, map directly onto abandoned tailings dams and spoil heaps, unless there are issues of public safety or environmental protection at stake. Whatever the occasion, it is good practice when addressing problems of mine water pollution to consider whether hydrological interventions might feasibly play a part in the overall solution being developed for a site. In some cases, hydrological interventions may be all that is needed to ensure a successful remediation effort. In most cases, hydrological controls will be only one element of a remediation scheme which will also include chemical interventions, either in the form of inhibition of pyrite weathering (see Section 2.4), or more likely in the form of active or passive treatment (Chapters 4 and 5).

   Hydrological interventions to prevent or minimise aqueous pollutant release from mine sites may be broadly classified into two categories:

(i)   Passive prevention of pollutant release, and
(ii)   Active mine water control

We will now consider both of these in turn.

3.12.2 PASSIVE PREVENTION OF POLLUTANT RELEASE

A formal definition of this approach has been developed by the European Commission's PIRAMID project (*www.piramid.org*), as follows:

> "Passive prevention of pollutant release from mine sites is achieved by the surface or subsurface installation of physical barriers (requiring little or no long-term maintenance) which inhibit pollution-generating chemical reactions (for instance, by permanently altering redox and/or moisture dynamics), and/or directly prevent the migration of existing polluted waters".

Specific techniques falling within this category are many and varied, though most fall into one of following four sub-categories:

• submergence techniques
• so-called "dry covers"
• subsurface impermeabilisation, and
• water diversion by gravity drainage.

We will now consider each of these in turn.

Submergence techniques
Possibly the simplest of all hydrological interventions for the remediation of polluted mine waters is to accept (or deliberately arrange) the flooding of mine voids in pyritic strata.

This serves to cut off the supply of atmospheric oxygen to submerged pyritic zones, thus largely halting pyrite oxidation in those zones (e.g. Fernández-Rubio *et al.*, 1987). Considerable research has gone into the design of water-tight seals for mine roadways to facilitate the flooding of zones which would otherwise be free-draining (Fernández-Rubio *et al.*, 1987; Ackman, 1987; Fuenkajorn and Daemen, 1996). The deliberate introduction of neutralising agents (e.g. slaked lime) or carbon sources for sulphate reducing bacteria (e.g. sewage sludge) into the deliberately flooded areas is also feasible in principle (Fernández-Rubio, *et al.*, 1987; Bowell *et al.*, 1999), though has not met with great success in field trials, mainly due to the difficulties of preventing the re-emergence of undesirable substances which are often present in these reagents (e.g. Younger, 2000d).

Some of the most successful applications of submergence techniques have been in the remediation of waste rock piles and abandoned tailings dams, in which context they are widely known as "water covers" or "wet covers". The use of water covers to inhibit the leaching of acidity from waste rock/tailings is predicated on the fact that the rates of oxygen diffusion in water are dramatically lower than in the atmosphere itself. The design of water covers is deceptively simple: all that is required is that some form of impoundment is arranged to ensure that a minimum depth of water is maintained above the surface of the flooded waste materials at all times. The minimum depth we would recommend is one metre, for two reasons:

(i)   where water covers are shallower than one metre, they are usually so well-mixed that the oxygen content at the sediment surface is little less than at the water surface (Li *et al.*, 2000); indeed, on these grounds, there are reasons to advocate water depths of several metres if this is consistent with other site constraints (access, safety etc).

(ii)   where water covers are shallower than one metre, the surface of the underlying tailings is prone to agitation by wind-blown waves of the magnitude which can develop in impoundments with areas up to several hundred hectares (i.e. of sizes appropriate to mine waste management situations). Re-suspension of tailings can result, leading to complications in water quality management (Catalan *et al.*, 2000).

In practice the topographic relief and other constraints of a given site (e.g. availability of runoff, intensity of evaporation) may make a 1 m minimum depth difficult to attain/maintain. Compromise solutions inevitably develop, with water covers as shallow as 0.3 m being developed in some cases. While re-suspension of tailings can be an issue with such shallow covers, observations of oxygen penetration into waste sediments below them show that they are still very effective in minimising sulphide oxidation. For instance, Li *et al.* (2000) used microelectrode techniques to measure dissolved oxygen profiles with depth through the standing water and underlying tailings in three water cover systems in Canada, in which water depths ranged between 0.3 m and 4.3 m. While the deepest covers worked best (restricting oxygen flux into the sediment surface to ≤ 0.35 mol·m^{-2}.y^{-1}, or about 1/2000th of the open-air flux rate into the same tailings), the shallowest of the covers (0.3 m) still restricted the flux to ≤ 16 mol·m^{-2}.y^{-1}, which was reckoned to be 1/200th of the open-air flux rate into the same tailings before flooding (Li *et al.*, 2000).

While minimisation of the downward diffusion of oxygen is the principal objective of water cover design, upward diffusion of metal contaminants is also an issue, particularly where previously dry tailings had to be newly flooded at the time of water cover establishment. In such cases, the process of flooding causes the dissolution of acid-generating secondary minerals (i.e. the 'vestigial acidity' (Section 2.2.5) comprising minerals such as those listed in Table 2.6). Gradual diffusion of these metals into the overlying standing water can give rise to problems in complying with environmental regulations. One possible mitigation method for this circumstance has been suggested by St-Arnaud (1994), who showed that addition of a layer of permeable quartz sand to the surface of the flooded tailings can help to minimise the diffusional release of metals to the water column. The process responsible is perceived to operate as follows: dissolved ferrous iron tends to oxidise briskly once in the aerated open waters, forming insoluble ferric hydroxide (ochre). The structural framework provided by the quartz sand acts as a locus for accumulation of the ochre. As ochre has a very high sorptive capacity for most other metals (see Section 2.4.2 and Dzombak and Morel, 1990), this sand-ochre layer can trap lots of the contaminants that would otherwise be released to the open water.

The problem of metal diffusion from recently-flooded tailings is one example of a wider drawback of submergence techniques in general, which is even more problematic where they are applied to deep mine voids (see Section 3.10.1). In other words, while flooding will prevent *future* oxidation of pyrite, it serves only to dissolve the existing products of former pyrite oxidation. Hence it can confidently be expected that the full water quality benefits of mine flooding (Fernández-Rubio *et al.*, 1987) will not be realised until the first flush is complete (Younger, 2000a).

Two other problems apply only to the use of submergence techniques for mine voids (and especially deep mine voids) as opposed to waste rock piles or tailings dams.

(i)   Since total submergence can often be effectively impossible to arrange in the case of large systems of deep voids, there is a tendency post-flooding for the water table to fluctuate some distance below ground. If the zone of fluctuation is in pyritic rocks, and is shallow enough to be in close connection with atmospheric oxygen, the seasonal rise and fall of the water table can lead to an ongoing release of 'juvenile acidity' (Section 2.2.5).

(ii)   Retention of ground water at high pressures behind subsurface impounds can be very difficult to ensure, both for reasons of the integrity of mine roadway plugs etc (see Section 3.7.5), but also because, as long as recharge continues, ground water head will continue to build up behind the impoundment until it overtops and decants via a natural outcrop or some other mining feature. For instance, as part of a pro-gramme of minimising acid drainage from the abandoned Summitville gold mine site in Colorado, the portal of one of the main drainage adits (Reynold's Adit) was plugged in January 1994. Four months later, the head of water within the workings had risen by about 50 m, giving rise to a major new discharge from the higher-lying portal of the Chandler Adit. In the words of Plumlee and Edelman (1995) "the leak underscores the fact that it is difficult to prevent leakage of ground water from a highly fractured and mined mountain. The plugging of the Reynold's Adit also resulted in the predictable reactivation of acid seeps and springs that had drained the site prior to underground mining, [and which had been previously identified by] mappable deposits of brown iron hydroxide minerals. Long-term leakage for groundwaters from these natural discharge points is inevitable".

"Dry covers": Barriers to moisture and oxygen ingress
As with water covers, the use of dry soil covers above reactive waste rock/tailings seems
deceptively simple upon first inspection. At their simplest, dry covers can be used to achieve
'dry entombment' of a body of mine waste by means of covering it with an impermeable
cap: if no water can contact the waste, no leachate can be generated. In this type of
application, dry covers do not differ at all from conventional landfill caps, the design and
construction of which has been widely discussed elsewhere (e.g. Suter *et al.*, 1993; Wing
and Gee, 1994), and will therefore not be discussed in any great detail here. At their
simplest, landfill caps consist of a layer of clay (usually about 1m thick, and rarely accepted
at less than 0.5m thick) which has been compacted such that it has so low a hydraulic
conductivity ($< 1 \times 10^{-9}$ m.s^{-1}) that it allows very little infiltration. Dry covers of this
relatively simple type (Figure 3.46a) have recently enjoyed widespread application in
Sweden, in the reclamation of tailings dams and waste rock piles (Gustafsson *et al.*, 1999).
These covers are typically in the range of 1.5 m to 2.0 m in thickness, and are composed
predominantly of clay-rich glacial till materials of Quaternary age, which are ubiquitous
in this northern land. Measured values of both hydraulic conductivity and oxygen diffu-
sivity in these covers typically fall in the range 5×10^{-9} m.s^{-1} to 1×10^{-10} m.s^{-1}. With such
low values, net infiltration through the covers is unsurprisingly low. For instance, studies
of the Steffenburg mine waste rock pile (Bersbo Mine, southern Sweden) revealed pre-
cover infiltration rates into bare waste rock of around 500 mm per year, compared with only
15 mm per year after the introduction of a 2.5 m till/compacted clay cover (Gustafsson *et
al.*, 1999). With recharge thus dramatically reduced, the ground water system within the
waste rock pile slowly adjusted over about four years, until a fairly steady residual discharge rate of only 0.02 l.s^{-1} was established (compared with about 0.5 l.s^{-1} prior to cover
installation).

In some Swedish dry covers, engineered low-permeability layers have been incorporated as alternatives to compacted clay layers. Materials used in these artificial layers
to date include a cement-PFA composite (e.g. at the Storgruve waste rock pile, Bersbo)
and paper mill bio-waste mixed with PFA (at Falun). Despite promising characteristics,
these artificial materials did not prove especially more effective than compacted clay.

In addition to their low permeability, paper mill wastes have two further potentially
desirable properties:

(i)   reactive organic matter within them tends to consume oxygen, thus further hinder-
       ing its ingress to the underlying waste rock.
(ii)   leachate formed by hydrolysis of the paper waste is often highly alkaline, and thus
       contributes to the neutralisation of acidity released by the sulphidic mine wastes.

Paper mill wastes have been successfully used as a capping material on a highly acidic
colliery spoil heap at Shilbottle (Northumberland, UK), where they eliminated acidic surface
runoff, leaving only acidic ground waters requiring passive treatment. Other organic waste
materials, such as sewage sludge (e.g. Metcalfe, 1994) and wood wastes from the forestry
industry (Germain *et al.*, 2000), have also been used as oxygen-consuming caps. Wood
wastes are particularly attractive where readily available as they do not carry the same
human health risks as sewage sludge. Often, the rate of oxygen consumption in organic
covers is so high that the oxygen concentration at the base of the cover (C_L) is always far

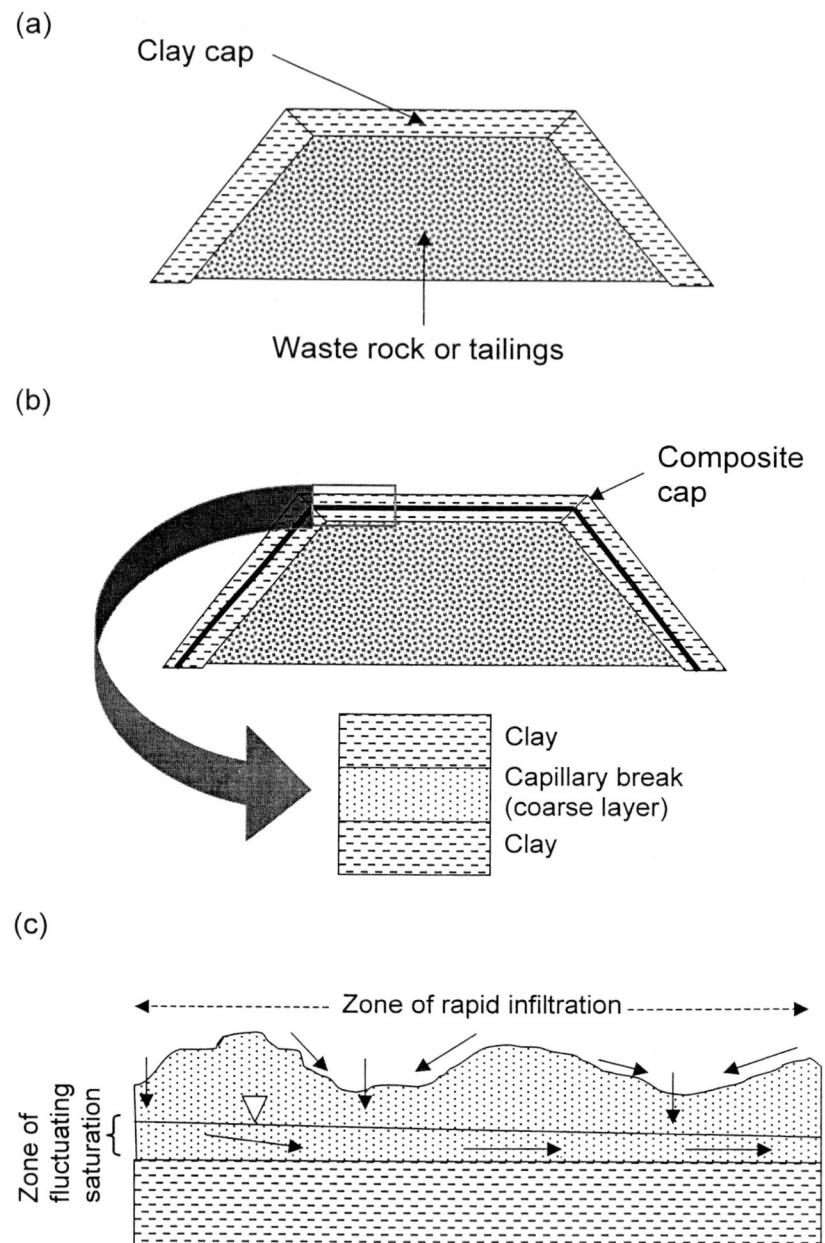

Figure 3.46: "Dry Cover" technology for abandoned waste rock piles/tailings deposits. (a) Simple low-permeability cover (b) Low permeability cover incorporating a capillary break (c) Detail of one edge of a storage and release cover. The overall layout will be similar to any other dry cover; the hummocky topography (to prevent surface runoff) and the deliberate lateral drainage above the compacted clay layer are the principal differences from (a) and (b).

smaller than the oxygen concentration at the ground surface (C_O), so that it is possible to calculate the oxygen flux at steady-state as follows (Nicholson *et al.*, 1989):

$$F_C = D_C (C_O/L) \qquad\qquad (3.57)$$

Where F_C is the flux of oxygen at the base of the cover ($M.L^{-2}.T^{-1}$)
        D_C is the effective oxygen diffusion coefficient for the cover material ($L^2.T^{-1}$) and
        L is the thickness of the cover (L)

While the reactivity of organic covers can effectively reduce oxygen fluxes into the under-lying waste, field trials with wood waste in Canada showed that the coarse, fibrous nature of such covers results in them actually *increasing* infiltration. This being the case, it is potentially important that some kind of impounding layer be introduced below the cover, or below the waste, to promote saturated conditions and thus gain some of the benefits of wet covers (Tassé, 2000). Nevertheless, even without impeding infiltration, wood waste covers remain highly effective, since they render infiltrating water sufficiently alkaline and anoxic that the resultant leachate from the waste rock pile is typically much lower in iron and higher in pH than had been the case before the cover was introduced (Germain *et al.*, 2000).

Slightly more elaborate dry cover designs (Figure 3.46b) incorporate a 0.5 m layer of coarse pebbles (cobbles) sandwiched between two clay horizons, forming what is commonly termed a 'capillary break'. The concept is that under unsaturated zone hydraulic conditions, the large pores of a cobble layer will not support continuous films of moisture passing from clast to clast, which would otherwise allow upward or downward migration of moisture in response to evapotranspiration or gravity respectively. Covers which incorporate a capillary break have been intensively studied in Canada, where they are termed CCBE (i.e. 'covers with capillary barrier effects'; Bussière *et al.*, 2000). The key to the successful application of CCBE technology is the attainment of such marked contrasts in the unsaturated hydraulic conductivity of successive fine- and coarse-grained layers within the cover that virtually permanent saturation is promoted in the fine-grained layers below the root zone (Nicholson *et al.*, 1989). This arises because the air-entry pressure of the coarse layer below a fine-grained horizon greatly limits infiltration, so that the fine-grained layer does not readily drain. (For this reason, CCBEs are sometimes termed "moisture-retaining covers", which is semantically incongruous with their inclusion in the category of 'dry covers'). Given the low diffusivity of oxygen through water, maintenance of permanently saturated conditions within the fine-grained layers can profoundly limit oxygen ingress, and thus oxidation of underlying sulphidic waste rock materials. Decreases in oxygen ingress and acidity release by up to four orders of magnitude are reported by Nicholson *et al.* (1989) from Canadian CCBE test sites. A range of Swedish CCBE case studies reveal comparable performance (Gustafsson *et al.*, 1999).

Simple diffusion models provide a credible basis for predicting and inter-comparing the efficiency of CCBEs as oxygen barriers (Nicholson *et al.*, 1989). The effectiveness of a cover (E_C) is defined as the ratio of the diffusive flux of oxygen into the waste which obtained before cover emplacement to the diffusive flux of oxygen through the cover, i.e. (Nicholson *et al.*, 1989):

$$E_C = F_O/F_C \tag{3.58}$$

Where F_O is the flux of oxygen before cover emplacement ($M.L^{-2}.T^{-1}$)

Because the rate of oxygen diffusion into uncovered mine waste is limited by the consumption of O_2 in sulphide oxidation reactions, Equation (3.58) can be expanded into the following parametric form:

$$E_C = (K_r D_W)^{\cdot}(L/D_C) + 1 \tag{3.59}$$

Where K_r is the reaction rate constant for sulphide oxidation (T^{-1})
         D_W is the effective oxygen diffusion coefficient for the uncovered waste rock pile ($L^2.T^{-1}$), (which typically takes values in the range 10–500 $m^2.a^{-1}$)
         L is, as before, the thickness of the cover (L), and
         D_C is, as before, the effective diffusion coefficient for the cover ($L^2.T^{-1}$), which for typical CCBE materials typically takes values in the range 0.09–0.9 $m^2.a^{-1}$.

Experimentation with Equation (3.59) rapidly reveals that, for all reasonable combinations of parameter values, the bulk of the reduction in acid leaching (by a factor of 10 to 50) occurs over the first 0.5 m of cover thickness (Nicholson et al., 1989). This result confirms 0.5 m as the minimum effective depth for such covers. Once cover thickness exceeds about 1.5 m, further gains of a factor of 10 in overall effectiveness turn out to require a further 10 m of cover in all cases. Hence in many situations, the cost-effective thickness of a CCBE will fall in the range 0.5–1.5 m.

More recently, Bussière et al. (2000) have used numerical models to demonstrate that CCBE efficiency is compromised where the cover has steep slopes. This is because lateral migration of the moisture within the fine-grained layers (as a resultant vector of vertical gravitational action) can lead to draining of the fine layers in dry periods, increasing the scope for oxygen ingress.

In arid and semi-arid climates, the use of dry covers made of fine-grained materials is inadvisable for two reasons:

(i)   Deep desiccation cracking during hot, dry periods is virtually unavoidable, and compromises the integrity of such covers.
(ii) The intense rainfall-runoff patterns characteristic of such regions can be expected to lead to severe erosion of any clay or silt rich covers, leading to the development of micro-scale 'badlands' topography (i.e. dendritic networks of drainage gullies which will eventually incise through the cover).

To overcome these problems, an alternative approach has been successfully developed, namely 'storage and release covers' (Durham et al., 2000). Figure 3.46c illustrates the concept. The cover comprises a layer of compacted clay (to ensure the development of a capillary break with the underlying waste rock), overlain by one or two metres of coarse-grained, geochemically inert materials (e.g. quartz -rich gravels). The ground surface developed on this upper gravel layer is deliberately shaped into hummocks. This micro-topography prevents direct surface runoff (and therefore surface erosion) and favours infiltration. The infiltrating water ponds above the compacted clay layer, temporarily functioning as an oxygen diffusion barrier in the same way as a conventional water cover. This ponded ground water can later be released by slow, lateral flow to a carefully-engineered seepage zone, or else retained *in situ* until such time as evapotranspiration by plants (such as eucalyptus) growing in the gravel can remove the moisture.

The problems which storage and release covers seek to avoid are not entirely restricted to arid and semi-arid regions, and vigilance must be maintained in all climatic zones to ensure the long-term integrity of dry covers. Even where erosion by surface runoff and/or desiccation cracking are negligible and vegetation establishment is vigorous, piercing of covers by deep tree roots can introduce macropore infiltration routes, allowing water to access the waste more readily during periods of surface runoff. In addition, where the reactivity of the cover material is important, eventual exhaustion of reactive components can be problematic where these cannot easily be periodically augmented.

Subsurface impermeabilisation approaches
One logical approach to the prevention of acid leachate generation within mine wastes is to destroy their permeability: if moisture cannot access the acid-generating parts of the waste, then acid cannot be generated. Implementation of this approach requires at least four steps:

(i)   Assay the body of waste for pyritic sulphur content, by means of closely-spaced boreholes and lab analysis of cuttings
(ii)   Identify the zones with a pyritic sulphur content over some critical threshold (we suggest: > 0.5 weight % pyritic sulphur, or > 1 weight % total sulphur), possibly using geostatistical interpolation methods
(iii)   Inject a grout under pressure into the most pyritic zones (taking care to use a sulphate-resistant grout)
(iv)   Monitor the leachate afterwards to ensure that acid generation is minimised; otherwise re-evaluate/repeat steps (i) through (iii).

Despite the simplicity of this approach, it has apparently not been implemented (or its implementation has not been reported) at large scale. However, it is a technique with considerable promise for dealing with leachate release from spoil heaps which have previously been successfully afforested or otherwise revegetated, such that the installation of a new dry or water cover is unlikely to be a popular option.

Control of polluted mine drainage waters by gravity drainage
Also known as "flow balancing", the use of simple gravity drains and impoundments can (in the right circumstances and with adequate design) play a useful role in the mitigation of mine water pollution.

For instance, the strategic diversion of surface waters away from known zones of infiltration to the mine voids may be possible in some cases. A good example of this approach in practice was undertaken by British Coal in the Dailly Coalfield (Ayrshire, Scotland) as part of their remedial efforts in relation to water pollution by a discharge from the abandoned Dalquharran Colliery (Younger, 2000c,e). In 1982–83, investigations of the hydrogeology of the mine complex revealed that most of the water entering the workings was flowing through the highest worked seam in a relatively clean condition, before falling to greater depths within the workings where it picked up very high acidity loadings. A gravity drain was therefore constructed to intercept the clean water in the upper seam at Dowhail, and this was successful in reducing the recharge to the Dalquharran system by about 95%.

A similar strategy was proposed by Ahmad (1970) in relation to the chronic pollution of Lake Hope in the Appalachian Coalfield. By careful identification of the aquifers and aquitards in the mined sequence, and their relative piezometric heads, it was deemed feasible to drill vertical wells which would divert clean waters into deeper aquifers, thus preventing these clean waters entering the mine workings where they were becoming contaminated.

Another gravity drainage approach to pollution mitigation was also implemented at the Dalquharran site: this took the form of a 'valve' (actually a small impoundment tower) at the discharge point, which served to trap discharging waters while head built up in the entire hinterland of partly-flooded workings. The system was designed such that overflow from the impoundment tower would only occur in very wet weather when dilution would be available in the receiving watercourse.

3.12.3 USE OF PUMPS IN MINE WATER CONTROL

Where gravity drainage is not feasible because of topographic relief and piezometric constraints, it might still prove feasible to use pumps to control the distribution of mine water, for purposes of minimising pollutant release and/or facilitating active or passive treatment (see Chapters 4 and 5). The main use of pumps in this context is in "pump-and-treat" applications, where boreholes or old mine shafts are used as pumping stations, creating sufficient drawdown that:

(i)   problematic gravity discharges of mine water which would otherwise occur in remote locations are prevented, and

(ii) the pumped water is readily accessible on a well-appointed site for treatment and release to the environment as appropriate.

This approach has been successfully implemented at a number of sites around the world, perhaps most notably in the UK, where at the time of writing around fifteen pump-and-treat systems are in operation (Younger, 2000c, 2001). The largest system of this type is in County Durham, England, where a total of 105 Ml.d^{-1} is pumped from nine old mine shafts, effectively preventing uncontrolled discharges of polluted water from deep mines which would otherwise occur at various points in a total catchment area of some 5,000 km^2 (Younger and Harbourne, 1995). A cost-benefit analysis of this pumping system concluded that the net present value of pumping costs (which were running at around £1M per annum in 1995) was less than that associated with permitting the mined system to flood and overspill, mainly because of the cost of replacing a £26M water treatment works on a river which would certainly receive uncontrolled mine water discharges if pumping were abandoned (Younger and Harbourne, 1995).

Rigorous appraisal of costs and benefits is an important part of the design for a proposed pump-and-treat system. In some cases, in which particularly sensitive receptors exist downstream of the zone of likely mine water emergence (e.g. a high value salmon fishery, or a public water supply abstraction), the outcome of the cost-benefit appraisal might be a foregone conclusion. However, in the majority of cases, the justification for a pump-and-treat scheme may be far less clear and a detailed technical-economic study will be necessary. Such a study must be conducted within an overall risk assessment framework, and will require input from a multi-disciplinary team of specialists, ranging from hydrogeologists and geochemists on the one hand to specialist environmental cost–benefit analysts on the other. From recent experiences with five such studies in the UK, it is possible to identify the steps which will be necessary in any such study:

1. A technical appraisal of the water quantities, qualities, and treatment requirements associated with the pump-and-treat option.
2. A technical appraisal of the observed (in the case of a retro-fitted system) or predicted locations, quantities and qualities of uncontrolled discharges which will occur in the absence of pumping. This must include an evaluation of likely treatment requirements and the feasibility of satisfying these requirements given the constraints of site access etc.
3. Environmental impact assessments in relation to 1 and 2 above.
4. Economic valuation of the costs associated with both options, including both use and non-use values for the receiving watercourses. Examples of use values include fisheries, water supply, navigation, recreation (formal and informal) and hydropower gener-ation. The impacts of alternative options on these values vary according to the location and the type of activity, in a manner likely to be unique to a given catchment. Non-use values might include ecological conservation goals and strategic values, such as the presence of a river of a given flow and quality as long-term insurance against failure of existing sources of public water supply.
5. Discounted cash flow analysis of the principal alternatives, so that all may be compared on a like-for-like basis, even though they may have different temporal characteristics in terms of timing of significant changes in water quality etc.

Resolution of all of the technical issues which arise in relation to the specification of an efficient mine water pump-and-treat system requires detailed, multi-disciplinary investi-gations involving mining engineers, hydrogeologists and geochemists. Some of the key technical issues which typically require resolution in the design of pump-and -treat schemes include:

(a) The quantity of water which must be pumped. The greater the drawdown which must be caused in order to prevent uncontrolled surface outflows, the greater the pumping rate which will be required. In the majority of cases, it will be necessary to pump at a greater rate than the combined flow rates of the uncontrolled surface discharges which would occur in the absence of pumping.
(b) The quality of water to be pumped, and how this might change over time. For instance, Chen *et al.* (1999) have described how the quality of water in two different parts of the same rebounding coal mine pond in Scotland was sufficiently different that it became an issue in deciding on the optimal location for a pump-and-treat operation. In the event, a site with a greater depth to water but better quality of water was chosen in preference to an alternative site near the coal seam outcrops, where pumping heads

were less but the water considerably more polluted (Sadler and Rees, 1998). After some months of pumping, however, the water at the chosen site gradually changed until it closely resembled the poorer quality water initially found near outcrop, no doubt because of later migration of the poorer quality water. Thus the pump-and-treat system ended up having to cope with the "worst of both worlds", i.e. the greatest pumping head and water of as poor quality as had been found anywhere else in the mine system.

(c) How best to ensure efficient pumping arrangements in relation to the mine voids. With surface mines this is relatively easy to arrange, but where deep mine voids must be controlled by pumping then a number of issues must be considered, such as:
- the structural stability of old shafts and adits if these are used for long-term pumping
- the likelihood of a newly-drilled borehole intersecting open voids which are well-connected to the rest of the workings
- the mechanical and electrical plant considerations, such as access to an adequate source of power supply, and the selection of pumps capable of coping with poor quality water and the peculiar hydraulics of large-diameter shafts.

With regard to the latter point, it is worth noting that many pump suppliers will be unused to providing electric submersible pumps capable of withstanding conditions of very low pH. In the Wheal Jane tin mine in Cornwall, UK, for instance, the pumps initially supplied to control a water with a pH of 2.5 had to be continually dosed with hydrated lime, delivered by a pipe right on their intake screens, in order to achieve pump component longevities on the scale of weeks rather than merely days. Another problem peculiar to the use of pumps in old mine shafts is that electric submersible pumps are designed to be used in relatively narrow diameter wells, such that the rush of water in the narrow annulus between the pump body and the well casing will serve to cool the motor. Where such pumps are used in shafts, they must be fitted with external "shrouds" (i.e. narrow tubes extending the length of the pump to an opening below the motor) to ensure the necessary rush of water past the motor.

CHAPTER FOUR

ACTIVE TREATMENT OF POLLUTED MINE WATERS

4.1. WHAT IS ACTIVE TREATMENT?

Before about 1980 (e.g. Barton, 1978), the only proven technologies for mine water treatment were those which we now term "active treatment". Active treatment is conventional waste water engineering applied to mine waters. The following formal definition of active treatment emphasizes the ways in which it differs from passive treatment (see Chapter 5):

> "Active treatment is the improvement of water quality by methods which require ongoing inputs of artificial energy and/or (bio)chemical reagents"

From the above definition, it is clear that "active treatment" embraces virtually all conventional mine water treatment technologies save those using wetlands and other "passive" unit processes (as described in Chapter 5). The "artificial energy" referred to in the definition given above can be provided in various forms, such as electrical power for pumping, mixing, aerating etc, heat to change reaction rates or pressure to control gas-liquid exchange rates. The "reagents" used in active treatment are usually alkaline liquids or solids (e.g. calcium hydroxide, sodium hydroxide), organic polymers (for coagulation/flocculation etc) or, less commonly, pressurized gases. One of the key advantages which active treatment has over passive treatment is the scope for precise process control: dose rates of reagents, speeds of pumps and mixers and other aspects of the treatment system can be adjusted instantaneously in response to changes in influent loadings or receiving watercourse conditions. As such, active treatment methods are especially suited to the management of water quality during mining operations, when the exigencies of dewatering and mineral extraction can lead to abrupt changes in the quantity and quality of the water make. Furthermore, it is invariably possible to actively treat mine waters on sites much smaller than those which would be required for passive treatment, which means that active treatment will always retain an important niche for large flows where land is scarce.

The fact that active treatment infrastructure comprises conventional technologies means that its detailed design does not differ significantly from the design of infrastructure for similar unit processes in ordinary wastewater treatment plants. Since guidelines for conventional plant design are well-established and widely available (e.g. Casey, 1997), the presentation of active treatment given in this chapter is far more concise than the presentation of passive treatment in Chapter 5 (where we deal with new technologies which in the main have not yet been thoroughly described in print).

Because of the variety of mine water chemistries encountered in nature (Chapter 2), and because of the familiarity of the mining sector with the physical and chemical processes necessary to separate metals and water (many of which were initially developed for mineral recovery purposes), there is a wide range of conventional, active treatment technologies for mine waters. Selection of the most appropriate technique for treatment of a given mine water should be based on consideration of the raw mine water quality and the desired final effluent quality. Figure 4.1 presents the decision-making logic for selection of mainstream active treatment processes.

In this chapter, practical guidance is given on the selection and use of the most widespread and most promising mine water treatment technologies. Because of its predominance in the mining industry, rather more emphasis is given in this chapter to the conventional "ODAS" (*o*xidation, *d*osing with *a*lkali and *s*edimentation) approach than to other more specialized techniques. Under virtually all regulatory regimes, treated effluents will have to comply with standards such as the following:

- $6.5 < pH < 8.5$
- Total Fe <1 mg.l^{-1} (increasingly, this limit is set at 0.5 mg.l^{-1})
- Total suspended solids < 50 mg.l^{-1}

Conventional ODAS is well suited to achieving these standards.

Where more stringent emission standards must be met, it may necessary employ enhanced ODAS techniques, or alternative techniques (e.g. sulfidization, reverse osmosis). Examples of more stringent standards which might be imposed depending on circumstances include:

$Zn < 0.5$ mg.l^{-1}
$Cu < 0.2$ mg.l^{-1}
$Al < 0.1$ mg.l^{-1} ⎫ Typically demanded where the receiving
$Cd < 0.5$ µg.l^{-1} ⎭ watercourse is ecologically sensitive

$SO_4^{2-} < 250$ mg.l^{-1}
$Mn < 0.5$ mg.l^{-1} ⎫ Typically demanded where the
Total dissolved solids < 1500 mg.l^{-1} ⎭ receiving watercourse supports water
supply abstractions downstream

In essence, there is no technical limit to the quality of water which can be achieved using existing technology, albeit the cost may be astronomic. In practice, therefore, the selection of a treatment approach to achieve the most stringent of effluent requirements comes down to an economic–environmental cost benefit analysis. For instance, demanding that a highly energy-intensive treatment method such as reverse osmosis be used to treat a mine water might cause more environmental damage (through the generation of the required power and disposal of highly concentrated residual brines) than it actually solves. While it is beyond the scope of this volume to consider how to manage such decisions, it is important to realize that technology selection and treatment plant design is ultimately a multidisciplinary task. Here we focus only on the technical issues of mine water treatment.

The disposal of residual brines from reverse osmosis plants is one example of a problem which is common to all active treatment methods: the management of the sludges which arise from the precipitation or sorption of metals. The closing section of this chapter focuses on this issue.

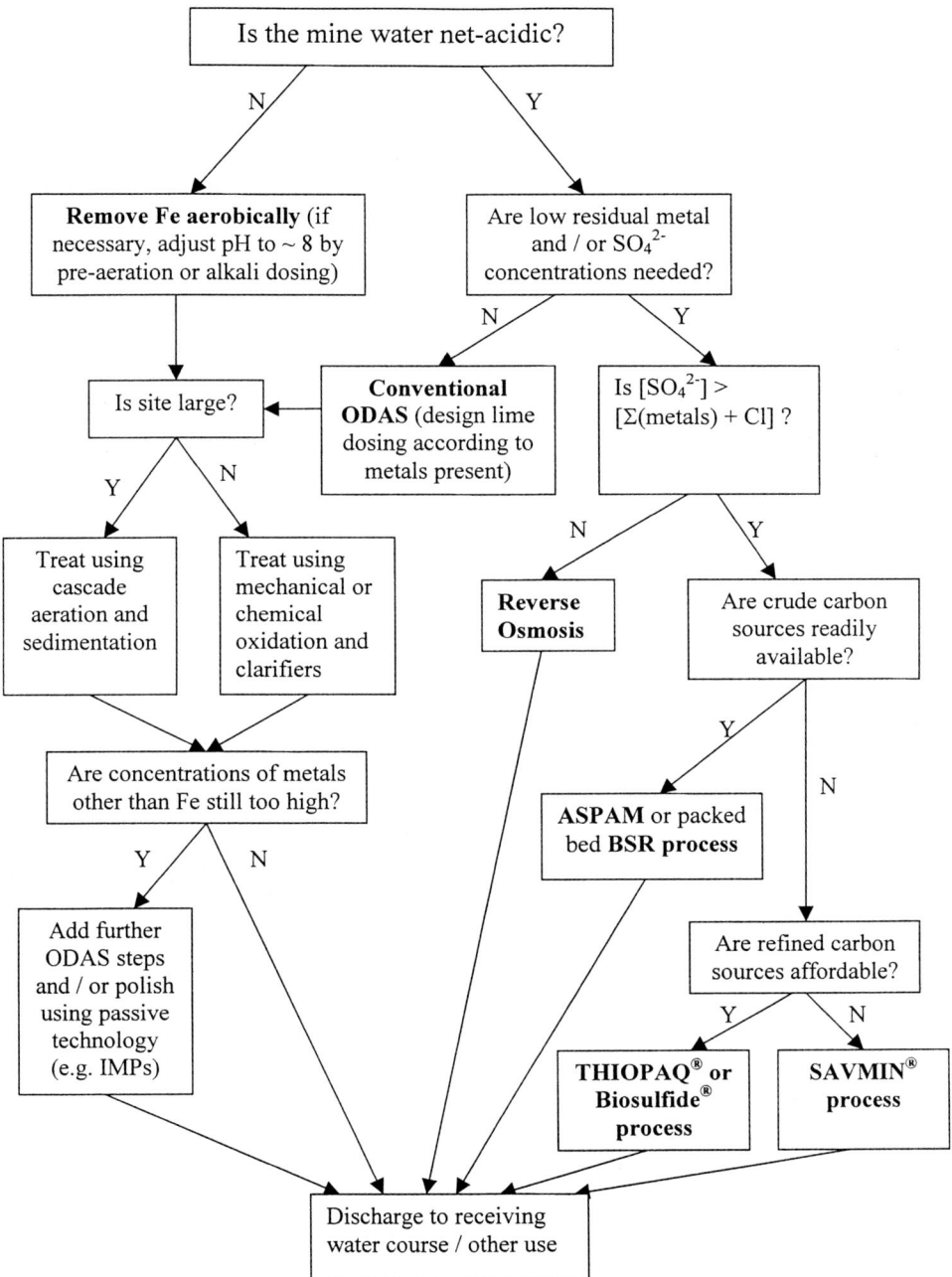

Figure 4.1: Flow-chart for selection of an appropriate active treatment process for a given mine water.

4.2 OXIDATION, DOSING WITH ALKALI AND SEDIMENTATION (ODAS)

4.2.1 INTRODUCTION

The techniques which we classify under the acronym ODAS are based on the same basic strategy:

- Oxidation, to transform relatively soluble ions such as Fe^{2+} and Mn^{2+} into their less soluble oxidized forms (Fe^{3+} and Mn^{4+} respectively).
- Dosing with alkali, to raise the pH of waters that are already acidic, and to counteract any further lowering of pH which would be expected to result from the release of protons which occurs when Fe^{3+}, Mn^{4+}, Al^{3+} and certain other pollutant metals hydrolyze to form hydroxide solids (see Section 2.2.4).
- Sedimentation, to remove these metal hydroxide solids from suspension.

It is important to note that the three steps (O, DA and S) will not always be arranged in precisely this order; rather the sequence of the three processes is usually varied according to the particular situation, as follows:

$DA \rightarrow O \rightarrow S$ } suitable for very acidic waters (raise pH first, to increase reaction rate of oxidation step)

$0 \rightarrow DA \rightarrow O \rightarrow S$ } pre-aeration step for highly carbonated waters, to de-gas CO_2 and thus raise pH, speeding oxidation

$S \rightarrow DA \rightarrow O \rightarrow S$ } where inert solids need to be removed first (common in active mines)

As a very widespread technology, ODAS can be implemented using various equipment to achieve the same ends. In the sections which follow we will describe some of the most common approaches to the various steps in the process. After discussing these we will consider some of the limitations of conventional ODAS processes, and then describe some enhanced ODAS systems which have been designed to overcome these limitations.

4.2.2 OXIDATION

4.2.2.1 Purpose of oxidation in mine water treatment

Oxidation is an essential time- and money-saving component of most mine water treatment systems. Where the oxidation step is missed out, or inefficiently implemented, dissolved Fe^{2+} can move right through the treatment plant, only to oxidize downstream, in the receiving watercourse, causing unnecessary staining and (if it is present in excess of the alkalinity content of the water) even lowering pH once more (e.g. Johnson, 1998). The primary objective in designing an oxidation feature is to ensure sufficient oxygen is transferred to the water to oxidize all of the Fe^{2+} and Mn^{2+} present in solution. In the case of certain strongly net-alkaline mine waters, the oxidation step can also serve to release excess CO_2 from the water to the atmosphere, which will usually cause pH to rise, thus increasing the reaction rate of the Fe^{2+} oxidation reaction. Some of the most common ways of achieving oxidation are described in the following sections.

4.2.2.2 Cascade aeration

This is the most popular option for achieving oxidation, at least on reasonably spacious sites where sufficient head exists to allow the water to cascade. The theoretical basis for oxygenation of water on cascades has been examined in detail by Novak (1994), from whose work fundamental guidance can be obtained. More empirical guidelines for cascade construction are provided by Wegelin (1996) amongst others.

Most aeration cascades are simply flights of steps over which the water is spread in a thin film. In any one cascade, three or four steps will typically be included, each around 70 cm high. Ideally each step should have a vertical 'lip' ($\leq$ 4 cm high) at its downstream edge, to promote the establishment turbulent eddies which favor efficient oxygen transfer (Novak, 1994).

To ensure that the entire water column has continuous access to atmospheric oxygen, the flow should be kept shallow (1–2 cm on average) and agitated. An appropriate depth can generally be ensured by allowing 10 cm width across the steps for every 60 l.min^{-1} of water requiring aeration. (For instance, a flow of 450 l.min^{-1} would require steps at least 75 cm wide). An example of an efficient mine water aeration cascade is shown in Figure 4.2, which illustrates the extent of ochre accretion which can be expected on an aeration cascade. As it will occasionally be necessary to scrape this ochre off the cascade, it is good practice to design the tail of the cascade such that access with cleaning equipment can be readily arranged (e.g. instead of having a deep lagoon section immediately at the foot of the cascade, leave an apron of hard standing 3–4 m wide between the foot of the cascade and the edge of any deep water.

A typical aeration cascade will be able to introduce sufficient oxygen to the water to oxidize around 50 mg.l^{-1} of Fe^{2+}; hence successive cascades and settling basins may be necessary to remove concentrations in excess of this amount.

4.2.2.3 Trickle filter aeration

Trickle filters (also known as "aerated biological filters", "percolating filters" or "static mixers") are aerated, unsaturated flow systems in which the water to be treated is introduced to the top of a chamber loosely packed with media of high specific surface area. Such units are commonly used in conventional wastewater treatment, and design guidance is therefore widely available elsewhere (e.g. Casey, 1997). Most trickle filters are circular in plan view and up to 2 m deep, though other shapes and dimensions can be as effective. Applications of these kinds of filters to mine water treatment are described by Best and Aikman (1983), Ackman (2000) and Jarvis and Younger (2001). Further details of the use of trickle filters as 'stand-alone' semi-passive treatment unit processes ('SCOOFI reactors') are given in Section 5.4.2.5, and will not be discussed further here.

4.2.2.4 In-line venturi aeration

Some of the highest rates of oxidation achievable by purely physical means have been

Figure 4.2: Cascade aeration of ferruginous mine water at a major coal mine dewatering station (Woolley, West Yorkshire, UK). This cascade is one of two in parallel, each aerating half of a total flow of around 200 l.s^{-1}. After cascade aeration, the water is treated by settlement in three lagoons and a 1.4 ha wetland. For further details see Laine (1997). (Photo: P. Younger).

accomplished using in-line venturi devices. Figure 4.3 illustrates a simple venturi device. As shown in the figure, the venturi is nothing more than a small pipe, open to the atmosphere at one end but submerged within a pipe flowing at full-bore at the other, and angled so that the venturi orifice faces downstream. As water flows through the pipe, air is entrained via the venturi, resulting in rapid oxygenation of the water. The degree of oxygenation which the water receives is a function of the hydraulic conditions within the water pipe, and optimal conditions occur when the pipe is flowing under an operating head of about 15 m.

The US Department of Energy have patented a mine water treatment system which has no internal moving parts (the "In-Line System" or ILS) of which a venturi aeration device is a key component (see Ackman and Place, 1987; Ackman and Kleinmann, 1993; Ackman, 2000). ILS applications of venturi devices have proven capable of oxidizing as much as 900 mg.l^{-1} of Fe^{2+} in a single aeration step (Ackman, 2000). Obviously, for such oxidation to be beneficial, it is important that the venturi injection occurs near the end of the water pipe line, so that this does not become clogged with ochre. (In the ILS application, alkali dosing is also undertaken via dripping of lime into the venturi, and the resulting oxidized and dosed effluent is passed through a static mixer of some kind to ensure complete reaction of the added reagent; Ackman and Kleinmann, 1993).

4.2.2.5 Mechanical aeration
A further popular option is to aerate the mine water by mechanical means, such as jetting the water at high pressure into mechanically-stirred tanks. There is a wide range of equipment suitable for such purposes, resulting in a very wide range of particular arrangements. Advice on suitable designs is readily available from suppliers of standard wastewater treatment equipment, and will therefore not be discussed further here.

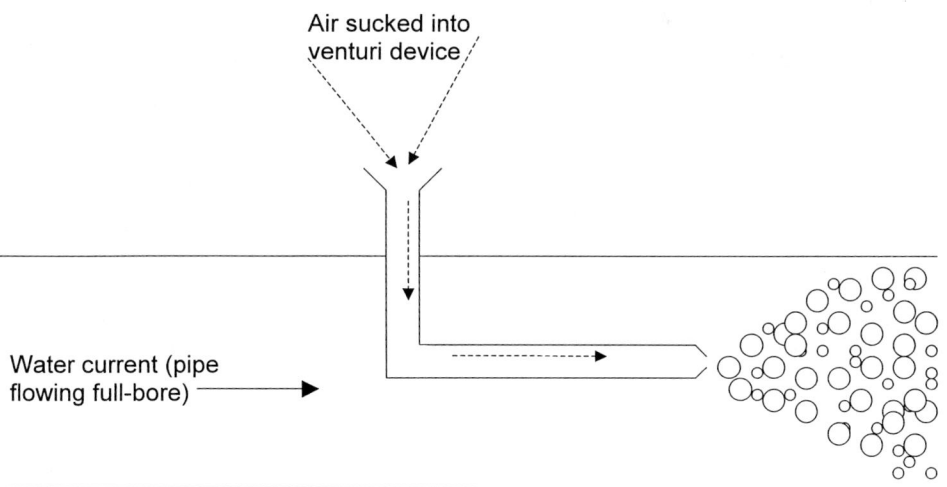

Figure 4.3: Sketch of a basic venturi device, as used to rapidly aerate mine water in the ILS (in-line system) of Ackman and Kleinmann (1993).

4.2.2.6 Biochemical oxidation by RBC

Rotating biological contactors (RBCs) are a proven technology for microbial oxidation of sewage BOD in wastewater plants serving small communities (Casey, 1997). RBCs depend on the establishment of bacterial colonies on the surfaces of sets of slowly-turning disks, mounted on a horizontal spindle such that at least half of each revolution exposes each point on every disk to the air. Olem and Unz (1977, 1980) undertook pilot-scale investigations of RBC applicability for ferruginous mine waters. Since bacterial oxidation of Fe^{2+} is only significant where pH is low (Section 2.1.2), it follows that RBCs are only appropriate for net-acidic mine waters. Although they have not enjoyed widespread uptake, RBCs could satisfy a niche market for treatment of small volume discharges which are too rich in ferric iron, aluminum or DO for treatment using anoxic limestone drain technology (see Section 5.4.2.2). Field trials show that a RBC can remove Fe from solution at efficiencies in excess of 90% under the following conditions:

- 2.3 < pH < 4.8
- 44 mg.l^{-1} < total Fe concentration < 200 mg.l^{-1}
- the system has been running long enough to develop dense bacterial colonies (>20 days typically)
- hydraulic loading rates ≤ 0.1 m^3.d^{-1}.m^{-2} of disk surface area
- retention time ≥ 1 hour in a RBC unit containing 40–50 disks.

4.2.2.7 Chemical oxidation

Where none of the physical and biochemical oxidation techniques listed above is feasible, which will most often be because of cramped conditions on site, chemical oxidation may provide an alternative with a much smaller 'footprint'. Hydrogen peroxide is the most common chemical oxidant used in practice by the mining industry. Dosing with hydrogen peroxide is physically simple: a container of 35% H_2O_2 is fitted with a dripping tap which doses directly into the mine water stream (in pipework or open channel) at a rate of around 0.35 millilitres of H_2O_2 (35% solution) per gramme of dissolved ferrous iron in the mine water. Some examples of successful hydrogen peroxide drip-feed dosing operations are given in Worked Example 4.2.

Besides peroxide, sodium hypochlorite (NaOCl) solution is frequently used. A 10% solution is usually selected (i.e. about twice as strong as conventional household bleach), and is typically applied to oxidize and Mn remaining in solution after Fe has already been removed by pH adjustment and aeration. Potassium permanganate (KMnO$_4$) is also used in some cases, and despite the apparent paradox, it successfully achieves low residual manganese concentrations if dosing is properly controlled.

Conventional disinfection oxidants used in potable water treatment works, such as chlorine and ozone, will also oxidize ferrous iron, though cost usually makes their use on mine sites unappealing. However, the fact that a few mg.l^{-1} of ferrous iron can be readily removed by chlorination or ozonation should be borne in mind when assessing the water resources impacts of mine water discharges, since this means that many municipal water treatment works will be able to cope with a few mg.l^{-1} of iron without adaptation.

4.2.3 DOSING WITH ALKALI

4.2.3.1 Rationale for dosing with alkali

There are numerous reasons why alkali dosing is widely practiced in active treatment:

- Under most regulatory regimes, a pH in the range 6.5–8.5 is a key requirement in every discharge permit for treated effluent.
- At low pH, ferrous and ferric iron are soluble and mobile, whereas both begin to become insoluble (i.e. their aqueous concentrations begin to be limited by the solubility of hydroxide phases) once pH exceeds 3.5 (for ferric iron) and 8.5 (for ferrous iron; see Table 4.1). Removal of iron as a hydroxide is therefore favored by high pH.
- The oxidation of ferrous iron to ferric iron occurs much more rapidly at high pH; for this reason, even though a mine water may contain plenty of alkalinity at pH 6.5, it may still be worth adding alkaline reagents to achieve a pH of around 8.5, which will favor far faster oxidation of Fe^{2+}. Similarly, Mn^{2+} oxidation is very slow at pH < 8, and proceeds most briskly where pH is held at about 10.2.
- Flocculation of ferric hydroxide particles proceeds more rapidly at pH in the range 7.5–8.5 (though at much higher pH, zero-charged particle formation may hinder flocculation once more).
- Aluminum, which is highly ecotoxic in the dissolved form, is highly soluble where pH is below 4, but virtually insoluble in the pH range 5.5 to 8.5.
- Where there are other potentially eco-toxic metals in the mine water besides iron, sedimentation at neutral pH will often not be sufficient to achieve metal removal. This is because the aqueous concentrations of many other metals (Cd, Ni, Pb, Zn, Mn) do not become limited by the solubilities of their respective hydroxides until pH is raised considerably (Table 4.1). Once hydroxide solids containing these metals have been precipitated, they can be removed from the mine water by sedimentation (aided by coagulation/flocculation in most cases) as described in the preceding section.

4.2.3.2 Pre-aeration to strip excess CO_2 before alkali addition

In cases where the mine water contains elevated concentrations of dissolved CO_2 it will usually be worthwhile incorporating a simple pre-aeration step prior to alkali dosing. The rationale for pre-aeration is two-fold:

(i) Some mine waters contain dissolved CO_2 well in excess of the amount they could hold at equilibrium with the atmosphere (for instance, pCO_2 of $10^{-1.0}$ in the water compared with $10^{-3.5}$ in the atmosphere). This is frequently the case, for instance, in waters pumped from flooded underground coal mines. Such waters tend to be buffered to a pH around 6.5 by the $H_2CO_3^*/HCO_3^-$ couple (see Section 2.2.2). As the excess CO_2 is driven of by aeration, the pH usually rises, promoting more rapid Fe^{2+} oxidation and thus lessening the amount of alkaline reagent which must be added to achieve the same goal

(ii) Excess dissolved CO_2 can also hinder the raising of pH by addition of alkaline reagents (which provide hydroxyl ions). This hindrance arises from buffering reactions, such as:

$$OH^-_{(aq)} + CO_{2\ (aq)} \Leftrightarrow HCO_3^-{}_{(aq)} \qquad\qquad (4.1)$$

In other words, the hydroxyl alkalinity added in the form of alkali dosing agents is consumed by reaction with dissolved carbon dioxide to form bicarbonate alkalinity, thus preventing the rise in pH which would correspond to an increase in dissolved OH^-. Pre-aeration helps to minimize this effect. Pre-aeration can be achieved using cascades or mechanical aeration tanks as described in Section 4.2.2.

4.2.3.3 Alkaline reagents for mine water dosing

Skousen *et al.* (1990) have discussed the main alternative reagents used in active mine water treatment, and provide a useful comparative cost analysis for specific scenarios. Here we discuss the technical attributes of the various reagents, merely noting in passing that the selection of a particular substance for a specific context is a key design task, which will usually demand some cost-benefit analysis

Calcium-based reagents
The most popular alkaline reagent used for mine water treatment is calcium hydroxide ($Ca(OH)_2$), which is commonly known as "hydrated lime" or "slaked lime". Calcium hydroxide is widely-used for pH adjustment in the water industry, and can be obtained in powdered form or as a pre-prepared slurry. The addition of calcium hydroxide to acidic ferruginous mine water causes the following reaction:

$$2Fe^{3+} + 2H^+ + 4Ca(OH)_2 \rightarrow 2Fe(OH)_3 + 4Ca^{2+} + 2H_2O \qquad (4.2)$$

Calcium hydroxide dosing of mine water is generally undertaken as follows:

- the powdered calcium hydroxide is stored in one or more silos (capacity 20 to 40 tonnes; see Figure 4.4), whence it is dispensed by a simple screw feeder
- dispensed powder enters a sealed "batching tank" of 5 to 10 m^3 capacity, in which it is mixed with water to produce a slurry with 5 to 10% solids by weight
- the slurry from the batching tank is added to the mine water by a metering pump at a rate sufficient to ensure attainment of the "target pH" (see 4.3.3 for definition). In most modern plants, a continuous pH monitor a few tens of meters downstream of the dosing point (at a point where mixing and initial precipitation will be complete) will be used to control the dosing pump.

Where suppliers are based within reasonable distance, it may be possible to obtain calcium hydroxide in the form of a pre-prepared aqueous slurry. Although unit reagent costs are higher than for powdered calcium hydroxide, the capital cost of the plant will be significantly less, as the silo-and-batching-tank arrangement can be replaced by a simple storage tank (storage volume 20 to 40 m^3) with metering pump. The simplicity of this dosing plant also reduces the level of inspection and maintenance required, which results in a reduction in labor costs. In some settings, pre-prepared slurry may prove to be cost-effective over the long-term, particularly where long discounting periods can be justified and where the visual impact of the tall silos (associated with powdered lime) must be avoided.

Figure 4.4: Lime silos at a high flow-capacity (8–15 Ml d^{-1}) conventional ODAS mine water treatment plant (former 'temporary plant' at Wheal Jane, Cornwall, UK). For a discussion of the evolution of treatment technology at this site, see Lamb *et al.* (1998) and Worked Example 4.3. (Photo: P. Younger).

Calcium oxide (CaO), which is also known as "quicklime" or "caustic lime", is cheaper than calcium hydroxide as a bulk reagent. However, calcium oxide must be hydrated on site to form a slurry before dosing the mine water. The cost of the hydration plant usually means that the economic advantage of calcium oxide is lost for all but the largest of operations.

Fragmented limestone typically costs less than a third of the price of calcium hydroxide, but it will not readily raise pH above 7.5, and requires long contact times with the mine water (around 12 hours). It is also prone to armoring with ferric hydroxides, unless used in a fluidized-bed reactor (e.g. van Tonder et al., 1994), which is sometimes termed a 'diversion well' (e.g. Arnold, 1991). Improved efficiency in fluidized-bed limestone reactors can be achieved by keeping the reactor gas-tight and injecting CO_2 gas under pressure, which favors more rapid dissolution kinetics and counteracts armoring (Sibrell et al., 2000). Nevertheless, successful applications of limestone as the principal alkaline reagent in full-scale active treatment applications have yet to be documented. In passive treatment, by contrast, it enjoys pre-eminence (see Section 5.4.2).

Sodium-based reagents
Perhaps second only to calcium hydroxide in popularity is sodium hydroxide (NaOH), also known as "caustic soda". Sodium hydroxide can be supplied in pelleted form for on-site batching, or as a concentrated aqueous solution. Being extremely caustic, sodium hydroxide is even more hazardous to handle than calcium hydroxide or calcium oxide, and it is more prone than hydrated lime to freezing in winter (although this can be controlled by keeping the solution strength within the 10–15% range). NaOH is also considerably more expensive tonne-for-tonne than hydrated lime. However, it has several important advantages over hydrated lime which ensure it will retain an important market share. These include:

(i)   A given mass of NaOH will achieve a higher pH in the water than will the same mass of hydrated lime. This makes it particularly suitable for raising pH high enough to remove Mn and other metals such as Zn and Cd (Table 4.1).
(ii)   It requires little infrastructure beyond a holding tank and a sedimentation pond.
(iii)   It is easier to handle than hydrated lime in terms of ease and reliability of delivery.

These advantages make it very attractive for rapid, short-term deployment in active mining situations, such as dewatering of voids known to be flooded with acid water. In longer-term applications, the main niche for NaOH is for treatment of mine water sources with a relatively low metals loadings (i.e. relatively low metal concentrations and/or low flows ($< \sim 30$ l.s^{-1}). In these circumstances the relatively high reagent cost is offset by the simplicity of application.

Another sodium-based agent, which is usually supplied in solid form, is sodium bicarbonate (Na_2CO_3, also known as "soda ash"). Although considerably more expensive than powdered calcium hydroxide, it achieves a higher pH on a weight-for-weight basis. It also has better caking properties than calcium hydroxide and can therefore be supplied in the form of briquettes, which may be stacked in simple gravity dispensers at remote sites (Dvorak, 1996). As each briquette dissolves, the overlying one in the stack falls into the water in its place. As dosage control is difficult to arrange with this approach, it should not be used where precise control is essential.

Other reagents
Magnesium hydroxide (which is often marketed as *Neutramag*) has two advantages over calcium hydroxide:

- it fosters the formation of denser sludges, which occupy less waste disposal space than sludges obtained by calcium hydroxide dosing
- it is far less hazardous than calcium hydroxide, which means that it is cheaper to handle

However, as a raw material magnesium hydroxide is usually more expensive than calcium hydroxide, except where it is produced locally.

Ammonia gas can be bubbled through the water, prompting protonation of the ammonia gas to form NH_4^+ and thus consuming protons:

$$NH_{3(g)} + H^+_{(aq)} = NH_4^+_{(aq)} \qquad (4.3)$$

As long as pH does not need to exceed 8.5, ammonia can be almost as cheap as calcium hydroxide (and, depending on location, as much as 50% cheaper than sodium hydroxide; Skousen *et al.*, 1990). The upper limit on pH effectively means that ammonia can be cost-effective for removing iron, copper and zinc, but not cadmium and manganese (cf. Table 4.1). The fact that ammonia can be dispensed from a simple gas cylinder gives it a distinct advantage on cramped sites, where large lime silos could not be installed. However, frequent cylinder changes may be necessary for high loading uses. One major drawback of using ammonia is the generation of NH_4^+ in the final effluent, since this compound is hazardous to aquatic life in its own right and is the subject of increasingly strict regulatory control. While relatively simple treatment processes can be used to convert NH_4^+ to innocuous N_2 gas, which readily vents into the atmosphere, accommodating such unit processes on a cramped site would defeat the object of using NH_3 dosing to save space.

Sometimes a local manufacturing process exists which produces alkaline waste products. For example, some oil refining plants yield 'alkali liquors' which are essentially weak (~10%) NaOH solutions. Subject to detailed chemical characterization and permitting, such liquors can be approved for use in mine water treatment. Two solid waste materials which have been occasionally used in mine water treatment are pulverized fuel ash (PFA) and cement kiln dust. While both are highly alkaline due to high CaO contents, their caking and compacting properties make them difficult to use in standard dosing equipment. Furthermore, many PFAs and kiln dusts contain toxic metals, which can be released to the water during treatment. Consequently, careful evaluation of specific sources of such materials must be made on a case-by-case basis before they can be recommended for use in a given mine water treatment application.

4.2.3.4 Quantities of alkali needed for dosing

Properties of a given water which will influence the amount of alkali required to precipitate dissolved metals and raise the pH to a target value include:

- the efficiency of the mixing and dissolution processes for the alkaline reagent under the specific circumstances of a particular treatment plant

- the total acidity of the water
- the initial pH of the water
- the content of SO_4^{2-} and HCO_3^-, since both of these can consume lime by reacting with the calcium which it contains
- the concentrations of 'non-target' metals in the water which can react with OH^- ions to form hydroxide solids. (The most common metal in this category is magnesium, which is frequently present at high concentrations in mine waters (especially coal mine waters) but is never a target for removal by active treatment).

Of these properties, the first two tend to dominate design considerations in practice. The efficiency of mixing and dissolution processes is a function of reagent type (solid vs liquid etc), plant configuration and process operation procedures. As such it is potentially a sensitive topic, as plant design and management procedures touch on matters of professional pride for treatment engineers. Studies have been made of the percentages of various reagents which effectively react in full-scale plants (US EPA, 1983). However, it is difficult to derive very specific rules from these findings, beyond noting that efficiencies tend to range from 50% to 95%, with most plants in the 70% to 80% range. With regard to total acidity, as it is a 'collective parameter' which reflects the sum of concentrations of a range of hydroxide-forming metals (see Table 2.5), the acidity-consumption reaction kinetics for a specific water will depend on the particular cocktail of metals contributing to the total acidity.

Given these complications, it is best to determine the amount of alkali needed to treat a specific mine water by undertaking laboratory tests in a pilot-scale treatment rig. Nevertheless, in advance of such testing (but not *instead* of it), it is possible to roughly calculate the amount of alkali which a particular mine water will require in the following manner (using both $Ca(OH)_2$ and NaOH as example reagents):

(i)   Calculate the amount of alkaline reagent which will be consumed in raising the pH to 7 and reacting with all of the contaminant metals to form hydroxides. This quantity is calculable from the total acidity, as follows:

If the total acidity is expressed in $mg.l^{-1}$ as $CaCO_3$, then
the number of milligrams of $Ca(OH)_2$ needed per liter = total acidity $\times 0.74$, or
the number of milligrams of NaOH needed per liter = total acidity $\times 0.8$.

If the total acidity is expressed as $meq.l^{-1}$ then:
the number of milligrams of $Ca(OH)_2$ needed per liter = total acidity $\times 37$, or
the number of milligrams of NaOH needed per liter = total acidity $\times 40$

(ii)   Calculate the 'extra' alkaline reagent required to raise the pH to a "target value" (pH_t) which will ensure sufficiently rapid reaction to precipitate the particular metals found in the mine water in question. Selection of a value for pH_t can be undertaken by reference to Table 4.1. For instance, if the only metals of concern are Fe and Zn, then Table 4.1 suggests that a pH of around 8.5 will generally be sufficient to ensure their rapid precipitation as hydroxides. If Mn must also be removed, then a pH of about 10.2 will be necessary. (Caution should be exercised in raising the pH so high if aluminum hydroxide solids remain in suspension at this point in the treatment plant, as these will dissolve again at such high pH). Having decided upon a pH_t value, the extra alkaline reagent needed to raise the pH from 7 (achieved in step (i) above) to pH_t is calculated as follows:

$$\text{mg Ca(OH)}_2 \text{ needed per liter} = ((1000 \times 10^{-pHt}) - 0.001) \times 37, \text{ or}$$
$$\text{mg NaOH needed per liter} = ((1000 \times 10^{-pHt}) \times 0.001) \times 40$$

(iii)   Account for the likely efficiency of the full-scale plant using local knowledge, or drawing upon the information given by US EPA (1983). The efficiency can be expressed as a percentage factor E, which will fall somewhere in the range 0.5–0.95.

The overall expression for the total amount of alkali required is (for the case of $Ca(OH)_2$) therefore given by:

$$\text{Total (mg.l}^{-1}) = \{[\text{total acidity (meq.l}^{-1}) + ((1000 \times 10^{-pHt}) - 0.001)] \times 37\}/E \qquad (4.4)$$

(N.B. where NaOH solutions are used, the final expression must also take into account the strength of the avalable solution (e.g. 50%)). An example application of Equation 4.4 is given as Worked Example 4.1, which demonstrates that it certainly yields estimates of the right order of magnitude for outline plant design purposes.

Worked Example 4.1 Predicted and actual alkali dosing requirements for an acidic mine water.

The Ynysarwed mine water is one of the largest and most damaging uncontrolled abandoned coal mine discharges to have emerged in Britain in recent years (Ranson et al., 1997). The proposed remediation scheme for this discharge advocates active treatment for a decade or more until iron concentrations reach their lower, asymptotic level, followed by passive treatment in perpetuity. Active treatment has to be achieved on a very small roadside site, and incorporates mechanical aeration, lime injection, sedimentation using lamellar plate thickeners and sludge thickening by scroll centrifuge.

To estimate the budget needed for this system, design calculations proceeded using Equation 4.4. The mine water exhibited the following mean characteristics during the year preceding the design activities (1996): Flow rate = 28.5 l.s^{-1}; Fe^{2+} = 300 mg.l^{-1}; Mn^{2+} = 6 mg.l^{-1}1; pH = 5.6. Using the expression given in Table 2.5, these values were used to calculate a total acidity of 547 mg.l^{-1} as $CaCO_3$ equivalent. As Mn removal is not required by the regulator in this case, the target pH for the treatment process is set at 8.5.

Applying Equation 4.4, with E values applied across the observed range (0.5–0.95), we obtain the following estimates for $Ca(OH)_2$ required:

Minimum (for E = 0.95): $0.95 \cdot [10.9-0.001] \cdot 37$ = 424 mg.l^{-1}, or 1037 kg.d^{-1}
Maximum (for E = 0.5): $0.5 \cdot [10.9-0.001] \cdot 37$ = 806 mg.l^{-1}, or 1985 kg.d^{-1}
Median: 508 mg.l^{-1}, or 1511 kg.d^{-1}

Using the simple rule-of-thumb given at the end of Section 4.2.3.4, we estimate the alkali requirement as follows: $1.25 \times (547 + 100)$ = 808 mg.l^{-1}, or 1990 kg.d^{-1}.

Since commissioning of the treatment plant in 1999, the actual $Ca(OH)_2$ usage has averaged 1500 kg.day^{-1}. This is strikingly close to the median estimate obtained using Equation 4.4, but only 75% of the quantity estimated by the rough rule-of-thumb.

Worked Example 4.2 Examples of oxidation by hydrogen peroxide dosing.

(a) External dewatering shaft, surface coal mine, north-east England.
The largest surface coal mine in England at the start of the new millennium is dewatered largely by means of a single deep shaft in a tight corner of the site, outside the peripheral overburden pile. The pumped mine water has around 5 mg.l^{-1} of Fe and a pH of 7.2. The pumping rate is very high (around 13.6 Ml.d^{-1} for 3 days per week). Treatment was traditionally provided by manual dosing with powdered lime, followed by sedimentation in two ponds in series. Although the effluent has always complied with regulatory limits, it is now considered that these were set a little too liberally, for the bed of the stream receiving the mine water has regularly been stained with iron hydroxide precipitates. A brief study concluded that the persistence of iron in the effluent was due to incomplete oxidation of ferrous iron in the existing treatment system. Full oxidation is now provided by drip-feeding the mine water with 35% H_2O_2 (Solvay Interox) at a rate of around 1 litre per hour. The lime powder dosing has been discontinued. Staining of the receiving stream has been eliminated. The peroxide treatment costs around $20 per day in reagents.

(b) Abandoned coal drift mine, northern West Virginia.
One of the worst abandoned mine discharges in West Virginia emerges on a very cramped site in the woodlands southwest of Morgantown, where there is insufficient space for aeration cascades. The pH of the water is raised from the initial 2.5 by dosing with ammonia gas, and oxidation is achieved by drip-feeding hydrogen peroxide. Ferrous iron in the mine water is about 2500 mg.l^{-1}, and the flow rate averages about 1.4 Ml.d^{-1}. The operation uses around 13 litres of H_2O_2 solution per hour, at a cost of around $200 per day.

A simpler rule of thumb for estimating the total alkali requirement which enjoys some currency in the eastern USA is based on the supposition that attainment of a pH within the usual regulatory limits is likely to require a net alkalinity on the order of 100 mg.l^{-1} as $CaCO_3$. If we add this amount to the initial net-acidity (= total acidity – total alkalinity) of the raw water, and multiply this total by 1.25 (to account for inefficiencies), we obtain a rough estimate of the total alkali requirement.

4.2.4 SEDIMENTATION

4.2.4.1 Sedimentation to remove inert solids from suspension

Many mine waters from areas of active working will require treatment by sedimentation, i.e. the removal of turbidity by the simple settling of rock particles from suspension. At its simplest, sedimentation in active treatment systems can occur in much the same way as in passive treatment systems (see section 5.4.3.2). In other cases, limitations of space or the cost of sludge disposal will demand a more reagent- or energy-intensive sedimentation process which begins to accord more with the definition of active treatment.

Almost all waters draining areas of active working will carry sufficient suspended solids that they would cause unacceptable silting up of natural watercourses if they were released to them directly. Typical suspended solids in active mine drainage may vary in grain size from mud to fine gravel, and will basically comprise comminuted fragments of ore and overburden/country rock, much of which will be relatively inert quartz and clay minerals. The removal of such chemically-inert material from water is a basic task in water and wastewater treatment world-wide (Smethurst, 1988).

In most mining situations, removal of inert suspended solids from mine water will require little more than the installation of ponds with sufficient surface area to provide sedimentation for the sediment load of the design flow. Generally, the design flow will be selected by analysis of the catchment area of the mine site and the local rainfall patterns, to determine the peak flow rate which has a probability of exceedance in any one year of between 0.05 and 0.01. (The reciprocal of these annual probabilities of exceedance equals the so-called "return period" in years, ie in this case $1/0.01 = 1$ year in 100). Values as low as 0.01 will typically be selected where the consequences of uncontrolled discharge are most serious, such as in nominally "zero discharge" mines (see Section 3.8.2). Guidance on the determination of design flows is available in standard hydrological texts, such as Shaw (1994) and Linsley *et al.* (1982).

Having obtained a design flow rate (Q_d, in $m^3.s^{-1}$), the required sedimentation pond area (A_{sp}, in m^2) required for settling inert suspended solids is given by:

$$A_{sp} = 100,000 . Qd \qquad (4.5)$$

(More exact expressions can be obtained by resorting to sedimentation theory, but the above expression is robust for most mining applications, as it is based on the settling velocity of 4 μm-diameter grains of shale or other sediment of similar density; NCB, 1982). In the majority of cases, Equation 4.5 is likely to yield pond areas giving a nominal retention time of 3 to 4 hours.

4.2.4.2 Sedimentation for removal of iron and other metals

Principle options in sedimentation

Having converted most of the incoming dissolved Fe^{2+} to suspended $Fe(OH)_3$ by means of alkali dosing and oxidation, the next step in the treatment process is to settle the $Fe(OH)_3$ from suspension in a sedimentation pond or clarifier. The choice between the two is usually predicated on land availability and costs: A sedimentation pond is invariably cheaper to construct and maintain than a clarifier, but clarifiers can achieve the same amount of sedimentation in a far smaller land area. There are many variations in sedimentation practice world-wide. We will begin by describing the simplest case, where treatment objectives can be met using one or more sedimentation ponds in series.

Sedimentation ponds

At many mines where iron is the only contaminant and the water is naturally net-alkaline, it proves possible to adequately treat the water using aeration and sedimentation alone. If the aeration is achieved without the use of energy, no flocculants/coagulants are added and the sedimentation ponds are large enough to require only infrequent maintenance, then such systems fall within the definition of 'passive treatment' given in Chapter 5. Very often, however, the aeration step and frequent pond de-sludging works will require electrical power, and/or starch-based flocculants will be added (at least occasionally) so that the treatment process can be considered 'active' (e.g. Laine, 1997).

In many cases, treatment of ferruginous waters using aeration and sedimentation will be successful, as long as the following criteria are observed:

- Only use aeration-sedimentation as the sole method of treatment for a ferruginous water if:
  - the water is net-alkaline, and
  - the pH of the water after the first aeration step is >6.5, and
  - iron is the only contaminant of interest
- If aeration is only to be achieved using cascades, then given that each cascade will provide enough dissolved oxygen to oxidize 50 mg.l^{-1} of Fe^{2+} (see Section 4.2.2.2) it will be necessary to allow one cascade and sedimentation pond for every 50 mg.l^{-1} of Fe^{2+} (or part thereof). For instance, if the mine water to be treated contains 165 mg.l^{-1} Fe^{2+}, then since 165/50 = 3.3, we round up to the nearest integer and design the system to have four cascades and ponds in series.
- The total retention time required in the system of sedimentation ponds depends on the settling velocity of ferric hydroxide particles. As these particles are naturally low in density and very small (initially < 2.5 μm diameter in many cases), settling velocities may be very slow. These sedimentation ponds should be constructed using the same loading criterion as is used for aerobic wetlands (i.e. assume that iron sedimentation will occur at a rate of about 10 g.d^{-1}.m^{-2}; see Chapter Five). Where sufficient land is available, experience has shown that virtually any net-alkaline ferruginous mine water can be treated using sedimentation alone to a residual iron concentration of around 1 mg.l^{-1} if sedimentation ponds have a nominal retention time in excess of 24 hours. It should be noted that this is retention time is some 8 times greater than that required to remove inert sediment from water. This reflects two factors:
  - (i) the relatively slow rate of formation of ferric hydroxide flocs in open water, and
  - (ii) the slow settling velocities of these low-density flocs.

Clarifiers, lamellar plate thickeners etc.

Where space is at a premium, it may be necessary for nearly all sedimentation to be achieved by coagulation and flocculation in conical tanks known as "thickeners" or "clarifiers" (Figure 4.5a). On the most cramped of sites, a lamellar plate thickener can be used (Figure 4.5b), or where the suspended solids are particularly coarse, by vortex solids separation in a hydro-cyclone. In some cases, it may be advantageous to use more advanced solids separation techniques, such as dissolved air flotation (in which millions of micro-bubbles of air per second are passed upwards through the water column, entraining and removing (as a floating froth) the suspended particles in their paths). However, full descriptions of such sophisticated techniques are beyond the scope of this text; interested readers are advised to consult Metcalf and Eddy Inc. (1991).

Coagulation and flocculation

Sedimentation can be enhanced in many cases by addition of coagulants to the mine water. Coagulants are substances which reduce electrostatic repulsive charges on particle surfaces, so that smaller particles coagulate with each other, producing larger particles ("flocs") which settle more readily under gravity. Coagulants include common starch, proprietary 'poly-electrolyte' compounds and reactive liquids such as aluminum sulfate. If the latter is added to a neutralized mine water, it rapidly reacts (consuming some of the alkalinity) to form frothy aluminum hydroxy-sulfate flocs, which in turn adsorb other metals and promote sedimentation. Other coagulants work in similar manners. Guidance

(a) Radial flow clarifier

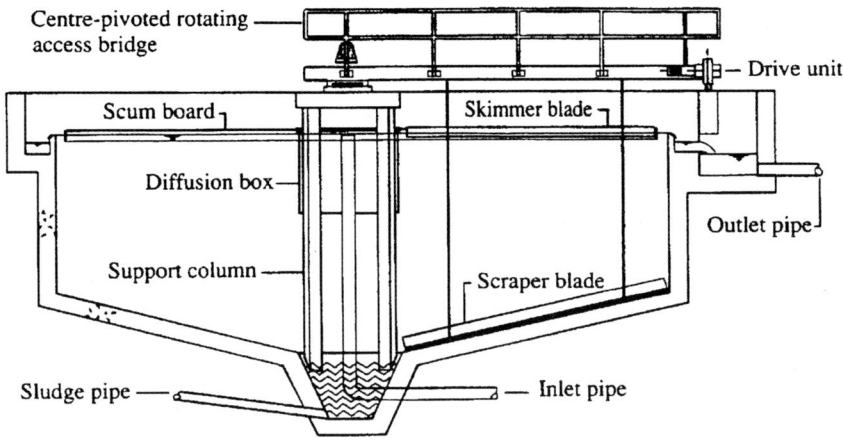

(b) Lamellar plate thickener

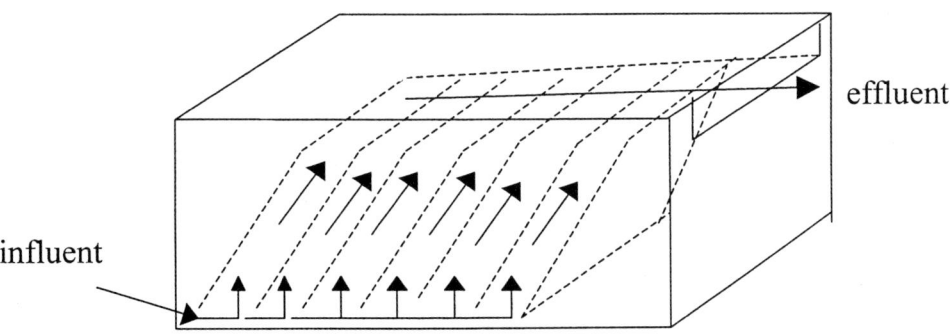

Figure 4.5: (a) Sketch section of a typical thickener suitable for mine water treatment purposes (courtesy of EA). (b) Sketch section of a lamellar plate thickener.

on other suitable coagulants for mine waters is given by US EPA (1983) and Skousen *et al.* (1990).

Having coagulated colloids to form flocs, it is often possible to further accelerate sedimentation by encouraging flocculation, in which flocs are caused to adhere to one another to form larger particles which will settle more readily from suspension. Deliberate promotion of flocculation in this manner is particularly common on cramped sites where clarifiers are being used. Flocculation can be achieved by physical and/or chemically-assisted means. Agitation of the water increases the frequency of floc collisions, thus speeding up the accretion of large particles. Such agitation is usually achieved by stirring with a rotating wand (or similar device) or by inducing vortex flow. Chemically-assisted flocculation involves dosing the water with proprietary substances (chiefly organic polymers), which are often designed to target specific metals, promoting the accretion of large, dense flocs, which in turn settle to form sludges of higher density than can be achieved using coagulation alone (Dudeney *et al.*, 1994; Clarke, 1995).

4.2.4.3 Further considerations in sedimentation basin design

Beyond ensuring adequate sizing according to whether the basin is intended to settle inert suspended solids or metal hydroxide flocs, there are three subsidiary matters to take into account when designing simple horizontal-flow sedimentation ponds (NCB, 1982; Smethurst, 1988):

1. It is important to cope with short-term fluctuations in inflow rates, particularly in areas prone to convective storms. This is best achieved by installing 'flow-balancing lagoons' or 'flow equalization ponds/basins' upstream of the main sedimentation ponds. Flow-equalization basins are typically designed to have a volume equal to several hours storage for the peak flow rate, with an adjustable orifice outlet at the base of the lagoon which can be used to throttle the release rate from the lagoon.
2. Horizontal flow ponds are prone to "streaming", i.e. the short-circuiting of the pond storage by discrete currents of water passing from the inlet to the outlet. There are a number of approaches to the minimization of streaming, none of which are entirely satisfactory. The first is to ensure that the total sedimentation pond area required is provided by *at least two ponds in series*. If by-pass pipe-work is incorporated, this has the further advantage of allowing one basin to be taken out of use for maintenance without totally losing treatment capacity. Larger ratios of pond length to width can also reduce the scope for channelization (Smethurst, 1988), although experience at mine sites suggests that this ratio should not normally exceed 5:1 (NCB, 1982). Within ponds, the flow should be spread as evenly as possible across the width of the pond. Although this is much easier to advocate than it is to achieve, it can usually be arranged by using crenellated weirs or multiple pipe installations for both influent and effluent ends of the pond.
3. The depth of sedimentation ponds is usually a compromise between the following three factors:
   (i)   the need to maximize storage space for settled solids, to maximize the time between pond cleaning operations
   (ii) the need to ensure that ponds are not so deep that it becomes difficult to clean the ponds using readily available equipment
   (iii) the constraints imposed by the geometry, topography and foundation conditions on the site.
   In most cases, depths of 3 to 4 m are suitable.

4.2.5 LIMITATIONS OF THE CONVENTIONAL ODAS APPROACH

The preceding sections have focused upon the standard industrial approach to active mine water treatment, based on the three processes of *o*xidation, *d*osing with *a*lkali and *s*edimentation (ODAS). The popularity of the ODAS approach in practice is testament to its predictability, flexibility and relatively low cost. It also introduces considerable hardness into the receiving watercourse, which is generally beneficial as most pollutant metals are far less ecotoxic in hard waters than in soft waters (e.g. Kelly, 1988).

Nevertheless, ODAS does have its limitations, including:

- The wider environmental costs associated with lime use, including vehicle movements for reagent deliveries, and the effects of limestone mining elsewhere in order to produce the alkali dosing agents.
- Equipment maintenance costs may be high due to clogging of fittings with ochre, calcium carbonate and/or gypsum. While anti-scaling agents can be used, these incur significant purchase and handling costs of their own.
- The high pH that is needed to remove metals such as manganese may cause remobilization of other metal hydroxides (e.g. aluminum).
- The usual failure to produce any net lowering of total dissolved solids. In many cases, the high dissolved solids content of mine waters is more detrimental to the use of receiving waters for water supply and irrigation than is the presence of elevated iron concentrations. This is true in many semi-arid regions, such as the coal-producing states of the American west (National Research Council, 1981) and the gold and coal-producing areas of South Africa (e.g. Smit, 1999).
- The low rate of sulfate removal associated with ODAS: Although precipitation of hydroxy-sulfates and gypsum may remove as much as 10% of the sulfate from solution in conventional ODAS operations, the residual concentrations may still be far in excess of the drinking water limit ($250 \ mg.l^{-1}$).
- Residual concentrations of some metals can still be relatively high after ODAS. Where regulators demand very low residual concentrations of certain metals, non-compliance can result. This is particularly likely where there are high initial concentrations of metals such as zinc, copper and cadmium, which will only precipitate as hydroxides in their own right at relatively high pH values.
- The chemical complexity of the sludge obtained from lime-dosing operations. These sludges differ from pure ferric hydroxides sludges in containing sulfate compounds, which alter the chemical and physical properties of the sludge and make it less useful for other purposes.
- The low density of the sludge results in large volumes which can make handling and disposal difficult and expensive (see Section 4.7.2).

In some cases, these drawbacks have proven sufficiently serious that they have prompted the development and application of alternative treatment technologies. Some of the most promising, proven alternatives to ODAS are reviewed below. While none of these techniques enjoys the wide currency of ODAS at present, they probably represent the future growth-areas in active mine water treatment, particularly for applications where high contaminant concentrations must be reconciled with strict environmental emission controls.

4.2.6 ADVANCED ODAS METHODOLOGIES

Some of the shortcomings of conventional ODAS have been overcome in recent decades by innovations in sludge handling and sorption/precipitation pathways for problematic metals and sulfates. Two of the most significant developments are briefly reviewed below.

(i) The High-Density Sludge Process (HDS)

Issues of sludge management are considered in some detail in Section 4.7.2 below. The key issue is that settlement of neutralized ferruginous waters under gravity typically results in the accumulation of a hydroxide sludge with a final water content in excess of 95% (e.g. Laine, 1997). Such low densities mean that water is being landfilled rather than undesired metals. It also means that the cleaning of sedimentation ponds requires the use of special pumps and tankers rather than conventional excavation machinery which is commonly available on active mine sites. Only when the water content can be reduced below 80% by weight will the sludge becomes easy to excavate (Laine, 1997). In Section 4.7.2 we discuss some of the mechanical means available to dewater existing sludges. Here, we are concerned with what can be done 'up front', during the hydroxide precipitation and sedimentation stages of treatment, to achieve higher sludge densities.

Sludge densities far greater than those achieved by conventional open-water sedimentation can be achieved using a so-called 'high-density sludge (HDS) circuit' (Bosman, 1983) (Figure 4.6). An HDS circuit operates by mixing about 80% of the sludge obtained from the underflow of a conventional thickener (Figure 4.5a) with the incoming mine water. The recycled sludge provides precipitation nuclei for further iron hydroxide formation. Resultant sludge densities from an HDS circuit are considerably higher than can be achieved by a conventional single-pass circuit. Even before dewatering, solids contents of around 15 to 20 wt % can be achieved. If frame-and-plate dewatering or filter pressing is applied to the HDS sludge, final solids contents in the range 50–80 wt % are attainable.

HDS mine water treatment plants have been developed by a number of organisations including the Tetra Corporation, Norando and Unipure Environmental. Although the specific forms of these plants differ they all share the same fundamental objective of achieving precipitation without the formation of the voluminous sludge created by conventional precipitation/sedimentation (Zinck and Griffith, 2000). A key decision in HDS design, and a key distinction between the processes marketed by different companies, is whether to:

(i)   mix the recirculated sludge with lime prior to introducing the mine water, or
(ii)   mix the sludge with the mine water prior to adding the lime.

It is not yet clear which is the best option overall. Indications are that the second option uses less lime and therefore has lower operating costs, but that this may be at the expense of producing a sludge which more readily releases metals to leaching in the long-term (Zinck and Griffith, 2000).

Worked example 4.3 summarizes the design and operation of a large HDS plant in Europe.

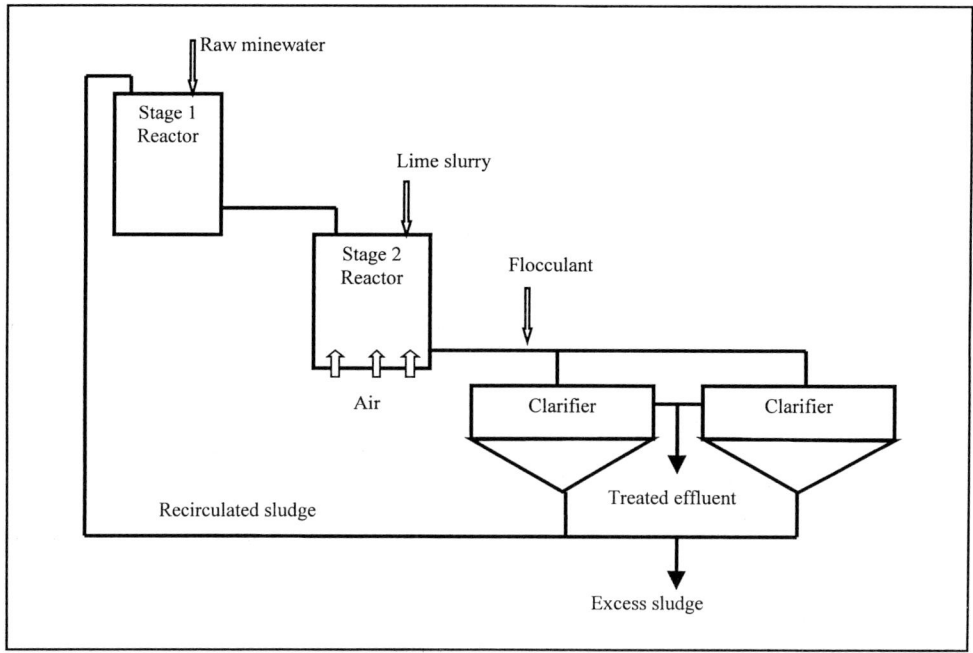

Figure 4.6: Process flow-chart for a high-density sludge (HDS) mine water treatment plant. Based on one of two parallel streams at Unipure Environmental plant at the Wheal Jane mine, Cornwall, UK. (See Worked Example 4.3 for further details.) (Diagram courtesy of Dr Richard Coulton, Director, Unipure Environmental).

(iii) Desalination by sulfate removal after lime dosing: the SAVMIN™ Process[1]

It was previously noted (Section 4.2.5) that one of the main drawbacks of conventional ODAS treatment systems is their inability to significantly lower the total dissolved solids (TDS) content of mine water; indeed, many ODAS plants actually increase the TDS by the addition of large amounts of Ca^{2+}. In semi-arid areas, such as the main mining districts of South Africa, this is a serious problem. On the other hand existing desalination technologies (such as reverse osmosis; Section 4.5) are very expensive in terms of both construction and operating costs. Within the last few years, South African hydro-metallurgists have devised a highly innovative process which can treat waters with several thousand mg.l^{-1} of sulfate to well below the permitted levels for drinking water (Smit, 1999). Furthermore, the process yields gypsum as a by-product, which is marketable to plaster manufacturers. This process, the SAVMIN process, uses conventional ODAS to remove the main problematic metals (Fe, Mn, Zn, Cd etc) as hydroxides, and to super-

---

[1]SAVMIN is a patented process and registered trademark belonging to MINTEK (S. Africa), Savannah Mining and the Wren Group

---

Worked Example 4.3 Design and performance of a large HDS plant, Wheal Jane Tin/Zinc Mine, Cornwall UK.

The Wheal Jane mine water treatment plant, which was commissioned in 2000 following an 8-year study (Lamb *et al.*, 1998) is a typical example of a HDS plant. Designed by Unipure Environmental, the plant treats up to 350 l.s^{-1} of acid mine water (pH 3.8) contaminated with the following metals (mean concentrations in mg.l^{-1}): Fe (160), Zn (44), Mn (5), Cd (0.06), Ni (0.4) and Cu (0.4).

A process flow-sheet for the plant is given in Figure 4.6. It comprises two streams, each with a nominal rated capacity of 240 l.s^{-1}. Each stream comprises:

- *Stage 1 reaction vessel*, in which the influent mine water is mixed with recirculated sludge pumped back from the clarifier. The pH in this vessel is not formally controlled but due to alkalinity present in the recirculated sludge, it typically varies between 7.5 and 8.5.
- *Stage 2 reaction vessel*; the pH of the mine water/sludge mixture overflowing from the stage 1 reaction vessel is raised in the stage 2 reactor to 9.3 by the addition of a 5% lime slurry. This target pH has been found to be sufficient to lower the dissolved Mn concentration to less than 1 mgl^{-1} in this particular water.
- *In-line static mixer*; treated water from the Stage 2 reaction vessel is passed through an in-line static mixer where a slightly anionic flocculant is added to assist solid/liquid separation.
- *Two lamellar clarifier/thickener units*. Solids/liquid separation is achieved in two 7 m square lamellar clarifier/thickener units operated in parallel. The clarified final effluent is discharged off site without tertiary treatment. Thickened solids from the clarifier/thickener are either recirculated to the Stage 1 reactor or pumped from the system at a solids concentration of between 15 and 25%.

Excess sludge from the plant is pumped to a nearby disused tailings dam where the solids rapidly settle to 30% solids (weight by weight) and subsequently consolidate to a greater density. Full-scale filter press trials demonstrated that the clarifier/thickener underflow can be dewatered to a solids concentration of up to 70 wt %.

During the 2000/2001 winter period the plant treated 4,500,000 m^3 of water removing some 1000 tonnes of metal. Effluent pH was typically 9 (which was acceptable for discharge to a stream already affected by other acidic discharges), and mean metal contents after treatment were as follows (mg.l^{-1}): Fe (1.3), Zn (0.4), Mn (0.6), Cd (< 0.001), Ni (< 0.1) and Cu (< 0.01).

---

saturate the mine water with respect to gypsum. After removing as much SO_4^{2-} as possible in the form of gypsum, the remaining water (still containing as much as 2500 mg.l^{-1} SO_4^{2-}) is reacted with aluminum hydroxide under carefully controlled high-pH conditions (pH in the range 11.6–12.0) to precipitate the mineral ettringite ($3CaO.3CaSO_4.Al_2O_3.31H_2O$) which is highly insoluble only in this narrow pH range. Ettringite is a very rare mineral in nature; it has the very favorable property of being at equilibrium in alkaline waters containing only a few mg.l^{-1} of dissolved SO_4^{2-}. The result is a final treated water of potable quality. The ettringite sludge is then decomposed by sulfuric acid, releasing the aluminum for recycling within the SAVMIN process, and yielding a concentrated Ca-SO_4 solution which is simply added into the gypsum precipitation channel. Hence, besides good quality water, the only other products of SAVMIN are metal hydroxide sludges (as in all ODAS operations) and potentially saleable gypsum (Smit, 1999).

4.3 SULFIDIZATION AND BIODESALINATION

4.3.1 PRINCIPLES OF ACTIVE SULFATE REDUCTION TECHNOLOGY

Many mine waters contain high concentrations of sulfate (Chapter Two). Under reducing conditions, sulfate can be converted to sulfide. This can react with dissolved metals to precipitate sulfide minerals, which are generally insoluble as long as reducing conditions persist. This process is known as "sulfidization", and it is becoming increasingly popular (in various forms) for the treatment of acidic mine waters where it is necessary to achieve strict compliance with very low effluent metals limits. (The principle of sulfidization is also applied in passive treatment systems using compost wetlands; see Sections 5.3.2 and 5.3.6).

The basic principle of sulfidization is to reduce sulfate (SO_4^{2-}), which is abundant in most mine waters (particularly in the most problematic acidic waters) to form various sulfide species (HS^-; H_2S), which react rapidly with divalent cations to precipitate sulfide solids, by means of a number of reactions such as:

$$H_2S + M^{2+} + 2\,HCO_3^- \rightarrow MS + 2\,H_2O + 2CO_2 \qquad (4.6)$$

Where "M^{2+}" represents a generic divalent cation. (It should be noted that trivalent cations such as Al^{3+} and ferric iron (Fe^{3+}) do not form sulfide solids at earth surface temperatures and pressures. Nevertheless, as many sulfate reduction reactions consume protons, and thus raise pH, trivalent ions can still be removed from solution in sulfidisation systems by precipitation of hydroxides, which is primarily a pH-dependent process (see Chapter Two)).

The reduction of sulfate to sulfide is catalyzed by bacteria; in the absence of bacteria the process is usually immeasurably slow. Consequently, sulfidisation is essentially a biological treatment process, in which sulfate-reducing bacteria (SRB) are brought into contact with sulfate-rich mine water. The various forms of sulfidisation technology differ principally in the manner in which they achieve the sulfate reduction and metal precipitation steps.

4.3.2 VARIETIES OF SULFIDIZATION TECHNOLOGY

A number of active treatment technologies based on bacterial sulfate reduction (BSR) processes have been patented and marketed in recent years. In the simplest BSR-based processes, bacteria are provided with a carbon source and a substrate for adhesion in the form of compost or some other organic matter (e.g. Tuttle et al., 1969; Dvorak et al., 1992; Hammack et al., 1994; Dvorak, 1996). Polluted mine waters are pumped through packed beds of such material, or else mixed with it in suspension. As metal sulfides are precipitated, these accumulate in the reactor, usually in a mixture including spent organic matter. Disposal of the resultant composite organic/sulfide sludge may prove costly.

One way to avoid this problem is to use a "light" carbon source which does not contribute bulky organic waste to the final sludge. Such an approach has been tested by Dill et al. (1994), who used producer gas (a mixture of CO_2, CO and H_2 and N_2, which is evolved in a number of industrial gasification processes) as a carbon source for SRBs in an anaerobic packed-bed reactor. The packed bed comprises a cylindrical chamber filled with clay rings to provide bacterial growth sites. Inoculation of the packed-bed was undertaken with sewage sludge. Consequently, the final sulfide sludge does contain *some* organic waste, but relatively little compared to conventional BSR-reactor sludges.

Gaseous mixtures of hydrogen and carbon dioxide are also used in the proprietary THIOPAQ® process[2], which can remove both metals and sulfate from mine waters down to extremely low residual concentrations. Ethanol is another common carbon source in this process. THIOPAQ produces low-carbon, sulfide-rich sludges, which in some cases will be suitable for use as smelter feedstock. Addition of a secondary oxidation step to the basic THIOPAQ process facilitates generation of elemental sulfur as a saleable by-product. While it is possible to operate the THIOPAQ® process using waste organic materials as a carbon source, this is at the expense of a concomitant increase in the organic content of the sulfide sludge.

Organic wastes can be eliminated entirely from the sulfide sludge if the reduction of sulfate is physically separated from the precipitation of sulfide minerals. This is achieved in the "biosulfide process" (Rowley *et al.*, 1994). The biosulfide plant layout is shown schematically in Figure 4.7. As mine water enters the treatment plant, hydrogen sulfide gas is bubbled through it, causing the precipitation of sulfide minerals as a sludge (which is removed by under-drainage). Typically three precipitation tanks are necessary to maximize metals removal. Upon leaving the third precipitation tank, the mine water is virtually free of ecotoxic metals, but still rich in sulfate. The mine water is therefore passed to the bioreactor vessel, in which the sulfate is bacterially reduced to hydrogen sulfide, which is then stripped from the water in the gaseous state and routed to the precipitation tanks.

Sulfidisation need not be seen solely as an alternative to hydroxide precipitation; rather, the two processes can be used conjunctively. For instance, Diaz *et al.* (1997) have described the laboratory-scale application of sequential hydroxide/sulfide precipitation steps. The hydroxide precipitation step removes non-valuable iron and arsenic, whereas the sulfide precipitation step recovers potentially-valuable copper and zinc as a dense sulfide sludge which could, in principle, be sold to smelters as feedstock. The sulfidisation step in the process they developed is based on a conventional upflow anaerobic sludge blanket (UASB) reactor, in which the mine water is made to flow upwards through a previously installed blanket of organic matter hosting SRBs.

4.3.3 BIODESALINATION

The term 'biodesalination' was coined by Professor Peter Rose of Rhodes University (Grahamstown, South Africa) to describe efforts to remove so much sulfate (and associated metals) from highly polluted mine waters that they are effectively desalinated. Most of the major mining districts in South Africa lie in semi-arid parts of the country, where they have spawned large conurbations and numerous spin-off industries. One such industry is leather production, which is predictably vigorous in a country famed for its wildlife resources and cattle ranches. Many major tanneries lie in the urban areas of the major mining districts. Tanneries are well-known for producing poor quality effluents, which typically require extensive treatment prior to discharge to natural watercourses. These

---

[2]THIOPAQ® is a registered trademark of Paques Bio-Systems BV of the Netherlands

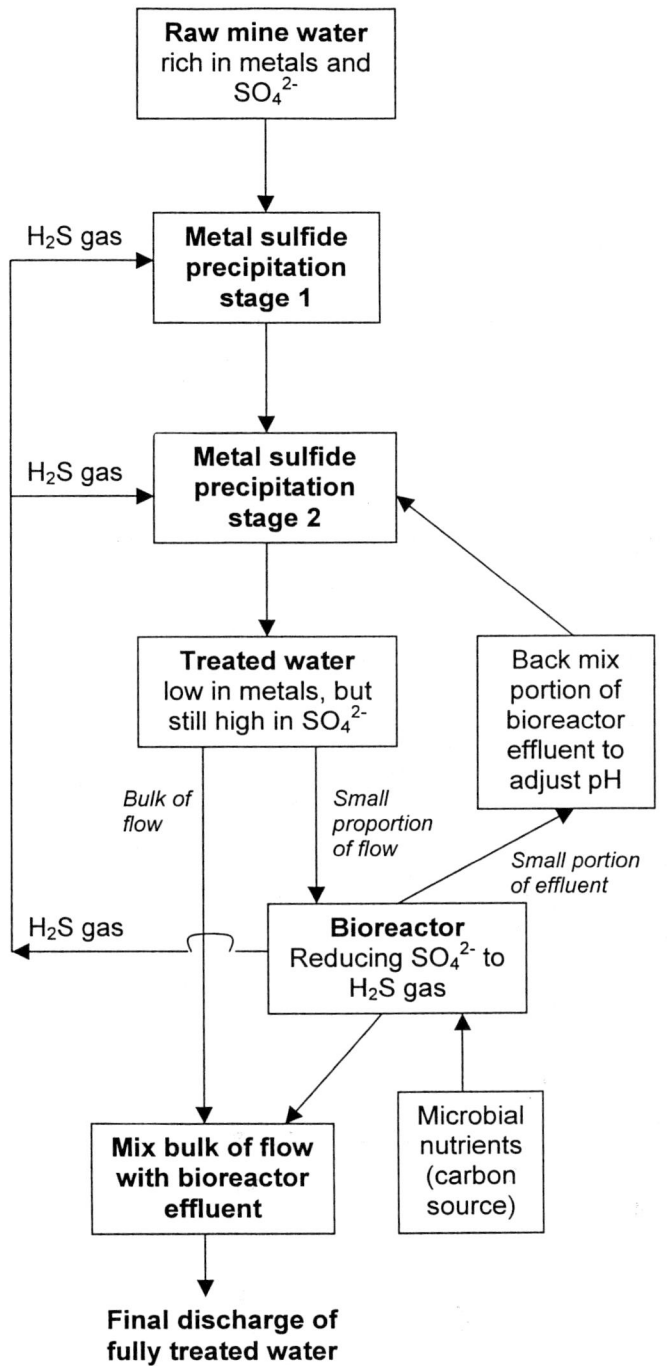

Figure 4.7: Process flow-chart for the Biosulfide Process® (developed from the concepts of Rowley *et al.*, 1994).

effluents are usually very rich in dissolved sulfur species, and usually contain very active SRB populations. Removal of dissolved sulfur from these wastewaters could be achieved by precipitation of metal sulfides, were metal concentrations not usually limiting in most tannery effluents.

Given the problematic super-abundance of metals in many mine effluents, the possibility arises of mixing the two waste streams, so that the metals from the mine waters can be used to assist sulfide precipitation from the tannery effluents, thereby potentially achieving treatment objectives for both waste streams. Taking advantage of the warm climate of South Africa, it ought also to be possible to implement the co-treatment of these waste streams by means of relatively simple waste stabilization (facultative) pond technology. Successful laboratory and pilot-scale field trials of this notion yielded sulfate reduction rates approaching 80%, and metal removal rates which were as much as eight times greater than those anticipated on the basis of stoichometry (based on iron mono-sulfide precipitation). Full-scale implementation of this process is now underway, as a patented technology named ASPAM (*A*lgal *S*ulfate Reducing *P*onding Process for the Treatment of *A*cidic and *M*etal Wastewaters). The ASPAM process route is shown in Figure 4.8. Results to date are extremely encouraging, and appear to offer a feasible option for returning formerly brackish mine waters to the rivers of semi-arid South Africa in a sufficiently dilute state that they do not degrade their water resource potential. Further details of these developments are given by Rhodes *et al.* (1998) and Boshoff (1999).

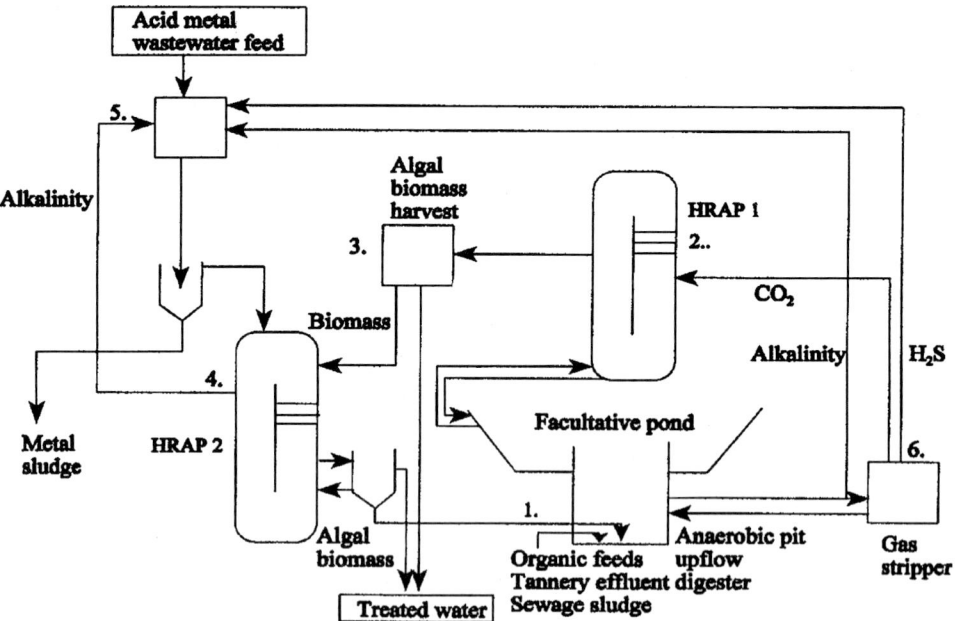

Figure 4.8: Process flow-chart for the 'ASPAM' process (after Rose *et al.*, 1999).

4.4 SORPTION AND ION EXCHANGE

The underlying scientific principles of the sorption of metals in aqueous systems are outlined in Section 2.4.2. Since sorption is an integral step in many hydrometallurgical processes (e.g. gold recovery by heap leaching) it is a well-known phenomenon in mineral processing circles. It is therefore natural that the possible use of sorption in mine water treatment should have received consideration. Particularly attractive is the proposition that sorption might be used to recover potentially valuable base metals (such as Pb, Cu and Zn) from mine waters, thus offsetting the costs of treatment against the value of the metal recovered. Partly in pursuit of this hope, a number of sorption-based mine water treatment techniques have been developed in recent decades.

Laboratory studies have long since demonstrated the feasibility of recovering potentially valuable base metals from mine waters by sorption (Barton, 1978), yielding metal concentrates which are of sufficient purity that they could be sold to smelters. Two problems have typically beset developers seeking to transform these laboratory findings into industrial processes:

1. The difficulties of scaling-up the lab-based reactors to cope with large mine water flows, and
2. The high costs of the most appropriate sorbing agents.

Where mine waters are particularly rich in base metals, the economics might make the installation of a sorptive metal recovery plant a worthwhile proposition, particularly where it is certain that metal recovery can be continued for periods of one to two decades. However, over such time-scales it is usual for the quality of mine water to improve naturally due to hydrological flushing processes (see Section 3.11.4). Most sorption-based systems will thus become uneconomic for all but the most persistent of grossly-polluted mine waters.

Recent developments have gone some way to avoiding the problems arising from the high cost of the best sorbent materials. It has been found that inexpensive sorbents can be used in conjunction with electrochemically-assisted winning of the metals from the sorbent surfaces. Termed the "BioElecDetox" process (Butter *et al.*, 1998), the procedure occurs in three stages:

1. The metals are sorbed onto dead biomass, and the biomass is removed from the aqueous phase by flocculation. (In trials to date, the biomass has comprised a waste product from the brewing industry, i.e. spent yeast cells).
2. The metals are leached from the biomass using a strong electrolyte, and the biomass is then recycled.
3. The metals are electrolytically recovered in their zero valent (native) forms, and the electrolyte re-cycled.

While this process has been successfully demonstrated at pilot scale, treating water draining from an abandoned Pb–Zn mine in northern England, it has relatively high capital costs compared with other technologies, and is therefore best suited to situations where saleable quantities of valuable metals can be recovered from mine waters with high concentrations of relatively valuable metals such as Zn and Cd.

4.5 MEMBRANE PROCESSES

It is feasible to remove solutes from water by forcing the water to pass at high pressure through membranes containing very small pores. Suspended and dissolved solid particles with effective diameters larger than the pore size will be trapped on the upstream side of the membrane, leaving more-or-less "pure" water to pass through to the downstream side. A highly concentrated liquor (which can require careful disposal) is left behind on the upstream side. Depending on the pore size of the membrane, this approach goes by various names:

- "Microfiltration", where pores are ≥ 0.1 µm and <0.45 µm. This is typically used for the removal of bacteria from water.
- "Ultrafiltration", where pores are ≥ 0.01 µm and <0.1 µm. This is typically used to remove colloids from water.
- "Nanofiltration", where pores are ≥ 0.001 µm and <0.01 µm. This is typically used to remove colour from water.
- "Reverse osmosis", where pores are < 0.001 µm. This is typically used to remove solutes from water.

The latter process is widely used in desalination plants producing drinking water, as an alternative to flash distillation. Reverse osmosis also has the potential to remove undesirable solutes from polluted mine waters, and since 1973, several attempts have been made to do this (Wilmoth, 1973; Barton, 1978; Clarke, 1995; Wilson and Brown, 1997). In general, the costs of reverse osmosis (which relate mainly to the cost of the membranes and the energy required to obtain the high operating pressures) preclude its use in routine mine water treatment. However, under certain economic conditions, reverse osmosis can prove attractive for mine water treatment. For instance, the highly-saline mine waters of Upper Silesia have economic value as a source of salt, and two reverse osmosis plants have been constructed near Katowice for this purpose (Clarke, 1995).

The largest low-pressure reverse osmosis mine water treatment plant in the world was designed, installed and brought into service to treat 20 Ml.d^{-1} of water for public supply in just eight weeks during a severe drought in Yorkshire, England, in 1996 (Wilson and Brown, 1997). The raw water for the reverse osmosis plant was obtained from an inactive surface coal mine (St Aidan's Opencast Mine) which had spectacularly flooded with river water seven years previously when the River Aire burst its banks and formed a waterfall down the highwall of the open pit (Hughes and Clarke, 2001). Dewatering of the St Aidan's Mine simultaneously achieved the rehabilitation of the mine (making 2.5 milllion tonnes of coal available for exploitation once more) and provided an invaluable addition to public water supplies in the Leeds area, for a critical period of eight months during the drought.

4.6 OTHER ACTIVE TREATMENT PROCESSES

Many other processes have been proposed for removing metals (and certain anions, especially sulfate) from mine waters. Where the value of the mined product (or the health hazards associated with it) demand element-specific processing (e.g. gold and uranium respectively), these techniques can come into their own. However, most mining engineers are unlikely to find them economic for the control of base metals in mine waters. Examples of these elaborate processes include:

- *Solvent extraction* (Barton, 1978), in which an immiscible organic solvent is agitated with the mine water (usually under highly-controlled pH conditions) until the metals of interest are preferentially partitioned into the organic solvent. The organic solvent is then separated from the mine water, and the metals transferred back into another aqueous solvent for subsequent concentration by precipitation, electrolysis or bio-concentration ("solvent extraction bio-winning"). Although widely used in the processing of uranium and gold, this technique is so metal-specific that it is rarely likely to be useful for most common types of mine water.
- *Electrochemical extraction*, in which electrodes of various compositions are suspended in the water and subjected to electrical currents appropriate for the attraction and precipitation of target metals. Although such techniques are central to various hydrometallurgical mineral beneficiation processes, there are no known cases in which mine waters have been sufficiently rich in valuable metals to make electrochemical extraction economically attractive as a water treatment technique.
- *Biochemical extraction* techniques are many and varied. Possible biochemical extraction techniques include:
  - biosorption using specific strains of microbes
  - algal uptake and bioconcentration of metals
  - The use of genetically-engineered "hyperconcentrators" (i.e. plants which take metals into their tissues at very high concentrations) to strip metals from water
  Extravagant claims continue to be made about the potential of such techniques for mine water treatment, but most methods in this category have yet to move beyond the test-tube scale.
- *The barium sulfide process* (Adlem et al., 1994) provides a means of removing sulfate from mine waters. Barium sulfide is added to the mine water under anoxic conditions. The barium rapidly reacts with the sulfate in the mine water, precipitating barite ($BaSO_4$), which is removed by gravity drainage as a sludge. The dissolved sulfide is then reacted with injected CO_2 gas, which allows recovery of elemental sulfur (a marketable commodity). This elegant process reliably yields residual SO_4^{2-} concentrations of only a few mg.l^{-1}. However, the process has very high energy requirements, for the barium sulfide used in the process is generated by thermal reduction of the recovered $BaSO_4$ at 1050°C, which requires maintenance of a substantial furnace. Given the expense associated with this step, it seems likely that SAVMIN process (Section 5.2.6) will be able to achieve sulfate removal more economically.
- Barium sulfate precipitation is also used for *removing barium and radium* from sulfate-poor mine waters (mostly highly saline waters) encountered in the Upper Silesian Coal Basin of Poland (Lebecka et al., 1994). Phosphogypsum, a waste material arising from phosphate fertiliser production, is added to the mine water, in which it slowly dissolves, releasing sulfate. The sulfate then reacts with the barium to precipitate barium sulfate (in which radium is also trapped by co-precipitation), which then settles from suspension. The process is applied not only in surface treatment plants, but is also used

underground, by addition of phosphogypsum (in slurry form) to areas of goaf through which the mine waters pass (Lebecka *et al.*, 1994).

4.7 SLUDGE MANAGEMENT – A KEY ISSUE IN MINE WATER TREATMENT

4.7.1 SLUDGE – A RESOURCE OR A WASTE PRODUCT?

Active treatment by the most popular methods (alkali dosing, sulfidisation, reverse osmosis) generally results in the production of a semi-solid waste material (sludge) rich in metals. Under the most favorable conditions, the sludge produced from a mine water treatment operation can be viewed as a resource rather than as a waste. For instance sludges arising from the treatment of ferruginous net-alkaline coal mine waters *without the use of lime or other alkaline reagents* will typically consist of very pure ferric hydroxides. Given favorable commercial circumstances, such pure sludges can have economic value in the following contexts:

- as pigments in a range of industries (Hedin, 1998, 1999; Hedin and Weaver, *in press*; see also *http://www.hedinenv.com/environ.htm*), including the production of colored concrete and bricks (Dudeney, 1997)
- as a soil amendment, to prevent excessive leaching of soil phosphorous (e.g. Evenson and Nairn, 2000)
- as a reagent for removal of P from sewage and agricultural runoff (e.g. Lamont-Black *et al.*, 2001).

In the longer term, stored ochre may eventually be transformed into an iron ore resource. It has been observed that retention of ochre in a dry impoundment for many years results in a gradual change in the nature of the material, from a low-density X-ray amorphous form to crystalline iron minerals. This occurs through slow structural changes which lead to the expulsion of chemically-bound water. The earliest crystalline forms to develop are relatively low-density hydroxide minerals, such as goethite ($FeOOH$) and ferrihydrite ($Fe(OH)_3$). Eventually, even these will lose their hydrogen content to form dense oxides such as hematite (Fe_2O_3), which is the principal economic ore mineral of iron. Observations by the authors at mine sites in northern England suggest that the transformation of fresh ochre to solid hematite requires one or two decades. However, once the iron is in the form of hematite, it is eminently marketable as long as the economic conditions are right. In the fullness of time, the presence of such shallow deposits of iron ore in well-known locations may well prove to be valuable to future generations of miners (see Hammer, 1992, p. 101).

Sulfide sludges derived from anaerobic treatment processes are less promising from the point of view of economic end-uses. For instance, iron sulfides are never likely to be worth processing for recovery of either the iron or the sulfur. Nevertheless, sulfide sludges rich in Cu and/or Zn may prove to have some commercial value as smelter feedstock (Diaz *et al.*, 1997), although it will generally be necessary to guarantee large quantities of high-grade sludge before the interest of a smelting operation can be assured.

Cases in which mine water treatment sludges have economic value are few and far between. Far more commonly, ODAS treatment of acidic coal mine drainage, and many base metal mine waters, normally yields sludges of variable physical and chemical stability, which are highly unlikely to have any economic value and which must be handled as waste. In extreme cases, for instance where the sludge contains arsenic or cadmium, they may even be classed as hazardous wastes, and will require special handling in line with local regulations.

4.7.2 WASTE SLUDGES ARISING FROM MINE WATER TREATMENT

4.7.2.1 General considerations

The practical and financial burdens associated with the disposal of mine water treatment sludges have long been recognized (e.g. Ackman, 1982; US EPA, 1983). In pump-and-treat schemes, sludge handling and disposal commonly accounts for 25–50% of the total treatment cost. At active mine sites, it is commonplace to dispose of these sludges by tipping/slurry pumping into mine voids. At abandoned mine sites, disposal into mine voids may no longer be an option for technical and/or regulatory reasons. Where disposal into mine voids is not an option, landfilling is the usual recourse. However, landfilling is becoming increasingly costly as regulatory pressures grow for waste minimization, typically pursued by specifying high tariffs for receipt of wastes, and/or by imposing taxes on every tonne of waste sent to landfill.

In planning a disposal route for mine water treatment sludges, it is often necessary to make a case for regarding the sludge as non-hazardous. Factors to be taken into account when preparing or considering such a case include the following:

- Sludges formed in lime dosing operations may be chemically unstable, and prone to release Zn, Ni, Cd and other metals to solution upon leaching after disposal.
- Any sludge containing large quantities of Pb, Zn, Cd, As etc is likely to be deemed a "hazardous waste", and will thus require special handling, and disposal at the most expensive tariffs.
- Sludges derived from sulfidization processes will only be stable if maintained in anaerobic conditions (e.g. below water covers (Section 3.12.2), or when mixed with reductants such as organic waste). Under oxidizing conditions, they may become net sources of pollutant metals once more.
- As noted above, ODAS sludges are typically of low density and of little or no economic value, necessitating much greater disposal volumes per tonne of metal than other methods.

Under most circumstances, the mine operator has little control over the overall metals content and chemical stability of waste sludges: if these contaminants have to be removed from the mine water, they must end up in the sludge. In rare cases it might be possible to preferentially remove different metals at different stages of the treatment process, for instance by carefully varying pH in accordance with Table 4.1. In this way, a non-hazardous component of the sludge may be generated separately from a 'hazardous' component.

The final factor mentioned above, that of low sludge density, is more amenable to solution by cautious process design.

Table 4.1: Target pH values for rapid removal of various metals as their hydroxides

Metal	pH
Fe^{3+}	3.5
Cu^{2+}	6.8
Zn^{2+}	8.2
Fe^{2+}	8.5
Cd^{2+}	9.8
Mn^{2+}	10.2

4.7.2.2 Increasing sludge density

The first step to take in maximizing sludge density is to operate the hydroxide or sulfide precipitation step in the treatment process in such a manner that the sludge density is maximized from the outset. For ODAS systems, this means that a high-density sludge circuit should be used (Section 4.2.6). For sulfidization systems, a process should be used which spatially separates the precipitation of metal sulfides from the production of HS⁻ by microbial activity in organic matter, such as the biosulfide process (Section 4.3.2). At the time of writing, there are few full-scale treatment plants using sulfidization processes. Consequently few experiences with the handling of sulfide treatment sludges (as opposed to sulfidic tailings) have as yet been published. In the following paragraphs, therefore, we focus on hydroxide sludges.

Having obtained the most dense sludge possible in the initial precipitation–sedimentation process, further reductions in the water content of the sludge can be achieved in a number of ways. Firstly, the sludge can be stored in *dry impoundments* and allowed to drain under gravity. Even low-density sludges can have their water contents reduced from about 95 wt % to less than 80 wt % in several weeks, at least in summer weather (Best and Aikman, 1983; Laine, 1997). However, the high specific retention of fine-grained iron hydroxides means that gravity drainage alone is unlikely to achieve water contents much less than 70% on time-scales of several months. As water contents as low as this can be achieved by HDS sedimentation processes (Section 4.2.6), there is little incentive to place HDS sludges in dry impoundments with the hope of further densification.

In many cases it will be necessary and/or desirable to increase sludge density beyond that attainable by gravity drainage. There are four major techniques to achieve this:

- Vacuum filtration
- Continuous pressure dewatering
- Frame-and-plate pressing
- Centrifuge separation

The first three of these rely on the filtration of sludge through finely-woven filter fabrics, and differ only in the strategy used to force the filtrate through the fabric.

Vacuum filters operate by first attracting a sludge layer onto the surface of a filter fabric by vacuum suction, drying the sludge, and then blowing and/or scraping the dried sludge off the fabric. In the rotary variety of vacuum filter (Figure 4.9), the vacuum is created by suction inside a cylindrical, mesh drum, around which the filter fabric is tightly wrapped. Slow rotation of the drum repeatedly immerses different belts of fabric in the feed tank containing the sludge. The vacuum suction attracts a layer of sludge onto the fabric surface, and the wetted area is then lifted free of the sludge surface, and dried rapidly as air is pulled into the central vacuum. As the fabric begins to turn once more towards the feed tank, it passes over the exhaust end of the vacuum machine, which serves to blow the dried sludge off the fabric. Scrapers may assist this process of sludge removal. The falling sludge (around 20 wt % solids) is caught in skips and transported to its final destination (waste dump or, less commonly, re-cycling facility).

Continuous pressure dewatering machines (Figure 4.10) also use filter fabrics, though in this case these are mounted on rollers rather than stretched over a drum. Sludge is fed onto the upper surface of the filter fabric, and gravity drainage through the fabric achieves

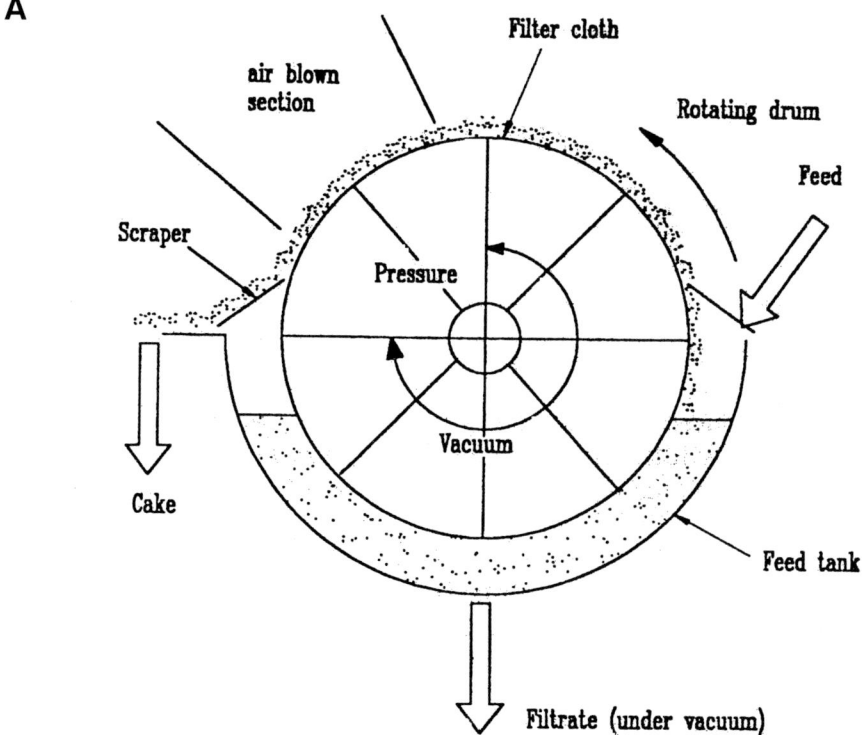

Figure 4.9: *(See next page for 4.9B)* **Rotary vacuum filter for sludge dewatering (a) Process schematic (courtesy of Environment Agency (EA)) (b) Example in use at an arsenic treatment plant in Brasil. (Photo: P. Younger, with the permission of Mineraçao Morro Velho).**

B

Figure 4.9B: (*See previous page for 4.9A*).

both a small increase in density *and* prompts gentle adhesion of the sludge to the fabric. The adhered layer of sludge is then squeezed between the top fabric and a lower fabric, firstly in a low pressure "wedge zone", and then in a high pressure zone. The dry, relatively dense sludge cake falls of the end of the belt into skips. Sludge densities on the order of 22 wt % solids can be obtained in this manner.

Frame-and-plate presses (Figure 4.11) operate by forcing a fabric-wrapped mesh "plate" downwards into a "frame" containing the sludge. Filtrate is forced through the fabric under pressure, leaving a compressed sludge cake behind in the frame. The compressed sludge cake remains in the frame during successive pressings of sludge aliquots, until the frame is full. At this juncture, the plate is removed, the sides of the frame lowered, and the compressed sludge cake removed by mechanical excavation and washing. Frame-and-plate presses produce the most dense sludges (25–30 wt % solids) of any common dewatering device.

Scroll centrifuges are the only full-scale sludge dewatering devices *not* relying on the use of filter fabrics. They are essentially scaled-up versions of the familiar bench top centrifuge (Figure 4.12), modified to include scraping devices to aid in flocculation and transport of solids. Scroll centrifuges produce sludges with a solids content of 22–25 wt %.

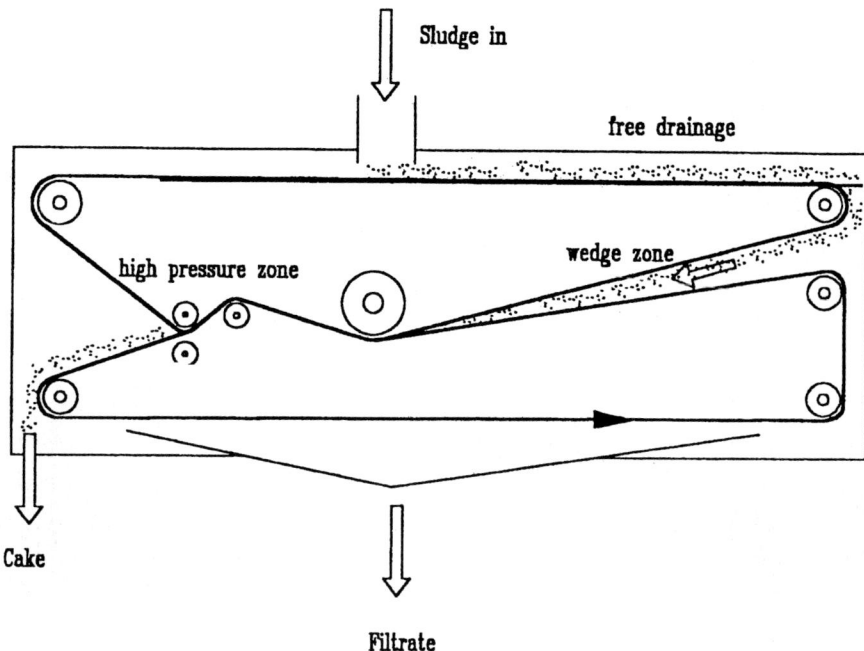

Figure 4.10: Schematic diagram of a continuous pressure sludge dewatering machine. (Courtesy of EA).

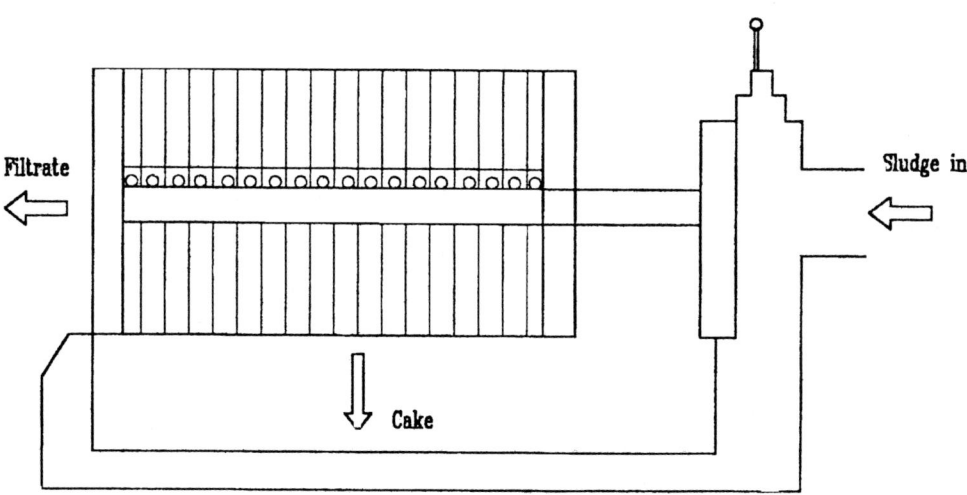

Hydraulic piston or screw compresses the press, filtrate passes through the filter cloth into plates and is removed. Cake remains in the frame until full, when it is disassembled and washed.

Figure 4.11: Schematic diagram of frame-and-plate presses for sludge dewatering. (Courtesy of EA).

A

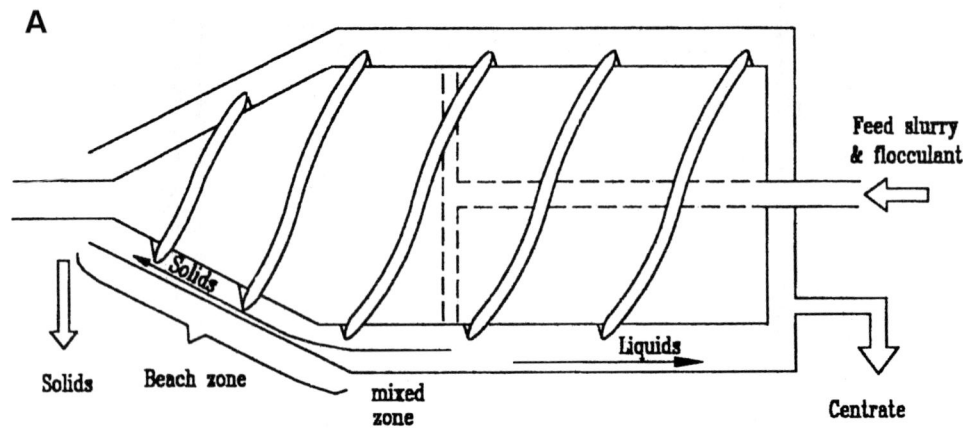

Feed slurry
& flocculant

Solids

Solids   Beach zone   mixed   Liquids
                               zone
                                                        Centrate

Typically, case can rotate at 3000 to 5000rpm
inner scroll at just a few revs slower (5 to 20rpm)

Figure 4.12: (*See opposite for 4.12B*) **Scroll centrifuge for sludge dewatering (a) Process schematic (courtesy of EA) (b) Example in use at Ynysarwed coal mine water treatment plant (see Worked Example 4.1). (Photo: P. Younger).**

B

Figure 4.12B: (*See opposite for 4.12A*).

CHAPTER FIVE

PASSIVE TREATMENT OF POLLUTED MINE WATERS

5.1 WHAT IS PASSIVE TREATMENT?

As contaminated mine waters flow into and through receiving systems (streams, rivers, wetlands and lakes), their toxic characteristics commonly decrease. Many studies (e.g. Tuttle *et al*., 1969, Wieder and Lang 1982; Huntsman 1978; Stark *et al*., 1990; Herlihy and Mills 1985) have shown that this amelioration in quality occurs as a result of:

- natural chemical and biological reactions and
- dilution with uncontaminated waters.

The net result is that potentially ecotoxic metals are removed from solution, and therefore placed largely out of harm's reach, by precipitation as hydroxide, oxide, sulfate, sulfide and (less commonly) carbonate minerals. Acidity is neutralized by:

- mixing with waters which already possess alkalinity and/or
- liberation of alkalinity from minerals (by *in situ* dissolution of pre-existing carbonate minerals, especially calcite, but to a lesser extent dolomite also) and/or
- generation of alkalinity from organic matter by heterotrophic microbial processes.

During the last 15 years, the possibility that natural ameliorative processes might be harnessed in mine water treatment systems has developed into a practice referred to as *passive treatment*. Although the term 'passive treatment' has been in use for almost a decade at the time of writing (e.g. Cohen and Staub, 1992), formal definitions of the term are not easy to find. The following definition of the term was coined by William Pulles of South Africa, and subsequently formally adopted by the European Union's PIRAMID R&D project (www.piramid.org):

> "Passive treatment is the deliberate improvement of water quality using only naturally-available energy sources (e.g. gravity, microbial metabolic energy, photosynthesis), in systems which require only infrequent (albeit regular) maintenance in order to operate effectively over the entire system design life"

As the definition given above makes clear, passive treatment technologies use natural materials to promote natural chemical and biological processes. Cost-effective treatment is obtained by manipulating environmental conditions in the treatment system so that particular contaminant removal processes are optimized. The principal materials utilized in passive treatment systems are locally-sourced carbonate rocks and organic substrates. Neither the materials nor the products of the vast majority of passive treatment are hazardous. It is often possible to design passive systems to operate for years (even decades) with minimal requirement for operator intervention and/or costly maintenance. Because most passive treatment systems include constructed wetlands, they may also provide wildlife habitat and can have substantial 'ancillary' values of social and ecological natures (e.g. Hawke and José, 1996; Younger *et al.*, 1998; Campbell and Ogden, 1999).

Passive treatment systems are now sufficiently numerous and sufficiently mature in North America (and increasingly in Europe) that it is possible to identify some of the key pros and cons of the technology. These may be summarised as:

Advantages...
- Low operating costs, and usually low capital costs also (at least for small- to medium-sized mine water discharges)
- Use non-hazardous materials
- If suitably designed and well-constructed, passive systems can work for long periods of time unattended
- Passive systems can often be directly integrated with surrounding ecosystems
- In many cases they will be more pleasant in appearance than active treatment systems

Disadvantages...
- Passive treatment technology is still relatively new, and hence reliable expertise is still scarce
- Because day-to-day intervention in treatment processes is precluded, precise control of treatment effluent quality is not feasible
- A large land-take is likely to be necessary foir high-flow and/or highly contaminated discharges
- Relatively high capital (construction) costs

The large land demand is probably the principal drawback of passive treatment (Figure 5.1). It derives from the fact that contaminant removal processes in passive systems generally occur at low or modestly-high pH, under which conditions most important reaction rates (oxidation, hydrolysis etc) are considerably slower than in the artificially high-pH environments typical of conventional active treatment systems (Chapter 4). As a consequence, passive systems require longer retention times and larger areas in order to achieve similar results.

The goal of passive treatment systems is to enhance the natural amelioration processes so that they occur within the treatment system, rather than in the receiving water body. As such, passive treatment can be considered to be an example of 'enhanced natural attenuation' (cf. Fetter, 1999). Two factors which determine whether this goal can be accomplished at a given site are the kinetics of the contaminant removal processes and the retention time of the mine water in the treatment system. At many sites, only a limited amount of land will actually be available, and thus an upper limit is imposed on the hydraulic retention time, which becomes a function only of flow rates. However, the kinetic rates of several contaminant removal processes can be enhanced by manipulating the environmental conditions which exist within the passive treatment system. Efficient manipulation of contaminant removal processes requires that the nature of the rate-limiting aspects of each removal process be understood. This chapter therefore focuses on the processes that underlie passive treatment and methods which may be used to manipulate their kinetics.

Figure 5.1: An extreme example of the large land take a passive mine water treatment system can demand: the UK Coal Authority's Taff Merthyr aerobic wetland system; South Wales. The photograph was taken shortly after commissioning of the system in the year 2000. Mine water emerges at surface from an old mine shaft in the center of the treatment area. It is then cascade-aerated and split three-ways, into each of three conspicuous sedimentation ponds, each of which is followed by a series of aerobic wetlands. The size of the system in this case made a major contribution to the clean-up of derelict land associated with the former colliery, and is integrated with a newly-established country park and adjoining commercial premises. (Photo: UK Coal Authority).

It is perhaps worth noting that passive treatment can be regarded as one of two sub-categories of a broader theme, which has been termed 'passive remediation' (see *www.piramid.org* for further discussion). The other sub-category is 'passive prevention of pollutant release', which encompasses techniques such as the installation of water covers or dry covers on waste rock piles. Passive prevention is discussed at length in Section 3.12.2 and will not be considered further here.

5.2 A BRIEF HISTORY OF PASSIVE MINE WATER TREATMENT

The first steps in the development of what is now termed "passive treatment" are ascribed by Campbell and Ogden (1999) to Dr Kathe Seidel, who, whilst working at the Max Planck Institute (Germany) in the 1950s, investigated the ability of various aquatic plants to absorb and break down common organic pollutants. Two more decades were to elapse before research into wastewater treatment using wetlands really began to flourish, in the USA during the mid-1970s (Campbell and Ogden, 1999). By the late 1980s, wetland treatment of wastewaters, stormwaters and agricultural runoff had become a well-established field of endeavour in many parts of the world (e.g. Hammer, 1989; Cooper, 2001).

Possibly because of their distinctive chemistry, polluted mine waters began to be considered candidates for passive treatment somewhat later than (and somewhat independently from) 'mainstream' wastewater treatment wetland technology. To this day, many passive mine water treatment systems are distinctly different in both form and function to analogous systems treating urban wastewaters. The roots of current passive mine water treatment technologies can be traced to two independent observations in the eastern USA (Huntsman, 1978; Weider and Lang, 1982), to the effect that *Sphagnum* bogs naturally improved the quality of coal mine waters flowing into them. These reports, and promotion of the concept of passive treatment by the US Bureau of Mines, resulted in the first attempts to construct wetlands specifically to treat polluted mine waters. Initially, most of these wetlands were constructed to mimic *Sphagnum* bogs. However, *Sphagnum* moss was not always readily available, proved difficult to transplant, and tended to accumulate metals to toxic levels after several months of exposure to mine drainage (Weider *et al.*, 1985; Spratt and Weider, 1988). Instead of abandoning the concept, researchers experimented with different kinds of constructed wetlands. Eventually a wetland design evolved that proved tolerant of years of exposure to contaminated coal mine drainage and was effective at lowering concentrations of dissolved metals. Most of these treatment systems consisted of a series of small wetlands (< 1 ha) that were vegetated with *Typha latifolia* (Girts *et al.*, 1987; Brodie, 1990). (*Typha latifolia* is a very well-known rush, which is colloquially named 'cattails' in the USA, and 'reedmace' in the British Commonwealth). Some wetlands were constructed with a compost substrate in which the *Typha* rooted, while others were constructed with whatever soil or spoil substrate was available. (Brodie, 1990). A common feature of these early passive systems was their general similarity to natural wetlands.

Recently, passive treatment technologies have been developed that do not rely on the wetland model that the early systems were designed to mimic (see, for instance, Younger, 2000d,e). At one extreme, various aerobic reactors (in the form of ponds, ditches, rock-filled basins and trickle filters) have been constructed which are not planted with emergent plants and, in some cases, contain no soil or organic substrate (e.g. Hedin and Nairn, 1993; Jarvis and Younger, 2001). At the other extreme are anoxic subsurface flow reactors packed with organic substrates, which have no 'daylight' portion and hence support no macrophytes (e.g. Staub and Cohen, 1992; Benner *et al.*, 1997). Similar in concept are a variety of passive 'pretreatment' systems, in which acidic waters contact limestone or organic substrates under anaerobic conditions, before flowing into ponds or wetland systems where aeration finally takes place (e.g. Hedin *et al.*, 1994).

The present status of passive treatment may be summarised by saying that techniques for removal of iron from acidic and alkaline mine waters are now well-established and can be implemented with confidence (e.g. Hedin *et al.*, 1994), whereas systems for removing most other contaminants from either type of water have been implemented less frequently and remain a subject of active research, with full-scale implementation still prone to significant uncertainties (e.g. Cohen and Staub, 1992; Younger, 2000e). At the time of writing, North America retains a leading role in the development of passive treatment technologies, with major research projects now underway in South Africa (Pulles, 2000), Europe (Younger, 2001b) and Australia, which are rapidly yielding technological advances for some of the more challenging mine water types.

Before considering the principal attributes of contemporary mine water passive treatment system designs, it is necessary to become familiar with the chemical and microbial processes upon which these designs are based.

5.3 CHEMICAL AND MICROBIAL BASIS FOR PASSIVE TREATMENT

5.3.1 SCOPE OF THIS SECTION

Chapter 2 gives comprehensive coverage of the main processes affecting the quality of mine waters in the natural environment. Most of the principles and guidelines presented there are just as valid in the context of the built environment of passive treatment systems. Nevertheless, at the risk of a little repetition, we consider it worthwhile providing here a succinct summary of the key chemical and microbial processes relevant to the design (and interpretation of performance) of passive mine water treatment systems, for two reasons:

(i)   to highlight those processes which are currently considered to be the most important ones to consider when designing passive treatment systems, and

(ii)   to allow this chapter to be used in "stand-alone" mode by readers who do not wish to deepen their understanding of the fundamental processes to the level of coverage afforded by Chapter 2.

Before continuing it is worth noting that a typical passive treatment facility is essentially a highly complex biogeochemical system (e.g. Walton-Day, 1999). Complete and precise description of the full range of processes occurring in any one system over a given period of time is not feasible. Rather, we have to content ourselves with identifying the *predominant* processes which are influencing water quality evolution in a particular system. Thus most 'aerobic' wetlands contain significant anaerobic zones within their bed sediments (and *vice versa*). Yet the designation of a given system is generally an adequate simplification for most engineering purposes. If we add hydrodynamic fluctuations to biogeochemical complexity, we might quickly despair of the likelihood of ever deriving sufficient system understanding that we could develop a predictive model suitable for design purposes. As will be seen, the reality is not quite as daunting as these considerations might suggest, but it certainly always pays to bear in mind the inherent complexity of passive systems.

The broad distinction between active and passive treatment systems given in Section 5.1 focuses on the prohibition of ongoing inputs of external energy and reagents in true passive systems. In purely chemical terms, there are two further distinctions between passive and active systems which need to be borne in mind:

(i)   The principal alkaline reagent in passive systems is bicarbonate (HCO_3^-), as opposed to hydroxyl (OH^-) in active systems (Section 4.2.3.3). Since bicarbonate tends to buffer the pH of mine waters between about 6 and 8.5 (Section 2.2.2), successful metal removal demands the utilization and optimization of reactions which operate briskly in this pH range. This differs substantially from the circumstances in active treatment systems, in which alkali dosing commonly results in pH ranging between 9 and 12, in which range most metals of interest will precipitate as hydroxides. Thus, in active treatment systems, Zn and Cu are commonly removed from solution as their hydroxides, whereas in passive systems their carbonate minerals are more feasible sinks (see Section 5.3.7 and Table 5.1 below).

(ii)   In passive systems, metal removal and alkalinity generation processes are commonly separated in space and time, whereas they tend to be synchronous in active treatment systems. This has important implications for the layout of passive treatment plants, as should become evident in Section 5.4.

Bearing these considerations in mind, we now turn to the description of the predominant processes operative in water quality amelioration in the principal types of passive treatment system.

5.3.2 REMOVAL OF Fe IN PASSIVE SYSTEMS

Aerobic Fe removal processes
Concentrations of dissolved iron decrease in natural aquatic systems by a variety of processes. From a treatment perspective, the most important process is the oxidative removal of Fe as an oxyhydroxide. The remediation process can be conceived as occurring in the following three steps:

ferrous iron oxidation $Fe^{2+} + 1/4\ O_2 + H^+ = Fe^{3+} + 1/2\ H_2O$ (5.1)
(i.e. the reaction previously introduced in Equation 2.2)

ferric iron hydrolysis $Fe^{3+} + 2H_2O = FeOOH(sus) + 3H^+$ (5.2)
(which is essentially a variant of Equation 2.3)

sedimentation of ferric $FeOOH(sus) = FeOOH(sed)$ (5.3)
oxyhydroxides solids

where "sus" indicates a suspended solid and "sed" indicates a sedimented (settled) solid.

Table 5.1: Possible removal processes for less common contaminants, which might be applicable in future passive system design

Contaminant	Possible removal process for passive systems	Examples/sources of further information
Arsenic	Oxidation in the presence of iron to As^{5+}, forming AsO_4^{3-}, which sorbs to Fe oxides; can also precipitate as ferric arsenate (scorodite)	McRae et al. (1999) (see also www.piramid.org for oxidation pond studies)
	Reduction of As^{5+} to As^{3+} in compost-based systems, forming sulfides such as AsS and As_2S_3	Cohen (1996)
Cadmium	Precipitation as a sulfide (CdS) in compost-based anaerobic systems	Cohen (1996)
Chromium	Reduction of Cr^{6+} to Cr^{3+} in compost-based systems, with hydrolysis to form $Cr(OH)_3$	Cohen (1996)
Copper	Oxidation in alkaline solution to form carbonate minerals (azurite/malachite etc)	Brown (1997)
	Reduction in compost-based systems to form sulfides	Cohen (1996); Thompson (1996)
Cyanide	Photolysis (in tropical regions) in open ponds	Young and Jordan (1996)
	Reduction to form CO_2 and NH_4^+ in compost-based systems	Thompson (1996) (see also www.piramid.org)
	Bacterially mediated oxidation to ammonia and nitrogen gas	Thompson (1996)
Lead	Oxidation in alkaline solution to form carbonate minerals	Thompson (1996)
	Precipitation as a sulfide in compost-based anaerobic systems	Cohen (1996)
Nickel	Precipitation as sulfides in a compost-based system	Ettner (1999)
Thallium	Reduction in compost-based systems to form sulfides	Mueller (2001)
Zinc	Precipitation as sulfides in a compost-based system	Cohen (1996) Lamb et al. (1998)
	Precipitation as a carbonate in aerobic ponds or limestone drains	Kalin (1998) Nuttall & Younger (2000)

As described in Section 2.1.1, the oxidation of ferrous iron in passive systems (Equation 5.1) can occur by abiotic (i.e. purely chemical) and/or microbial processes, with the relative dominance of one or other process being dependent on pH. Abiotic dominates over biotic at pH 6–7, while the converse is true at low pH (e.g. Athay et al., 2001).

The chemical oxidation of ferrous iron at pH values between 4 and 8 is usually described by the model developed by Singer & Stumm (1970) (see Section 2.1.2):

$$\frac{d\ Fe^{2+}}{dt} = K\ \frac{[Fe^{2+}]\ [O_2]}{[H^+]^2} \qquad\qquad (5.4)$$

Reaction 5.4 is first order with respect to Fe and dissolved oxygen and second order with respect to pH. (The strong non-linear dependence on pH explains the virtual absence of abiotic iron oxidation when pH is less than 5). At pH values greater than 9, the rate of Reaction 5.4 is so fast it becomes primarily dependent on the transfer rate of oxygen into the water (Hustwit et al., 1991). In the pH range 6 to 8, which is very commonly encountered in passive treatment systems, the rate of Reaction 5.4 is fast enough to make passive treatment feasible, but still slow enough that the kinetics remains a very important consideration in the design and sizing of passive systems.

As previously noted, microbial oxidation of ferrous iron predominates at pH values less than 5 (e.g. Kleinmann and Crerar 1979). Iron oxidizing bacteria, such as *Thiobacillus ferrooxidans* are often invoked as the responsible taxa, though recent work has shown that other bacteria and even archaea (i.e. non-bacterial microbes) can play a significant role (e.g. Bond et al., 2000). The iron-oxidizing microbes appear to be optimally active at pH values between 2 and 3. While there is little doubt that bacteria significantly catalyze iron oxidation at low pH, there is limited information on the kinetics of the process, save the oft-quoted axiom that it may be as much as 10^6 times faster than the abiotic rate at the same pH (e.g. Singer and Stumm, 1970). In particular, it is unclear whether microbial iron oxidation rates at low pH are faster than abiotic iron oxidation rates measured at circumneutral pH (e.g. Watzlaf et al., 2001).

The kinetics of the process by which ferric iron hydrolyses to form suspended (oxy) hydroxide solids (i.e. reaction 5.2) has not been studied in detail. However, preliminary data presented by Singer & Stumm (1970) and numerous field observations indicate that the rate is rapid at pH values greater than 4. Certainly, *dissolved* ferric iron is rarely detected in waters with pH > 4, suggesting that any dissolved ferric iron at this pH is very rapidly removed from solution by (oxy)hydroxide formation. Conversely, the rate of the hydrolysis reaction appears to be very slow at a pH value less than 3, where solutions maintain their ferric iron content for days despite over saturation with respect to numerous ferric (oxy)hydroxide and oxide minerals.

The sedimentation of iron (oxy)hydroxide solids in passive systems has not been widely investigated. Nevertheless, common observations and some experimental data demonstrate that there are at least four major processes which perform the transformation represented by Equation 5.3. In order of increasing rapidity, these :

(i)   Settlement of ferric hydroxide flocs from aqueous suspension. This process apparently predominates where neutralized and thoroughly oxidized mine waters are stilled in ponds.

(ii)   Physical filtration of colloidal ferric hydroxide from solution by fixed solids (e,g, plant stalks and roots, fibrous wetland substrate materials, or non-biological filter media). Though difficult to quantify by field measurements, this process has been shown to be more important than the other three in some experimental trickle filter reactors (Burke and Banwart, 2001).

(iii) Iron plaque formation on roots and rhizomes of wetland plants. It has been found that the roots of common wetland plants (such as *Typha latifolia* and *Phragmites australis*) prolifically 'leak' oxygen into the surrounding wetland sediments. Ferrous iron dissolved in the sediment pore waters is therefore subject to vigorous oxidation in the immediate vicinity of the roots and rhizomes, and precipitates as ferric hydroxide coatings (known as 'plaques') on the plant material (e.g. Batty, 1999).

(iv)   *In situ* accretion of ferric hydroxide, by a process termed *surface-catalysed oxidation of ferrous iron* (SCOOFI). SCOOFI may be conceived as a three-step cyclical process (Jarvis and Younger, 2001), as follows:

1.   Dissolved Fe^{2+} is adsorbed by existing ferric hydroxide present as a substrate
2.   The ferric hydroxide surface itself acts as a powerful catalyst for the oxidation of the adsorbed ferrous iron to the ferric (Fe^{3+}) form (Equation 5.1)
3.   This ferric iron then hydrolyses *in situ* (as per Equation 5.2), forming a new surface layer of fresh ferric hydroxide (which can then adsorb more Fe^{2+} and reinitiate Step 1 above)

All four iron sedimentation processes can (and probably generally do) occur within close proximity to each other in any one passive system. However, some of the processes are more characteristic of one situation than another. For instance, in naturally oxidized mine waters with a pH less than about 5, particulate iron concentrations rarely exceed 1 mg/L. This is probably because ferric hydroxide solids are precipitated directly onto the substrates to which the iron-oxidizing bacteria are also attached (essentially a microbially-enhanced SCOOFI process). At pH values greater than 6, SCOOFI predominates where flows are rapid and turbulent (e.g. Demspey *et al.*, 2001), but where water velocities are sluggish, particulate iron frequently forms in suspension, giving the mine water the dense "tomato soup" appearance so damaging to the appearance of watercourses. In these circumstances, the suspended ferric hydroxide probably precipitates abiotically near the water surface, where dissolved oxygen concentrations are continually renewed. The suspended particles then slowly sink and coalesce into larger settleable solids, at a rate which is presumed to be a function of iron concentration.

The term "ochre" is commonly used as a collective term for the red, orange and yellow iron salts produced by hydrolysis of ferric iron. The precise composition of ochre varies with pH and the availability of dissolved anions such as SO_4^{2-}. Under circum-neutral conditions (pH 6–8), a mixture of X-ray amorphous iron hydroxide and goethite (α-FeOOH) precipitates (Equation 5.2). At more elevated pH (> 8) ferrihydrite ($Fe(OH)_3$) is more commonly precipitated in a reaction (Equation 2.3) which also yield three protons (H^+) for every mole of ferric iron which hydrolyses. At lower pH conditions, substitution of SO_4^{2-} for OH^- occurs, resulting in the formation of "oxyhydroxysulfate" minerals such as schwartmanite (Winland *et al.*, 1991; Bigham *et al.*, 1992). If left to consolidate over many years (or perhaps decades) the various hydrated ferric iron salts which comprise ochre will gradually dehydrate and recrystallise to form haematite (Fe_2O_3), which is a major iron ore mineral. Haematite is far more dense than the hydrated ferric iron salts, a fact which has potentially important implications for the long-term performance of passive systems in which ochre is presently accumulating.

The rate-limiting aspect of the oxidative iron removal process in passive mine water systems varies primarily with pH, and also with iron and dissolved oxygen concentrations (e.g. Watzlaf *et al.*, 2001; Dempsey *et al.*, 2001). At pH values less than 3, microbial processes oxidize Fe^{2+} to Fe^{3+}, but the slow kinetics of hydrolysis limit the formation of a solid. At pH values between 3 and 6, iron removal is limited by the oxidation step. At pH values greater than 6, iron removal appears to be limited by the oxidation step at concentrations greater than 10–20 mg/L, but becomes limited by the solid precipitation step at concentrations less than 10 mg/L.

Anaerobic iron removal
Under reducing conditions, iron can be removed from mine waters in passive treatment systems by the formation of iron sulfides and iron carbonates.

$$Fe^{2+} + HS^- = FeS + H^+ \qquad\qquad (5.4)$$
$$Fe^{2+} + HCO_3^- = FeCO_3 + H^+ \qquad\qquad (5.5)$$

The formation and stability of both of these solids requires circum-neutral pH conditions. As the production of the necessary HS^- and HCO_3^- anions relies on microbial alkalinity-generating processes, further discussion of reactions 5.4 and 5.5 is reserved until Section 5.3.4 below.

5.3.3 REMOVAL OF Al^{3+} IN PASSIVE SYSTEMS

Mine waters with pH less than 4 commonly contain high concentrations of Al (>10 mg/L) while waters with pH between 5 and 8 generally have dissolved Al concentrations less than 1 mg/L. As a highly ecotoxic metal in its aqueous form, Al removal is often a matter of considerable importance. Al removal processes in passive systems are still the subject of active research (e.g. Jarvis, 2000), and much remains to be investigated in detail. It is nevertheless already clear that hydroxide and hydroxy-sulfate solids are major sinks for Al in passive systems.

Given the pH-dependence of Al concentrations in natural waters noted above, it is logical to propose that formation of aluminum hydroxide will follow a reaction similar to that for ferric iron (Equation 2.2). The following reaction is commonly proposed (e.g. Hedin *et al.*, 1994; Younger *et al.*, 1997; see also Equation 2.56):

$$Al^{3+} + 3H_2O = Al(OH)_3 + 3H^+ \qquad\qquad (5.6)$$

Since Al is invariably present in the Al^{3+} form in solution, reaction 5.6 requires no prior oxidative step, unlike in the case of ferric iron (i.e. equations 5.1 and 5.2). Hence, while formation of ferric hydroxide normally occurs under aerobic conditions, the formation of aluminum hydroxide can occur under either aerobic or anaerobic conditions, if only the pH is high enough. (This is precisely why care must be taken in introducing Al-rich waters to anoxic limestone drains, as will be explained in Section 5.4.2).

The freshly-precipitated aluminum hydroxide formed by reaction 5.6 (and similar reactions) is usually X-ray amorphous, white and of very low density (e.g. Younger, 1995). If waters saturated with it are agitated (e.g. by an aeration cascade) the aluminum hydroxide usually forms as an unsightly froth, which can become an airborne eyesore (e.g. Younger *et al.*, 1997). Once captured and sedimented, however, the low-density amorphous aluminum hydroxide will slowly crystallise to one of the mineral varieties of the compound (Section 2.2.4), such as gibbsite and/or boehmite. These minerals are extremely common, non-toxic components of many natural soils, and therefore represent an environmentally suitable long-term store for Al in wetland sediments.

Although the existence and precise nature of low-temperature aluminum hydroxy-sulfate minerals has attracted some controversy over the years (see Appelo and Postma, 1993, for a good summary) recent field investigations have demonstrated the existence of X-ray amorphous aluminum hydroxy-sulfates in precipitates forming naturally from mine waters. For instance, in observations of two streams heavily impacted by Al-rich mine waters, Younger (1995) observed synchronous downstream decreases in dissolved sulfate and dissolved aluminum. In the absence of other possible explanations, the incorporation of sulfate into aluminum hydroxy-sulfate solids was inferred. Subsequent studies of the elemental composition of white deposits collected from the streambed (by means of electron back-scattering analysis using a scanning electron microscope) revealed them to include substantial quantities of sulphur (Younger, 1995). Similar hydroxy-sulfate phases have subsequently been found throughout the sediments of a treatment wetland receiving aluminum-rich mine drainage (e.g. Jarvis and Younger, 1999).

5.3.4 REMOVAL OF MANGANESE IN PASSIVE SYSTEMS

Aerobic Mn removal
Manganese is primarily removed from solution in passive systems by oxidation and hydrolysis reactions.

$$Mn^{2+} + 1/2\ O_2 + 2H^+ = Mn^{4+} + H_2O \qquad (5.7)$$
$$Mn^{4+} + 2H_2O = MnO_2 + 4H^+ \qquad (5.8)$$

Because of the need for oxidising conditions, most passive treatment strategies for Mn removal are predicated on aerobic processes. Nevertheless, these processes are not particularly well understood.

Because abiotic oxidation rates of Mn^{2+} are very slow at pH values less than 8, Mn^{2+} oxidation in passive systems is usually considered to be microbially-catalysed (cf Lind and Hem, 1993). Supporting this hypothesis is the virtual absence of Mn precipitation from filter-sterilized circum-neutral mine water samples (in laboratory experiments), and the tendency for Mn removal to improve significantly when a passive system is inoculated with Mn-oxidizing microorganisms (e.g. Brant *et al.*, 1999). Besides bacteria, both algae and fungi have been implicated in the oxidative precipitation of MnO_2. Algal mats have been shown to promote MnO_2 precipitation under summer conditions (Phillips *et al.*, 1995), presumably due to their ability to photosynthesise so vigorously that they super-saturate the surrounding water with dissolved oxygen. White rot fungus releases an extra-cellular enzyme (manganese peroxidase) which greatly catalyses the oxidation of Mn^{2+} (I Singleton, University of Newcastle, personal communication, 2000). Where manganese dioxide precipitation is especially prolific, thick coatings of black mineral matter are usually observed (e.g. Thornton, 1995; Brant *et al.*, 1999). This black material is termed 'wad' in mineralogical circles, and it typically comprises a mixture of X-ray amorphous and crystalline MnO_2 (the latter being the mineral 'pyrolusite'). Over time, all of the wad will eventually crystallise to form pyrolusite.

Sustainable manganese removal in aerobic passive systems only occurs under circum-neutral conditions. Several factors may explain this treatment limitation (Hedin *et al.*, 1994):

- The catalysing microbes may be intolerant of pH less than 6
- Formation of MnO_2, by biotic or abiotic processes, may require formation of a $MnCO_3$ precursor, which will not precipitate at low pH
- Manganese oxides are inherently unstable at lower pH because of their increasing solubility
- Many low-pH systems contain high concentrations of ferrous iron which readily reduce and solubilize manganese oxides.

In highly alkaline systems, it is possible that some Mn may be removed from solution as a carbonate. The best-known Mn carbonate is the attractive pink-white mineral rhodocrosite ($MnCO_3$). However, this pure phase is a less likely carbonate sink for Mn than certain dolomite-like Mn minerals (with a composition $(Ca,(Mg),Mn)(CO_3)_2$; Goldsmith and Graf, 1957) which have been found to be a significant sink for Mn in certain mine-water impacted streams (e.g. Lind and Hem, 1993).

Anaerobic Mn removal
Under anaerobic conditions, Mn^{2+} is usually quite mobile. While a manganese sulphide mineral does exist in nature (i.e. alabandite; MnS), it is very rare, only stable at very high pH, and at equilibrium with several mg L^{-1} of dissolved Mn^{2+}; it therefore holds out little promise as a sink for Mn in anaerobic passive systems. Nevertheless, research has shown Mn removal rates as high as 99% in anaerobic passive systems (e.g. Staub and Cohen, 1992. It has been suggested that the solid-phase sink for Mn in these systems could be rhodocrosite (e.g. Staub and Cohen, 1992); the dolomite-like Mn minerals described by Goldsmith and Graf (1957) and Lind and Hem (1993) are also feasible candidates. However, given that the few data which exist on the aqueous geochemistry of manganese carbonate minerals suggest they are only slightly less soluble than calcite, it may well be that the major sinks for Mn in anaerobic systems are as-yet-unidentified organic-manganese complexes. Such organic complexes may not be permanent sinks for Mn, however, to judge from sustained monitoring of the Jennings Environmental Center RAPS in Pennsylvania, which revealed high initial rates of Mn removal in the anaerobic portion of the system, followed by a gradual decline and, a year later, virtually no Mn removal. The implication is a slow, reversible sorption reaction involving the organic substrate.

5.3.5 GENERATION OF ALKALINITY IN PASSIVE TREATMENT SYSTEMS

There are two major processes of alkalinity generation in passive systems:

(i)   the dissolution of carbonate materials and
(ii)   microbial metabolism (see Section 2.2.6, and especially Table 2.7)

The most common carbonate material used in passive systems is limestone rich in calcite ($CaCO_3$). Calcite reacts with water to neutralize proton acidity (Equation 5.9) and generate bicarbonate alkalinity (Equation 5.10).

$$CaCO_3 + 2H^+ = Ca^{2+} + H_2O + CO_2 \tag{5.9}$$
$$CaCO_3 + H_2CO_3 = Ca^{2+} + 2HCO_3^- \tag{5.10}$$

The consumption of protons by reaction 5.9 is rapid, typically reaching equilibrium in less than a day of contact between the water and the calcite, and it is the more important of the two above reactions at pH values less than 5. The alkalinity-generating reaction (5.10) occurs at pH values greater than about 5 (see Section 2.2.2), and is slower than reaction 5.9, being limited, eventually, by the solubility of $CaCO_3$. Contaminated mine waters commonly contain high CO_2 partial pressures which provide the potential for significant alkalinity production under appropriate environmental conditions. CO_2 partial pressures for contaminated mine waters are commonly $10^{-1.0}$atm or greater (Hedin and Watzlaf, 1994; Hedin et al. 1994). When mine waters are exposed to limestone in a closed environment that maintains the high CO_2 partial pressures, bicarbonate alkalinity concentrations of 150–350 mg/L commonly result. When the same water is exposed to limestone in an open environment (CO_2 partial pressure of $10^{-3.5}$atm), the CO_2 degasses to the atmosphere and bicarbonate alkalinity concentrations of 40–60 mg/l result.

The overall effect of carbonate dissolution on mine water chemistry is the combination of proton-consuming (5.9) and alkalinity-generating (5.10) reactions. For waters with low pH and high concentrations of hydrolyzable metals (Fe^{3+} and Al^{3+}), proton-consuming reactions can dominate changes in acidity. For mine waters with pH > 4 and containing only Fe^{2+} and Mn^{2+}, changes in mine water acidity are primarily associated with the generation of bicarbonate alkalinity.

Table 2.7 lists a range of biological processes which can affect the acidity-alkalinity balance of soils and natural waters. Amongst the processes listed there, by far the most important in passive mine water treatment systems is dissimilatory bacterial sulfate reduction. In a very simplified form, this may be written:

$$SO_4^{2-} + 2CH_2O = H_2S + 2HCO_3^- \tag{5.11}$$

Sulfate reducing bacteria (SRB) utilize low-carbon number compounds (represented above by CH_2O), most notably acetate, which are in turn produced by the hydrolysis of ligno-cellulose materials, itself mediated by the action of a diverse range of heterotrophic microbes. SRB require anoxic environments with an absence of Fe^{3+} and at least 100 mg/L sulfate. The conditions favored by SRB are readily satisfied in a pond or wetland that contains an alkaline, fertile organic substrate and which is fed by contaminated mine water rich in sulfate.

The majority of SRB cultured to date require circum-neutral pH in addition to the other environmental conditions listed above. A few studies have indicated that as long as calcite dissolution maintains pH high enough for these 'neutrophile' SRB to thrive, sulfate reduction will be vigorous in systems receiving acid waters, but that once the calcite is depleted SRB activity declines concomitantly (e.g. Watzlaf, 1997; Eger and Wagner, 2001). Nevertheless, many observations exist of persistent microbial sulfate reduction in substrates receiving low pH, highly oxidized waters (e.g. Tuttle et al., 1969; Herlihy and Mills 1985; Hedin et al., 1988; McIntire and Edenborn 1990; Elliott et al., 1998; Jarvis and Younger, 1999). It may be that the alkalinity produced by SRB helps to produce and maintain circum-neutral microenvironments, so that the bacteria can remain active in systems that receive mine water with pH lower than would generally be considered toler-able for them. Another possible explanation lies in the recent identification by Johnson (1998) of previously-undescribed acidophilic SRB genera. Since it is increasingly evident that such acidophile microbes are widely distributed in nature (Norris and Johnson, 1998), they might provide the "pioneer" consortia which reduce sulfate in the most acidic zones, facilitating a rise in pH into the range tolerable by the better-known neutrophile SRB.

5.3.6 INTERACTION OF METAL REMOVAL AND ACID NEUTRALIZATION PROCESSES

Successful passive treatment of mine water requires attention to both metal removal and acid neutralization reactions. Mine waters commonly contain *both* alkalinity and potential acidity. Potential acidity results from dissolved metals that, upon hydrolysis, produce proton acidity (H^+). Indeed, as was seen in Section 2.2.4, the contributions of dissolved metals to total acidity are so important in most mine waters that total acidity can be accurately calculated using the expressions given in Table 2.5. These can be simplified to the following form:

Total acidity (mg.l^{-1} as $CaCO_3$ equivalent) =

$$50[2(Fe^{2+}/56) + 3(Fe^{3+}/56) + 2(Mn^{2+}/55) + 3(Al^{3+}/27) + 2(Zn/65) + 1000(10^{-pH})] \qquad (5.12)$$

(Other cations such as Cu and Cd can be added to the above expression if they are present at significant concentrations, using their atomic masses and valencies as shown). If the total acidity represented by Equation 5.12 is not matched by sufficient alkalinity (or will not be so matched after treatment), then hydrolysis of the various metal ions will lead to development of a low pH, which will slow further metal removal reactions and degrade water quality. Where Fe predominates in a mine water, the total acidity – alkalinity balance is well expressed by the following further simplification of Equation 5.12 and Table 2.5:

"*Hopkins' Rule*": If the ratio (Total Fe (mg.l^{-1})/Alkalinity (mg.l^{-1} as $CaCO_3$)) > 1.1 then the mine water is net acidic.

(The logic of Hopkins' Rule, which we named after the engineer who suggested such a simple rule ought to exist, is that each mole of Fe^{2+} which eventually oxidizes and hydrolyzes will release two moles of H^+. Hence there must be two moles of alkalinity present for every mole of Fe in order to ensure overall net-alkalinity. Substituting the relevant relative atomic/molecular masses (Fe = 55.85, HCO_3^- = 61; $CaCO_3$ = 100) we find the following 'threshold' values:

55.85/(61/2) = 1.83, if the alkalinity is expressed as HCO_3^-
55.85/(100/2) = 1.117, if the alkalinity is expressed as $CaCO_3$ equivalent).

During the 1980's dozens of wetlands were constructed on mine sites in the eastern USA by designers who failed to take cognizance of the spirit of Hopkins' rule (as detailed in Table 2.5). These wetlands decreased iron concentrations to some degree, but also decreased pH, so that the net effect on total acidity was negligible. Figure 5.2a shows water chemistry for a constructed wetland in Pennsylvania (USA) that decreased iron concentrations from 95 mg/L to 15 mg/L, but also decreased pH from 5.5 to 3.1. Flow through the wetland had little effect on acidity (5% decrease). Concentrations of Al increased from 0.5 mg/L to 3 mg/L, apparently due to the dissolution of the clay substrate by the low pH water. The effluent of this passive system required supplemental chemical treatment before it could be discharged to the receiving stream.

When mine waters contain sufficient alkalinity, metal removal occurs without a decrease in pH. Figure 5.2b shows Fe and pH concentrations for a mine water that contains sufficient alkalinity. The mine water was aerated and passed through two sedimentation ponds that precipitated iron oxide. Iron concentrations decreased from 97 mg/L to 6 mg/L, while pH increased from 6.4 to 7.1. The combined iron oxidation, hydrolysis and neutralization reaction may be summarized as follows:

$$Fe^{2+} + \tfrac{1}{4} O_2 + 2HCO_3^- = FeOOH + 2CO_2 + \tfrac{1}{2} H_2O \qquad (5.13)$$

The increase in pH, a common occurrence for passive systems that are adequately buffered with bicarbonate alkalinity, is predominantly attributable to the degassing of dissolved CO_2 (see also Sections 4.2.2 and 4.2.3):

$$HCO_3^- = CO_2(gas) + OH^- \qquad (5.14)$$

Mine waters that contain an excess of alkalinity are referred to as *net alkaline* waters (see Sections 2.2.1 and 2.2.3). A fundamental aspect of the efficient passive treatment of mine drainage is the recognition of net alkaline conditions when they exist naturally, and the passive creation of net alkaline conditions when they do not exist.

A second common interaction of metal removal and alkalinity-producing reactions occurs in an anoxic organic substrate where microbial activity precipitates metals as sulfide and carbonate solids. Representative reactions for iron are shown below.

$$Fe^{2+} + SO_4^{2-} + 2CH_2O = FeS + 2H_2O + 2CO_2 \qquad (5.15)$$
$$Fe^{2+} + SO4^{2-} + 2CH_2O = FeCO_3 + H_2S + H_2O + CO_2 \qquad (5.16)$$
$$Fe^{2+} + CaCO_3 = FeCO_3 + Ca^{2+} \qquad (5.17)$$

Substrate analyses of organic substrates in constructed wetlands often reveal the presence of mixtures of metal sulfide and metal carbonate compounds (Wieder *et al.*, 1990; Faulkner and Richardson, 1989; Younger, 2000d; Batty *et al.*, in press). Microenvironmental conditions presumably influence whether metal carbonates or sulfides form at any one point in a substrate. When metals are precipitated *in situ* as sulfides and carbonates, bicarbonate alkalinity is consumed, which might be deemed to be counter-productive. On complete analysis, these reactions are found to consume protons by ferric iron reduction (Vile and Wieder, 1993) and by incomplete re-oxidation of HS-, such that it is arrested with S at zero valency (i.e. as S^0) rather than proceeding to hexavalency (the valency of S in SO_4^{2-}), which would release protons (e.g. Younger, 2000e). These reactions are particularly important in passive systems in which acidic water is constrained to flow through alkaline organic substrates (e.g. Cohen and Staub, 1992; Staub and Cohen, 1992; Machemer and Wildman, 1992; Kepler and McCleary, 1994; Pulles, 2000). In such systems, decreases in acidity often greatly exceed the production of measurable alkalinity (e.g. Kepler and McCleary, 1994; Younger, 2000e).

a) Aerobic system receiving net-acidic mine water

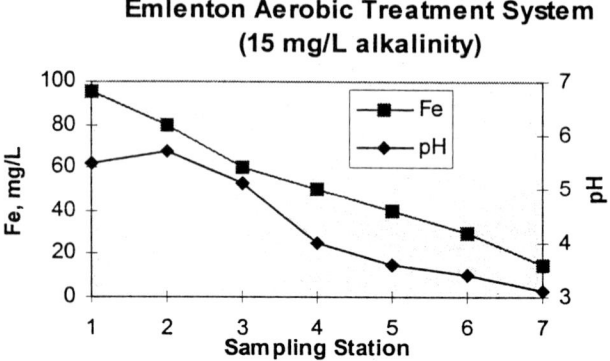

b) Aerobic system receiving net-alkaline mine water

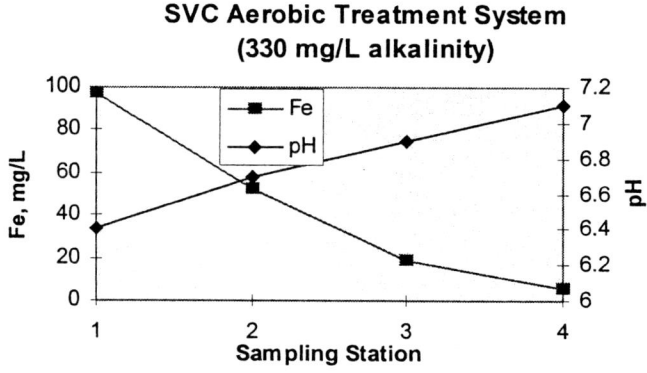

Figure 5.2: Changes in pH and Fe concentrations in aerobic systems treating (a) net-acidic water, and (b) net-alkaline water.

5.3.7 REMOVAL PROCESSES FOR OTHER CONTAMINANTS

The foregoing sections have focused on the most common contaminants in acid-releasing mined environments (i.e. Fe, Mn, Al and proton acidity). However, in many mined systems other contaminants are at least as important. For instance, at the Tar Creek Superfund Site, Oklahoma, abandoned Pb-Zn mines in the carbonate-hosted Tri-State Orefield, which began to overflow in the mid-1980s, discharge alkaline waters containing 300 mg.l^{-1} Fe and 150 mg.l^{-1} Zn (Nairn *et al.*, 2001). Similar instances have been noted elsewhere in the world (e.g. Kelly, 1988; Nuttall and Younger, 1999), and copper, cadmium, nickel, lead, chromium, selenium, mercury and other ecotoxic metals are also encountered in certain cases (e.g. Banks et al., 1997). The metalloid arsenic can be released both from sulphide ore-bodies (in which it commonly occurs in the form of arsenopyrite (FeAsS)) and by desorption from haematite and other iron oxides in iron ore-bodies. Since arsenic is usually mobilized as an oxy-anion (AsO_4^-), it is often still fairly mobile at neutral pH. Most other anions commonly found in mine drainage waters are relatively innocuous compared to the metals. However, sulfate (which has strong, temporary laxative effects in most mammals) is frequently present at such high concentrations that it accounts for a very large proportion of the total dissolved solids. In many arid and semi-arid areas, the removal of sulfate from mine waters is desirable if salinization of receiving waters is to be avoided.

Besides these naturally occurring contaminants, certain substances used in mineral processing can also contribute substantially to mine water pollution, for instance mercury (widely used by informal sector gold miners), cyanide (used by large-scale gold mines) and flotation agents such as xanthates.

While specialised active treatment processes are available to deal with all of the contaminants listed above, passive treatment options are still very limited; in most cases the development of suitable passive processes is still at the experimental stage. Up-to-date information on technologies for such contaminants can be found at the web-site of the European Commission's 'PIRAMID' project (*Passive In-situ Remediation of Acidic Mine/Industrial Drainage*), which may be consulted at *www.piramid.org*. A brief overview of some of the more promising removal processes currently being investigated for the contaminants mentioned above is given in Table 5.1. As with Fe, Mn and Al, the basic choice is usually between oxidative and reductive strategies, with carbonate precipitation being an attractive option for Cu and Zn (see sections 5.4.2.4 and 5.4.3.3). The reaction mechanisms are in most cases closely analogous to those presented in detail above for iron, and will not be discussed in depth here. If more detail is required, the references cited in Table 5.1 should be consulted.

One important *caveat* must be given with regard to potential passive treatment solutions for some of the more toxic elements (Cd, As, Hg etc): passive treatment systems (like wetlands) which make such substances readily accessible to plants and animals in the surface environment may well be ecologically inadvisable, and probably regulatorily unacceptable. It is likely, therefore, that large-scale passive treatment of these toxic elements will best be pursued in subsurface-flow reactors, in which the pollutants can be kept isolated from all higher plants and animals, and be biologically accessible only to the relevant 'extremophile' microbes which thrive in their presence and help to immobilize them geochemically.

5.3.8 RATES OF CONTAMINANT REMOVAL IN PASSIVE SYSTEMS: TOWARDS DESIGN CRITERIA

The preceding sections have shown that acidity and metal removal in passive systems can occur by a variety of processes under a wide range of pH and redox conditions, and subject to a range of surface and subsurface flow conditions. Ideally, the design and sizing of passive systems should be based upon a sound understanding of the relative rates of contaminant removal for each possible combination of chemical and hydraulic circumstances. This is a very demanding ideal, for it would require powers of measurement and mathematical modeling way beyond the present state-of-the-art. It might be considered that a logical first step towards this ideal would be to equate the overall kinetics of the removal process with the kinetics of the predominant chemical reaction. For instance, we have already noted that the oxidation of ferrous iron by oxygen obeys first order kinetics with respect to dissolved Fe^{2+} concentration (Equation 5.4; Section 5.3.2). Similarly, Staub and Cohen (1992) found bacterial sulphate reduction in experimental passive systems to be first order with respect to SO_4^{2-} concentration. On the basis of such information, Tarutis *et al.* (1999) proposed that the sizing of mine water treatment wetlands could be undertaken using a conventional first order kinetics model (based on Equation 5.4), which can be written:

$$A = \frac{Q_d \, (\ln C_i - \ln C_t)}{K_l} \qquad (5.18)$$

Where: A = the required wetland area (m^2)
Q_d = the average daily flow which the wetland will receive ($m^3.d^{-1}$)
C_i = the average daily influent concentration of the relevant contaminant ($mg.l^{-1}$)
C_t = the required concentration for this contaminant after treatment ($mg.l^{-1}$), and
K_l = first order kinetic reaction constant for the reaction under consideration ($m.d^{-1}$)

Tarutis *et al.* (1999) suggested that the above expression could be used to design treatment wetlands for mine waters, if only suitable values for K_l were available. They tentatively proposed a K_l value of 0.18 $m.d^{-1}$ for iron removal in aerobic wetlands (though this value was derived from data for natural wetlands rather than constructed wetlands). Jarvis (2000) used data from a constructed compost wetland in the UK to develop a K_l value of 0.12 $m.d^{-1}$ for total acidity removal, which appears to be first order in acidity in such systems. (It is worth noting that there is counter evidence to suggest that bacterial sulphate reduction, which is the driving process for acidity removal in compost wetlands, is actually zero order in SO_4^{2-}, and may be the limiting step in profoundly anaerobic passive systems; Willow and Cohen, 1998).

For cases in which first order kinetics can reasonably be assumed, how well can Equation 5.18 be expected to perform in practice? We can gain some insight by examining its track record in the design of sewage treatment wetlands. In that case, the relevant reaction is consumption of Biochemical Oxygen Demand (BOD), which is widely held to be first order with respect to BOD (with estimated K_l values in the range 0.06 to 0.10 $m.d^{-1}$; Cooper, 2001). However, the use of first order expressions (such as Equation 5.18) as the main design criteria for sewage treatment wetlands has been trenchantly criticized by Kadlec (2000) on the grounds that it ignores:

• variations in influent flow rates
• variations in influent contaminant concentrations (which are rarely in phase with the flow variations)

- the fact that flow through wetland systems is always subject to significant hydrodynamic dispersion, whereas expressions such as Equation 5.18 assume simple plug-flow hydraulics
- the likelihood that contaminant removal processes differ somewhat between the fast-flowing and stagnant zones within a treatment wetland.

The net result of these factors is that the overall rate of contaminant removal in a treatment wetland is unlikely to be well-modelled by comparison only with the first order kinetics of one of the main chemical reaction processes.

There is no *a priori* reason why the situation should be any better in the case of mine water treatment wetlands. Indeed, as was shown in Section 5.3.2, oxidation of ferrous iron (Equations 5.1 and 5.4), which is the crux of the design model proposed by Tarutis *et al.* (1999), happens only to be the best-known of the sequence of three reactions (Equations 5.1 through 5.3) which are altogether involved in Fe removal in aerobic systems. Since the kinetics of reactions 5.2 and 5.3 under field conditions have not been determined, we cannot even account fully for the relevant reaction chain, before we even consider issues such as fluctuating inputs and the internal hydraulic complexity of wetlands. While it is theoretically possible to develop comprehensive, process-based mathematical models which incorporate direct representations of all of the relevant chemical and hydraulic processes (a prototype for such a code is described at *www.piramid.org*), full parameter-ization of such models for real field systems is never likely to be feasible. In the absence of such a respectable scientific tool, the mine water engineer must work instead with empirically-derived design rules. Water quality data and general system design and construction information exist for dozens of passive mine water systems (Wieder 1989; Stark *et al.*, 1990, 1994; Girts *et al.*, 1987; Hellier, 1989; Hellier *et al.*, 1994; Faulkner and Skousen, 1993; Hedin *et al.*, 1994; Younger, 2000d,e); the basis therefore exists for the derivation of robust design rules. But how should these rules be framed?

One possible formulation in common use is "treatment efficiency" (e.g. Wieder, 1989; Girts *et al.*, 1987; Stark *et al.*, 1990). This is calculated from changes in contaminant concentration (X) as follows:

$$\text{Treatment efficiency} = 100 \cdot [(X_{inf} - X_{eff})/(X_{inf})] \qquad (5.19)$$

Where the subscripts "inf" and "eff" refer to the influent and effluent sampling positions respectively. Because Equation 5.19 does not incorporate flows or the dimensions of the treatment system, it has very limited value for comparing different systems. A more robust approach is to deal with pollutant *loadings* (dimensions: $M \cdot T^{-1}$), rather than concen-trations ($M.L^{-3}$), since loadings combine data on both concentrations and flows. If changes in contaminant loadings across a treatment system are related to some measure of the system size (such as surface area, volume, substrate tonnage, or retention time) a rule can be obtained which incorporates all relevant factors. A common performance calculation used in recent analyses is the area-adjusted contaminant removal (R_A).

$$R_A = (\text{Flow}_{inf}*X_{inf} - \text{Flow}_{eff}*X_{eff})/(\text{surface area}) \qquad (5.20)$$

Once a reliable value for R_A is known (by study of a range of similar systems), an expression analogous to Equation 5.18 can be used to determine the size of a wetland needed to obtain a desired decrease in contaminant loading:

$$A = \frac{Q_d \, (C_i - Ct)}{R_A} \qquad\qquad (5.21)$$

Where all terms have their previous meanings.

The use of surface area to represent system dimensions in Equations 5.20 and 5.21 is frankly a matter of convenience, resulting from its ease of measurement and from the fact that surface area is often the key limitation on passive system design in practice (Section 5.1). Surface area may well be a reasonable metric of overall system size for surface-flow, aerobic reed beds. However, as the design of passive systems advances, it is likely that volume-based or retention-time based measures of performance will prove to be more appropriate in many circumstances, particularly for passive systems in which flow occurs predominantly in the subsurface (Cohen and Staub, 1992).

When measuring the performance of a treatment system it is important to consider the effect of dilution by uncontaminated water on contaminant concentrations. Because passive systems are typically very large, such non-target water can be important. Sources of dilution include direct precipitation, surface runoff, and inflows of shallow, uncon-taminated groundwater. The latter source is particularly important to bear in mind, as it is usually not visible. However, its importance is underlined by the results of studies of wetland systems without significant surface water inputs, for which it was found that dilution averaged 5% and reached 10–20% at some sites (Hedin *et al.*, 1994). Before data can be interpreted or compared, dilution must be accounted for. The data necessary to construct a balanced hydrologic budget (see Section 3.3.1) are rarely available for operational passive treatment systems. An alternative method for estimating the importance of dilution is to monitor concentrations of a conservative ion, i.e. one which is transported largely without participating in chemical reactions. A good conservative ion will have the following characteristics:

- typically be present in contaminated mine waters at sufficiently high concentrations (say, > 50 mg.L^{-1}) that it can be measured accurately
- be present in potentially diluting waters only at low concentrations (say, < 5 mg.L^{-1})
- be chemically unreactive at the pH and Eh conditions found in passive treatment systems.

Chloride is the classic conservative ion, and in some deep mine waters in Europe it is present at concentrations well in excess of those in shallow ground waters. In the eastern USA, magnesium is very often a good alternative in waters where Cl is scarce. Sodium is also a suitable ion at some sites, particularly flooded underground coal mines that have overburdens influenced by marine conditions. Once a suitable conservative ion has been selected, changes in its concentration across a passive system can be used to calculate a 'dilution adjustment factor', which can then be used to adjust contaminant concentrations prior to making removal calculations, so that reaction processes are quantified rather than simple dilution.

Table 5.2 shows area-adjusted Fe and Mn removal rates (calculated using Equation 5.20) for 22 sites in the eastern US. Significant removal of iron occurred at every site studied. The highest rates of Fe removal were associated with circum-neutral, bicarbonate buffered, oxidizing environments (cf Figure 5.3). When mine waters contain more than 20 mg/L Fe and are alkaline, iron is typically removed at rates of 10–30 g m^{-2}d^{-1}. At iron concentrations less than 20 mg/L, the rate of removal decreases to 5–10 g m^{-2}d^{-1} or less. This could either be a simple manifestation of the first order kinetics of ferrous iron oxidation (Equations 5.1 and 5.4), as postulated by Tarutis *et al.* (1999), or else it could be due to the sedimentation rate of iron oxide particles. As the concentration of iron falls, the density of oxide particles decreases, which results in decreased rates of particle aggregation and settlement. For instance, Table 5.3 shows cell-by-cell iron removal data for an aerobic passive system that effectively oxidized iron, but was not effectively removing suspended iron oxides solids. (The final discharge was a turbid orange). In the first two cells, where ferrous iron concentrations exceeded 7 mg.L^{-1}, iron oxidized at an average rate of 23 g m^{-2}d^{-1}. This rate is consistent with iron removal in systems with higher concentrations of iron. In the last two cells oxidation rates slow, but the feature primarily limiting treatment performance was the poor sedimentation of suspended iron oxides. This problem with final "polishing" of passively treated water is a common problem for passive systems designed to optimize treatment efficiency. Designs that optimize clarification of the mine water appear to be necessary in future systems.

In the preceding paragraphs, we have focused our attention on Fe removal. However, the same system evaluation and design logic applies to other contaminants, with appropriate adjustments being made according to circumstance. For instance, manganese removal, though mechanistically similar to Fe removal, occurs at much slower rates in passive treatment systems. After dilution adjustments are made, many sites that decrease Mn concentrations moderately display no significant removal of Mn. Significant removal of Mn only occurs at sites with circum-neutral aerobic conditions and low concentrations

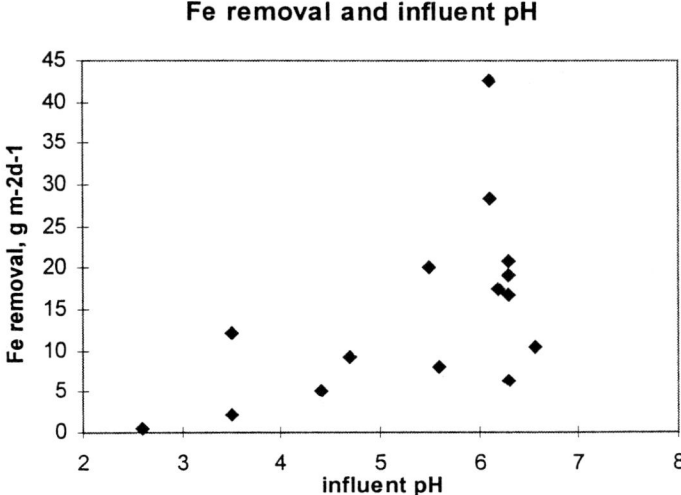

Figure 5.3: Relationship between influent pH and the rate of iron removal at twenty passive mine water treatment systems in the eastern USA.

Table 5.2: Summary Fe and Mn removal statistics for constructed wetlands in the eastern US. Fe and Mn concentrations are Mg.l⁻¹. Removal rates are g.m⁻²d⁻¹. "sig" indicates the whether the rate is significantly greater than zero (t-test, 0.05 level)

Site	pH in	pH out	Fe in	Fe out	Fe removal ave (sd)	n	Sig?	Mn In	Mn out	Mn removal ave (sd)	N	sig?	flow	surface	design area
Donegal	7.1	7.4	5	<1	>>0.4	9	yes	8	2	0.50 (0.25)	9	yes	501	8100	four constructed wetlands
Simco	6.5	6.4	111	42	12.4 (5.6)	167	yes	2	2				451	4138	four constructed wetlands
SVC1	6.4	7.0	119	9	17.0 (1.1)	2	yes	50	47	0.06 (0.07)	2	no	965	8100	ALD; two sediment ponds
Mor-low	6.4	6.6	56	<1	>>0.6	24	yes	37	11	0.20 (0.18)	24	yes	7	1015	sediment pond; two constructed wetlands
SR 114D	6.4	6.8	40	14	9.5 (na)	1	na	2	2				590	2311	ALD, two sediment ponds, two aerobic wetlands
Cedar	6.3	6.4	92	41	6.3 (2.2)	7	yes	2	2	na	na	na	156	1360	sediment pond, three constructed wetlands
Keystone	6.3	6.4	37	32	20.7	27	yes	1	1	na	na	na	8606	4200	deep ditch
Mor-up	6.3	6.4	151	56	19.2 (10.6)	24	yes	42	37	0.17 (0.41)	24	yes	7	60	ALD, shallow ditch
Blair	6.2	6.4	52	<1	>>0.7	6	yes	30	5	0.43 (0.37)	6	yes	11	1080	
SVC2	6.2	7.6	46	<1	>>5.0 (1.7)	4	yes	32	6	2.10 (0.52)	4	yes	2650	52,664	series of sediment ponds and volunteer wetlands
StVincent3	6.2	6.5	96	2	17.4 (na)	1	yes	2	2				833		five sediment ponds
REM-left	6.1	3.8	190	84	28.3 (5.7)	20	yes	50	48	-0.05 (0.13)	20	no			
Howe-up	6.1	5.6	265	185	42.7 (8.2)	13	yes	37	34	-0.43 (0.49)	13	no	130		
Shade	6.0	6.8	2	<1	na	na	na	23	10	0.72 (0.64)	17	yes	10	880	ALD; aerobic limestone cell
Piney	5.8	6.1	1	<1	na	na	na	15	11	1.07 (0.34)	33	yes	468	2500	constructed wetland
Howe-low	5.6	6.2	185	68	8.1 (1.9)	13	yes	34	33	0.06 (0.16)	13	no	130		
REM-right	5.5	3.3	473	338	20.1 (4.0)	18	yes	130	113	0.10 (0.33)	18	no			
Emlenton	4.7	3.2	89	15	9.1 (3.3)	40	yes	77	73	-0.09 (0.19)	40	no	55	643	nine constructed wetlands
Somerset	4.4	5.5	162	18	5.0 (4.9)	40	yes	50	33	-0.01 (0.54)	40	no	47	1005	two constructed wetlands
REM-low	3.5	2.9	246	115	12.0 (3.4)	9	yes	92	88	-0.05 (0.14)	9	no			
Latrobe	3.5	3.7	125	56	2.1 (1.0)	43	yes	32	29	0.03 (0.09)	21	no	86	2800	three constructed wetlands
Friendhill	2.6	2.9	153	137	0.5 (0.50)	73	yes	10	10	0.00 (0.02)	73	no	15	667	six constructed wetlands

Table 5.3: Changes in dissolved iron (Fe^{dis}) and total iron (Fe^{tot}) at the SR 114-D system on June 11, 1997. The flow rate was 590 L.min^{-1}. Removal rates in g.m^{-2}.d^{-1}.

	Surface Area, m^2	pH	Alk	Fe^{dis}	Fe^{tot}	Fe^{dis} removal	Fe^{tot} removal
Influent		6.4	141	40	40		
Cell 1 Effluent	470	6.8	158	23	33	29.7	13.2
Cell 2 Effluent	844	6.6	91	7	19	16.6	14.1
Cell 3 Effluent	523	6.6	84	4	17	4.2	2.0
Cell 4 (Final) Effluent	474	5.8	67	2	14	3.9	6.6

of Fe. Even at these sites, the removal rates are generally only 0.5–2.0 g m^{-2}d^{-1} (Table 5.2). Thus Mn removal in the present generation of passive treatment systems occurs an order magnitude more slowly than iron removal.

5.4 PASSIVE TREATMENT TECHNOLOGIES

5.4.1 A TYPOLOGY OF PASSIVE MINE WATER TREATMENT SYSTEMS

The contaminant removal processes described in Section 5.3 are deployed in a wide (and rapidly widening) range of technological forms. At the start of the 21st Century, the passive treatment repertoire for mine waters encompasses three main types of system:

1. *Inorganic media passive systems (IMPs)*. This category includes a range of technologies based on the dissolution and/or precipitation of inorganic, mineral substances in surface or subsurface flow reactors. There are currently two main types of IMPs:
   - carbonate-dissolution based IMPs, such as anoxic limestone drains (ALDs), oxic limestone drains (OLDs), closed-system Zn removal reactors and siderite-calcite reactors for Cd and As removal
   - systems in which inorganic media provide surfaces on which mineral precipitates can accrete, such as 'SCOOFI' reactors and 'pyrolusite process' reactors
   These are all described in Section 5.4.2 below. IMPs are a rapid growth area in passive treatment, and the variety of systems available is likely to expand greatly over the next decade.
2. *Wetland-type passive systems*. These can in turn be subdivided into three varieties, which differ radically from one another in form, function and applicability, as will be seen in Section 5.4.3. The three types are:
   - aerobic wetlands (reed beds)
   - compost wetlands (sometimes rather inaccurately labeled 'anaerobic' wetlands)
   - *R*educing and *a*lkalinity-*p*roducing *s*ystems (RAPS). These systems were originally termed 'SAPS' (Successive Alkalinity Producing Systems) by their originators (Kepler and McCleary, 1994), but were recently re-named by Watzlaf and co-workers to more accurately reflect their functioning.

The latter are sometimes also referred to as 'vertical flow ponds' (e.g. Demchak *et al.*, 2001). This terminology arose from experiments with compost wetland systems in the mid-1990s, from experiments aimed at maximizing subsurface flow through compost substrates. Attempts to achieve this by horizontal flow failed because the water eventually developed flow paths on the compost surface. By contrast, systems with a *vertical* flow path worked. Such systems are essentially the same as 'SFBs', described below. However the term 'vertical flow pond' became established in some circles and found ready transfer to RAPS when they were invented. However, as the term 'vertical flow ponds' has very specific connotations in relation to passive treatment of sewage (e.g. Cooper, 2001), and as that connotation is synonymous with *aerobic* treatment, its use with another meaning in the mine water is context is probably best avoided.

3. *Subsurface-flow bacterial sulfate reduction systems (SFBs)*. These further sub divide into two sub-categories:
   - *in situ* permeable reactive barriers (PRBs) treating contaminated ground water within an aquifer, and
   - SFBs constructed to treat contaminated mine water discharges

   Apart from their mode of construction and the locations of system influents and effluents, there is no scientific difference between these two sub-categories (Section 5.4.4).

It should be stressed that the above typology is somewhat artificial, in that certain elements of IMPs technology are integral to most wetland-type systems, and there is considerable commonality in form and function between the SFBs and the subsurface portions of RAPS wetland systems. Furthermore, the treatment of any given mine water may require deployment of more than one of the above systems in series. The issue of selection and staging of 'unit processes' within an overall passive treatment system is addressed in Section 5.5 below, illustrated by examples of actual systems.

5.4.2 INORGANIC MEDIA PASSIVE SYSTEMS (IMPS)

5.4.2.1 The concept of IMPs

IMPs are conceptually simple: they use dissolution and/or precipitation of one or two mineral species to effect major changes in mine water chemistry. The predominant examples of the genre are limestone-dissolution based IMPs (ALDs and OLDs) in which the dissolution of calcite ($CaCO_3$) is used to raise pH, neutralize acidity and add bicarbonate alkalinity to mine waters. Limestone has many virtues as an alkaline reagent: it is inexpensive, non-hazardous, and is often readily available in close proximity to major coal fields. However, limestone has been little used in active treatment for two main reasons:

(i)   its tendency to "armor", i.e. for clast surfaces to become covered with metal hydroxide solids that result in the material becoming less reactive, and

(ii) the slowness of its dissolution reactions when compared to strong alkalis such as hydrated lime (itself produced by thermal treatment of limestone) and sodium hydroxide.

While fluidized-bed reactors have been developed to overcome the first limitation in active systems (e.g. van Tonder *et al.*, 1994; Sibrell *et al.*, 2000), little can be done to accelerate the dissolution rate of unarmored limestone clasts. In most passive systems, provided the problems of armoring can be circumvented, the slow rate of limestone dissolution is a valuable asset, since it permits the design of systems which can persistently treat acid waters for periods of several decades before requiring major refurbishment (see Sections 5.4.2.2 and 5.4.2.3 below). In other carbonate-based IMP systems (Section 5.4.2.4), relatively little calcite dissolution is needed, and the focus is on subtle alterations of mineral saturation balances for key metals, particularly Zn (e.g. Nuttall and Younger, 2000) Cd and the metalloid, As (e.g. Wang and Reardon, 2001) .

Another IMP concept is to use inorganic media (rock clasts or some other solid, such as blast furnace slag or even specially shaped plastic media) to provide stationary frameworks upon which mineral precipitates can accumulate (e.g. Thornton, 1995; Brant *et al.*, 1999; Jarvis and Younger, 2001). In some cases, the medium can be utterly inert (e.g. for SCOOFI reactors; Section 5.4.2.5), whereas in others a high-pH medium and/or a medium rich in magnesium carbonate (Section 5.4.2.6) is recommended.

5.4.2.2 Anoxic Limestone Drains (ALDs)

Principles of ALDs:
ALDs are very simple engineered structures, consisting of a buried bed of limestone aggregate sited such that it intercepts acidic mine water. The limestone neutralizes proton acidity and generates bicarbonate alkalinity by means of calcite dissolution. The limestone aggregate is emplaced such that anoxic conditions are maintained throughout the system when in operation, so that Fe^{2+} and Mn^{2+} will remain in their reduced states, and will thus not precipitate hydroxide/oxide solids within the ALD. The maintenance of gas-tight anoxic conditions also prevents release of CO_2 derived from calcite dissolution to the atmosphere. This results in the dissolution reaction occurring under 'closed-system' conditions, which are well-known to result in a higher pH and bicarbonate alkalinity than would occur where dissolved CO_2 was allowed to equilibrate with atmospheric CO_2 (e.g. Freeze and Cherry, 1979).

Where Fe^{2+} and Mn^{2+} are allowed to oxidize, or where the incoming mine water already contains dissolved Fe^{3+}, the pores of the ALD will gradually clog with $Fe(OH)_3$ and/or MnO_2. To ensure that this will be the case, it is unwise to deploy ALDs for waters containing more than about 1 $mg.L^{-1}$ of dissolved oxygen (DO), for at concentrations much higher than this oxidation and hydrolysis of Fe within the drain is likely. Another contra-indication for ALD use is the presence of more than about 2 $mg.L^{-1}$ of dissolved aluminum in the mine water (Section 5.3.3). In contrast to Fe, aluminum occurs in only the trivalent form (Al^{3+}) under ambient environmental conditions, and can therefore hydrolyse readily under anoxic conditions (Equation 5.6). $Al(OH)_3$ precipitates on contact with the high-pH conditions near the limestone surfaces, gradually filling the pore space.

Several cases are known (though for obvious reasons not published!) in which ALDs fed with waters containing high concentrations of Al^{3+} or Fe^{3+} have ceased to accept water after about six months. Excavation of these failed ALDs reveals that the clogging is usually limited to the first 10–15% of the total length of the ALD. Hence, in cases where removal of the first portion of the ALD can be afforded every few months, such clogging might be viewed as a maintenance frequency issue rather than a matter of complete failure. However, in most passive treatment settings, this will be unacceptable.

To summarize, therefore:

- an adequately designed and constructed ALD should only alter the pH, total acidity and alkalinity of the mine water passing through it, with Fe^{2+} and Mn^{2+} being transported conservatively through it
- ALDs are appropriate only for waters containing less than 2 mg.L^{-1} each of Al^{3+} and Fe^{3+}, and for other ferruginous mine waters containing less than 1 mg.L^{-1} of DO.

Design practice for ALDs:
Studies of existing ALDs have revealed that alkalinity generation plateaus after 8–14 hours of retention time, at a value that is usually only 10–20% of theoretical calcite saturation (Hedin *et al.*, 1994). This behavior probably reflects a reduction in the rate of calcite dissolution as saturation is approached (Berner and Morse, 1974).

The use of the word "drain" in the name of this technology reflects the physical resemblance of many early ALDs to conventional 'french drains', i.e. aggregate-filled trenches which feed water into a perforated drainage pipe. Such drains are commonly used in the mining industry (as elsewhere) to gather seepage from the toes of spoil heaps or tailings dams. However, the high length-to-width ratio typical of french drains is not necessary in order to promote calcite dissolution, and most of the more recently constructed ALDs have been rectangular in shape.

In terms of practical design, the principal considerations are as follows:

(i)   In order to obtain the maximum alkalinity generation, the ALD should be sized to ensure a *minimum* retention time (t_R) on the order of 14 hours. (While it may still be beneficial to use ALDs with shorter t_R, systems in which this is less than about 10 hours are unlikely to be cost-effective). Sizing for a 14-hour retention time can usually be achieved by assuming that loose-tipped, single-size limestone aggregate will end up with a porosity on the order of 50%. Taking the design flow (Q_d) for the system, we can readily calculate the volume of void space needed to store a 14-hours worth of water (V_v) as follows:

$$V_v = Q_d \cdot \varphi \qquad (5.22)$$

Where Q_d is the design flow in $m^3.d^{-1}$, V_v is in m^3, and the factor φ is in this case equal to 14.

(ii)   Calculate the *minimum total volume* (V_t) of the active part of the ALD (i.e. voids plus limestone clasts) as follows:

$$V_t = V_v/n_e \qquad (5.23)$$

Where n_e is the effective porosity (Equation 3.3) of the aggregate after emplacement, expressed as a decimal (i.e. as 0.3 instead of 30%). In tests undertaken by the authors, n_e has been found to fall always in the range 0.38–0.5. However, care should be taken in specifying the aggregate properties to a supplier, for 'single size' aggregate can be misconstrued to mean 'all sediment passing a sieve of the stated size' (i.e. 'single size and down') rather than true single size (i.e. all particles pass the specified sieve, but are retained by the next standard sieve size down). If 'single size and down' is obtained by mistake, n_e may well prove to be as low as 0.2.

As well as providing the minimum ALD volume for design purposes, V_t also gives a measure of the amount of limestone aggregate you need to purchase for installation in your ALD. Usually, your local aggregate supplier will be able to advise you on how much of their particular product you will need to fill a space of volume V_t. For design and costing purposes, it is helpful to estimate the required weight of limestone aggregate (W_l) directly, which can be done as follows:

$$W_l = f \cdot V_t \qquad (5.24)$$

For V_t in cubic yards and W_l in US tons, f tends to range between 1.3 and 1.5 (US tons per cubic yard). With V_t in m^3 and W_l in tonnes, f ranges from 1.71 to 1.98 (tonnes per m^3). (N.B.: In view of the density of the mineral calcite (2.7 tonnes per m^3), these ranges of f values imply n_e values in the range 0.6 to 0.7. After emplacement and compaction of the aggregate, n_e values in the range 0.4 to 0.5 are to be expected).

(iii) Obviously, dissolution of limestone ensures that V_v initially increases as a proportion of V_t, so that equation is only valid during the earliest stage of ALD operation. Once sufficient limestone has dissolved that settlement of the ALD begins to occur, V_v will begin to decrease as a proportion of V_t. At this stage, t_R will also decrease, leading to a decline in ALD performance. In order to postpone this evil day as long as possible, it is recommended that the sizing derived using steps (i) and (ii) above be regarded as being precisely what we have called it, i.e. a *minimum* size for the system. If at all possible, the ALD should be increased in size, to increase the longevity of its useful life. Skousen (1991) suggested that the best way to design an ALD for a specific lifetime was to estimate the total amount of acidity which the drain will receive over this lifetime, and then provide an equal amount of limestone. The total amount of acidity over the ALD lifetime is estimated from observed acidity load per unit time, which is then multiplied by the total length of the lifetime in the same time units. This approach assumes that the acidity load will be constant over time; if this is not anticipated, more sophisticated predictions of time-variant acidity loads could be used (see Sections 2.3.3 and 2.3.4). In practice, the attainable ALD lifetime will be as much a matter of available land area, the driving head achievable on the site and the costs of materials and construction as of considerations of long-term acidity loads.

(iv) Once the size of the reactive portion of the ALD has been established as described above, it is necessary to design the precise layout of the system. Flow through an ALD occurs in accordance with Darcy's Law (Equation 3.4). Reflection on that law reveals that the best way to ensure a given flow (Q_d) will pass through an ALD is to maximize three factors:
  • the hydraulic conductivity (K) of the ALD
  • the cross-sectional area (A) through which flow occurs
  • the hydraulic gradient across the ALD (i)

Maximization of K is a matter of ensuring that the limestone aggregate supplied is well-sorted (i.e. a single-size aggregate) and coarse in grain size. With regard to the latter, there is a conflict between the hydraulic requirement for high K (for which large clasts, and therefore large pore-necks, are favored) and the geochemical requirement for maximum reactivity (for which small grain size gives a higher surface area in contact with the water). If there is plenty of available head on the site in question (i.e. several metres drop from the point where the water can be captured by the ALD to the point where it must exit the ALD, then a relatively fine-grained limestone aggregate can be used (say 10 mm diameter clasts). Where head is limited (< 1.5 m) then it is probably unwise to use clasts much less than 50 mm diameter. Best practice is to perform constant-head permeameter tests (see, for instance, Freeze and Cherry, 1979) on the various aggregates available, so that a reliable K value can be obtained and combined with site topography information and the design flow to select the finest-grained aggregate which is still sufficiently permeable to transmit the quantities of water in question. As an example, 28 measurements of bulk hydraulic conductivity for 20 mm single-size limestone aggregate in our own labs yielded a mean K value of 20.5×10^3 m.d^{-1}, within 95% confidence limits (2σ) of 5.8×10^3 to 35×10^3 m.d^{-1}. This is a very high value compared with natural fluvial gravels (e.g. Morgan-Jones et al., 1984), which generally include considerable quantities of interstitial sand. Were we to use this mean value for an ALD design, we would proceed as follows:

- note our design mine water flow rate (Q_d)
- calculate the available hydraulic gradient (i) on our site, by surveying the elevation difference between the point of mine water capture and the most distant point on the area proposed for our ALD, and dividing this elevation difference between the lateral distance between the two points
- calculate the cross-sectional area (A) required for our ALD:
  e.g. for Q_d = 2,500 m^3.d^{-1} (a substantial mine water discharge), i = 0.01, and with the mean K value cited above (20.5×10^3 m.d^{-1}), we can re-arrange Equation 3.4 to obtain A = $2,500/(0.01 \times 20.5 \times 10^3)$ = 12.2 m^2, an area which could be accomplished by constructing an ALD 2 m deep by 6 m wide (dimensions which are likely to be attainable on most sites). If an even larger area can be accommodated on site, then this is a bonus.

Maximization of cross-sectional area is achieved by passing the water through the broadest edge of the ALD wherever possible. Figure 5.4 illustrates the point for a rectangular prism: Given that the values of K and total available head tend to be fixed on any one site, the greater the surface area presented to the incoming mine water, the greater will be the flow accommodated without building up unwanted back-pressure. A further advantage of passing water through the largest face of a prism is that any clogging of pores which does occur (due to unanticipated variations in the DO, Al^{3+} or Fe^{3+} content of the mine water in question) will be far less likely to close off the entire influent area than would be the case for water passing through the smallest face of the prism.

Optimization of the hydraulic gradient generally arises naturally from optimal design of other elements of the ALD. Field measurements by the third author at several ALD sites in the eastern USA show that where ALDs receive appropriate water (i.e. very low Al, DO, Fe^{3+}) and are constructed with clean (true single-sized) 25–50 mm diameter aggregate, and designed with 20–40 hours of retention time, the head loss across the system is usually less than 25 mm. Where a site is particularly cramped, so that maximization of retention time is difficult, valuable extra driving head can be gained by impounding the point of mine water emergence (so long as this does not cause the mine water to emerge at some other, less-convenient point), or by deep excavations at the foot of the proposed ALD (provided this does not leave the effluent point below the base level of site drainage).

(a)

(b)

(c)

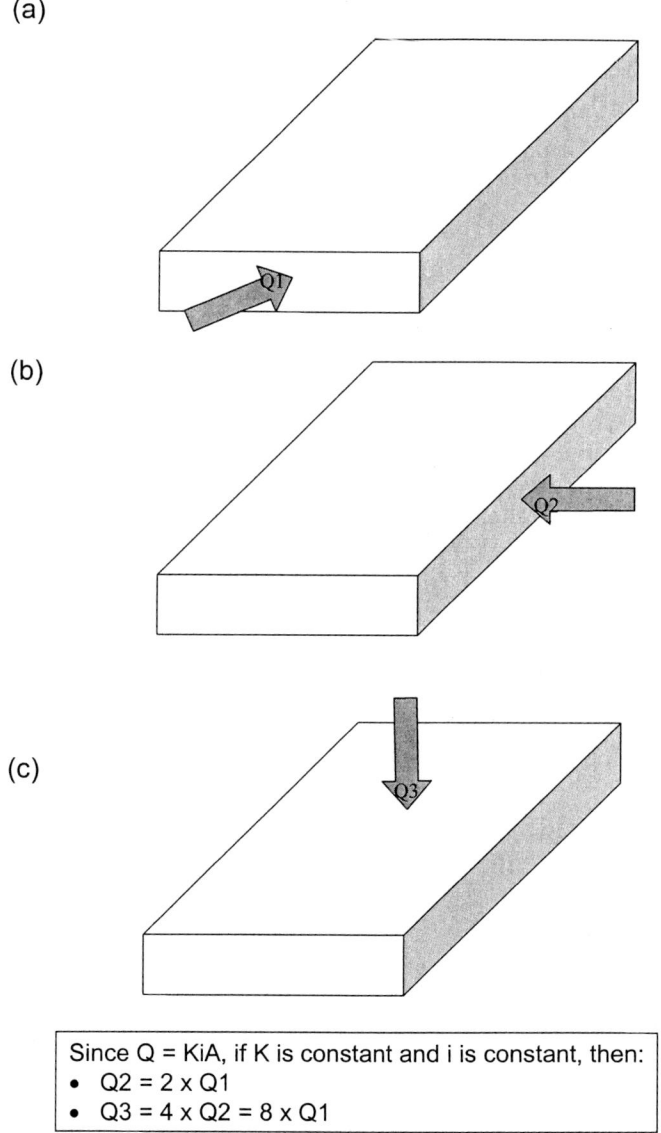

Since Q = KiA, if K is constant and i is constant, then:
- Q2 = 2 x Q1
- Q3 = 4 x Q2 = 8 x Q1

Figure 5.4: Rectangular prism diagram illustrating the maximization of cross-sectional area through which flow occurs to optimize the design of anoxic limestone drains (ALDs).

(v)   The final consideration is the 'encasement' of the limestone aggregate (Figure 5.5).
      Key objectives here are:

- to minimize oxygen ingress
- to entrap CO_2 evolved as calcite dissolves (to ensure that $CaCO_3$ dissolution occurs under closed-system conditions)
- to ensure that the ALD is hydraulically confined (which is beneficial in terms of minimizing gas exchange with the atmosphere, and ensures no transient storage phenomena affect retention times)
- to prevent invasion (and clogging) of the ALD by surrounding mud.

    In many cases, the local soil will be sufficiently clay rich that the ALD can be
constructed simply by excavating a trench, emplacing the influent and effluent
pipework and back-filling the trench, first with the limestone aggregate then with the
previously-excavated soil as a cap. Occasionally (experience to date suggests 5–10%
of cases) it will prove necessary to introduce a heavy-duty PVC, HDPE or geotextile '
liner, forming a sealed bag enclosing the limestone aggregate. Particular attention
needs to be paid to ensuring that any seams between separate sheets of liner
material are well caulked, and that the points where the influent and effluent
pipework pass through the liner are very well sealed. This is usually achieved by
careful installation of rubber gaskets (tightly-fitting rings) on the pipes, which are
glued to the liner and the pipes with long-lasting adhesive. Finally, it is as well to
arrange matters such that they ensure that the influent pipes are positioned in the
bottom half of the limestone aggregate bed. This is because most dissolution occurs
near the influent zone, and hence subsidence of the ALD will be most pronounced
in this area, and might compromise inflow if the influent pipe is positioned near the
top of the bed.

5.4.2.3 Oxic limestone drains (OLDs) and open-channel limestone reactors (OCLRs)

Principles of OLDs and OCLRs:
The limitation of ALDs to waters with less than 1 mg.l^{-1} each of DO, Fe^{3+} and Al^{3+} has
proved to be burdensome in practice. For instance, despite many initial proposals to use
ALDs in the UK, not one full-scale ALD has actually been installed (Younger, 2000d,e),
for in every case the waters to be treated have violated the above requirement. One
response to this eventuality is to install a RAPS instead (see 5.4.3.5), and this has recently
become common practice in the eastern USA (e.g. Kepler and McCleary, 1994) and in the
UK (Younger, 1998d; 2000d,e). RAPS have the drawback that they require even more
driving head than ALDs. So what if head is sufficiently limited that a RAPS is not
feasible? Is there an alternative to a counsel of despair? Cravotta and Trahan (1999) have
documented an alternative approach for such circumstances, in which a system physically
identical to an ALD is installed to treat oxic mine waters. Termed an "oxic limestone drain"
(OLD), such a system is confidently expected to provoke the precipitation of ferric and/or
aluminum hydroxides within itself. By maintaining rapid flow velocities (> 0.1 m. min^{-1})
within the OLD, it has proved possible to retain a large proportion of these hydroxides in
suspension, so that they exit the OLD without contributing to clogging. (The suspended

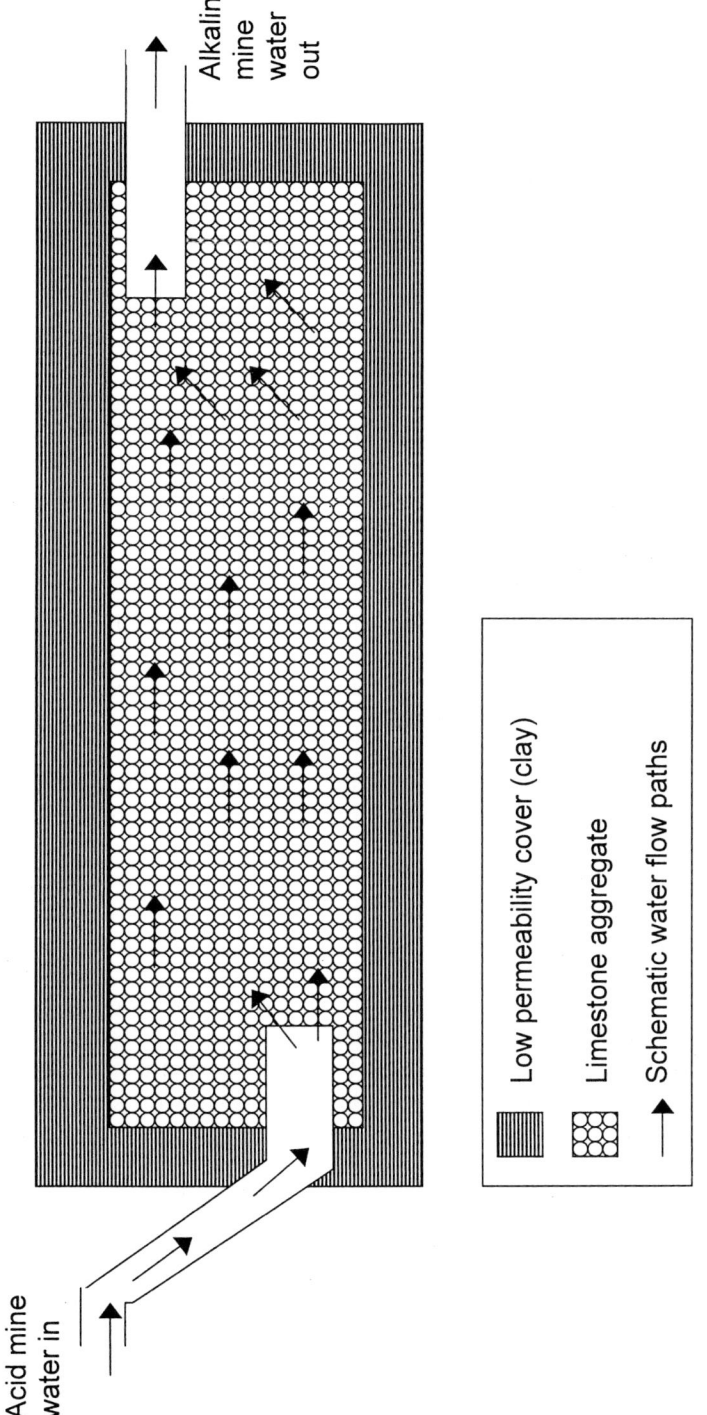

Figure 5.5: Cross-section showing a typical layout for an anoxic limestone drain (ALD).

hydroxides can then be settled in a sedimentation pond/aerobic wetland). In detailed experiments, Cravotta and Trahan (1999) demonstrated generation of up to 120 mg.l^{-1} (as CaCO$_3$ equivalent) of alkalinity in only 2–3 hours residence in OLDs. Like ALDs, these OLDs take advantage of closed-system CaCO$_3$ kinetics to achieve these elevated alkalinities.

Even where open-system conditions are maintained, and limestone becomes heavily armored with ferric hydroxide, it has been found that calcite dissolution still occurs at significant rates (albeit less vigorously than in an ALD or OLD). For instance, Ziemkiewicz *et al.* (1997) reported that armored limestone was only 2 to 45% less effective than unarmored limestone in neutralizing acidity. However, this straightforward comparison did not take into account the differences in hydraulic residence time between the open-channel systems they studied and the ALDs with which they made their comparison. When this is taken into account, the armored limestone is seen to be between 17% and 67% less effective than the unarmored variety (Jarvis, 2000). Nevertheless, the observation that armoring hinders calcite dissolution rather than preventing it altogether is very encouraging, and has prompted Ziemkiewicz *et al.* (1997) to advocate the installation of limestone aggregate into the beds of open channels through which acid water is flowing. Though hardly recommendable instead of ALD technology where the latter is feasible, the use of such open-channel limestone reactors (OCLRs) may well be worthwhile in a range of peculiar circumstances, such as:

- Places where passive treatment must be accomplished in areas of hard bedrock, particularly in mountainous terrain, where excavation of an ALD may be prohibitively expensive.
- In remote, upland tributaries of streams afflicted by acidic mine drainage, where all treatment options would be infeasible save the "installation" of limestone aggregate into the streambed by helicopter lift and tipping (Ziemkiewicz *et al.*, 1997). (It is easily imagined that wilderness enthusiasts might not be too enthusiastic about this well-intentioned option, implying as it does the pulverisation of channels and riparian areas).

Design practice for OLDs and OCLRs:
OLDs constructed to date have been sized and installed essentially in the same manner as ALDs (see Section 5.4.2.2), with the proviso that hydraulic conditions are arranged such that the interstitial velocity is everywhere in excess of 0.1 m.min^{-1} (and preferably > 0.5 m.min^{-1}). To comply with this criterion, it is likely that a large limestone clast size will have to be specified, so as to attain a high K value. For instance, Cravotta and Trahan (1999) used tabular limestone clasts with long axes of about 100 mm and short axes of 30 mm. With such high velocities, total retention times (t_R) in excess of about 3 hours are unlikely to be achievable on most reasonably-sized sites, so that the design will imply a value of $\varphi = 3$ for Equation 5.22. Besides this consideration, OLDs differ only from ALDs in the quality of the water they receive. While this water can be saturated with respect to DO, experimental results presented by Watzlaf (1997) and Cravotta and Trahan (1999) suggest that OLDs are most likely to be effective where the mine water has relatively modest concentrations of dissolved Fe^{3+} and Al^{3+} (10 to 20 mg.l^{-1} of each) and a total acidity ≤ 90 mg.l^{-1} as CaCO$_3$. Notwithstanding the transport of colloidal hydroxides out of the OLD under ordinary operational conditions, Cravotta and Trahan (1999) recommend the inclusion of a 'scour pipe' in the OLD substrate, comprising a large-diameter perforated pipe, which will usually be kept sealed, but can be opened periodically (every 2–3 months)

to flush larger hydroxide flocs from the OLD, thus helping to prolong the periods between system overhaul.

No specific guidelines have yet been proposed for OCLRs, though it is likely that a channel reach would need a water transit time on the tens of hours to days for major benefits to become evident.

5.4.2.4 Carbonate-based reactors for removal of Mn, Zn, Cd and As

Principles of Mn, Zn, Cd and As removal using carbonate media:
The last few years have seen considerable advances in the use of reactors containing carbonate minerals (principally calcite, but also dolomite and siderite) for the precipitation of some of the less common (bit often more toxic) mine water contaminants. Most of the development to date has been at laboratory or field pilot scale, so that few of these techniques can be regarded as 'proven technology' at field scale as yet. Nevertheless, we consider it is worth mentioning these techniques here since they are likely to enter the repertoire of full-scale passive treatment techniques during the shelf-life of this book.

Perhaps the best-developed of these methods is the removal of Mn using an OCLR installed in the path of a mine water from which the iron has already been removed. A thick black coating of 'manganese wad' generally develops on the surfaces of the limestone clasts. The wad comprises a biofilm of Mn-oxidizing bacteria heavily impregnated with X-ray amorphous MnO2, which gradually matures *in situ* to form its crystalline equivalent, pyrolusite (e.g. Thornton, 1995; Brant *et al.*, 1999). A patented variant of this natural process (termed the 'Pyrolusite Process®'), which has a growing track-record of implementation in the eastern USA. The Pyrolusite Process® involves site-specific laboratory-culturing of aerobic bacteria, which are then introduced to a serpentine OCLR by means of pre-installed 'inoculation ports'.

Geological observations reveal that pyrolusite is more commonly present as a surface coating on dolomite than calcite in ancient rocks. Bedding planes in dolomite sequences often host beautiful dendritic accumulations of pyrolusite, while nearby calcitic bedding planes do not. This observation has prompted experiments into the use of dolomite as an alternative substrate for MnO_2 precipitation. Initial results from lab and pilot-reactor experiments are very encouraging, showing rapid and comprehensive removal of Mn from solution in residence times of only a few hours (K L Johnson, University of Newcastle, *personal communication*, 2001). Also very encouraging in these experiments has been the complete removal of Zn, Cu and other contaminants, probably by means of irreversible sorption onto the MnO_2, which is known to be a powerful sorbent at circum-neutral pH (e.g. Pretorius and Linder, 2001).

Closed-system limestone dissolution reactors have been successfully piloted by Nuttall and Younger (2000) for the removal of Zn from certain hard, circum-neutral mine waters in which the Zn is the major contaminant. The principle of the field reactors used by Nuttall and Younger (2000) is to use closed-system calcite dissolution to raise pH sufficiently (to around 8.2) that Zn becomes insoluble as its carbonate, smithsonite ($ZnCO_3$). Upon degassing of the water as it leaves the closed-system limestone bed, the water can also attain saturation with respect to hydrozincite ($ZnCO_3.3Zn(OH)_2$).

Most recently, Wang and Reardon (2001) have devised a method for using siderite ($FeCO_3$) and calcite to remove As and Cd from iron-poor waters in which these two contaminants are present at high concentrations. The process is implemented in two steps:

(i)   The water is passed through a saturated-flow reactor packed with siderite, causing siderite dissolution and the release of $\leq$ 15.2 mg.l^{-1} Fe to solution.

(ii)   Next, the water is sprinkled onto an aerated column packed with limestone clasts, through which the water percolates by unsaturated flow. The de-gassing of CO_2 and dissolution of O_2 prompts the precipitation of Cd as its carbonate (otavite, $CdCO_3$) (or by incorporation into calcite which also precipitates under these conditions), and also prompts precipitation of ferric hydroxide, which strongly sorbs AsO_4^{3-}.

Design practice for Mn, Zn, Cd and As removal using carbonate media:
With the exception of the Pyrolusite Process®, which is a proprietary procedure, all of the systems described above remain at the research stage of their evolution, and firm guidelines for their implementation at full scale are not yet forthcoming. Hence the notes given below can be regarded only as provisional guidance at this stage. Nuttall and Younger (2000) suggested that an 8-hour retention time of water in a closed-system limestone reactor should be sufficient to remove 20–40% of the Zn present in a given mine water before the water becomes too super-charged with CO_2 for calcite dissolution to become sluggish. Hence each reactor is designed in a manner identical to an ALD, save that φ is set equal to 8 in Equation (5.22). An aeration and de-gassing step after each closed-system reactor can be used to restore sufficient aggressivity to the water with respect to calcite that returning the water to another closed-system reactor will then remove a further quotient of Zn.

The removal of Mn by formation of wad in OCLRs is climate-dependent, being brisker in warm areas. Although reliable removal rates for such reactors have yet to be determined with statistical rigour, the limited data available to date suggest that a mature, efficiently functioning OCLR may remove Mn at rates of 0.5–3 g.d^{-1}·m^{-2} (K L Johnson, *personal communication*, 2001).

The sequential siderite and calcite reactors used by Wang and Reardon (2001) achieved removal of Cd and As to below detection limits with t_R values of less than two hours in both reactors. The saturated flow siderite reactor can therefore be sized using Equations 5.22–5.24, with φ set equal to 2.

5.4.2.5 SCOOFI reactors

Observations at natural mine water discharges commonly reveal that iron is leaving solution in fast-flowing waters to form a thick substrate of relatively dense ochre ($Fe(OH)_3$). Given the velocity of these waters, settlement of flocs of $Fe(OH)_3$ from suspension is clearly not the responsible process, rather (as explained in Section 5.3.2) it is 'SCOOFI' (surface-catalysed oxidation of ferrous iron). The first application of SCOOFI for mine water treatment is arguably found in work reported from Scotland by Best and Aikman (1983), albeit they did not recognize the process responsible for the rapid removal of Fe which they achieved by percolating deep mine water through brushwood trickle filters. Recent large-scale experiments (Jarvis and Younger, 2001) have consciously examined the potential for harnessing SCOOFI to provide efficient treatment for ferruginous mine waters in areas where lack of space would preclude otherwise-feasible alternative treatment options (such as aerobic wetlands).

The principle is to expose oxygenated mine water to a porous medium characterised by a very high specific surface area (i.e. a high surface area per unit volume of material). Such a porous medium can be constructed by stacking various kinds of material in a container through which the mine water is constrained to flow. Materials suited to the purpose include proprietary plastic 'trickle filter media' (as used in conventional biological aeration systems for wastewaters), or (less expensively) blast furnace slag, fragments of which have large surface areas due to the presence of innumerable fractured bubbles. The high-surface area media provide sites to which ochre can bind, forming a highly reactive surface of ferric hydroxide in contact with the water. The SCOOFI process leads to continuous sorption and *in situ* oxidation of Fe^{2+}, and hence the ever-increasing accretion of ochre to the media surfaces. Because the surface-catalysed process of iron oxidation is more rapid than open-water oxidation of Fe^{2+} by dissolved oxygen, once a thin coating of ochre is present on the supporting media iron removal proceeds very rapidly. Eventually, the accumulation of ochre begins to block the media pore space to such a degree that the full flow will no longer pass through the reactor. At this point (or ideally some time shortly before) the media will need to be removed and replaced. Options for media replacement and ochre disposal constitute one of the key matters for consideration in design practice for SCOOFI reactors (see below).

There are two possible configurations for SCOOFI reactors (Figure 5.6):

• saturated flow SCOOFI reactors (e.g. Younger, 2000d), and
• unsaturated flow SCOOFI reactors (e.g. Jarvis and Younger, 2001)

The former achieve much more intimate contact between the mine water and the media, but can only be recommended where sufficient dissolved oxygen has already been added to the water (e.g. by cascade aeration or other means; see Section 4.2.2.2) to oxidize all of the dissolved Fe^{2+} given sufficient reaction time. (In practice this means that mine waters with more than 50 mg.l^{-1} Fe^{2+} are unlikely to be amenable to comprehensive treatment using a single-step saturated flow SCOOFI reactor). Saturated flow reactors (Figure 5.6a) are hydraulically designed in a manner similar to ALDs (Section 5.4.2.2), in that media hydraulic conductivity and cross-sectional area perpendicular to flow are key considerations. Unlike ALDs, no upper seal is used in SCOOFI reactors, since free exchange of gases with the atmosphere is desirable.

Where site conditions preclude pre-aeration of the mine water, a saturated flow reactor is unlikely to be successful, as the rate of flow through the reactor will always be far more rapid than the rate of DO dissolution through the free surface of the water. In such cases, vertical, unsaturated flow SCOOFI reactors (Figure 5.6b) may find a niche, since these can simultaneously aerate a mine water and oxidize its Fe^{2+} content (e.g. Jarvis and Younger, 2001). Nevertheless, unsaturated flow reactors achieve far less contact between the mine water and the media surfaces than do saturated flow SCOOFI reactors. This probably explains why iron removal rates in unsaturated flow reactors have been found to increase with the rate of throughput of water (Jarvis and Younger, 2001): as more water is passed through an unsaturated flow system, more of the medium surface area is wetted.

A

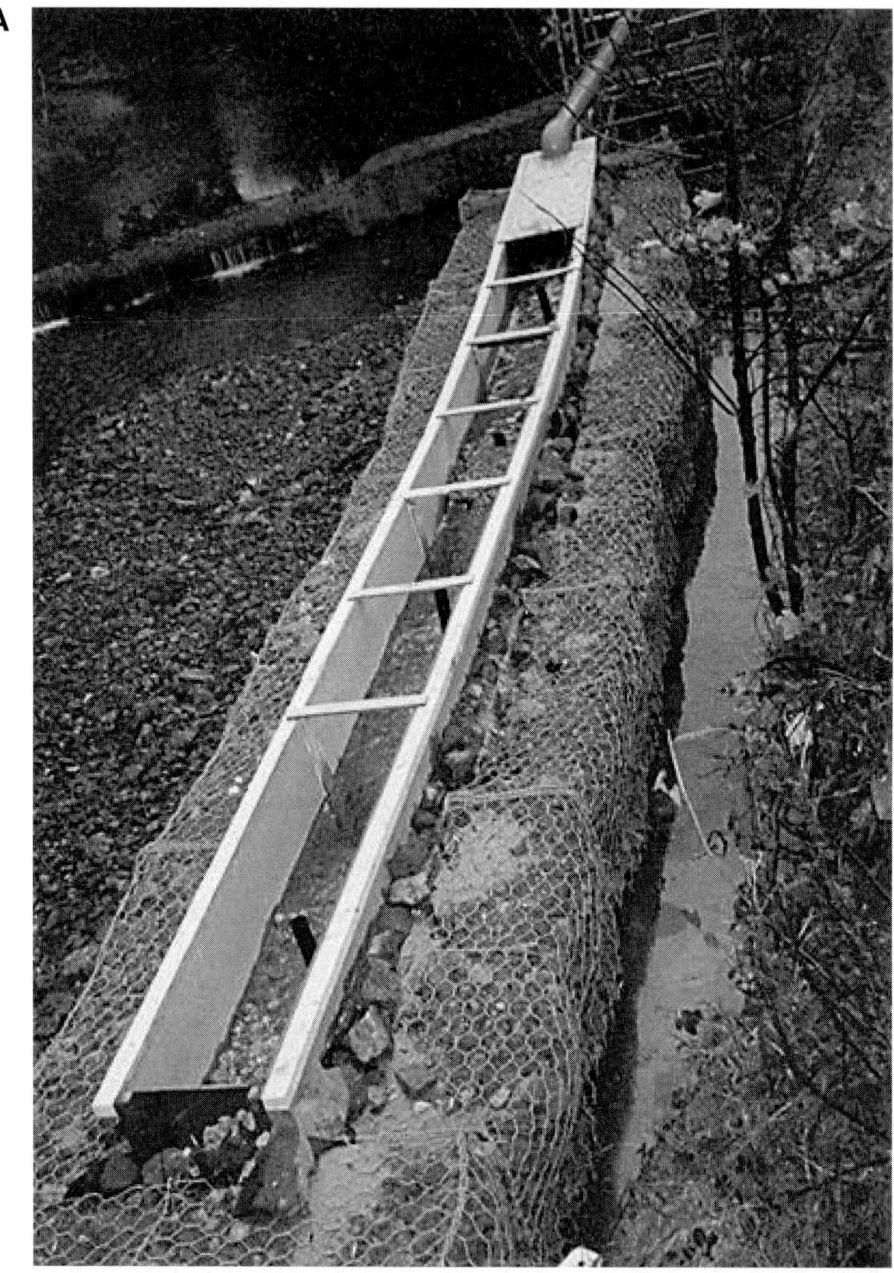

Figure 5.6: (*See opposite for 5.6B*) A – Pilot-scale saturated-flow SCOOFI reactor at Skinnigrove, Cleveland (UK), as described by Younger (2000d). The mine water moves through the blast-furnace slag held in the gabions, and iron is removed by surface-catalyzed oxidation of Fe^{2+}. B – The prototype unsaturated flow SCOOFI reactors at Kimblesworth, County Durham (UK), as described by Jarvis and Younger (2001). The reactor comprises two towers filled with lumps of plastic and/or blast furnace slag (both of which have high specific surface areas).

Figure 5.6B: *(See opposite for 5.6A).*

Reactor sizing criteria for both unsaturated and saturated flow SCOOFI reactors remain tentative at present, being based on the systems constructed to date (all of which have been built in the UK, comprising four pilot plants and two full-scale plants; Younger, 2000d; Younger *et al.*, 2001). As already mentioned, saturated flow SCOOFI reactors are intrinsically more efficient than the unsaturated flow variety, and this is reflected in reported Fc removal rates which equate to hundreds of grams of Fe per $m^3.d^{-1}$ (Younger, 2000d). By contrast, unsaturated flow SCOOFI reactors achieve far lower Fe removal rates, amounting to only $12 \ g \cdot m^{-3} d^{-1}$ (Jarvis and Younger, 2001). (While these removal rates are currently being used in the design of full-scale SCOOFI reactors of the respective types, it has to be stressed that both values are likely to be amended substantially in the light of future experience; Younger *et al.*, 2001).

It was mentioned above that the motivation for developing SCOOFI reactors was to provide solutions for sites in which there is insufficient available land area to accommodate a conventional wetland. But how do SCOOFI reactors compare with aerobic wetland treatment systems (Section 5.4.3.3) in absolute terms? If the Fe removal rates for saturated and unsaturated flow SCOOFI reactors are normalised by the surface area of media in the reactors from which they were derived, the following figures are obtained:

- $25 \ g.m^{-2}.d^{-1}$ for saturated flow SCOOFI reactors
- $0.05 \ g.m^{-2}.d^{-1}$ for unsaturated flow SCOOFI reactors.

The areally-adjusted removal rate for the saturated flow reactor exceeds the areally-normalised Fe removal rates generally cited for aerobic wetlands (i.e. $10–20 \ g.m^{-2}.d^{-1}$; Hedin *et al.*, 1994a). However, the rate for unsaturated flow reactors is much lower than for aerobic wetlands. Before jumping to conclusions, it is important to note that the mean residence time of water in the aerobic wetlands studied by Hedin *et al.* (1994) was 3.5 days, whereas the residence time of water within the unsaturated flow SCOOFI reactors studied to date has been in the range of only 90–120 *seconds*. If we therefore divide the area-adjusted removal rates for SCOOFI reactors and aerobic wetlands by their respective residence times, we obtain the following figures:

- Aerobic wetlands: $2.9–5.7 \ g.m^{-2}$
- Unsaturated flow SCOOFI reactors: $36 \ g.m^{-2}$
- Saturated flow SCOOFI reactors: $4000 \ g.m^{-2}$

Notwithstanding the *caveat* that these performance figures will no doubt be amended substantially in future (Younger *et al.*, 2001), the values presented here at least serve to underline the process intensification which SCOOFI reactors represent in comparison to land-hungry wetland treatment methods. They also illustrate why SCOOFI has significant potential as a 'polishing' process for waters with $\leq 50 \ mg.l^{-1}$ Fe, at least where land availability is at a premium.

Probably the most important design considerations for full-scale SCOOFI reactors relate to media replacement and ochre disposal. As SCOOFI reactors are predicated on the accumulation of ochre within a porous medium, that porous medium must be cleaned occasionally if it is not to clog completely. With this in mind, current SCOOFI reactor designs incorporate modular blocks of plastic media which can be easily removed from the reactor for cleaning. Ochre can be removed from the plastic media by brushing or jetting, before returning the de-clogged media elements into the reactor. If a market can be found for the ochre, then SCOOFI reactors have the advantage over wetland treatment systems that the ochre can be obtained in a 'clean' form, in that it won't be mixed with plant and soil debris which might compromise its use in other industries. At present, strenuous research initiatives are underway in the USA and Europe to develop markets for ochre in the pigment, sewage treatment, agricultural drainage and composting industries. Some of these efforts are summarized in Section 4.7.1. For examples relating particularly to passive treatment systems, see Hedin (1998, 1999) and Hedin and Weaver (*in press*).

5.4.3 WETLAND-TYPE PASSIVE SYSTEMS

5.4.3.1 General considerations

Constructed wetland systems are currently the most widely-used passive mine water treatment technology, and are likely to remain so. There are several reasons for this, including:

1. The excellent track record of constructed, aerobic wetlands in treating net-alkaline mine waters in which the only pollutant of concern is iron. These systems are now so widespread, and invariably successful when designed and constructed in accordance with established guidelines (e.g. Hedin *et al.*, 1994; Younger, 1997b, 2000d), that they fully merit the tag of 'proven technology'.
2. The generally low running costs of wetland systems in comparison to active treatment systems.
3. The inherent ability of large wetland systems to cope with unforeseen fluctuations in environmental conditions, by providing flexible storage volumes etc.
4. The environmental attractiveness of wetlands, as prime habitats for birds and other animals, and as landscape amenities of appeal to human visitors.

Notwithstanding these virtues, the use of wetlands for passive treatment of mine waters has not been without its detractors (e.g. Wieder, 1989; McGinness *et al.*, 1997). Most of the mis-givings which have been expressed in the literature arise from cases where the technology has been misapplied and/or the data misinterpreted. For instance two cases may be cited from England alone in which disappointing early applications of wetlands technology gave rise to denigration of the technology as a whole. In both cases, the reason for the poor performance was that simple aerobic reed-beds were inappropriately constructed to receive extremely acidic spoil leachates, for which other technologies (compost wetlands or RAPS) are actually recommended. By contrast, 25 other UK systems were constructed around the same time in accordance with the guidelines of Hedin *et al.* (1994), all of which proved very successful (Younger, 2000d). These successes nullified the previous bad publicity, and both of the early, failed systems are now being retro-fitted with alkalinity-generating variants of wetland systems.

Despite the numerous success stories, some wetlands-based technologies remain less certain in their applicability than others. In particular, wetlands treatment is still challenging for highly acidic waters, and may even be inadvisable for waters containing significant concentrations of toxic metals (such as Hg and Cd). Hence we do not wish to give the impression that the techniques described below represent a *panacea*, and we will attempt to highlight current limitations to the technology as we see them.

5.4.3.2 Sedimentation ponds

The design of simple sedimentation ponds for the removal of Fe from mine waters has already been described in Section 4.2.4, and will therefore not be discussed in any detail here. In the context of their use in passive treatment systems, sedimentation ponds can be very useful for removing high concentrations of Fe in an easily-cleaned basin, upstream of less-easily cleaned wetlands. In the UK, for instance, the government's Coal Authority now recommends that a sedimentation pond be the first element in any passive treatment system receiving mine waters with more than about 50 mg.l^{-1} Fe. As with SCOOFI reactors, the ochre recovered from these primary sedimentation ponds is far more likely to be marketable than the ochre rich in plant debris which is typically dredged from treatment wetlands.

5.4.3.3 Aerobic Wetlands (Reed beds)

Principles of aerobic wetlands (reed beds):
We have already noted that aerobic wetlands can legitimately be regarded as 'proven technology' when applied to the treatment of ferruginous, net-alkaline mine waters. They have also been successfully used to remove Mn from net-alkaline waters (Hedin *et al.*, 1994), albeit at 5% of the rate of Fe removal. Faster rates of Mn removal can also occur in certain aerobic wetlands in which algal mats develop during the summer months (e.g. Phillips *et al.*, 1995). Similar algal mats in aerobic ponds have been accredited by Kalin (1998) with fostering the removal of Zn from solution as a carbonate (most likely as hydrozincite, as explained in section 5.4.2.4).

A common factor in all of these successful applications of aerobic wetland technology is the dominance of oxidative hydrolytic processes (see sections 5.3.2, 5.3.4 and 5.3.7), coupled with an array of sedimentation mechanisms (see 5.3.2). Because the net result of the predominant oxidation and hydrolysis reactions is a release of protons (see Equations 5.2, 5.6 and 5.8), aerobic wetlands are *only* recommended for the removal of metals from waters which are net-alkaline. If net-acidic waters are subjected to oxidative hydrolysis in an aerobic wetland the pH will drop, usually to such a level (< 5) that most metals become soluble once more (see Section 2.2.4). Aerobic wetlands can, of course, be used to treat originally acidic mine waters which have already been neutralized by some other process (e.g. by some form of active treatment, or using an ALD or RAPS etc).

Four processes of iron sedimentation were outlined in Section 5.3.2, all of which occur in aerobic wetlands, and these may be generalized to cover other metal species as follows:

(i)   Settlement of precipitated solids from aqueous suspension
(ii)   Physical filtration of colloidal precipitates by plants stalks etc
(iii)   Formation of hydroxide/oxide plaques on plant roots and rhizomes
(iv)   Sorption and *in situ* oxidation (analogous to SCOOFI)

All four processes occur in an analogous manner for Mn. Since aluminum and zinc each display only one valence state under environmentally-relevant conditions (i.e. Al^{3+} and Zn^{2+}), *in situ* oxidation does not apply to them, though all other processes listed (including the sorption step in (iv); see Section 2.4.2) most certainly do. The central role of plants in sedimentation processes (ii) and (iv) above explains why wetlands are preferred to ordinary sedimentation ponds, particularly for removing the last few tens of milligrams per liter of Fe from a mine water.

Natural and constructed aerobic wetlands are often referred to as "reed beds" in the British Commonwealth (e.g. Hawke and José, 1996). The term "reed bed" is also specifically applied to 'horizontal subsurface flow reed-bed treatment systems' (HFRBTS) which are widely used for tertiary treatment of sewage effluents in the UK and elsewhere (e.g. Cooper, 2001). Substantial differences in both form and function exist between HFRBTS and aerobic mine water treatment wetlands, such that the term 'reed bed' is probably best avoided in this context. Nevertheless, reeds and rushes are important components of aerobic mine water treatment wetlands. In addition to promoting sedimentation of metals, the stalks of emergent wetland plants serve to baffle surface flows, and provide so much frictional resistance to moving water that they significantly alter the profile of velocity over depth. This can foster removal of metals by relatively slow reactions. The seasonal die-back of some of the more prolific reeds (especially *Typha latifolia*) contributes substantial quantities of labile organic matter to the wetland substrate, within which microbial degradation processes may turn it into a significant further source of alkalinity (see Section 2.2.6 and especially Table 2.7).

It is often stated that uptake of metals by wetland plants, though often dramatic in terms of metal concentrations in plant biomass (e.g. Batty, 1999), is not a significant process in water quality terms (e.g. Cohen and Staub, 1992; Hedin *et al.*, 1994). This is almost certainly true in the majority of cases where the wetland is treating water with high concentrations of Fe, Mn etc. However, in the more distal parts of wetlands, or in wetlands receiving waters with only a few $mg.l^{-1}$ of Fe/Mn, plant uptake may well be an important polishing process. Figure 5.7 illustrates that where aqueous concentrations of Fe are around $1\ mg.l^{-1}$, uptake by *Phragmites australis* can account for nearly 100% of overall iron removal. (Similar results have been obtained in relation to Mn). Bearing in mind that inorganic oxidation of Fe^{2+} is a first order reaction (Equation 5.4), so that abiotic removal of iron at such low concentrations of Fe is very slow, the results shown in Figure 5.7 have important implications for the design of wetlands to achieve stringent Fe emission limits.

Design practice for aerobic wetlands (reed beds):
Aerobic wetlands are simply shallow water bodies supporting more-or-less dense stands of reeds and/or rushes (Figure 5.8). Several practical manuals have been produced on the creation of freshwater wetlands in general. We particularly recommend the guidelines of Hammer (1992) and Hawke and José (1996). These two books contain a wealth of practical information on the design, construction and management of wetlands, with particular emphasis on measures to maximize the wider ecological value of a wetland which may be under construction for utterly utilitarian reasons. Reflection on these published recommendations, and our own experience, leads us to propose the following general guidelines for wetland design (not specific to mine waters):

(i)   Ensure that the wetland substrate has an uneven surface. After the water is intro-
       duced, this undulating surface will be manifest in islands, spits, heavily-vegetated
       shallows and deep plant-less pools. This range of habitats will maximize the wildlife
       value of the wetland, and it also serves to baffle the flow of water through the
       wetland, thus improving treatment efficiency.

(ii)   Make sure that most (if not all) of the water margins have slopes of 1:3 or less.
       Slopes much steeper than this will be of little use to most wading birds, amphibians
       and mammals. Very steep side-walls to a wetland are also potentially hazardous to
       young children and domestic pets, who can easily fall into the wetlands, but may be
       unable to climb out again unaided.

(iii) Try to avoid having tall trees flanking the very margins of the pond, as shading will
       result in the wetland being devoid of emergent plants. On the other hand, low shrubs
       around the wetland margin are a positive advantage for terrestrial wildlife, enabling
       riparian mammals and amphibians to access open waters with least risk of predation.

(iv)   Avoid steep, bare concrete sidewalls and hydraulic control features wherever possi-
       ble. Such structures detract from the amenity value of the wetland, frequently entail
       the steep-ness hazards already mentioned, and provide no cover for animals moving
       to and from the wetland to the surrounding land. Bitter experience shows that the

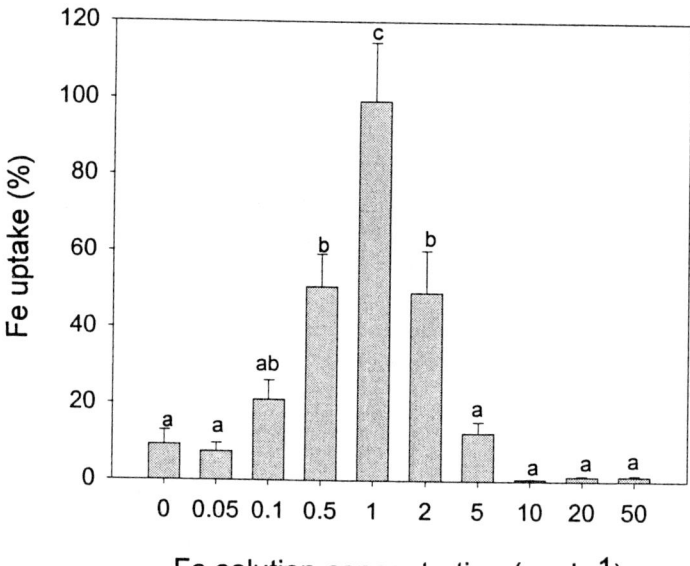

Figure 5.7: Uptake of Fe by seedlings of *Phragmites australis*, expressed as a percentage of the Fe supplied in solution. (Different letters indicate a significant difference with respect to Fe concentration at $P < 0.05$). The majority of the iron taken up by the plants is found in the root tissues, primarily as iron plaque deposits. Although absolute Fe concentrations within plant tissues are higher at higher Fe supply rates, as a percentage the Fe uptake is significantly lower above 1 mg.l^{-1}. This suggests that there is a limit to the amount of Fe that can be removed by this plant and that the optimum removal rate is achieved when Fe is supplied at 1 mg.l^{-1}. (Data courtesy of the European Commission's PIRAMID project (*www.piramid.org*), provided by Dr L.C. Batty).

A

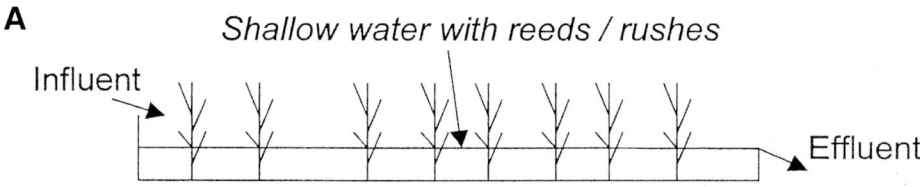

B

Figure 5.8: Aerobic wetlands: (a) Generalized and simplified cross-section illustrating the basic design concept (b) Photograph of a real aerobic mine water treatment wetland: St Helen Auckland (Co Durham, UK). (Photo: P.L Younger).

use of concrete is often inevitable if a stable bund is to be created. However, in mine water situations, the concrete must either be based on sulfate-resistant cement (which is slow to set and more expensive than many other cements), or it must be painstakingly painted with an acid-resistant bitumen coat. If surficial concrete must be used, consider the extra expense of covering the surface with geo-textile covers which will support plant growth.

(v) Think carefully before specifying an artificial liner for a treatment wetland: not only will it greatly increase the costs and difficulties of construction, but it can very easily be damaged during or after construction, thus wasting the major investment. Where it emerges above the water surface, an artificial liner is often ugly and usually slippery and therefore hazardous. In many cases, compaction of the existing soil, or importation and compaction of clay will be preferable to installing a liner. Conduct *in situ* geotechnical tests and take expert advice before jumping to the conclusion that an artificial liner is necessary.

(vi) Avoid including heavily-engineered rectilinear corners when laying out a wetland bund. Not only do they look ugly and detract from wildlife habitat, but square corners usually result in large stagnant zones which contribute little to the treatment process.

(vii) Try to distribute influent water evenly across the width of the wetland. This can be readily achieved using a line of adjustable t-pieces on the influent pipe work (Figure 5.9). (As long as there is plenty of internal baffling of flow within the wetland, distribution of effluent points is not usually necessary).

Turning to the specifics of designing aerobic mine water treatment wetlands, there are two key issues to consider:

(a) Sizing of the wetland (including area, depth and freeboard)
(b) Selection, introduction and nurturing of wetland plants

With regard to sizing the required wetland area, the two main alternative models for this purpose (i.e. the model of Tarutis *et al.*, 1999 (Equation 5.18), and that of Hedin *et al.*, 1994 (Equation 5.21)), have already been discussed in Section 5.3.8. In that Section we concluded that the model of Hedin *et al.* (1994) currently remains the most useful of the two models. To apply this method, we need to substitute site-specific values of flow (Q_d) and the design pollutant concentration into Equation 5.21, together with a generic value of the area-adjusted removal rate (R_A) for the pollutant of interest. Values of R_A have been derived for a range of metals likely to be encountered in net-alkaline mine waters, and these are summarized in Table 5.4, together with notes on their origins and guidance on their applicability. Thus Table 5.4 and Equation 5.21 together provide the basic tools necessary for wetland sizing. We currently recommend the R_A values proposed by Hedin *et al.* (1994) in preference both to the previous (more conservative) criteria of Brodie *et al.* (1988) and to the more recent (and more optimistic) criteria of Phillips *et al.* (1995).

Before proceeding to apply the values in Table 5.4 in practice, a few nuances need to be taken into account:

(i)   Where a wetland must remove both Fe and Mn, the total area required is obtained by first sizing for Fe only, then sizing for Mn separately and adding the two areas together. (The reason for this practice is that Mn oxidation is inhibited in the presence of significant dissolved Fe^{2+}, as explained in Section 5.3.4).

(ii)   It will be noted that no criterion for removal of aluminum is given in Table 5.4. This is because aluminum is never present in dissolved form in net-alkaline waters, being highly insoluble as its hydroxide where pH exceeds 4.5 (see Section 5.3.3). However, it is possible for recently-neutralized mine waters to contain significant suspended aluminum. In such cases, we would advocate the use of simple sedimentation ponds (see 5.4.3.2), sized according to the criteria for Fe removal (section 4.2.4), or else to allow a small additional area of aerobic wetland (say, 1 m² for every 1.s⁻¹ of flow).

(iii) If it is desirable that an aerobic wetland also remove some zinc, then it is important to realize that the value of R_A for Zn given in Table 5.4 is not a 'consensus' value, for the predominant processes of Zn removal in aerobic systems remain controversial. For instance, Brown (1997) observed that Zn was only removed from solution in an aerobic wetland during the summer months, being returned to solution in winter. The process deemed responsible was a reversible sorption reaction involving leaf litter. Similarly Kalin (1998) has documented significant Zn removal in aerobic ponds which contained floating colonies of algae. In this case, the Zn was removed

Figure 5.9: The use of adjustable t-pieces on the influent water pipe to achieve even spreading of water across the width of a newly-constructed treatment wetland, Sherburn, Co Durham (UK). (Photo: P.L. Younger, with the permission of the site owners, Northumbrian Water Ltd).

Table 5.4: Area-adjusted removal rates (R_A) for Fe, Mn and Zn in aerobic wetlands. These values are substituted into Equation 5.21 to derive design areas for aerobic wetlands (reed beds) for the treatment of net–alkaline mine waters

Target metal	R_A value (g.d^{-1}.m^{-2})	Suggested conditions for which applicable	Source
(a) Values which we recommend for routine design purposes			
Fe	10	Wherever the receiving stream is sensitive/regulatory compliance essential.	Hedin *et al.* (1994)
Mn	0.5	Wherever regulatory compliance is deemed essential.	Hedin *et al.* (1994)
(b) Other values proposed in the literature			
Fe	20	Suitable where receiving stream not especially sensitive; reliable for well-oxygenated waters with pH > 6.5, but unlikely to be achieved where pH and/or DO are depressed.	Hedin *et al.* (1994); Watzlaf *et al.* (2001)
Mn	1	Suitable where regulatory compliance not critical.	Hedin *et al.* (1994)
Fe	2–4	*Proposed* for aerobic wetlands receiving acidic waters (pH < 5). N.B.: we *do not recommend* using aerobic wetlands for such waters.	Lamb *et al.* (1998)
Fe	5–11	*Proposed* for waters in pH range 5–7 (marginally net-alkaline)	Lamb *et al.* (1998)
Fe	0.72	*Proposed* for aerobic wetlands receiving acidic waters (pH < 5.5). Although we do not recommend using aerobic wetlands for such waters, this is nevertheless a conservative sizing criterion.	Obtained by re-arranging sizing criteria given by Brodie *et al.* (1988), which were said to be capable of ensuring the following residual concentrations in wetland effluents: Fe ≤ 3 mg.l^{-1}, and Mn ≤ 2 mg.l^{-1}
Fe	1.92	For waters with pH > 5.5.	
Mn	0.205	*Proposed* for aerobic wetlands receiving acidic waters (pH < 5.5); again, aerobic wetlands no longer recommended for such waters, but this is a very conservative sizing criterion.	
Mn	0.72	For waters with pH > 5.5.	
Mn	2	Suitable where microbial-algal mats growing in shallow aerobic pond; normally operative only in summer months, though difficult to sustain.	Phillips *et al.* (1995); Lamb *et al.* (1998)
Zn	7	Aerobic ponds/wetlands with floating algal colonies. Prone to major seasonal fluctuations from 0 to 15 g.d^{-1}.m^{-2}.	Calculated by the authors from data presented by Kalin (1998)

by precipitation of hydroxycarbonate solids. However, removal rates varied from effectively zero on some occasions in winter to almost 15 g.d^{-1}.m^{-2} in summer. If these experiences are adopted to aid wetland design, then it is probably advisable to size for Zn removal as an area additional to that needed for Fe removal (though *not* additional also to the area needed for Mn removal, as Zn is invariably found to be less mobile than Mn in mine waters containing both). At first sight, this recommendation may seem too conservative: after all, given that Zn^{2+} is essentially non-oxidizable under the pH and Eh conditions relevant to wetlands, there is no *a priori* reason to suppose that formation of hydroxycarbonates of Zn^{2+} will be inhibited by competition for O_2 by the oxidation of dissolved Fe^{2+} (in contrast to the case of Mn^{2+}, where such inhibition is usually very marked). However, the rate of precipitation of the hydroxycarbonates of zinc is apparently significantly slower than the rate of ferric hydroxide precipitation. For instance, the removal of Zn from solution in aerobic pond systems described by Kalin (1998) took place in ponds with nominal hydraulic retention times on the order of 1 to 2 weeks, compared with the average of 3.5 days in the iron-removing wetlands studied by Hedin *et al.* (1994). Hence, more time is needed to remove Zn than Fe, so that provision of additional wetland area is almost certainly warranted.

Before leaving the issue of sizing wetland areas, it is worth briefly comparing the areas predicted by using Equation 5.21 (i.e. the model of Hedin *et al.*, 1994) with those predicted by Equation 5.18 (i.e. the first order reaction model of Tarutis *et al.* (1999)). To make this comparison, we apply the formulae to an existing aerobic treatment wetland, at Edmondsley (County Durham, UK), the construction of which was completed in the summer of 1999. The Edmondsley mine water flows from an old coal drift at a rate of about $10 \, l.s^{-1}$, and contains some $30 \, mg.l^{-1}$ of Fe. The designers of the wetland used Equation 5.21 (with an R_A value of $10 \, g.d^{-1}.m^{-2}$; see Table 5.4) to estimate the minimum wetland area required. The calculation is as follows:

$$A = (864 \, m^3.d^{-1} \times (30\text{--}1))/10 = 2505 \, m^2$$

Fortunately, there was more than twice as much suitable land area available for purchase at the site. Hence, the system was constructed as four aerobic reed bed cells in series, totaling $4000 \, m^2$ in area. This design allowed one or two of the cells to be taken out of operation at any time for maintenance purposes without compromising the treatment ability of the system as a whole. Monitoring of the system since it was commissioned has shown performance in line with expectations, with virtually all of the Fe (down to a residual $< 0.5 \, mg.l^{-1}$) being removed in the first two cells (i.e. the first $2000 \, m^2$ of wetland). By contrast if the wetland had been designed using Equation 5.18 and the K_1 value of $0.18 \, m.d^{-1}$ suggested by Tarutis *et al.* (1999), very different conclusions might have been reached: the wetland area predicted by Equation 5.18 to achieve an Fe residual $< 0.5 \, mg.l^{-1}$ is:

$$A = (864 \, m^3.d^{-1} \times (\ln(30) - \ln(0.5)))/0.18 = 19652 \, m^2$$

Had this been the design figure used, then the costs of acquiring 2 ha of land in this scenic area would have precluded wetland treatment as a serious option. An active treatment system would have been developed, with huge cost implications for long-term operation. This example underlines the point made in Section 5.3.8 that Fe removal in aerobic wetlands is a function of far more than just the rate of oxidation of Fe^{2+} (which is the basis of the model of Tarutis *et al.*, 1999).

Besides area, the other key aspect of wetland design is depth. There are two aspects to this: the depth of water during wetland operation and the amount of freeboard in the system. With regard to water depth, experience has shown that preferential germination and growth of major wetland plant species occurs where the soil surface is submerged by 0.15 m–0.25 m. If the water is much shallower than 0.15 m, invasion by grass species is common. At the other extreme, where water depths are much in excess of 0.25 m, sprouting of all but the hardiest wetland plants will be inhibited, and mature plants will tend to be stunted compared to neighbors growing in shallower water.

Freeboard is the height between the water surface and the lowest point on the crest of any water-retaining structures (bunds, dams etc). The provision of adequate freeboard in a wetland system is important for at least two reasons:

(i)   It provides a volume for potential storage of water in the wetland during particularly wet periods, thus helping to minimize local flooding problems.

(ii)   It provides scope for a gradual raising of the sediment surface over time as plant debris, clastic sediments and metal precipitates accumulate in the wetland. Periodic raising of the water level (by adjusting outflow structures such as weirs etc using stop-logs; see Hawke and José (1996) for guidance on these details) can ensure that an adequate water depth is always retained, thus preventing desiccation of the wetland and its colonisation by terrestrial plants (*hydroseral succession*).

In the design of most wetland systems constructed to date, a nominal freeboard of 0.5m to 2 m has typically been sought. However, selection of freeboard must always be made on a case-specific basis, for insistence on a certain minimum freeboard may well compromise other desirable wetland features, such as gently sloping banks around the water margins. Experience to date suggests that designs should make allowance for the accumulation of ferric hydroxide sludge to consume between 25–50 mm of freeboard per year.

The selection of suitable plant species for use in mine water treatment wetlands is frequently a focus of deeply earnest discussion by newcomers to the technology. With the benefit of hindsight, we would advocate the following maxim: the ideal wetland plant is one which thrives in the system in question. Whether or not a given wetland plant will thrive depends on its physiology, in particular its tolerance to high concentrations of potentially phytotoxic metals. The science of metal uptake, exclusion and excretion in wetland plants is an area of rapid development. One key process currently receiving considerable research attention is the development and functions of metal 'plaques' (i.e. dense mineral coatings) which commonly occur on plant roots and rhizomes in metalliferous wetlands (e.g. Batty *et al.*, 2000). Ferric hydroxide and manganese oxide plaques appear to develop due to oxygen leakage from plant roots. Aluminum phosphate plaques have been more recently described from more acidic mine water wetlands (Batty *et al.*, in press). Apart from representing significant sinks for their constituent metals, and other metals which sorb to them, these plaques might also limit the passage of sorption-prone nutrients such as phosphate into the aerial parts of the wetland plants, thus limiting the growth potential of these plants.

Not all wetland plants are equally robust in the face of these threats to prosperity. Experience has shown that there are three wetland plants which are particularly capable of flourishing in metal-rich wetlands:

Scientific name	Common name (USA)	Common name (British Commonwealth)	Figure in which illustrated
Phragmites australis	Giant reed	Common reed	Figure 5.10a
Typha latifolia	Cattails	Reedmace	Figure 5.10b
Juncus effusus	Soft rush	Common rush	Figure 5.10c

Other common wetland plants have also proved to be adaptable to the mine water environment (e.g. Laine, 1997; Batty, 1999). This is certainly true of *Scirpus lacustris* (bulrush), *Eriophorum angustifolium* (bog cotton grass), Yellow-Flag Iris (*Iris pseudacorus*), and various *Juncus* species in addition to *J. effusus*. Inclusion of these less common plants in a planting campaign can both help to diversify the habitat offered by a wetland and improve its appearance (for instance in early summer when Yellow-Flag Iris blooms spectacularly). Nevertheless, the predominance of the 'big three' listed above is well illustrated by the case of a large mine water wetland in West Yorkshire, UK, in which seven different species of wetland plant were initially planted (Laine, 1997): only four growing seasons passed before the three plants shown in Figure 5.10 dominated the wetland, with *Typha* and *Phragmites* in deeper waters and *Juncus effusus* around the margins of the wetland.

Other more specialised wetland plants (such as *Sphagnum* and other plants typical of ombrotrophic (rain-fed) mires) have died during long exposure to elevated (> 10 mg.l^{-1}) concentrations of iron for more than one growing season (Spratt and Wieder, 1988). Hence the selection of plants is not an arbitrary process, and expert advice should be sought in a particular geographical region. Moving beyond mere survival of plants to wider ecological issues, it is also wise to consider the natural ranges of certain wetland plants before introducing them to the harsh conditions of a mine water treatment system. For instance, *Phragmites australis* is (as the name 'australis' implies) naturally most abundant in the lower latitudes. However, it has been repeatedly introduced to constructed wetlands far to the north of its natural limit (e.g. in Scotland and northern England). While it can become established from pot-grown plants, and stands of the reed will spread by rhizome migration, it will never go to seed so far north and is therefore arguably an artificial element in the wider ecosystem.

The introduction and early nurturing of wetland plants can be approached in several ways. Where a wetland is being created without external pressures for early performance, it may be best to simply create a water body with the appropriate dimensions and await colonization by reeds and rushes which find the habitat amenable. Another approach is to speed up the process of colonization by introducing a few mature specimens of the plants which are judged most likely to prove successful locally. These will subsequently colonize the remainder of the wetland system by rhizome migration. If this approach is taken, transplantation from a suitable local source can often be inexpensively arranged by offering to clear the ditches on adjoining farm land. When digging up mature wetland plants, it is desirable to dig a good 0.3 m down before scooping below the plant and removing a large block of soil containing the roots, taking care to avoid damaging most of the main roots. If the wetland construction is completed in the late summer or fall, then transplantation of plants is likely to leave the new wetland looking more like a flooded compost heap, full of dead plants. In those circumstances, it is preferable to sow seeds, which are usually abundant on *Juncus* and *Phragmites* at this time of year, and which are embedded in the panicles of *Typha* (i.e. the long brown seed cases which develop at the

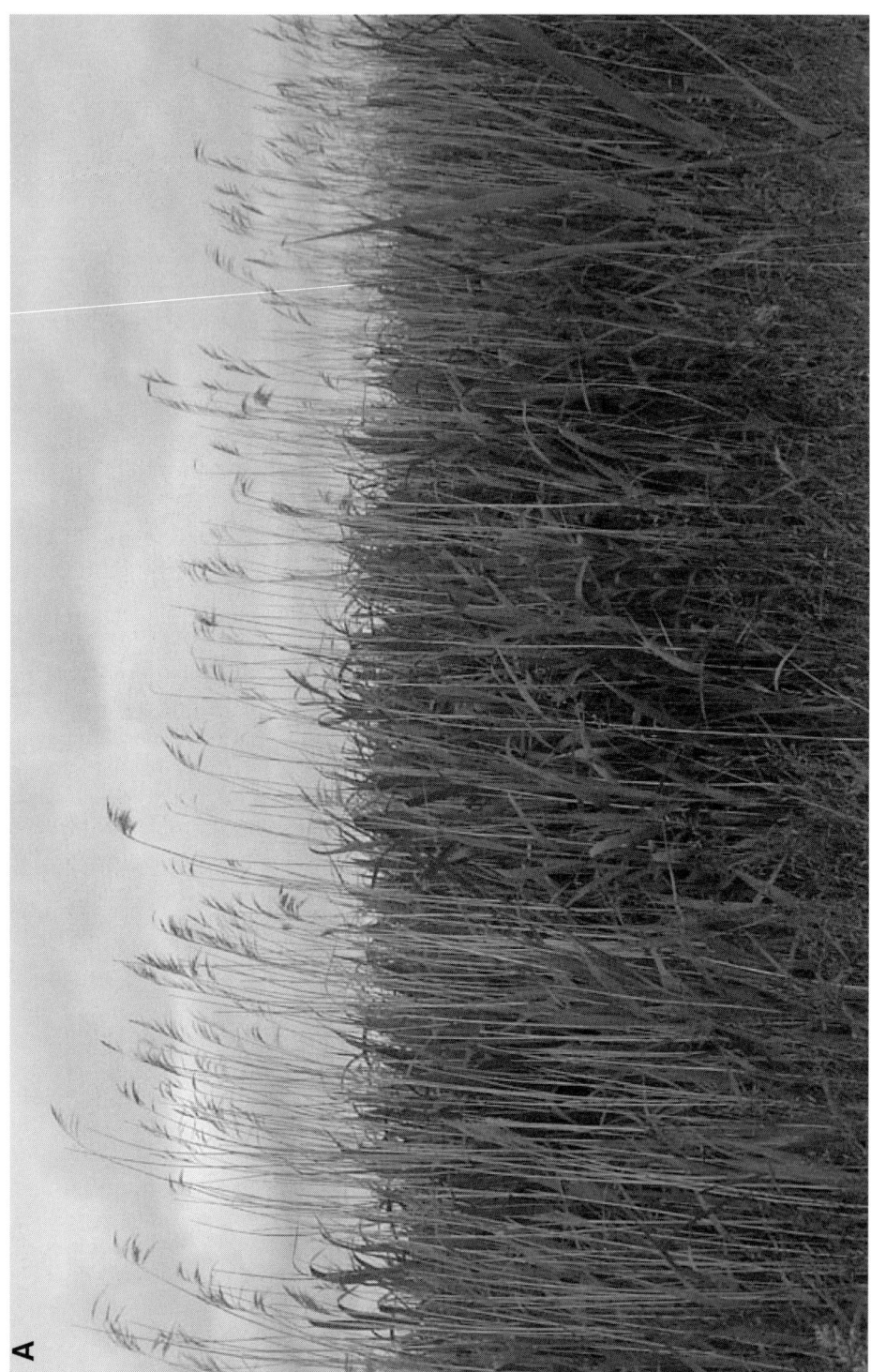

Figure 5.10: *(See following pages for 5.10b & c)* **Photos of common mine water wetland plants.** *A: Phragmites australis, B: Typha latifolia, C: Juncus effusus.*

B

Figure 5.10B.

C

Figure 5.10C.

top of *Typha* plants in the fall). Hawke and José (1996) provide valuable advice on such autumnal pursuits. In high-profile wetland creation projects, it will probably be necessary to formally plant the wetland to a specified density (usually one plant to every 0.5 m^2). Early liaison with specialist wetland plant nurseries is advisable, to ensure they will have a sufficient quantity of the desired plants by the time planting commences. The nurseries will usually arrange planting themselves. When ordering plants from a nursery, it is usually better to specify that the supplied plants be pot-grown rather than bare-rooted. If a DIY approach to wetland planting is preferred, then the practical guidance of Hawke and José (1996) will once more be found to be very useful.

After planting, careful management of water levels in the wetland is necessary to avoid drowning of the newly-emerging plants. Small plants will be best grown in a few centimetres of water, which is sufficient to deter grazing rabbits but not so deep as to drown a sprouting shoot. There is also growing evidence that it would be best to limit iron concentrations to only a few mg.l^{-1} while plants are germinating and sprouting (Figure 5.11). As the plants grow, so the water depth can be increased, to ensure that terrestrial grasses do not gain a foothold. Figure 5.12 gives a good example of vegetation establishment from seed, with careful control of water levels aiding sprouting. The "grasses" standing in the water are *Typha* seedlings. These were established from seeds sown in March 2001. Water depths were maintained at less than 50 mm until June, when water depths were increased to between 150 to 300 mm.

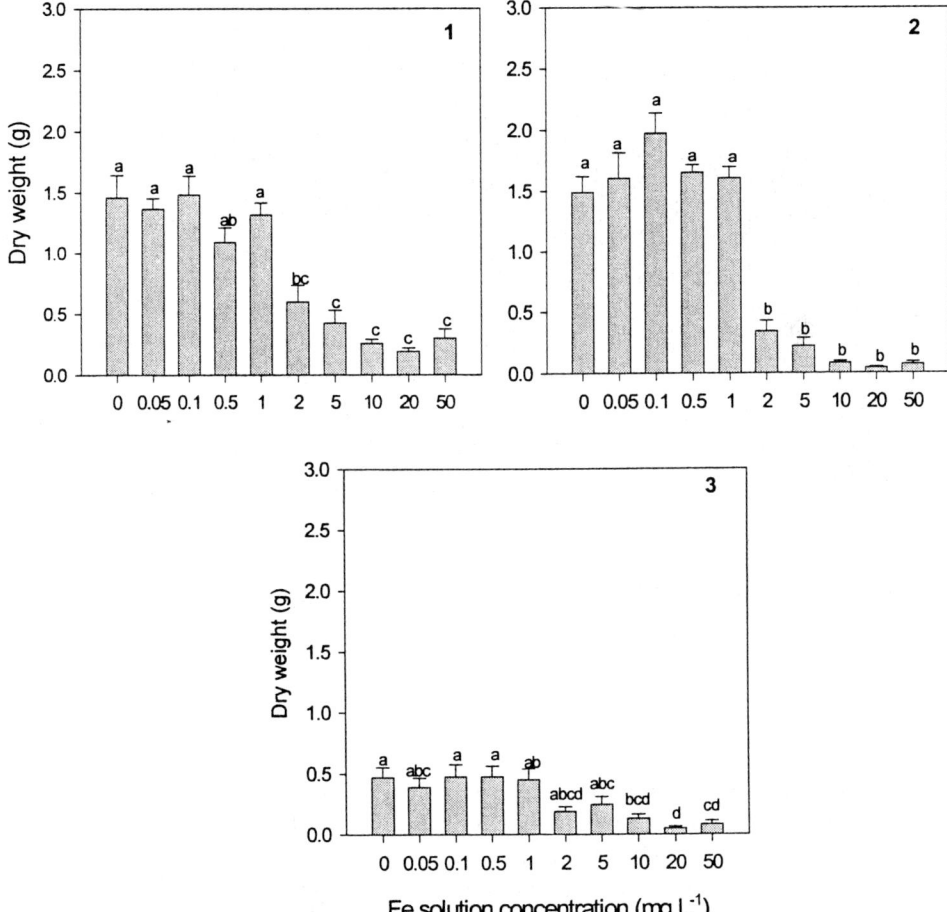

Figure 5.11: The influence of iron concentration in the water to the growth rate of seedlings of the common reed, *Phragmites australis*. The three plots show dry weight Fe concentrations in (1) shoots (2) roots and (3) rhizomes. (Different letters indicate a significant difference with respect to iron concentration at p < 0.05). As for Fe uptake (Figure 5.7) there appears to be an optimum around 1 mg.l^{-1}, with significant stunting above this concentration. This suggests that it is good practice to nurture seedlings at lower Fe concentrations if possible before subjecting them to elevated Fe. (Data courtesy of the European Commission's PIRAMID project (*www.piramid.org*), provided by Dr L.C. Batty).

Figure 5.12: An example of raising *Typha latifolia* directly from seed, in a newly-established wetland in the eastern USA. See text for discussion. (Photo: R.S. Hedin).

Generally speaking, harvesting of reeds is not practiced in aerobic mine water treatment wetlands, though there is no *a priori* reason why it should not undertaken in the unlikely event that a market exists for the reeds. (N.B. The same may not be true in the case of compost wetlands (see below), in which the seasonal addition of plant leaf debris to the substrate is generally considered desirable).

5.4.3.4 Compost Wetlands

Principles of compost wetlands:
In the evolution of passive treatment technologies for mine waters, aerobic wetlands were the first invention. It soon became clear that aerobic wetlands often serve to further lower the pH of net-acidic mine waters, due to the release of protons during the hydrolysis of Fe^{3+} and Al^{3+} (see Figure 5.2a). Those wetlands least prone to this problem were those in which substantial beds of compost had been incorporated to aid plant growth. Further investigation identified sulfate reduction and calcite dissolution in the compost as processes which counteracted acidification (Hedin *et al.*, 1988). Thus was the concept of compost wetlands which incorporate acid-neutralization processes born (see Kepler and McCleary, 1994; Hedin *et al.*, 1994). Compost wetlands (sometimes inaccurately referred to as 'anaerobic wetlands') superficially resemble aerobic wetlands, in that they comprise surface impoundments, generally vegetated with *Typha*, *Phragmites* and/or other wetland plants (Figure 5.13). They differ from aerobic wetlands in having a thick, anoxic substrate of saturated organic material ('compost'), within which bacterial sulfate reduction consumes acidity and/or generates bicarbonate alkalinity (see Sections 2.2.6 and 5.3.5). Sulfate reduction also removes some of the Fe and Zn as their sulfides (see Section 5.3.2, Equations 5.4 and 5.5 and Table 5.1), and the rise in pH contributes to the removal of Al as its hydroxide (Section 5.3.3) and Mn as its oxide or carbonate (Section 5.3.4).

The surprising thing about compost wetlands is that they work as well as they do. This is because much of the flow is evidently surficial, and would be anticipated to by-pass the reactive substrate. In practice, the substrate is continually exchanging solutes with the surface water by means of molecular diffusion, and by 'hyporheic flow', which refers to groundwater flow through protuberances in the upper surface of the organic substrate, in accordance with Darcy's Law (Equation 3.4). The interface between the organic substrate and the open water is therefore the locus of the most dynamic chemical processes which attenuate metals and acidity in compost wetlands. Coring through compost wetland substrates usually reveals that the steepest concentration gradients occur in this interface zone, which acts as a buffer between largely oxygenated surface waters and anoxic waters deep within the substrate (Batty *et al.*, 2001). Adding to the complexity of the processes in the interface zone of most compost wetlands is the presence of the root systems of wetland plants such as *Typha* and *Phragmites*, which locally maintain oxic conditions around their roots by leakage of oxygen, thus inhibiting the activity of obligate anaerobe bacteria. On the other hand, the root exudates of these same plants often contain precisely those organic compounds (such as acetate) which are the ultimate metabolites of sulfate reducing bacteria.

The net result of these various influences is a complex, temporally dynamic array of (bio)chemical reactions, some of which serve to improve water quality, while others may actually degrade water quality. At any one time, the interface zone may be truly anoxic, micro-aerobic (0 mg.l^{-1} < DO ≤ 0.5 mg.l^{-1}) or aerobic. Consequently, SO_4^{2-} may be under-

A

Shallow water with reeds / rushes over an anoxic, compost substrate

Influent

Effluent

B

Figure 5.13: Compost wetlands: (a) Generalized and simplified cross-section illustrating basic design concept (b) Photograph of a real compost wetland: Quaking Houses (Co Durham, UK). This system is described in detail in Section 5.5.2.2. The open water lies over a 0.5 m thick layer of composted municipal waste and composted cow manure. The figures on the right are standing on the central weir (the position of which is marked on Figure 5.22). (Photo: © University of Newcastle, by permission).

going reduction to HS⁻, or HS⁻ might be oxidizing to elemental sulfur (S^0) or even SO_4^{2-} once more. Iron may be precipitating as $Fe^{3+}(OH)_3$, or this compound may be reduced (consuming protons) to liberate Fe^{2+} to solution, which may then further react with HS⁻ to form iron sulfide (Equation 5.4) or with HCO_3^- to form siderite (Equation 5.5). Whether the water leaving the compost wetland is of better quality than the water which entered it depends on the relevant predominance of these competing redox reactions. Fortunately, the beneficial anaerobic reactions seem to predominate most of the time in the majority of systems. Nevertheless, compost wetlands are very finely balanced (e.g. Younger, 2000e), as illustrated by the fact that they rarely remove more than 20% of the sulfate load they receive (e.g. Jarvis, 2000).

In view of the uncertainties which attend water quality improvement in compost wetlands, it is probably best to restrict their use to circumstances in which alternative technologies such as RAPS (Section 5.4.3.5) or SFBs (5.4.4) are impractical. Thus, in most cases, compost wetlands are indicated where net-acidic mine waters require passive treatment, but there is insufficient topographic relief on the site to provide enough hydraulic head to force the water through the subsurface flow medium of a RAPS or SFB.

Design practice for compost wetlands:
Design practice for compost wetlands shadows that for aerobic wetlands in most details. The two substantial differences relate to the sizing of wetland area and the provision of sufficient space within the wetland area to install the compost substrate.

The sizing of compost wetlands is currently undertaken using Equation 5.21 with an R_A value of 3.5 to 7 g.d⁻¹.m⁻² of *total acidity* (expressed as $CaCO_3$ equivalent). As with sizing aerobic wetlands, the lower figure should be used where the receiving watercourse is particularly sensitive. Total acidity can be accurately calculated for most mine waters using Equation 5.12.

As yet, there is no alternative sizing model for compost wetlands. Although Jarvis (2000) considered the overall acidity removal reaction to be first order in total acidity, and used data obtained from three years of monitoring the Quaking Houses Wetland (Jarvis and Younger, 1999) to propose a K_1 value of 0.12 m.d⁻¹ for use in Equation 5.18 to describe the process, the underlying sulfate reduction is apparently zero order in sulfate (Willow and Cohen, 1998). It therefore seems unlikely that Equation 5.21 will be supplanted as a serious design model in the near future.

Since total acidity includes not only the proton acidity (H^+) but also the mineral acidity arising from dissolved metals such as Fe, Mn, Al and Zn (see Section 2.2.4, Table 2.5 and Equation 5.12), its use in Equation 5.21 for compost wetland makes a lot of sense. Nevertheless, experience shows that the various metals which contribute to the total acidity display differing tendencies to leave solution (e.g. Lamb *et al.*, 1998), so that a compost wetland will preferentially remove metals in the following order:

$$Al > Fe > Cu > Zn > Cd > Mn$$

With regard to the installation of the compost substrate, a thickness of about 0.5 m is recommended. There is no great virtue in installing a much thicker layer of compost, for at least three reasons:

- the thicker the compost layer, the more expensive the excavation and bunding and the more expensive the compost
- since much of the vertical transport of metals and sulfur species is by the very slow process of molecular diffusion, compost buried at great depth is unlikely to contribute greatly to the dynamics of the all-important interface zone, and
- the thicker the layer of compost, the more hazardous it is to anyone who (accidentally or deliberately) happens to stand in the wetland.

The compost should be installed by loose tipping. Avoid compacting the substrate (e.g. by driving over it in construction vehicles), so as to maintain as high a permeability as possible. (Hydraulic conductivities in the range 0.01 to 1 m.d^{-1} should be achievable in this manner). Ridges and furrows (with amplitudes of a few cm) oriented perpendicular to the future flow direction will help to baffle surface flow to some degree, and will also promote the occurrence of hyporheic flow.

In selecting a suitable material of the substrate, a 'compost' or similar organic waste material is sought which has a reasonable combination of the following properties:

(i)   It should be of a sufficiently fibrous nature that it will retain a reasonable permeability when saturated and self-loaded to a depth of 0.5 m.
(ii)   It should contain some sulfate-reducing bacteria (SRB). (This ought to be easily achievable if the compost contains at least some mammalian fecal material).
(iii) It should preferably be alkaline in nature, or at least not be prone to release strong organic acids to solution.
(iv)   It should not contain potentially harmful viruses (e.g. avoid using cattle manure from farms affected by diseases such as BSE or Foot-and-Mouth disease).

Organic materials measuring up to the above criteria are widespread, and usually available for no more than the cost of haulage. If alternative media are available at equal cost, it may be worth performing laboratory microcosm tests in order to determine which of the media best promotes sulfate reduction (e.g. Younger et al., 1997; Jarvis, 2000). Examples of materials which have successfully been used in compost wetlands include:

- Spent mushroom compost (e.g. Hedin et al., 1994), an alkaline organic material which is widely available in the eastern USA. $CaCO_3$ accumulates in mushroom compost during the composting process, by reaction of Ca^{2+} released from gypsum (a standard ingredient in mushroom compost production) with biogenic carbonate (resulting from very high CO_2 partial pressures which develop within the compost pile).
- Horse manure and straw (e.g. Younger et al., 1997)
- Composted conifer bark mulch (Younger, 1998d), which has excellent permeability characteristics and provides a good long-term source of ligno-cellulose materials from which SRB metabolites can be released by hydrolysis; however it needs to be mixed with manure to provide sufficient SRB for substrate colonization to occur
- Cow manure and straw (e.g. Cohen and Staub, 1992)
- Composted municipal waste (e.g. Jarvis and Younger, 1999), particularly vegetable wastes and 'green waste' (i.e. grass and hedge cuttings etc)
- Sewage sludge cake (D Laine, personal communication, 2000), and

- Organic wastes arising from paper making (e.g. Chang *et al.*, 2000; D Laine, *personal communication*, 2000).

As a counter-example, during construction of the Pelenna III wetlands in South Wales (Younger, 1998d), the contractor proposed chicken litter as an inexpensive alternative to cow manure and composted bark mulch. On testing the chicken litter in laboratory microcosms, however, it was found to *lower* the pH of the mine water placed into contact with it, presumably through release of strong organic acids.

In areas where compost wetland technology is new, regulatory resistance to the use of such materials in a water body may well be encountered. This is understandable, given that most pollution control officers spend much of their time trying to *prevent* the introduction of such materials to watercourses from agricultural premises and sewage works. (Accidental releases of such organic wastes to watercourses are often devastating due to their very high biochemical oxygen demands (BOD), which strip oxygen from the water and cause fish to suffocate). Where such hesitancy is encountered, the best way to proceed is to build a pilot-scale demonstration wetland prior to installing a full-scale system. Monitoring of the pilot wetland will demonstrate that the more labile organic substances are washed from the wetland in one or two weeks (e.g. Younger *et al.*, 1997), leaving behind the more refractory organic matter which slowly yields SRB metabolites by *in situ* hydrolyis. (The labile organic wastes which are flushed out of the wetland in the first few weeks account for most of the BOD, since this parameter is measured over five days and therefore relates only to readily available organic materials). Given that compost wetlands tend to be constructed to improve the quality of streams which are currently heavily polluted by acidic mine drainage, two weeks of elevated BOD inputs are not likely to be viewed with alarm, as it is impossible to kill aquatic animals which are already dead!

Since compost wetlands rely on water-substrate interactions for their key water treatment processes, wetland plants are not strictly necessary. Many compost wetlands work well from their first week of commissioning, when there may not be a single plant in the system. However, in the longer-term, emergent wetland plants can serve to counteract the tendency to channelization which besets all shallow water ponds, and through seasonal die-back they potentially provide a long-term source of organic matter to replenish that which is gradually consumed by SRB metabolism in the substrate (although it remains to be established whether this process is as efficacious as has generally been assumed). On the negative side of the balance sheet, the root systems of wetland plants tend to oxygenate the surrounding substrate, potentially turning anoxic zones into micro-aerobic zones. In some cases, these processes have been found to hinder acidity removal efficiency (e.g. Gusek, 2000), though there are many well-vegetated compost wetlands in which acidity removal remains substantial (e.g. Jarvis and Younger, 1999). Indeed, anaerobic and aerobic processes clearly occur in close proximity to each other in many natural wetland systems: wade out into almost any mature stand of *Phragmites australis* or *Typha latifolia* and as your footsteps churn the substrate you will see that it is black with sulfide solids (just below the surface), and you will smell the odor of H_2S gas released by the disturbance. Eventually, supposedly aerobic constructed wetlands develop a thick, partially anoxic substrate also. Hence, compost wetlands may be viewed as precocious wetlands, mature beyond their years by dint of being fitted with a thick organic substrate from the outset, without having to wait years for it to develop naturally.

In finalizing a design for a compost wetland, it is necessary to ensure that the substrate remains flooded at all times, to avoid oxidation of the sulfide solids which accumulate in the compost. The sulfides which initially precipitate within wetlands are very fine-grained and X-ray amorphous, and as such belong to the most highly reactive category of sulfides (cf Caruccio, 1975). This fact has sometimes been seized upon by skeptics as an argument against the use of compost wetlands to treat acidic mine waters. There are at least four reasons for regarding this argument as too alarmist:

(i)   The maintenance of permanently flooded conditions is not especially difficult to achieve in most climate zones (see Hawke and José, 1996, for practical advice).

(ii)   The compost substrate, being a soft material only 0.5 m in thickness, is highly amenable to excavation for off-site disposal, should this ever be deemed necessary.

(iii) If the wetland is to be decommissioned *in situ*, a long-term water cover or dry cover may be installed in the same way as for tailings and waste rock piles (see Section 3.12.2).

(iv)   Even in the worst-case scenario, (where (i) through (iii) above all fail to be implemented), the total area of most compost wetlands is sufficiently small compared to the mine wastes drained by the discharge from which they have received their metals contents that they represent a tiny rainfall catchment, and hence will be unlikely to yield large volumes of leachate (provided external sources of surface water and ground water do not flow through them).

It is nevertheless clear that a sound compost wetland design ought to incorporate an outline decommissioning strategy which takes the above factors into account.

Once your system is constructed and water is added, be sure to organize your opening ceremony early: in this way you can take advantage of the 'honeymoon period' (McGinness *et al.*, 1997), typically two to three weeks in duration, during which the treatment performance of the wetland exceeds both previous expectations and future experiences. The explanation for this frequently-observed phenomenon is the vast sorptive capacity of the organic matter in the substrate, which mops up metals (and therefore attenuates acidity) comprehensively until all surface sites are saturated, after which desorption slowly returns a large proportion of these metals to solution (Machemer, and Wildeman, 1992; Cohen and Staub, 1992). Only after the honeymoon period is over will the long-term rate of acidity removal by sulfate reduction processes become clearly evident.

5.4.3.5 *Reducing and alkalinity producing systems (RAPS)*

Principles of RAPS:
These systems were invented in Clarion County, Pennsylvania, in the early 1990s (Kepler and McCleary, 1994), as a response to the major shortcoming of ALDs, i.e. their restricted applicability to waters with less than 1 mg.l^{-1} of DO, Fe^{3+} and Al^{3+}. Not only does this restrict the raw waters to which ALDs might be applied, but it also means that an ALD can only be used as the *first step* in a passive system, since waters which have been oxidized in a sedimentation pond or aerobic wetland will certainly have more than 1 mg.l^{-1} DO. But what if the water is still acidic after all of the alkalinity from the ALD has been used up in neutralizing the protons released from the hydrolysis of Al^{3+} and/or Fe^{3+}? This is where

the RAPS concept originated: as a means of incorporating *successive alkalinity producing systems* within an overall passive treatment plant. (The term *'successive alkalinity producing systems'* (SAPS) was the one first coined for this technology, by Kepler and McCleary (1994). This name arises from the fact that the passive treatment unit contains alternating alkalinity generating and aerobic metal-precipitating units. By arranging several such units in series, it is theoretically possible to treat any level of acidity. However, in view of the scarcity in practice of passive treatment plants which contain more than one 'SAPS', the more descriptive name of RAPS has since been introduced by G Watzlaf and co-workers).

In concept, a RAPS unit is an ALD overlain by a compost bed, the purpose of which is to strip DO from the water, and to reduce any Fe^{3+} to Fe^{2+} prior to the water contacting the limestone aggregate (Figure 5.14). In this way, ALD-type treatment can be applied to a wider range of waters. While the inventors of RAPS (Kepler and McCleary, 1994) were clearly fully aware of the likelihood of bacterial sulfate reduction occurring in the compost bed, a robust, conservative design philosophy for RAPS can be developed by making the naïve assumption that the compost bed will *only* remove DO and reduce Fe^{3+} to Fe^{2+}. In reality, the compost layer in the RAPS will also remove substantial amounts of Fe as its sulfide (and Al as its hydroxide), but if we ignore these processes, our design will err on the side of caution, which is comforting in the context of a technology which is (like compost wetlands) highly complex and prone to behave in a manner not fully amenable to deterministic predictions.

Because RAPS constrain all of the water to contact the compost and limestone, they are generally much more efficient in treatment terms than simple compost wetlands, and hence require far smaller land areas to provide the same level of treatment (see Kepler and McCleary, 1994; Younger *et al.*, 1997 and Younger, 1998d). Hence, wherever construction of a RAPS is feasible, it is the preferred option for the treatment of acidic mine waters. However, RAPS systems require substantial driving head and freeboard, and hence are unlikely to be a practical proposition where there is much less than about 5m of topographic relief on the proposed treatment plant site.

Aluminum removal in RAPS systems presumably occurs mainly by hydroxide precipitation, both in the compost substrate and in the limestone aggregate layer. In the latter, hydroxide accumulation can lead to clogging of the system, as in ALDs. The fact that this problem is not usually reported from RAPS suggests that much of the aluminum hydroxide does in fact leave solution in the compost substrate.

Design practice for RAPS:
In essence the design of RAPS systems should follow the same logic as for ALDs, i.e. to size the limestone aggregate bed for a minimum retention time of about 14 hours (see Section 5.4.2.2). Once the total volume of the limestone bed has been determined, the vertical and horizontal dimensions must be established. We generally seek a vertical thickness of at least 0.5 m or more for the aggregate layer, and fix the length and width accordingly, taking into account the land area available and the site layout. The limestone is best emplaced in a basin of compacted clay; if this is not possible an artificial liner can be used.

The installation of drainage pipes within the limestone aggregate bed of a RAPS is controversial. Some authorities argue that a dendritic network of perforated drain pipes should be installed throughout the lowermost part of the limestone aggregate layer (e.g. Watzlaf *et al.*, 2000; Demchak *et al.*, 2001), since this ought to favor subsurface flow down

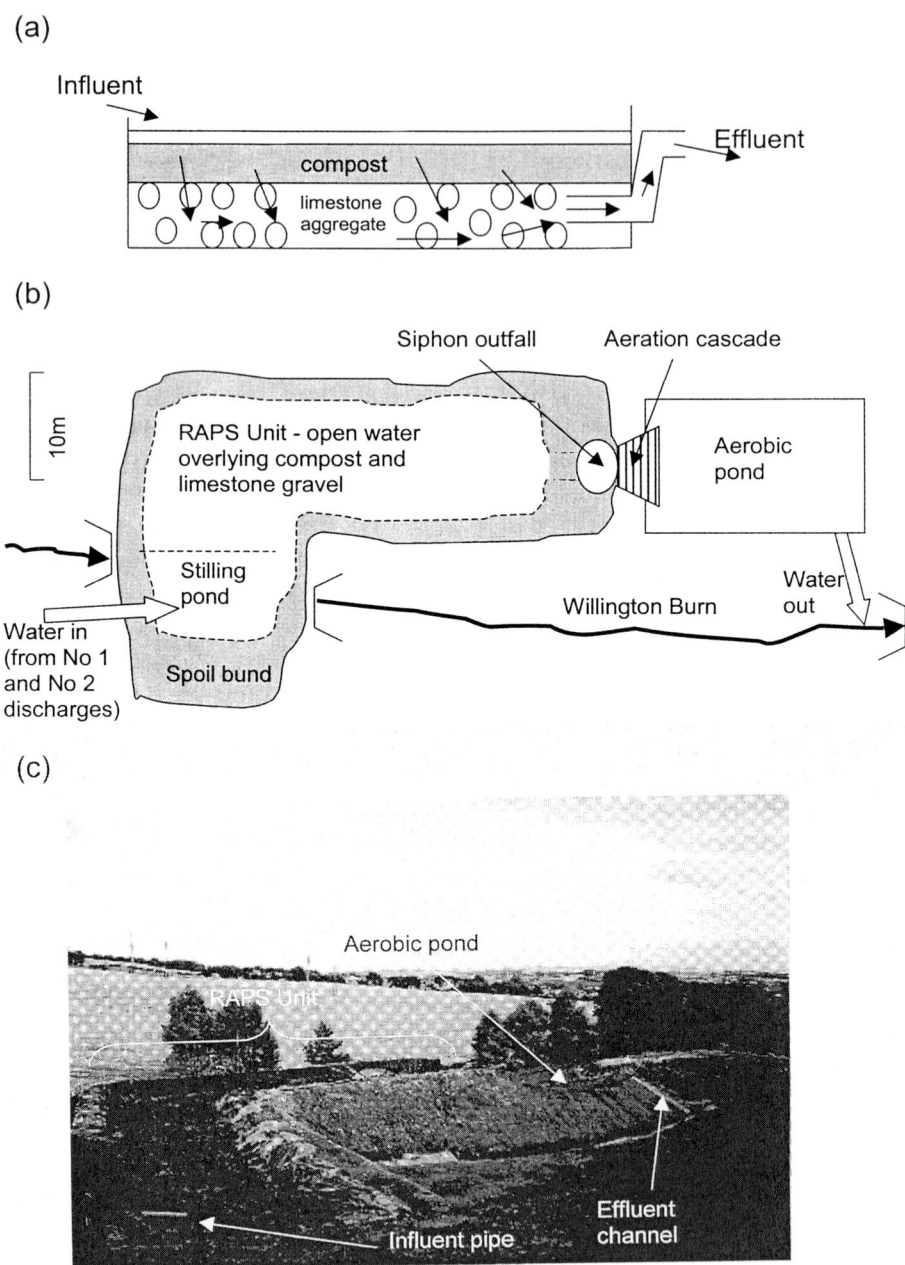

Figure 5.14: Reducing and alkalinity producing systems (RAPS) (a) Generalized and simplified cross-section illustrating the basic design concept (b) Sketch plan of the Bowden Close Pilot RAPS unit (Co Durham, UK) (c) Photograph of Bowden Close Pilot RAPS at the time of commissioning (September 1999), just before introduction of acid mine water. (Photo: P.L. Younger, with the permission of the site owners, Durham County Council).

through all parts of the limestone bed. On the other hand, the presence of pipes has the tendency to promote short-circuiting of the limestone, with water tending to sit in plastic pipes rather than pores in the limestone aggregate pores. Hence some RAPS designs include a short length of pipe-work in the final few metres of the aggregate bed only, where all drainage must ultimately be gathered together for passage through the main dam, and into the 'penstock' (the pipe-work which controls the head in the entire system).

Once the limestone unit has been sized, we allow for a minimum of 0.5 m of compost substrate completely overlying it, so that water entering the RAPS can only pass into the limestone aggregate layer after flowing downwards through the compost layer. The required properties for the compost substrate are the same as for compost wetlands (Section 5.4.3.4). Wherever possible, the retaining walls of the lagoon should slope gently (no steeper than 1:3, as in the case of other wetlands; see 5.4.3.3) for reasons of safety and ecological habitat provision. It is most important that the lagoon have as much freeboard as possible (and certainly not less than about 1.5 m in any circumstance) so that any clogging of the underlying compost and limestone can be auto-compensated by a build up of the extra head required to ensure continued flow of all of the mine water through the subsurface. Where the mine water source is highly variable in flow rate (which is commonly the case for spoil heap discharges), steps should be taken to maximize the flow coming into the system by means of storm overflow devices; for peace of mind, an emergency flood spill-way should also be fitted to the RAPS lagoon, so that head cannot rise to such levels that over-topping and erosion of the lagoon walls could occur.

The most important task in RAPS design relates to the impoundment of the lagoon and the satisfactory sealing of the penstock pipe-work (Figure 5.15). RAPS units inevitably violate one of the 'golden rules' of small dam design: don't have any pipe-work coming through the dam! For a RAPS unit to work, we have no option than to collect water at the base of the limestone unit and cause it to flow out at a higher level (determined by the crest of the penstock outfall). Making sure that the penstock pipe-work passes through the dam with a water-tight seal on its exterior can be very demanding, depending on the material available to construct the dam. However, failure to achieve a water-tight seal for this pipe-work will result in leakage, erosion and even complete failure of the dam, with potentially grave consequences. In many cases, adequate sealing of the pipe-work as it passes through the dam will not be possible without resort to the use of concrete (into which the clay base or artificial liner of the main lagoon must also be keyed very effi-ciently). On the downstream side of the dam, the penstock pipework must be arranged so as to maintain a water level in the lagoon above the surface of the compost layer. At its simplest, this can be achieved by making the pipework rise to a suitable level (ideally cresting at a level 0.5 m to 1 m above the surface of the compost, depending on the desired depth of standing water in the RAPS unit). This can be achieved using an adjustable elbow-bend in the pipe (Figure 5.15), which can be raised or lowered until the right head is achieved, or by including a pressure-sustaining valve in the pipeline.

As with all wetlands, the depth of standing water will exercise a major control on the presence and vitality of wetland plants. As wetland plants are in no way necessary for the success of a RAPS unit, it is perfectly reasonable to specify a depth of water which will preclude growth of most plants (> 1 m). On the other hand, if you would like to the RAPS lagoon to be vegetated (for aesthetic and / or habitat-creation reasons), then adjust the penstock in accordance with the guidelines for water depth given in Section 5.4.3.3.

Figure 5.15: Adjustable-angle elbow joints forming a simple but effective means of controlling water levels in a constructed wetland. If the angle of the decanting pipe is changed, the water level in the wetland upstream must also change to maintain the same flow rate. (Photo: P.L. Younger, with the permission of the site owners, Northumbrian Water Ltd).

A RAPS unit constructed in accordance with the guidelines given above should be capable of yielding between 150 and 300 mg.l^{-1} (as $CaCO_3$ equivalent) of alkalinity (or, put another way, of neutralizing the same amounts of acidity). If the mine water in question contains more net-acidity than this, several RAPS can be used in series, with sedimentation ponds between them to precipitate ferric hydroxides.

As noted above, aluminum hydroxide accumulation in the limestone aggregate layer is potentially a problem in RAPS. To counteract this problem, Kepler and McCleary (1997) proposed the incorporation of a 'scour pipe' in the limestone layer, which can be periodically opened to rapidly drain the pores of the aggregate layer, entraining and removing aluminum hydroxides in the process. This flushing facility is recommended wherever RAPS receive waters with more than 10 mg.l^{-1} Al (Kepler and McCleary, 1997; Demchak et al., 2001).

How do RAPS compare in terms of treatment efficiency when compared with conventional compost wetlands? If we normalize the vertical dimensions of the limestone and compost layers at 0.5 m each, then we can compare the overall performance of RAPS units with compost wetlands using an equivalent areal removal rate (RA) for acidity (e.g. Demchak et al., 2001). From studies of several long-established RAPS in Appalachia, Watzlaf et al. (2000) deduced an equivalent acidity R_A of 30 to 50 g.d^{-1}.m^{-2} (where the acidity is expressed as $CaCO_3$ equivalent). By comparison with the equivalent R_A values for compost wetlands (Section 5.4.3.4) this implies that a RAPS unit can be accommodated in a land area only 15 to 20% of that needed for a compost wetland. This tallies with our experience on the ground, although it has to be re-emphasized that a compost wetland can be constructed on a site with only a few centimetres of topographic relief, rather than the several metres of relief needed for RAPS installation.

5.4.4 SUBSURFACE-FLOW BACTERIAL SULFATE REDUCTION SYSTEMS (SFBS)

5.4.4.1 In situ permeable reactive barriers (PRBs)

Principles of PRBs:
Thus far in this book, our attention has mainly been focused on the problems caused by discrete discharges of polluted mine waters to surface water courses. In some cases, the principal cause for concern is migration of contaminated mine water by subsurface flow in an aquifer which is used for water supply purposes. In others, discharge to surface water courses is still the main concern, but it occurs in a diffuse manner by groundwater upwelling through a long length of streambed. In these cases, *in situ* treatment of polluted groundwater is the logical response.

In situ remediation of polluted groundwater is a vast field of research and engineering practice in its own right (e.g. Fetter, 1999). Nevertheless, most developments in that field to date have focused on the biodegradation of persistent organic micro-pollutants, and have thus led to practices which are largely inapplicable to typical mine water contaminants. However, one of the principal techniques of groundwater remediation is rapidly finding wide application at mining and mineral processing sites where the local groundwater has become contaminated: the use of permeable reactive barriers.

The concept of a Permeable Reactive Barrier (PRB) is very simple (Figure 5.16): a permeable medium of geochemically appropriate material is placed in the path of the polluted groundwater in the form of a 'barrier' across the flow path. As the groundwater flows through the barrier, beneficial (bio)chemical reactions take place which result in an overall improvement in water quality, so that the groundwater flowing out of the down-gradient face of the barrier is significantly less polluted than that which entered.

Unfortunately, the simplicity of PRBs ends at this point, for both the construction and long-term deployment of PRBs are beset with considerable uncertainties, which remain topics of active research. The key issues which must be addressed in any PRB design are nonetheless clear, and include:

(i)   the likelihood that water will flow *through* the PRB rather than by-passing it
(ii)   the degree to which geochemical reactions within the PRB will improve water quality, and
(iii)   the frequency with which the reactive medium must be replenished (or completely removed and replaced) to prevent either:
    • a loss of performance due to exhaustion of reactive components and/or
    • destruction of permeability by the clogging of pores by minerals which precipitate within the PRB substrate

In selecting a suitable substrate for a PRB to treat mine waters, the usual choice is a compost-based medium which will promote bacterial sulfate reduction (e.g. Waybrant *et al.*, 1998). As such, the desirable properties of the medium are essentially the same as in the case of compost wetlands and RAPS (see Sections 5.4.3.4 and 5.4.3.5), albeit with particular attention being paid to the permeability of the substrate, as we will see below.

Although most PRB applications have been (and are likely to continue to be) for shallow aquifers contaminated with sulfide oxidation products, a recent extension of the technology to abandoned deep mine systems has been described by Canty (2000). Organic substrates were emplaced in flooded and overflowing underground workings of the abandoned Lilly/Orphan Boy Mine in Montana, USA. Emplacement was achieved by two methods:

(i)   Suspension on platforms in the main shaft, and
(ii)   Tipping down two boreholes drilled into the main drainage adit.

The results from the first four years of operation have been very encouraging, with pH being raised consistently from 3 to 7, and removal efficiencies seasonally reaching > 99% for Cd, Cu, Zn and Al (Canty, 2000). Lower removal efficiencies corresponded to periods of high flow when retention times in the PRB zones must have been very brief. Interestingly, the technology appears to have resulted in greater mobility of Fe and As, presumably due to reductive dissolution of ferrihydrite which had previously accumulated in the mine. (The release of As suggests it was present as a sorbed/co-precipitated element within the ferrihydrite). Clearly, further work will be necessary before generic design rules can be proffered for this deep mine application of the PRB concept.

Design Practice for PRBs:
The prior investigation of the polluted body of groundwater is an essential pre-requisite

(a)

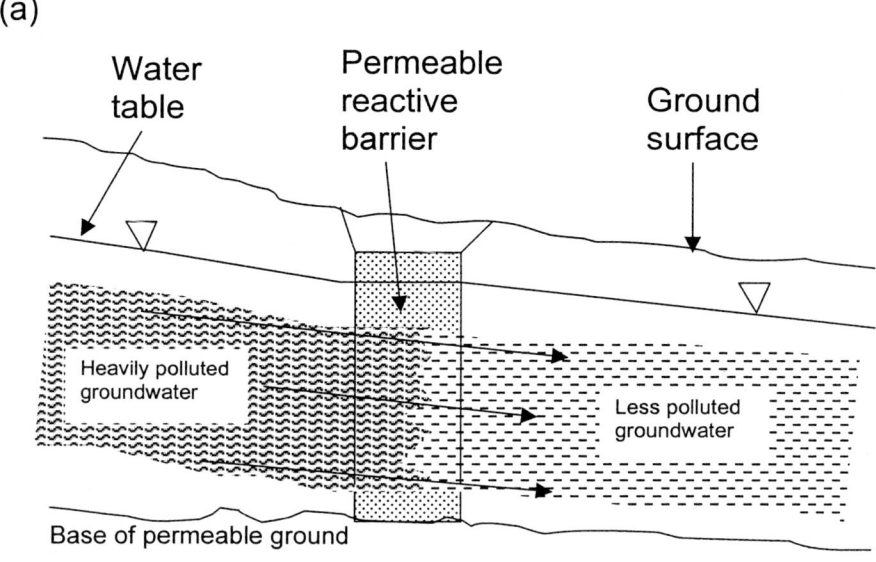

(b)

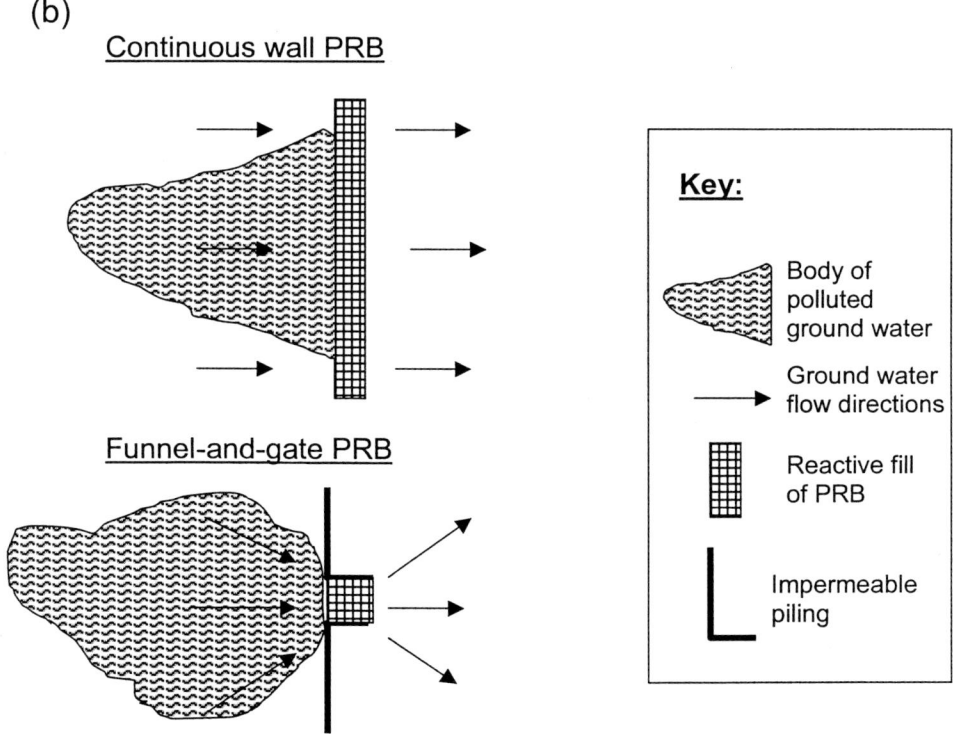

Figure 5.16: Permeable reactive barriers (PRBs) (a) Generalized and simplified cross-section illustrating the basic design concept (b) schematic plan layouts for the two most common PRB configurations.

for robust PRB design. Relevant parameters which must be determined (preferably over several seasonal cycles) include:

(i)   The hydraulic conductivity (or transmissivity) of the aquifer (see Section 3.3.3), which are best determined by test pumping of boreholes (using well-established techniques described by numerous authors, such as Freeze and Cherry, 1979, Fetter, 1994, and Kruseman and de Ridder, 1990).

(ii)   The hydraulic head distribution in the aquifer (measured using boreholes, as described in Section 3.4.4), from which groundwater flow directions can be deduced and groundwater velocities can be calculated (when head gradients are multiplied by hydraulic conductivity values, as per Equations 3.5 and 3.23).

(iii)   Groundwater quality as indicated by borehole samples, from which the extent of the 'plume' of polluted groundwater may be mapped. Pollutant concentrations are likely to be highest in close proximity to likely 'source areas' (such as bodies of flooded workings, spoil heaps or tailings ponds), and to decline down the direction of groundwater flow, due to mixing and reaction with unpolluted groundwater (cf. Equation 3.25 and corresponding discussion in Section 3.3.3).

Once the location and direction of flow of the body of contaminated groundwater is known, and the hydraulic conductivity of the host aquifer has been determined, substrate selection can begin in earnest. The first step in this process is to prepare an inventory of suitable organic materials within economic haulage distance of the site. Critical issues here will be the availability of sufficient quantities to match the volume of the PRB. In some areas, there may be so little organic material available locally that little choice may exist: all available material will be required. Usually, some element of choice will exist, and alternative media can be selected on the basis of an optimal combination of:

• their capacity for promoting sulfate reduction and metal removal, and
• their hydraulic conductivity.

Both of these properties can be evaluated using laboratory tests; suitable protocols for such tests are described by several authors (Waybrant et al., 1998; Alcolea et al., 2001; Amos and Younger, 2002). Laboratory microcosm tests, in which the organic material is placed in contact with a large sample of the polluted groundwater for a month or more, will allow the propensity for sulfate reduction to be evaluated. Lab permeametry can be used to evaluate the hydraulic conductivities of the various alternative barrier fill materials, as long as care is taken to impose weights on the materials in the permeameter to reproduce the effect of compaction by self-loading which will occur at depth in the field PRB (Amos and Younger, 2002). In many cases it will be necessary to create a mixture of organic material with some proportion of mineral clasts in order to obtain a suitable combination of reactivity and hydraulic conductivity (K). Recent field and modeling studies have revealed that the target hydraulic conductivity for a PRB substrate ought to be 5 to 10 times greater than that of the enclosing aquifer (Gavaskar et al., 1998). Examples of mixtures selected in this manner for use in mine water PRB installations include:

Site	Mixture chosen	Mean K (m.d^{-1})	References
Sudbury, Canada	20% municipal compost, 20% leaf mulch, 9% wood chips, 50% pea gravel, 1% limestone	345	Benner et al. (1997); Waybrant et al. (1998)
Vancouver, Canada	70% pea gravel, 30% compost	130	McGregor et al. (1999)
Shilbottle UK	50% limestone aggregate, 25% cattle slurry screenings, 25% green waste compost	6	Amos and Younger (2002)
Aznalcóllar, Spain	50% limestone aggregate, 40% compost, 10% river sediment	12	Carrera et al. (2001)

The installation of PRBs has been described in detail by Gavaskar et al. (1998). Wherever possible, the reactive media and any cut-off features (funnel walls) should be keyed into the underlying aquitard (i.e. a lower permeability layer which underlies the aquifer in question). This is achieved by digging 0.5 to 1 m into the underlying aquitard and installing the media/cut-off material as appropriate.

There are two PRB layouts in common use (Figure 5.16b):

(i)   Continuous barriers, in which the reactive material is installed across the full width of the groundwater contaminant plume (e.g. Benner et al., 1997), and

(ii) Funnel-and-gate barriers (Starr and Cherry, 1994), in which impermeable 'funnel walls' divert polluted groundwater from the full width of the plume into a 'gate' filled with a permeable reactive medium.

Funnel-and-gate designs are especially suitable where suitable reactive media are scarce and/or expensive, since they result in the installation of far smaller volumes of media than continuous barrier designs (Star and Cherry, 1994). Detailed modeling studies have revealed that funnel-and-gate PRBs are best designed with the funnel walls at 180° to the direction of groundwater flow (and the therefore to the long axis of the gate section); angled funnel walls are inefficient at capturing the bulk of the contaminated water (Gavaskar et al., 1998). Funnel walls are typically installed as sheet piling, but can also be formed from grout curtains (i.e. vertical barriers of impermeable, cement-based materials). The latter option is likely to be uneconomic in mine water settings, due to the need to use expensive sulfate-resistant cements. Although funnel-and-gate systems are increasingly popular for the treatment of synthetic organic compounds in mine waters (Gavaskar et al., 1996), none of the six full-scale PRBs known to have been installed for mine water treatment to date have used a funnel-and-gate configuration (Benner et al., 1997, 1999; McGregor et al., 1999; Laine, 2000; Carrera et al., 2001; Alcolea et al., 2001; Amos and Younger, 2002). This probably reflects the low costs of the kinds of organic media typically used in mine water PRBs.

Continuous barriers can be installed by a range of techniques. The simplest and most common is 'simultaneous cut-and-fill', in which one digging machine excavates a trench while another immediately backfills the cut with the reactive medium. The key advantage of this approach is that it avoids the need to temporarily support the side walls of the excavation. However, it is only likely to be feasible for relatively shallow PRB installations in cohesive soils. In practice, it has been found that simultaneous cut-and-fill without temporary wall support is feasible for barriers up to about 6 m depth, i.e. within easy reach of a conventional backhoe (Benner *et al.*, 1997; Laine, 2000). For deeper barriers, or for barriers excavated in non-cohesive aquifer sediments, temporary support of the excavation walls will be necessary. This has been achieved in recently-constructed mine water PRBs using linked steel sheet piling (to 8 m depth in the Aznalcóllar PRB, Spain; Alcolea *et al.*, 2001) and with dense, low viscosity, biodegradable support fluids (to 6.7 m depth in a PRB in Vancouver; McGregor *et al.*, 1999).

Where pollution from mine sites penetrates very deep aquifers, PRBs become more and more difficult to implement. The lower limit for excavation with a clamshell (i.e. a crane-mounted bucket scoop) is around 60 m (Gavaskar *et al.*, 1998), and wall support is likely to be both necessary and prohibitively expensive well before this depth is reached. Other alternatives included drilling lines of closely-spaced large-diameter boreholes and filling them with reactive media. This is by no means a cheap option either; for instance, a single 0.9 m diameter borehole to 200 m depth can be expected to cost in excess of $350K at 2001 prices. Other alternatives, none of which are likely to be cheap, are reviewed by Gavaskar *et al.* (1998).

There are as yet no established rules for sizing the width of PRBs in the direction of flow (i.e. the length of the groundwater flowpath through the reactive media). Systems installed to date have varied from 1.4 m to 4 m in width (Carrera *et al.*, 2001; McGregor *et al.*, 1999). Analysis of data presented by various authors indicates that groundwater residence times in successful mine water PRBs vary between 3 days (McGregor *et al.*, 1999), through about 12 days (Alcolea *et al.*, 2001) to a maximum of 90 days (Benner *et al.*, 2000). The PRBs with the longer residence times are not notably more efficient than that with the shortest residence time, suggesting that even three days is a sufficient residence time for most purposes. As will be seen in the following section, this is consistent with the findings of Cohen and Staub (1992), who found that residence times in the range 1.7 to 4 days were sufficient to ensure metal removal efficiencies of more than 98% in anaerobic reactors used to treat surface mine water discharges. In the light of these considerations, it is recommended that the width of a mine water PRB be designed as follows:

(i) Determine the hydraulic gradient in the aquifer in question (as described above)
(ii) Using this value, and the hydraulic conductivity and porosity of the particular reactive medium to be used in your PRB (determined by laboratory investigations; Waybrant *et al.*, 1998; Amos and Younger, 2002), calculate the average linear velocity of groundwater flow through the PRB (V_{PRB}) using Equations 3.5 and 3.23.
(iii) With V_{PRB} expressed in units of m.d^{-1}, calculate the required width of the PRB (W_{PRB}) in metres as follows:

$$W_{PRB} = \frac{V_{PRB}}{\varpi} \tag{5.25}$$

Where the factor $\varpi \geq 2$. In practice, as high a value of ϖ should be selected as possible, since the wider the PRB, the less prone will it be to bypass flows (Benner *et al.*, 2001). In reality ϖ values in the range 2–6 ought to be attainable at a reasonable cost for most sites; values in excess of 10 are highly unlikely to be economically feasible, due to the costs of materials, construction or both. However, if it proves difficult to accommodate a PRB on a given site without setting ϖ to be somewhat less than 2, then the PRB is not likely to function adequately, and thus may not be a cost-effective solution.

5.4.4.2 Anaerobic reactors for surface mine water discharges

Principles of anaerobic reactors for mine water discharges:
In most mine waters, there is a huge molar excess of SO_4^{2-} over dissolved metals (e.g. Younger, 2000b), mainly because there are many reactions which remove metals from solution in oxidizing environments, but very few which remove sulfate to any great extent. An important consequence of the molar excess of SO_4^{2-} is that if one desires to remove metals from solution by reacting them with sulfide (HS^-) (in a reaction such as that shown in Equation 5.4) then only a small fraction of the total SO_4^{2-} need be reduced to supply enough moles of HS^- to satisfy the demand from all of the relevant metals. This is the principal reason why compost wetlands are effective at removing acidity from mine waters, even though they tend to reductively remove only a small percentage of the SO_4^{2-} which they are fed (typically < 20%). If removal of metals is the only objective of treatment, then a compost wetland or RAPS may be all that is required. However, in arid and semi-arid areas, the total dissolved solids content of mine water is often more of a problem than the metals content (e.g. Boshoff, 1999; Smit, 1999). Sulfate typically accounts for a major proportion of the total dissolved solids, such that removal of sulfate is tantamount to desalination for many mine waters (e.g. Boshoff, 1999).

Anaerobic reactors have been under investigation for the last decade as a possible technology for the wholesale removal of sulfate and metals from contaminated mine waters. Investigations to date have been undertaken across a range of scales, from numerous laboratory column experiments (e.g. Hammack *et al.*, 1994; Christensen *et al.*, 1996; Elliott *et al.*, 1998; Willow and Cohen, 1998; Thomas *et al.*, 1999; Tsukamoto and Miller, 1999; Chang *et al.*, 2000; Johnson *et al.*, 2000), through several notable pilot-scale field reactors in Colorado (e.g. Cohen and Staub, 1992; Machemer and Wildeman, 1992; Staub and Cohen, 1992; Cohen, 1996) and South Africa (Pulles, 2000) to a very limited number of full-scale applications (e.g. Gusek, 1998, 2000; Gusek *et al.*, 2000; Zaluski *et al.*, 2000).

Key issues which emerge from these studies include:

(a) The need to identify a suitable source of carbon to sustain the bacteria in the long-term. (see Section 5.4.3.4).
(b) The need to arrange the hydraulics such that the water is heavily insulated from atmospheric oxygen, so that an Eh < –150 mV can develop and be sustained (Thomas *et al.*, 1999). Such deeply reducing conditions ought to be sufficient to assure an adequate SO_4^{2-} removal rate.
(c) Tentative design values for SO_4^{2-} reduction in suitably-constructed systems range from 300 millimoles per cubic metre per day ($mM.m^{-3}.d^{-1}$) (Gusek, 1998; Lamb *et al.*, 1998) to around 800 $mM.m^{-3}.d^{-1}$ (Willow and Cohen, 1998).

(d) The need to arrange the hydraulics of the system such that adequate retention times are realized to achieve goals (b) and (c) above, without inducing clogging at one extreme (e.g. Gusek *et al.*, 2000) or short-circuiting of flow at the other (e.g. Cohen and Staub, 1992; Cohen, 1996).
(e) The need to quantify the rate of consumption of organic matter, as a basis for planning its augmentation (either by subsurface injection of liquid organic matter (e.g. Tsukamoto and Miller, 1999) or by simple addition to the top of the reactor).
(f) How to control re-oxidation of HS⁻ upon leaving the reactor so that it halts at elemental sulfur (S^o) rather than continuing to SO_4^{2-} (which implies proton release and thus a possible lowering of pH).

All of these issues (and others) remain subjects of considerable ongoing research. Although guidelines for the construction and use of such systems have been published (Cohen and Staub, 1992; Wildeman *et al.*, 1993; Cohen, 1996) it is fair to say that there is as yet little in the way of consensus on design criteria. Hence the comments given below must be viewed with caution, as a necessarily subjective distillation of the findings of recent research.

Design practice for anaerobic reactors for surface mine water discharges:
One of the first issues to resolve in an anaerobic reactor design is the direction of flow, from among the three obvious options of:

(i)   horizontal
(ii)   vertically downwards, and
(iii) vertically upwards

Experience suggests that, where there is sufficient head available, the third option is usually the best (Cohen and Staub, 1992). Horizontal flow systems are prone to the same constraints as ALDs (see Section 5.4.2.2), in that it is easier to pass a large flow through the largest side of a prism (Figure 5.4), and yet this is inevitably difficult to arrange in horizontal flow. Gusek *et al.* (2000) have reported a number of clogging problems with horizontal flow anaerobic reactors in Missouri, ranging from physical clogging to blockage by large bubbles of H_2S gas which became trapped at the upper surface of the reactor and occupied a proportion of the cross-sectional thickness. In vertically-downward flowing systems, compaction of the substrate (and thus a reduction in its permeability) is promoted by the movement of water in the same direction as gravity. By contrast, in vertically-upward flowing anaerobic reactors, the flow of the water resists gravitational compaction and helps to maintain reasonable permeabilities (Cohen and Staub, 1992).
Once the orientation of the flow system has been established, the shape of the reactor can be established. A range of reactor geometries, from cylindrical through prismatic to inverted-pyramidal have all been tested (Figure 5.17a), without any firm conclusions being reached as to the optimum shape to minimize short-circuiting. Whatever geometry is decided upon, be prepared to have to dig the substrate back out and re-emplace it a number of times before your reactor will finally perform adequately (cf. Gusek *et al.*, 2000). Spreading the influent water widely by means of a distributory manifold (Cohen and Staub, 1992; Cohen, 1996), so that it enters the substrate at as many points as possible (Figure 5.17b), minimizes the velocity at any one point and thus limits the potential for

erosional development of pipes and other preferential flowpaths. A further measure which can help to arrest channelization of flow within the substrate is to install a layer of pea gravel every metre or so (Figure 5.17c), which will tend to re-distribute the flow more evenly before it enters the next layer of reactive substrate.

Maximization of hydraulic retention time is probably the most important objective in anaerobic bioreactor design, and yet the most difficult to achieve. Some earlier experiments with anaerobic bioreactors which reported disappointing results (e.g. Younger, 1997b) can now be re-interpreted with the benefit of hindsight as resulting from using far too brief residence times in the reactors. The latest research results (Pulles, 2000; R.R.H. Cohen, Colorado School of Mines, personal communication, 2001) suggest that a nominal retention time of 40 hours is the least that should be considered in designing full-scale systems. At the other extreme, a reactor with a residence time of four days proved capable of treating the very worst waters encountered in the abandoned mines of the Rocky Mountain mining districts (Cohen and Staub, 1992). In practice, there is nearly always external pressure to minimize retention time, since designing an anaerobic bioreactor so that it has a hydraulic retention time of only 40 hours is likely to result in a large system, for all but the most diminutive of mine water discharges.

Designing an anaerobic reactor for a specific hydraulic retention time is difficult because the hydraulic conductivity distribution of the substrate can never be known thoroughly in advance of installation (and remains very difficult to characterize even after installation, without installing so many piezometers that the hydraulic behavior is affected by the measurement infrastructure itself). The volume of substrate (V_s) needed to achieve a given hydraulic retention time (t_R) can be estimated by the following expression (derived from Darcy's Law, Equation 3.4):

$$V_s = \frac{(t_R \cdot Q_d)}{n_e} \tag{5.26}$$

Where n_e is the effective porosity (see Equation 3.3) and Q_d is the mean flow rate of the mine water. Effective porosity must be estimated at the design stage. As a guide, field testing of established compost-filled reactors typically returns values in the range 0.15 to 0.35. However, it is recommended that laboratory permeameters (falling-head variety) be used to test the permeability and porosity of specific substrates prior to final selection and system sizing (Waybrant et al., 1998; Amos and Younger, 2001). It should be noted that Equation 5.26 uses only an estimate of the mean retention time, whereas true saturated flow through the reactor will yield a distribution of retention times for individual water molecules which will tend to be normally distributed about this mean value. Given this and other uncertainties, precautionary assumptions should be taken in design of anaerobic bioreactors until such time as the canon of field experiences expands considerably.

5.5 PASSIVE TREATMENT PROCESS SELECTION AND STAGING

5.5.1 PRINCIPLES

In Section 5.4, we have described the various passive treatment technologies as if they stand in isolation from one another. In practice, this is not so: each of the technologies

(a)

(b)

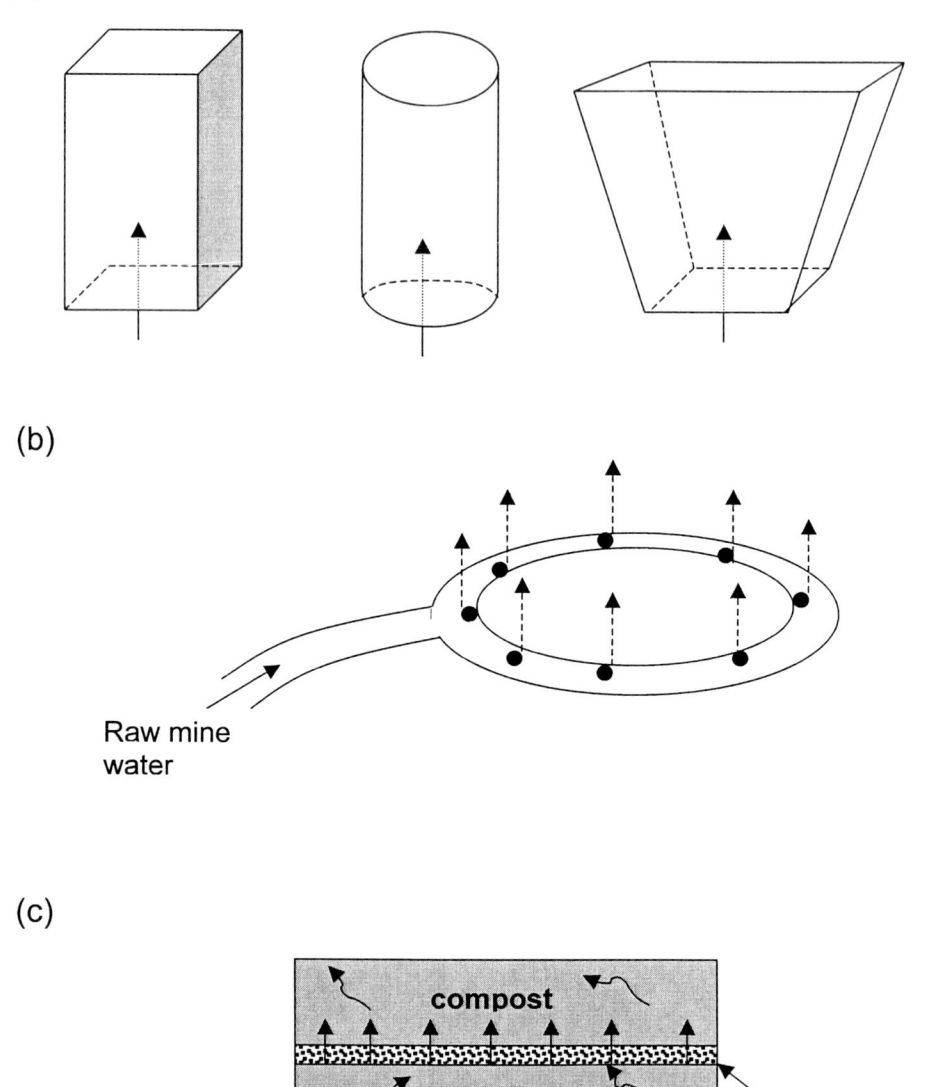

Raw mine
water

(c)

Flow re-spread
evenly by gravel

Flow tending to
preferentialize to
one or two paths

compost

compost

pea gravel
layers

compost

Figure 5.17: Some aspects of design for subsurface flow anaerobic bioreactors used to treat surface discharges from mines. (a) Various geometries which have been trialed for upflow reactors. (b) Distribution manifold to spread influent water evenly at the base of an upflow reactors (adapted after Cohen and Staub, 1992). (c) Use of pea gravel layers to redistribute preferential flows.

described above is essentially a 'unit process' which can be used in combination with other unit processes to contribute to an overall treatment system for a given polluted mine water. How do we decide what selection of unit processes to deploy, and in which order? The decision logic is summarized in Figure 5.18. Essentially, the guiding principle is to neutralize the mine water using sulfate reduction and/or carbonate dissolution, and then use aerobic processes to strip the remaining metals from solution.

Where Fe, Mn and Al are the only contaminants of concern, the above principle will always hold sway. However, where more toxic metals are present in the mine water, a number of *caveats* must be observed, including the following:

- mine waters containing mercury should never be subjected to sulfate reducing conditions, as this will promote the formation of methyl mercury, which is the most toxic form of the metal (e.g. King *et al.*, 2001).
- avoid discharging waters containing cadmium and/or arsenic into uncovered wetland-type passive treatment units, since exposure of wetland fauna to such toxins can be ecologically disastrous (e.g. Kelly, 1988).

To illustrate the application of the logic of Figure 5.18 in practice, we will close this chapter with some case studies.

5.5.2 ILLUSTRATIVE CASE STUDIES

5.5.2.1 *Aerobic wetland treatment of a net-alkaline water contaminated with Fe*

Loyalhanna Creek is a mid-sized stream located in Westmoreland County, in the southwest of Pennsylvania. The headwaters of the stream are highly valued for cold water (trout) and warm water (bass, panfish) fisheries. In the city of Latrobe, Loyalhanna Creek is polluted by the Monastery Run, whose problems can be traced further upstream to Four Mile Run. The stream is polluted by mine drainage flowing from underground coal mines that were mined and abandoned during the first half of the 20th century. Flows discharge from subsidence holes located near the coal crop and from bore holes that were drilled into the mine decades ago for a variety of reasons. The discharges occur on, or adjacent, to the campus of Saint Vincent College. In 1993, the College and Westmoreland County Conservation District initiated an effort to restore Loyalhanna Creek, and focused initially on the Four Mile Run discharges. A sampling program followed where all major discharges were collected into pipes or weirs so that flows could be measured and samples reliably collected. The results of the program were reported in 1994 (Jones *et al.*, 1994) and used to develop a restoration plan that consisted of three large wetland systems. One large discharge, flowing from an abandoned borehole, was targeted for treatment by an aerobic wetland named 'Wetland 3'. An analysis of the discharge destined for Wetland 3 is shown in Table 5.5. The flow of $7.5–30 \ l.s^{-1}$ is net alkaline, but contaminated with dissolved iron (all in the ferrous form). This chemistry is representative of other discharges in the Four Mile Run watershed and also similar to dozens of discharges from abandoned flooded underground coal mines in western Pennsylvania.

Wetland 3 was designed by the US Natural Resource Conservation Service and constructed with funding provided by the US Environmental Protection Agency. The system

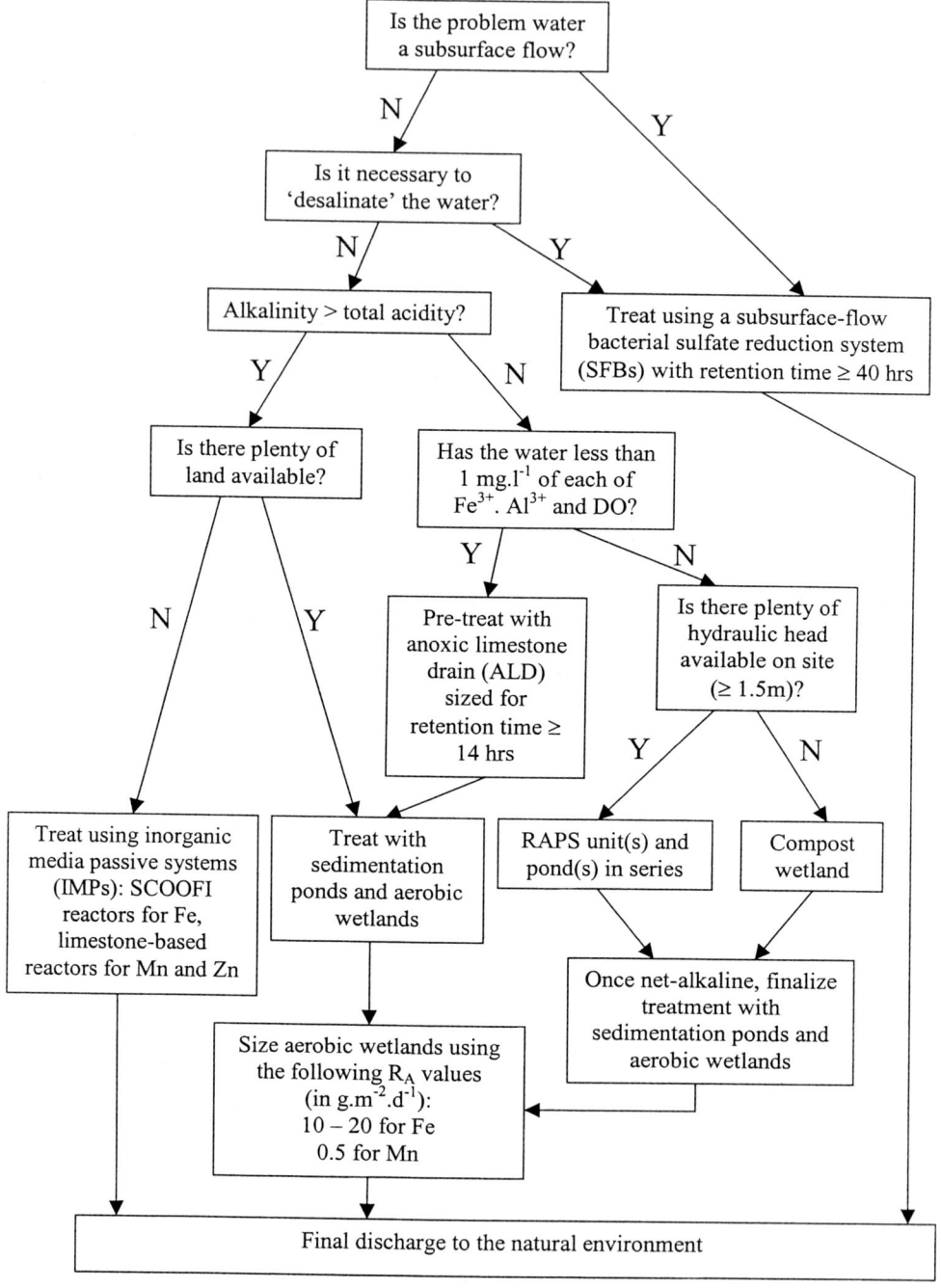

Figure 5.18: Flow chart giving the decision logic for selecting a passive treatment process for a given water.

consists of five cells. The first cell is a 1.2–1.8 m deep pond, the 2nd and 3rd cells pond/wetland mixtures (0.3–0.7 m deep) and the 4th and 5th cells are wetlands. The system was constructed in an existing wetland and efforts were made to protect existing vegetation during construction. Because of these efforts the wetland cells are densely vegetated with *Typha* and a variety of other woody and herbaceous species.

Wetland 3 is located within walking distance of the College science buildings and is adjacent to a restored grist mill that has been operated by the Saint Vincent Archabbey for over 100 years. The system is used for research purposes by College students and faculty and has developed into a popular educational stop for local schools and groups.

The system was designed based on a flow of 38 $l.s^{-1}$ containing 85 $mg.l^{-1}$ Fe and an expected iron removal rate of the aerobic system of 20 $g.m^{-2}d^{-1}$. The calculation was considered conservative because the 38 $l.s^{-1}$ flow rate was the highest measured during the monitoring period. The calculated target size of the system was 11,500 m^2. The installed system had a total wet surface area of 13,500 m^2.

The performance of the system is shown in Figure 5.19 and Table 5.6. The system has removed an average 97% of the iron contained in the mine discharge. The flow rate has averaged 18 $l.s^{-1}$ with high flows reaching approximately 26.5 $l.s^{-1}$. The performance of the system is flow-related (Figure 5.20). During periods when flows are less than 19 $l.s^{-1}$ (250 gpm), the system removes essentially all of the iron (< 1 $mg.l^{-1}$ is discharged). When flows are above this threshold, the effluent contains 3–7 $mg.l^{-1}$ Fe.

On average, the system has removed iron at a rate of 9 $g.m^{-2}.d^{-1}$. This value does not show the true capability of the system for Fe removal because on half of the days Fe loadings were less than 9 $g.m^{-2}.d^{-1}$ and essentially all the iron was removed. When only those days that the system was discharging Fe > 3 mg/L are considered, the average Fe removal rate is 13 $g.m^{-2}.d^{-1}$. Even this value is an underestimate of the system's capabilities. For a one year period between September 1997 and August 1998 the discharges of each of the cells was monitored. The results are shown in Figure 5.21. In low-flow months (summer and fall), 85% of the iron is removed in cell 1 and the concentrations are less than 5 $mg.l^{-1}$ at the discharge of cell 3. On these occasions the system looks (and is) oversized. During high-flow periods (winter and spring) much more iron is discharged from the cell 1 and the rest of the system becomes more important in the production of a good quality final effluent.

5.5.2.2 Compost wetland treatment of an acidic, Al-rich waste rock pile leachate

The Stanley Burn is a small headwater tributary of the River Wear, one of the principal rivers of north-east England. Since the mid-1980s, the Stanley Burn has suffered conspicuous pollution by acidic drainage emanating from a perched water table within a superficially restored waste rock pile appertaining to the former Morrison Busty Colliery (abandoned

Table 5.5: Composition of the borehole overflow that is treated by Saint Vincent College Wetland 3. (Data provided by R. Capo, University of Pittsburgh Program in Geological and Planetary Sciences)

pH	Alkalinity	Net Acidity	Fe	Mn	Al	Mg	Ca	Na	Cl	SO_4
6.1	216	−44	88.0	3.8	0.1	59.2	218.5	89.7	36.0	926

All concentrations are $mg.l^{-1}$. Alkalinity and net acidity are $mg.l^{-1}$ as $CaCO_3$

1975). The leachate contains up to 200 mg.l^{-1} total acidity (as $CaCO_3$ equivalent), with elevated concentrations of Fe (≤ 30 mg.l^{-1}), Al (≤ 30 mg.l^{-1}) and Mn (≤ 15 mg.l^{-1}). This polluted drainage constitutes a classic 'orphan discharge' for which no legally responsible party could be identified. In 1994, when residents of the nearby village of Quaking Houses finally accepted that no remedial action was ever likely to be forthcoming from elsewhere, a "do-it-yourself" remedial program was launched. This program drew together volunteers, professionals and charitable sponsors in the development of a complete solution to the pollution of the Stanley Burn. Drawing upon the inspiration of USA experiences (especially the work of Hedin *et al.*, 1994) passive treatment was soon identified as the most appropriate solution. As the water is acidic, a compost wetland or RAPS-based system was the obvious choice. However, given that compost wetland technology had no

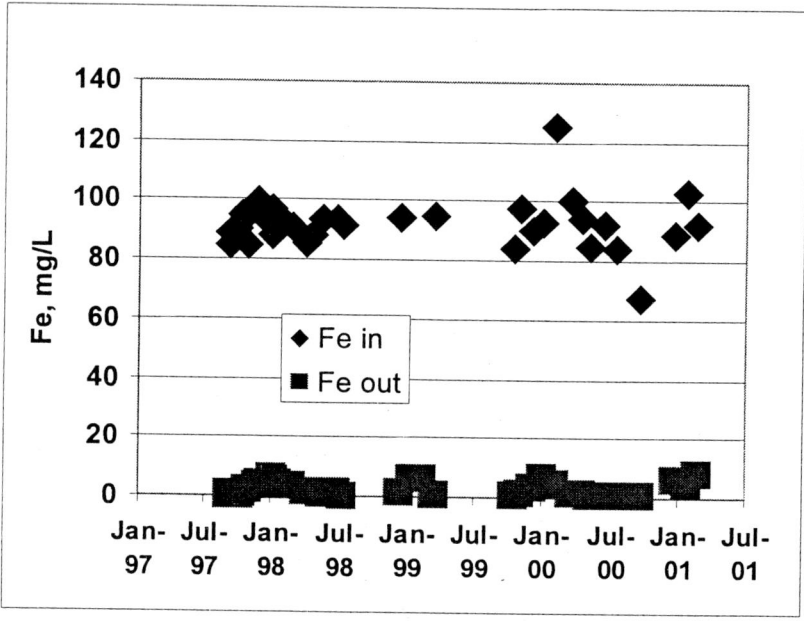

Figure 5.19: Saint Vincent College Wetland 3 performance: Concentrations of total Fe at the influent and effluent of Wetland 3. (Data provided by D. Fish, Chemistry Department, Saint Vincent College, Latrobe, PA).

Table 5.6: Average chemical characteristics of Saint Vincent College Wetland 3 influent and effluent, September 1997 – March 2001. (Data provided by D. Fish, Saint Vincent College Chemistry Department). Concentrations are in mg.l^{-1}

	Flow (l.s^{-1})	pH	Alk	Fe	Mn
Influent	} 18	6.1	191	92.1	3.3
Effluent	}	6.7	102	2.7	2.4

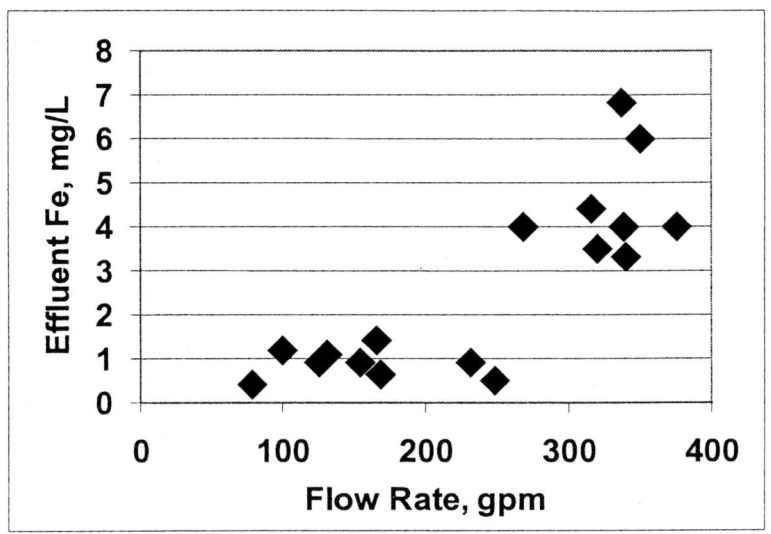

Figure 5.20: Relationship between flow rate and effluent concentrations of Fe at the outflow from Wetland 3. (N.B. The flow rates of 100, 200, 300 and 400 gpm (US gallons per minute) correspond to 7.6, 15.2, 22.8 and 30.3 l.s^{-1} respectively). (Data provided by D. Fish, Chemistry Department, Saint Vincent College, Latrobe, PA).

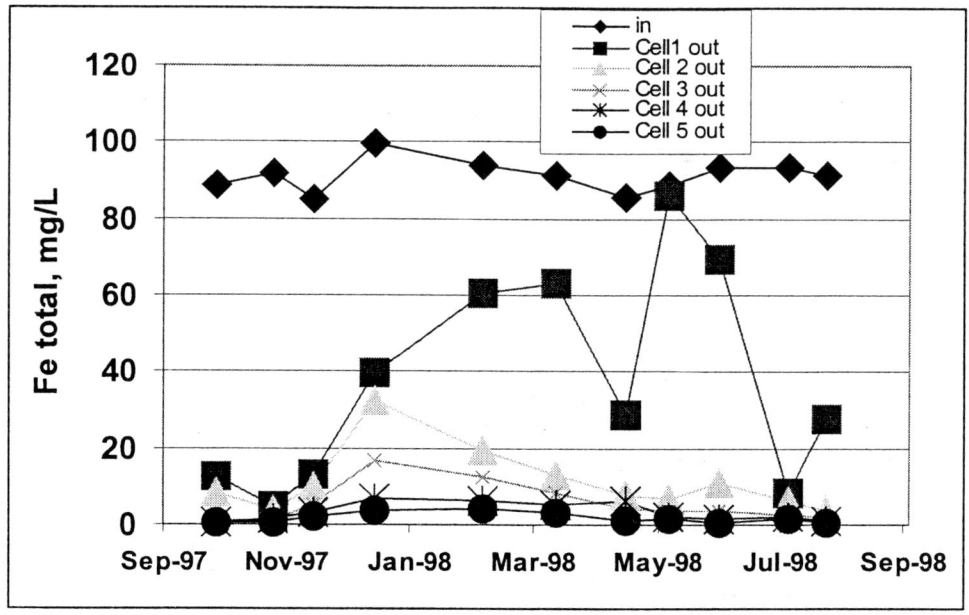

Figure 5.21: Concentrations of Fe at various points within Wetland 3. (Data provided by D. Fish, Chemistry Department, Saint Vincent College, Latrobe, PA).

previous track record in Europe at that time, the first step in the remedial program was to build and monitor a small-scale pilot wetland. This not only allowed the designers to gain valuable hands-on experience which would prove invaluable later, but also proved crucial in building confidence in the efficacy of the technology amongst regulators and potential sponsors. The design, construction and performance of this pilot wetland has been described in detail by Younger *et al.* (1997). In essence, it was a shallow pond, 45 m^2 in area, with a 0.3 m substrate of horse manure and straw from the Quaking Houses Village Stables. It was designed to treat around 5% of the average leachate flow, in accordance with the guidance given in Section 5.4.3.4. After 18 months of monitoring, this pilot wetland yielded an average removal rate of 9 g.d^{-1}.m^{-2} total acidity (as CaCO$_3$ equivalent), which compared favorably with the R$_A$ values for total acidity derived from studies of similar systems in the USA (i.e. 3.5 to 7 g.d^{-1}.m^{-2}; see Section 5.4.3.4 and Hedin *et al.*, 1994).

These encouraging performance figures, and the pleasant appearance of the pilot wetland, proved influential with potential sponsors, and by mid-1997 sufficient funding had been obtained from a range of charitable and philanthropic foundations to finance the construction of a full-scale system. The full story of the wetland creation is the subject of an entire book (Kemp and Griffiths, 2000), which recounts the ups-and-downs of community organization, volunteer participation, and professional inputs (both by engineers and by 'artists-in-residence'). Here we focus on the technical issues.

While the pilot plant had been in operation, the seminal work of Kepler and McCleary (1994) had begun to influence passive system design in the UK, and preliminary plans were laid to construct the full-scale system at Quaking Houses as a RAPS (Younger *et al.*, 1997). However, when the available plot of land was finally cleared of scrub vegetation and trial-pitted, it was found to be underlain by highly pyritic waste materials derived from a previously unrecorded coal washery finings pond. This meant that excavation of a suitable basin to install a RAPS would have entailed disposal of large volumes of highly reactive, acid-generating waste, the landfilling of which would have consumed much of the budget available for wetland construction. Without such excavation, there was a maximum of 1.0m of head available across the entire site, which is insufficient to drive water through a RAPS (see Section 5.4.3.5); hence the full-scale system was designed as a compost wetland, scaled-up from the original pilot-plant design.

Design and construction of the full-scale wetland system at Quaking Houses has been outlined by Jarvis and Younger (1999). Construction of the wetland commenced in August 1997 and took about 6 weeks of site work. The layout of the system is shown in Figure 5.22. The leachate was captured by construction of a concrete headwall across the outfall of the culvert from which the discharge emanates. This headwall gained some 0.5 m of head to help drive water through the system. Two sections of 100 mm diameter pipe were built into the headwall. The first carries water underground in an inverted siphon to the influent point of the wetland, discharging into a basin from where the water is distributed across the width of the wetland. The second section of pipe allows overflow back into the original watercourse when flow-rates exceed approximately 400 litres per minute. Because pollutant concentrations are lower at higher flow-rates due to dilution, and because of further dilution of the overflow water by the effluent from the wetland, the impact of this water on the receiving watercourse is minimal.

The heart of the Quaking Houses treatment system is a compost wetland unit (see Figures 5.13 and 5.22) occupying some 440 m^2. This is enclosed by a bund composed of

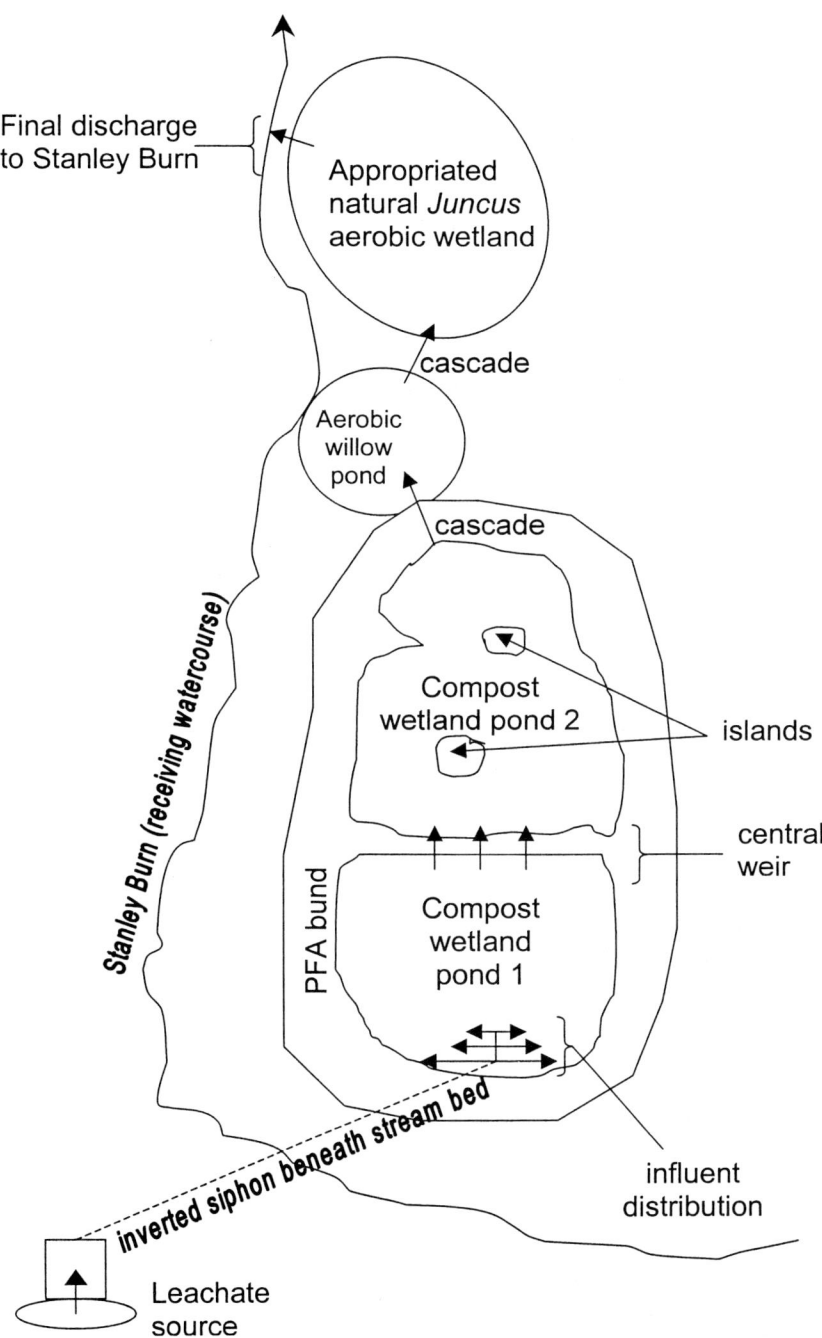

Figure 5.22: Layout of the Quaking Houses Community Wetland for Mine Water Treatment, County Durham (UK). The performance data in Table 5.7 relate only to the 'compost wetland' (ponds 1 and 2), sampled at the point where the water cascades into the aerobic willow pond.

pulverised fuel ash (PFA), which is both strong and highly impermeable after mechanical compaction, yet costs less than half the price of the main alternative material (clay). To avoid toe drainage, which may have affected the integrity of the bund, the base of the embankment was sunk approximately 0.2 m into the *in situ* soil. The bund had a minimum crest width of 1.5 m, with inner slope angles (i.e. facing into the ponded area) of 1:3 or less, in order to encourage wildlife. (Outer slope angles were made to be not more than 1:2). Baffles and islands were also constructed from PFA, both to help minimize hydraulic short-circuiting within each wetland cell and to improve the appearance and habitat diversity of the wetland. PFA was also used to construct a central weir of about 0.4 m height, which was incorporated into the design for four reasons:

(i)   To accommodate the natural slope of the site away from the influent point: if the bund crest had been maintained at the same elevation around the entire wetland, much more PFA would have been needed, construction would have taken longer, and less area would have been available for treatment unless the slopes were steepened significantly.

(ii) To minimize short-circuiting by re-spreading the water at the entrance to the second basin.

(iii) To allow one or other of the basins to be taken out of commission for maintenance at any one time without totally losing treatment capacity.

(iv) To provide an attractive waterfall feature in the middle of the wetland.

Because the horse manure used in the pilot-scale wetland was not available in sufficient quantity, additional sources of compost were sought: the final compost wetland substrate comprises horse manure, cattle manure and municipal waste compost in the following proportions: 30:40:30. Additionally 30 tonnes of limestone were deposited at the far end of the wetland, to facilitate final pH adjustment if it should be required. The system was designed such that the compost depth in the wetland would be 0.30–0.50 m, leaving 0.30 m of freeboard for future accumulation of material on the substrate surface.

When first constructed, the water leaving the compost wetland was routed directly to the Stanley Burn. Subsequently, an aerobic wetland unit was added to the end of the system, to polish residual iron concentrations down to below 0.5 mg.l⁻¹. This comprises a circular, ornamental 'willow pond' and an appropriated area of natural *Juncus* stands, totalling some 100 m² in area. It should be noted that, for illustrative purposes, the contaminant removal rates and efficiencies discussed below *relate only to the compost wetland unit (ponds 1 and 2)*, not to the passive treatment system as a whole, which achieves significantly more Fe removal than the compost wetland alone.

Monitoring of the compost wetland unit within the Quaking Houses system over its first 27 months of operation (Jarvis, 2000) revealed a mean acidity removal rate of 5.6 g.d⁻¹.m⁻². This mean value falls almost exactly in the middle of the range of R_A values (3.5–7 g.d⁻¹.m⁻²) derived from studying similar systems in the USA. However, acidity removal performance ranges from 1.3–46.1 g.d⁻¹.m⁻², with the higher rates corresponding to times when the influent was at its most acidic. This is consistent with other evidence (discussed by Jarvis, 2000) which suggests that the overall acidity removal reaction may be first order in acidity.

While total acidity removal was consistent with expectations, it is instructive to consider the removal rates of the individual metals which contribute to this total acidity. Table 5.7

summarizes these, calculated as both % treatment efficiencies and as R_A values. It is interesting to note that the removal rates for each of the individual metals are substantially lower than would be anticipated in an aerobic wetland receiving net–alkaline mine water (see Table 5.4), but are similar to the proposed removal rates for these metals in aerobic wetlands receiving acid waters (Brodie *et al.*, 1988).

One major benefit of the Quaking Houses passive treatment project deserves further mention here: once the long-term pollution from acidic mine site drainage had been abated, pressures mounted for the clean-up of other sources of pollution to the stream (particularly from combined sewer overflows and deicing salt store runoff). Previously, the organizations responsible for these two sources of pollution had justifiably claimed that there was little point treating their effluents to higher standards given that the acid drainage was clearly killing all aquatic life in its path. With that excuse removed, the other sources of pollution soon became priorities for clean-up, with the result that the Stanley Burn has now been thoroughly restored as a healthy stream ecosystem.

5.5.2.3 Use of RAPS and wetlands to treat an acidic discharge rich in Fe and Al

The Pit 431 treatment system was constructed to treat a surface mine discharge in Pennsylvania, which is highly contaminated with Fe, Mn, and Al. Previous efforts to

Table 5.7: Contaminant removal performance of the compost wetland unit within the Quaking Houses Passive Treatment System (after Jarvis, 2000)

	Mean	Range (n)[a]
Acidity treatment efficiency (%)[b]	53.4	11.8–95.5 (80)
Area-adjusted acidity removal (g/m²/d)[c]	5.6	1.3–16.3 (57)
Area-adjusted acidity removal with DF (g/m²/d)[c,d]	5.01	1.33–15.67 (29)
First-order acidity removal (m/d)[e]	0.118	0.34–0.296 (29)
Fe treatment efficiency (%)[b]	46.4	−23.8–96.6 (104)
Area-adjusted Fe removal (g/m²/d)[c]	0.58	−0.06–2.58 (76)
Area-adjusted Fe removal with DF (g/m²/d)[c,d]	0.53	−0.10–2.58 (34)
First-order Fe removal (m/d)[e]	0.105	−0.013–0.322 (34)
Mn treatment efficiency (%)[b]	26.4	−13.9–100 (102)
Area-adjusted Mn removal (g/m²/d)[c]	0.26	−0.03–1.18 (72)
Area-adjusted Mn removal with DF (g/m²/d)[c,d]	0.23	−0.41–1.15 (34)
First-order Mn removal (m/d)[e]	0.061	−0.197–0.282 (34)
Zn treatment efficiency (%)[b]	25.8	−107.2–81.0 (49)
Area-adjusted Zn removal (g/m²/d)[c]	0.090	−0.02–0.280 (39)
Area-adjusted Zn removal with DF (g/m²/d)[c,d]	0.065	−0.047–0.230 (29)
First-order Zn removal (m/d)[e]	0.062	−0.040–0.322 (29)
Al treatment efficiency (%)[b]	51.2	−2.9–100 (99)
Area-adjusted Al removal (g/m²/d)[c]	0.81	−0.02–8.76 (70)
Area-adjusted Al removal with DF (g/m²/d)[c,d]	0.60	−0.01–2.17 (34)
First-order Al removal (m/d)[e]	0.161	−0.002–0.563 (34)

[a]: n = number of observations.
[b]: calculated according to Equation 5.19.
[c]: i.e. R_A value, calculated using Equation 5.20.
[d]: DF = dilution factor; see Section 5.3.8.
[e]: K_1 value for the specified reaction, assuming it to be first order – see Equation 5.18

passively treat the water with ALDs had failed due to clogging and inadequate alkalinity generation. A design was developed that targeted Al and Fe for removal in a combined system of RAPS and constructed wetlands. The system was constructed in October 1996. Figure 5.23 is a plan view of the system. The RAPS units were expected to remove Al and Fe and generate alkalinity. The aerobic wetlands were expected to precipitate iron (as ferric hydroxide). The high acidity and Fe concentrations necessitated the construction of RAPS units. These were designed on the supposition that they would remove total acidity at a rate of 40 g $m^{-2}d^{-1}$, while the aerobic wetlands were expected to remove Fe at a rate of 10 g $m^{-2}d^{-1}$.

The RAPS units were constructed with 45 cm of organic substrate overlying 45 cm of limestone aggregate in which drainage pipes were installed. The organic substrate was amended with limestone fines to increase its neutralization potential and delay acidification.

Table 5.8 presents the average chemistry of samples collected from the system in January, March and July 1997. Samples from the end of Wetland #1 could not be collected because of stagnant, pooled conditions. The first RAPS unit decreased Al to less than 1 mg.l^{-1}, decreased Fe concentrations by 85–190 mg.l^{-1}, and discharged water with an alkalinity of 80–175 mg.l^{-1} as $CaCO_3$. The unexpectedly high retention of iron in the first RAPS unit resulted in large decreases in acidity and an acidity removal rate of approximately 60 g.m^{-2}.d^{-1}. The constructed wetlands, which are not receiving the iron loadings predicted, are removing iron at a rate of approximately 5 g $m^{-2}d^{-1}$.

The system was designed for a maximum flow rate of 200 l.min^{-1}. Because of very high precipitation in 1996, flows during the winter and spring of 1997 exceeded 250 l.min^{-1}

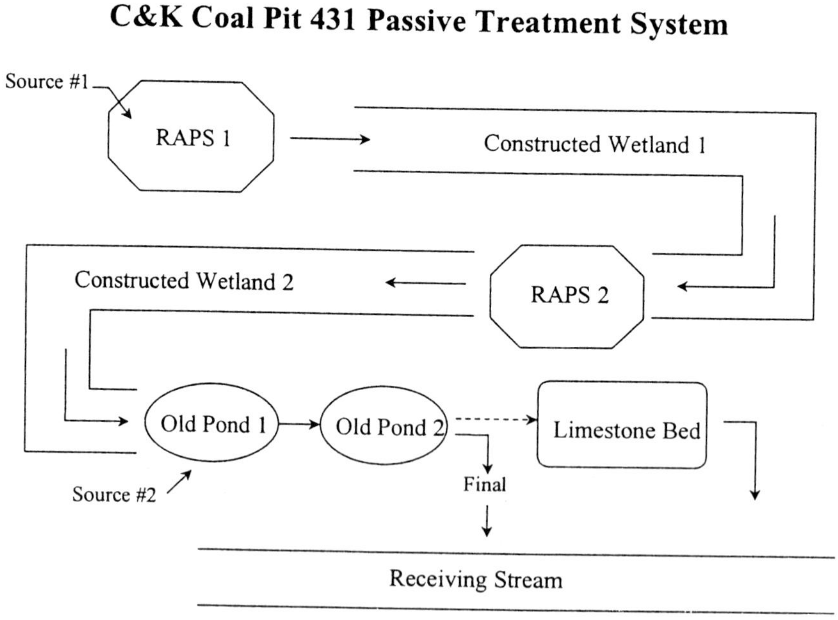

C&K Coal Pit 431 Passive Treatment System

Figure 5.23: Layout of the Pit 431 passive treatment system.

Table 5.8: Average water chemistry for the Pit 431 passive treatment system

	Source	RAPS1 eff	RAPS2 eff	Wet#2 eff
pH	4.05	5.98	6.11	6.11
alkalinity	0	117	113	47
acidity	947	455	223	218
Fe	354	225	108	79
Al	31	<1	<1	<1
Mn	143	131	100	93
Ca	219	344	320	320
Mg	337	295	235	249
Na	12	16	15	16
SO_4	2962	2684	2105	2056
cations	60.6	55.1	43.4	43.4
anions	61.7	55.9	43.8	42.8
flow (L/min)	na	250	na	na
n	3	3	2	3

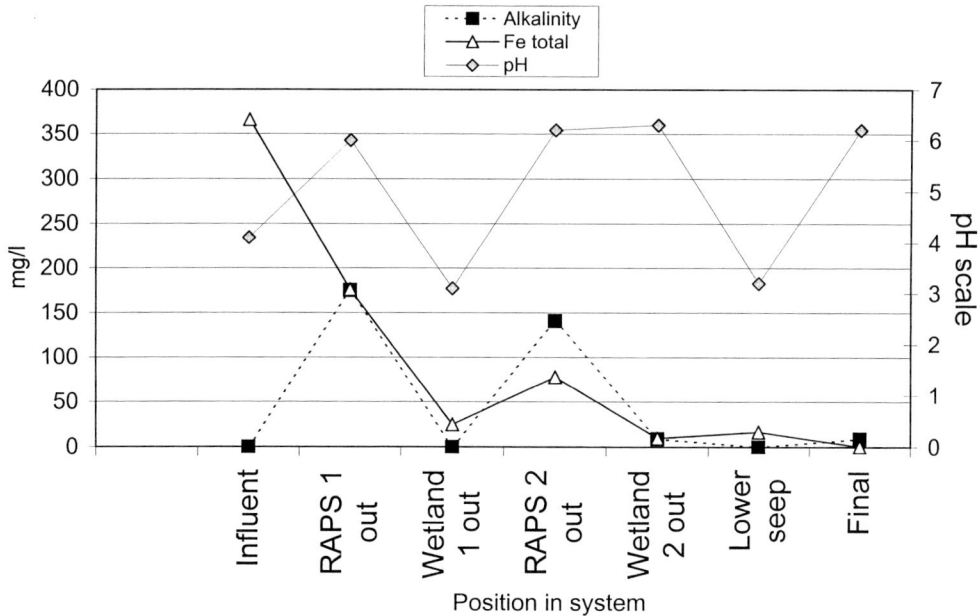

Figure 5.24: Changes in concentrations of Fe and alkalinity with position in the Pit 431 passive system.

continuously and 350 l.min^{-1} on occasions. In the summer of 1997 flows decreased and performance of the system under maximum design conditions could be evaluated. Figure 5.24 shows changes in alkalinity and Fe position within the Pit 431 system. Figure 5.24 illustrates well the performance of a RAPS-based passive system: Alkalinity is generated by the RAPS, then consumed in the aerobic wetland by oxidative iron precipitation reactions, then regenerated in another RAPS, before finally being lowered again by iron precipitation reactions in a further aerobic wetland.

APPENDIX I

GLOSSARY OF MINING AND RELATED TERMS RELEVANT TO MINE WATER SCIENCE AND ENGINEERING

The glossary which follows is drawn from the experience of the authors, and represents terminology which was current in the English-speaking mining world at the start of the 21st Century. The glossary is restricted to mining and geological terms likely to be heard in the context of mine water management. We *do not* provide a general geochemical or hydrological glossary. Neither is this a thoroughly exhaustive glossary of mining engineering. The reader seeking a general mining glossary is referred to works such as Greenwell (1888), Orchard (1991) or Rieuwerts (1998). The glossary is *not* intended to cover geological and chemical terminology, with the exception of a few terms restricted largely to mining applications of these sciences.

In the definitions which follow, words are underlined where they are themselves defined elsewhere in this glossary.

Acid mine drainage	Water which became acidic during its passage through pyrite-bearing mined strata. Synonym: *"acid rock drainage"*.
Acid rock drainage	See *"acid mine drainage"*. The term "acid rock drainage" is used predominantly in Canada, with the implication that the acidity problem is due to the rocks rather than the process of mining. There are at least two reasons why this usage is objectionable: (i) "Acid rocks" *sensu stricto* are a group of igneous rocks which are not noted for giving rise to problematic drainage. (ii) It is somewhat disingenuous to attempt to exonerate mining in this manner, as acidic drainage is relatively rare in unmined sulphidic strata, but is common where mining has increased the rock mass permeability and allowed oxygen to access previously reduced strata.
Adit	A roadway, horizontal or gently inclined, which connects subsurface mine workings to the ground surface. Synonyms: *day-level, day-drift, drift, level, Ingaun E'e.*
Advance dewatering	*Dewatering* an area to be mined, by pumping or gravity drainage (e.g. using inclined boreholes), prior to the commencement of mining.
Afterdamp	Carbon monoxide, formed by methane explosion or fire underground. (The name 'afterdamp' refers to its common presence as a hazard in a mine after an explosion, in which circumstances it poses a grave danger to unprepared rescuers).

397

Angle of draw	The angle formed by the edge of a *longwall panel* and the outermost limit of *subsidence* in the overlying strata caused by the *goafing* of this panel.
Auger mining	The use of horizontal, large-diameter boring machines to extract coal from beneath *highwalls* in strip mines, or (less commonly) from outcrops in hilly terrain.
Backfill	(i) (verb) To return previously-*stripped overburden* to an *opencast* mine *void* after *coaling* ceases.
	(ii) (noun) The broken waste rock material which occupies a restored opencast void.
Baffle banks	A mound of tipped *overburden* or stored soil placed around the perimeter of an *opencast* or *open-pit* mine to limit the external visibility of the mine *void* and/or provide a partial barrier to site noise and dust. Synonym: *bund* (meaning (ii).
Bank	(i) A slope.
	(ii) The ground surface, as a point of reference for underground workings. Typical usage: "Get the *ore* to bank", meaning "take the ore out of the mine".
Batch	The name given to a *waste rock pile* in the Somerset coalfield.
Batter	(i) An artificial slope in a *waste rock pile* or *surface mine*
	(ii) The slope angle of such an artificial slope.
Bench	A more-or-less horizontal surface of rock, developed within a surface mine as the *highwall* retreats in response to repeated blasting and excavation.
Blackdamp	See "*stythe*".
Block caving	A specialized form of *stoping*.
Bing	Scottish term for a *waste rock pile*.
Bord or Board	A mine *roadway* driven as one of a set of mutually perpendicular roadways for the purpose of extracting coal or other minerals by means of *bord-and-pillar* mining.
Bord-and-pillar	A method of extraction in which the coal or ore is removed by means of a set of mutually perpendicular roadways ('*bords*'), leaving *pillars* of intact coal or ore to support the roof. Synonyms: pillar-and-stall, room-and-pillar, stoop-and-room.
Box cut	A roughly rectangular opencast void sunk through overburden cover to exploit payable rock. A box cut typically has *sidewalls* on two sides, and a *highwall* on the third side. One or more box cuts are usually sunk in the first stages of excavation of a major opencast mine void, with extraction of the material between them resulting in development of the main opencast *highwall*.
Bund	(i) An embankment constructed to impound water (commonly used in constructed wetlands).
	(ii) A pile of *overburden* constructed around a *strip mine void* for amenity purposes (limiting the external visibility of the mine *void*, providing a partial barrier to blasting noise and dust). Synonym: *baffle banks*.

Canch	Headspace cut in a roadway above the roof of the seam, usually to obtain clearance for access. See also *"ripping"*.
Chokedamp	See *"stythe"*
Coaling	The extraction of coal from an *opencast void*.
Collar	The rim of a mine shaft and/or the structural features (in timber, brick, concrete or steel) used to support it.
Colliery	Coal mine.
Country rock	Non-mineralized bedrock enclosing an *orebody*.
Cross-measures drift (or raise)	See *"inter-seam drift"*.
Crown hole	A small (typically < 5 m diameter) hollow on the ground surface, usually roughly circular in plan, caused by collapse due to *void migration*, (or in some cases by winnowing of soil into an underground void, which may have migrated). Synonym: *pit fall, subsidence depression, subsidence hollow.*
Crown pillar	A block of intact rock left between the top of a *stope* and the ground surface.
Crut	See *"inter-seam drift"*.
Cross-cut	(i) A roadway driven between two separate *orebodies* (ii) See *"stenton"*.
Day-level	See *"adit"*.
Day-drift	See *"adit"*.
Decline	See *"inclined drift"*
Deep mine	Any mine in which the miner and/or his machinery work underground.
Deep mining	Mining underground.
Dewatering	(i) Preventing the flooding of mine workings by pumping or gravity drainage. (ii) Removing excess water from sludges generated during mineral processing or mine water treatment. Usually undertaken to make disposal easier and cheaper.
Dib	An inclined *adit* or *inter-seam drift*. (Term used mainly in northern English lead mines).
Downcast	Adjective applied to signify that a given shaft or adit conducts (or conducted) ventilation *into* the workings. (see also "upcast").
Drainage adit	An *adit* which is constructed such that it under-drains a volume of workings by gravity. Synonyms: *sough, watergate, water-level, waterloose.*
Draw-point	A short tunnel driven from a main haulage roadway into the foot of a *stope*, from which ore can be "mucked-out" after blasting in a *sub-level stoping* operation.
Drift	(i) A synonym of *adit*, particularly in coal mining. (ii) A mine accessed by an *adit* (*"drift mine"*). (iii) Sometimes used in coal mining to signify underground roadways other than an adit. For instance a roadway driven outside of a seam may be termed a "stone drift", and an inclined

roadway connecting two separate working horizons may be termed an *"inter-seam drift"*.

(iv) A verb, meaning "to drive a mine tunnel" as in: "They drifted into the Main Seam, but found it wasn't worth working".

(v) (Geol.) Unconsolidated strata, generally of Quaternary age, lying between the soil and rockhead.

Drift mine   A deep mine (usually a coal mine) accessed by a *drift* (i.e. an *adit*).

Dripper   Water entering an underground working by dripping through the ceiling. If the chemistry is favourable, drippers may form stalactites.

Egress   A route out of a *deep mine*. (See *"secondary (means of) egress"*).

Endless   A very long hoop of cable stretching the length of an *engine plane*, turning on pulleys installed at either end of the *plane*, and used to haul tubs *in-* or *out-bye*.

Engine plane   A main haulage *roadway* in a mine, along which *run-of-mine* ore/ coal is drawn by engines operating endless cable-hauled sets of tubs.

Engine shaft   A *shaft* used by a haulage engine.

Face   The active front of excavation of the coal/*ore* in a mine.

Fan drift   An *adit* fitted with a ventilation fan.

Fan pit   A *shaft* fitted with a ventilation fan.

Fault   (Geol.) A rock-mass discontinuity, on either side of which the strata have demonstrably been displaced vertically and/or transversally.

Feeder   A sustained, discrete, voluminous inflow of water to underground workings (analogous in appearance and behaviour to a natural spring).

Fines or Finings   See *"washery fines"*

Finings pond   A pond in which coal *washery fines* are disposed of, by sedimentation. Synonym: *tailings pond*.

Firedamp   An explosive mixture of methane and air.

Flat or Flatt   (i) A horizontal *orebody* flanking a generally vertical vein.

(ii) The place in which all of the coal won from a given district of *bord-and-pillar* workings was marshalled for haulage to *bank*.

Footwall   The country rock immediately underlying a dipping *orebody* or *fault*.

Furnace pit or   A *shaft* fitted with a furnace which heated air to drive a mine
Furnace shaft   ventilation circuit. (Rarely used after about 1900).

Gangue   Minerals of no economic value which occur in close association with *ore* minerals.

Gate   See *"roadway"*.

Gin   (i) A winding device used to haul materials and *run-of-mine* from the *workings*. (An antique term).

Gin shaft   A *shaft* fitted with a *gin*.

Gin whimsey   A North Pennine term for a *gin shaft*.

Goaf   (i) A shattered mass of rock, formed by the collapse of roof strata into a void from which coal (or, less commonly, other minerals) has been removed. (Synonym: *gob*)

(ii) (To goaf): A verb meaning "To collapse forming *goaf*".

Goafing	Present participle of the verb 'to goaf' (see *goaf* meaning (ii)).
Gob	See *"goaf"*.
Gob pile	The *waste rock pile* associated with a coal mine (only in North American usage).
Gunnis	An open, abandoned *stope* (especially in Cornwall).
Gurt	See *"Ricket"*
Gutter	See *"Ricket"*
Hanging wall	The strata immediately overlying a dipping *orebody* or *fault*.
Highwall	The artificially-cut cliff of intact strata which gradually retreats during the working of a *surface mine*.
Holing	(i) A connection between one body of mine workings and another (ii) The act of connecting (often accidentally) one body of workings into another, as in "the pit holed into flooded old workings that weren't marked on any plan".
Hopper	A bin in which *run-of-mine* ore is stored prior to removal from a *deep mine*. Ore is typically tipped into a hopper from above, via an *ore chute*, and is removed from below, by lifting a gate and allowing it to pour into a waiting vehicle or onto a conveyor belt.
Hush	An artificial channel on a hillside, created by repeated scouring with water to reveal/extract mineral deposits. (Term originated in northern England, but is also applied to similar remains of *hydraulic mining* elsewhere in the world).
Hydraulic mining	*Surface mining* by means of erosion with jets/torrents of water.
In-bye	A directional term used in underground mining, signifying the direction towards the innermost reaches of a mine. (See *"out-bye"*).
Incline	See *"inclined drift"*
Inclined drift	An underground *roadway* which dips at a considerable angle (usually > 5°), used to connect one area of workings to another, or as a mine access from the surface. Synonyms: *decline, dib, inter-seam drift, slant, slope entry.*
Ingaun E'e	A Scottish term for an *adit*.
Inrush	A sudden, unforeseen and usually devastating inflow of water into underground workings.
Inset	A point where a roadway leads off from a mine *shaft*.
Intake roadway	A *roadway* which conducts (or conducted) ventilation *into* the workings. Such a *roadway* will be connected upwind to a *downcast shaft* or adit. (see also *"return roadway"*).
Interburden	A pedantic term referring to *overburden* materials lying between two coal *seams*.
Inter-seam drift	An inclined underground *roadway* in a coal mine, which connects one coal seam to another. Synonyms: *cross-measures drift; cross-measures raise; crut.*
Ladderway	A *shaft* fitted with ladders.
Landing	The point where a *deep mine level* leads off from a major *inclined* access *drift* (equivalent to an *"inset"* on a vertical shaft).

Launder	A shallow channel used to carry mine water over uneven ground or *voids*. Most launders were made of wood, a few of steel. These were largely superseded by steel (and later HDPE) *pipe ranges* in the 20th Century.
Level	(i) Synonym for *adit* (in northern England). (ii) Major *roadway* in a deep mine (e.g. main haulage *level*).
Lift	(i) The process of developing a new *bench* within a surface mine. (ii) The haulage provided by one *staple shaft* or *main shaft*.
Lodge	See "*pumping lodge*".
Longhole	A borehole drilled from a *sub-level*, generally down towards a series of *draw-points*, used for blasting to create an open *stope*.
Longwall	A method of working flat-lying coal and similar deposits, in which rectangular '*panels*' (usually 100 to 200 m wide by 1000 m long) are removed by successively cutting away a strip along one of the short sides of the rectangle (the working *face*), allowing the roof overlying previous face positions to collapse, forming *goaf*.
Loose wall	The frontal slope of the main mound of *backfill* in an *opencast* mine which faces onto the highwall.
Main gate	The nearest to the *main shaft* of the two *roadways* which run along either edge of a *longwall panel*. (See also *tail gate*).
Main shaft	The principal access shaft to a *deep mine*.
Make	(i) See "*water make*". (ii) A verb meaning "to yield water", as in "The mine makes 8 Ml/d from the underlying dolomite".
Manway rise or raise	A narrow *shaft*, typically fitted with ladders, which allows miners local access to working areas within a *deep mine*.
Manriding	Adjective applied to any piece of mine infrastructure (shaft, railed set of tubs, conveyor belt) which is authorized for transport of personnel.
Mine	(i) A void excavated for purposes of *mineral extraction*. (ii) (To mine). To extract rock from a mine. (iii) A Lancashire synonym for *seam*.
Muck out	To remove shattered ore from a *stope*.
Opencast	*Surface mine* working coal (British usage). Also used in the terms "opencast coal site" (OCCS), and opencast colliery. Largely synonymous with the US term "*strip mine*".
Open-pit	(i) sensu lato: Any working (or worked) *void* of a *surface mine*. (ii) sensu stricto: A type of *surface mine* in which relatively little *overburden* is stripped before wholesale *benching* of the *orebody* is undertaken, with little back-filling of *spoil* until production has finally ceased.
Ore	An economically valuable rock.
Orebody	The geological structure/complex in which all of the ore occurs.
Ore chute	A shallow underground *shaft* within a *deep mine*, down which ore is tipped on its way to the main haulage *roadway* by which it will leave the mine. Synonym: *hopper*.

Out-bye	A directional term used in underground mining, signifying the direction towards the mine entrance and therefore the surface. (See "*In-bye*").
Overburden	Barren strata overlying the economic deposits (especially in surface mining).
Overhand stoping	See "*stoping*". Overhand stoping is undertaken by excavating upwards from an access *roadway*.
Portal	The entrance to an *adit*.
Panel	A rectangular area of coal worked by *longwall* methods.
Pillar	A volume of unmined rock left in place for structural support or as a barrier to water and/or gas.
Pillar-and-stall	See "*bord-and-pillar*"
Pipe range	A pipe line used to convey mine water or compressed air.
Pit	(i) *Colliery*
	(ii) *Shaft*
	(iii) *Surface mine void*.
Pit fall	A colloquial term for a *crown hole*.
Punch mine	A small-scale *drift mine* for coal, developed by short *roadways* driven in from outcrop. (Appalachian usage).
Pumping lodge	A *sump* or area of abandoned, partly flooded workings in which water is allowed to accumulate prior to pumping out as part of *dewatering* operations. See also "*standage*".
Quarry	(i) (British usage) Any *surface mine* other than an *opencast* coal mine.
	(ii) (US usage) A *surface mine* working dimension stone.
Return roadway	A *roadway* which conducts (or conducted) ventilation out of the workings. Such a roadway will be connected down-wind to an *upcast shaft* or *adit*. (see also "*intake roadway*").
Ribside	The line along which one or more *goafed longwall panels* meet a substantial area of unworked strata.
Ricket	A mine water gutter along the edge of a roadway. Synonyms: *gutter, gurt*.
Ring dyke (or dike)	A *tailings dam* constructed by enclosing it entirely in a circumferential bund.
Ripping	Increasing the height of a roadway in a coal mine by removal of stone above the seam roof. This creates '*canches*'.
Roadway	An underground tunnel in a *deep mine*. Synonyms: *drift, gate, level, way*.
Rockhead	The uppermost surface of solid rocks in the subsurface, below the soil and *drift* cover. (see "*drift*", meaning (v)).
Room-and-pillar	See "*bord-and-pillar*"
Run-of-mine	An adjective applied to *ore* to signify unprocessed *ore*, in the condition in which it leaves the mine en route for the *beneficiation plant*.
Seam	A bed of coal. (Less commonly applied to other lithologies, such as Ironstone in the Cleveland Ironstone Field).

Secondary (means of) egress	An *adit* or *shaft* maintained to provide an emergency escape route from a *deep mine* in the event the principal *adit* or *shaft* becomes unusable.
Shaft	A vertical or nearly-vertical tunnel used to access *deep mine* workings.
Shaft station	An *inset* at which a shaft *skip* or *cage* makes regular stops to take on personnel and/or materials.
Shrinkage stoping	An *overhand stoping* technique which is undertaken in the following cycle: (a) The *orebody* is blasted and the shattered rock falls into the *stope*. (b) Enough of the shattered rock is drawn off from below ("mucked out at the *drawpoints*") in order to that the pile of shattered rock "shrinks" sufficiently for miners to gain access to the roof of the stope by walking on top of the rock pile. (c) The miners on top of the rock pile set charges for the next blast, then withdraw without firing the charges. (d) The remaining rock is mucked out of the *stope*. (e) return to step (a) above.
Sidewall	A rock face developed parallel to the direction of driving of an *opencast cut*, typically forming a right-angle with the *highwall*.
Sink or Sinking	(i) In *deep mining*: the excavation of a *shaft*. (ii) In *surface mining*: excavating a *box cut* or forming a new *bench* lower than the last.
Slant	Welsh term for an inclined access *drift*.
Slimes	See "*tailings*".
Slimes dam	See "*tailings dam*".
Slope entry	USA term for an inclined access *drift*.
Sough	Central England term for a *drainage adit*.
Spoil	*Waste rock.*
Spoil heap	See "*waste rock pile*"
Standage	A section of *roadway* or abandoned workings which is used as a reservoir to store mine water during periods of high water make or pump failure. See also "*pumping lodge*".
Staple	An underground *shaft* linking one level of workings with another. (pronounced to rhyme with "apple"). Occasionally applied to a *ladderway shaft* to surface. Synonyms: *sump* (meaning (ii)); *winze*.
Staple shaft	See "*staple*".
Stenton	A short *roadway* inter-connecting two major, parallel *roadways* in a *deep mine*. Synonym: *cross-cut*. (meaning (ii)).
Stinkdamp	Hydrogen sulphide gas.
Stoop-and-room	The Scottish equivalent of the term "*bord-and-pillar*".
Stope	(i) A vertical or sub-vertical mine void, typically resulting from mining of veins or irregularly-shaped orebodies of significant vertical extension. (ii) A verb, meaning "to mine in the vertical plane", as in "They stoped up and worked the richest part of the orebody".
Stoping	Deep mining in the vertical or sub-vertical plane.
Stope-and-pillar	Stoping in which substantial pillars of intact *ore* and/or *gangue* are left in place to support the walls.

Strike	(Geol.) The orientation of a horizontal line projected across the surface of a dipping plane, such as a *fault, footwall, hanging wall vein* or similar geological surface.
Strip	To remove *overburden* in *surface mining*.
Strip mine	A *surface mine* for coal, in which stripped overburden is cast behind the advancing excavation works onto worked-out areas. Synonym: *opencast coal mine* (UK, S Africa usage).
Stythe	Oxygen-deficient air; air rich in carbon dioxide. Poses a suffocation risk. Synonyms: *blackdamp, choke damp*.
Sub-level	A *roadway* driven along *strike* within a sub-vertical or vertical orebody to facilitate *sub-level stoping*.
Sub-level stoping	*Stoping* according to the following cycle of activities: semi-permanent haulage roadways in the *country rock*, located in the *hanging wall* of the *vein*, communicate with *draw-points* driven into the vein (every 10 m or so) and *manway rises* which give access to higher positions within the *vein*. (b) *Sub-levels* driven off at intervals from the manway rises provide access from which *longholes* can be drilled within the *vein*. These holes are charged, and then fired at the end of each shift. (c) Broken *ore* which falls under gravity into the stope is extracted (*"mucked out"*) via the *draw-points* at the foot of the *stope* and loaded into tubs or onto a conveyor belt. (d) Drilling and firing of *longholes* is then repeated all the way back along the *sub-level* to the *manway rise* over the succeeding days, leaving an empty *stope* behind.
Subsidence	Lowering of the ground surface due to closure of underground voids.
Subsidence depression	A low-lying area of ground formed by *subsidence*. Synonyms: *subsidence hollow*. Small subsidence depressions may be synonymous with *crown holes*.
Subsidence hollow	A *subsidence depression*.
Sump	(i) A pond excavated at a low point within a body of surface or *deep mine* workings, to collect water which can then be pumped from the mine. (ii) A *staple shaft* in a metal ore mine (northern England usage).
Sump dewatering	Preventing the flooding of workings solely by pumping from one or more *sumps*.
Surface mine	Any mine in which the miner and/or his machinery works in the open air.
Surface mining	The process of *working* a surface mine.
Swally or Swilley	A body of ponded water in a low lying stretch of a *roadway* (from which the *roadway* rises in both directions).
Tail gate	The furthest from the *main shaft* of the two *roadways* which run either side of a *longwall panel*. (See also *main gate*).
Tailings	Finely crushed particles of *gangue* material, produced during the *beneficiation* of ore. Synonym: *slimes*.
Tailings dam	A large sedimentation pond in which *tailings* are allowed to settle

	from suspension in mineral processing effluents prior to discharge of the supernatant to natural watercourses. Synonym: *slimes dam*.
Tailings impoundment	Synonym of *Tailings dam*.
Tailings management facility	Euphemistic synonym of *tailings dam*.
Tailings pond	Synonym of *Tailings dam*.
Tailings storage facility	Euphemistic synonym of *tailings dam*.
Tip	See "*waste rock pile*".
Tubbing	Concentric rings, usually of steel or iron and timber, used to line shafts to prevent water ingress.
Underhand stoping	*Stoping* undertaken by excavating downwards from an access *roadway*.
Upcast	Adjective applied to signify that a given *shaft* or *adit* conducts (or conducted) ventilation out of the workings. (see also "downcast").
Valley dam	A *tailings dam* constructed by impoundment of a natural valley.
VCR	*Vertical Crater Retreat*, i.e. a specialized form of *stoping*.
Vein	(Geol.) A vertical or sub-vertical tabular *orebody*.
Vertical crater retreat	See *VCR*.
Void	(i) Any cavity left behind by mining (ii) The *open pit* in a *surface mine*.
Void migration	The vertical movement of a *void* towards the ground surface by means of successive failure of the roof strata. If a migrating void reaches the ground surface, it forms a *crown hole*.
Washery fines or finings	Fine-grained dirt left behind after the washing of coal. Analogous to *tailings* in metalliferous mineral processing, and typically disposed of in a *finings pond*.
Waste rock	*Overburden* or *gangue* material removed from the ore/coal during mining or mineral processing.
Waste rock pile	A pile of *overburden, gangue* or other *waste rock* produced during the working of a mine. Synonyms: *bing* (Scottish usage), *gob pile* (US coal mining usage), *spoil heap*, *tip* (UK usage).
Watergate	See "*drainage adit*".
Water-level	See "*drainage adit*".
Waterloose	See "*drainage adit*".
Water make	The amount of water entering a mine (surface of deep) from natural sources (i.e. all water in the workings except that introduced artificially for drill cooling, dust suppression etc).
Way	(i) A mine *roadway*. (ii) A rail track in a mine.
Whimsey	A shaft used for haulage. (See *gin whimsey*).
Wingwalls	The support walls typically constructed either side of the access route to a *portal* from the surface.
Winze	The Cornish term for a *staple shaft* in a metal ore mine. (Term also used in parts of the New World influenced by Cornish miners).

Winning	(i) The process of establishing a new *deep mine*
	(ii) A *deep mine*.
	(iii) The *working* of coal/mineral.
Working	(i) The extraction of coal or minerals from a *mine*.
	(ii) A mine *void*.
	(iii) The natural adjustment of strata in a *deep mine* to changes in stress caused by extraction.
Workings	A collective term for mine voids.

REFERENCES

Abbott, M.B., Bathurst, J.C., Cunge, J.A., O'Connell, P.E. and Rasmussen, J. (1986) An introduction to the European Hydrological System – Systéme Hydrologique Européen, "SHE", 2: Structure of a physically-based distributed modelling system. *Journal of Hydrology*, **87**, 61–77.

Ackers, P., White, W.R., Perkins, J.A. and Harrison, A.J.M. (1978) *Weirs and flumes for flow measurement*. John Wiley and Sons, Chichester, 327 pp.

Ackman, T.E. (1982) *Sludge disposal from acid mine drainage treatment*. US Bureau of Mines Report RI 8672, 25 pp.

Ackman, T.E. (1987) Conceptual operation and closure techniques to prevent perpetual treatment of acid mine drainage. In: White. J.M. (editor), *Proceedings of the Third International Conference on Innovative Mining Systems*, University of Missouri, Rolla, November 2–4, 1987, pp. 63–72.

Ackman, T.E. (2000) Feasibility of lime treatment at the Leviathan Mine using the In-Line System. *Mine Water and the Environment*, **19**, 56–75.

Ackman, T.E. and Place, J.M. (1987) Acid mine water aeration and treatment system. US Patent No. 4,695,378, Sept 22.

Ackman, T.E. and Kleinmann, R.L.P. (1993) An in-line system for the treatment of mine water. *International Mine Waste Management*, **1**(3), 1–4.

Adams, R. (2000) *Simulation of groundwater redound in abandoned coalfields: development, testing and application of a generic model*. PhD Thesis, Department of Civil Engineering, University of Newcastle. 211 pp.

Adams, R. and Younger, P.L. (2000) Simulating groundwater rebound in a recently-closed tin mine. *Proceedings of the 7th International Mine Water Association Congress*, Katowice, Poland, September 11–15th 2000, pp. 218–228.

Adams, R. and Younger, P.L. (2001) A strategy for modelling ground water rebound in abandoned deep mine systems. *Ground Water*, **39**, 249–261.

Adlem, C.J.L., Maree, J.P. and McCrindle, R.I. (1994) Water softening in the barium sulphide process for sulphate removal. In: *Proceedings of the 5th International Mine Water Congress*, Nottingham, UK. Volume 2, pp. 601–611.

Ager, D.V. (1993) *The nature of the stratigraphical record*. (Third edition). Wiley, Chichester, 151 pp.

Ahmad, M.U. (1970) A hydrological approach to control acid mine pollution for Lake Hope. *Ground Water*, **8**, 19–24.

Ahmad, M.U. (1974) Coal Mining and its Effect on Water Quality. *Water Resources Problems Related to Mining*. Proceedings No 18, American Water Resources Association, pp. 138–148.

Alcolea, A., Ayora, C., Bernet, O., Bolzicco, J., Carrera, J., Cortina, J.L., Coscera, G., de Pablo, J., Domenèch, C., Galache, J., Gibert, O., Knudby, C., Mantecón, R., Manzano, M., Saaltink, M. and Silgado, A. (2001) Barrera geoquímica. *Boletín Geológico y Minero*, **112**, 229–255.

Aldous, P.J. and Smart, P.L. (1988) Tracing ground-water movement in abandoned coal mined aquifers using fluorescent dyes. *Ground Water*, Vol. 26, pp. 172–178.

Aljoe, W.W. (1994a) Application of an analytical ground water flow model to a pseudokarst setting in a surface coal mine spoil. In: *Proceedings of the International Land Reclamation and Mine Drainage Conference and the Third International Conference on the Abatement of Acidic Drainage*. Pittsburgh, PA, April 1994. Volume 4 (US Bureau of Mines Special Publication SP 06D-94), pp. 190–198.

Aljoe, W.W. (1994b) Hydrologic and water quality characteristics of a partially-flooded, abandoned underground coal mine. In: *Proceedings of the International Land Reclamation and Mine Drainage Conference and the Third International Conference on the Abatement of Acidic Drainage*. Pittsburgh, PA, April 1994. Volume 2 (US Bureau of Mines Special Publication SP 06B-94), pp. 178–187.

Aljoe, W.W. and Hawkins, J.W. (1994) Application of aquifer testing in surface and underground coal mines. In: *Proceedings of the 5th International Mine Water Congress*, Nottingham (UK), September 1994. Vol. 1, pp. 3–21.

American Society for Testing and Materials (ASTM). (1996) *ASTM Designation: D 5744-96 – Standard Test Method for Accelerated Weathering of Solid Materials Using a Modified Humidity Cell*. ASTM, West Conshohocken, PA.

Amos, P. and Younger, P.L. (2002) Substrate characterisation for a subsurface reactive barrier to treat colliery spoil leachate. *Water Research* (in press).

Anderson, M.P. and Woessner, W. (1992) *Applied Groundwater Modeling. Simulation of Flow and Advective Transport*. Academic Press, San Diego, CA, 381 pp.

Appelo, C.A.J. and Postma, D. (1993) *Geochemistry, Groundwater and Pollution*. A.A. Balkema Publishers, Rotterdam, 536 pp.

Argall, G.O. and Brawner, C.O. (1979) *Mine Drainage*. (Proceedings of the First International Mine Drainage Symposium, Denver, Colorado, May 1979). Miller Freeman Publications Inc., San Francisco, CA, 848 pp.

Arnesen, R.T., Bjerkeng, B. and Iversen, E.R. (1997) Comparison of model predicted and measured copper and zinc concentrations at three Norwegian underwater tailings disposal sites. In: *Proceedings of the Fourth International Conference on Acid Rock Drainage* (held Vancouver, Canada, May 31–June 6, 1997). Volume IV, pp. 1833–1847.

Arnold, D.E. (1991) Diversion wells – a low cost approach to treatment of acid mine drainage. In: *Proceedings of the Twelfth Annual West Virginia Task Force Symposium,* held Morgantown, West Virginia, 3–4 April 1991.

Arze-Quintanilla, G. (1994) Un nuevo enfoque para el control ambiental. In: Montes de Oca, I., and Canedo, C., (Editors), *Mesa redonda: Impacto ambiental en Minería*. Asociación Boliviana Para El Avance de la Ciencia/Academía Nacional de Ciencias de Bolivia, La Paz, pp. 23–50.

Aston, T.R.C. and Whittaker, B.N. (1985) Undersea longwall mining subsidence with special reference to geological and water occurrence criteria in the north-east of England. *Mining Science and Technology*, **2**, 105–130.

Aston, T. and Lamb, T. (1993) An evaluation of groundwater at the Gays River Mine, Halifax County, Nova Scotia. *CIM Bulletin*, **86**, 46–54.

Athay, D., Nairn, R.W. and Strevett, K.A. (2001) Biotic and abiotic iron oxidation kinetics in net alkaline mine drainage. In: *Proceedings of the 2001 Meeting of the American Association for Surface Mining and Reclamation,* June 3–7, 2001, Albuquerque, New Mexico, USA, pp. 458–471.

Atkin, S.A. and Schrand, W.D. (2000) Limnology of the Sleeper Pit Lake, Humboldt County, Nevada. *Proceedings of the Fifth International Conference on Acid Rock Drainage*, (ICARD 2000), Denver, Colorado, May 21–24, 2000. Volume I, pp. 347–357.

Awberry, H.G. (1988) The protection of the Nottinghamshire Coalfield by the Bentinck Colliery minewater concentration scheme. *International Journal of Mine Water*, **7**, 9–24.

Azcue, J.M. (1999) *Environmental impacts of mining activities. Emphasis on mitigation and remedial measures*. Springer, Heidelberg, 300 pp.

Baird, A.J. (1997) Continuity in hydrological systems. In: Wilby, R.L. (editor), *Contemporary hydrology. Towards holistic environmental science*. Wiley, Chichester, pp. 25–58.

Baird, A.J. and Wilby, R.L. (1999) *Eco-hydrology. Plants and water in terrestrial and aquatic environments*. Routledge, London, 402 pp.

Banks, D. (2001) A variable-volume, head-dependent mine water filling model. *Ground Water*, **39**, 362–365.

Banks, D., Younger, P.L. and Dumpleton, S. (1996) The historical use of mine-drainage and pyrite-oxidation waters in central and eastern England, United Kingdom. *Hydrogeology Journal*, **4**(4), 55–68.

Banks, D., Younger, P.L., Arnesen, R.-T., Iversen, E.R. and Banks, S.D. (1997) Mine-water chemistry: the good, the bad and the ugly. *Environmental Geology*, **32**(3), 157–174.

Banwart, S. (1997) Aqueous speciation at the interface between geological solids and groundwater. In: *Modelling in Aquatic Chemistry* (Eds. I. Grenthe and I. Puigdomenech), pp. 245–287. Nuclear Energy Agency, OECD, Paris.

Banwart S. (1994) Surface processes in water technology. In: *Chemistry of Aquatic Systems: Local and Global Perspectives* (Eds. G. Bidoglio and W. Stumm), Kluwer, Dordrecht, pp. 307–335.

Banwart, S., Destouni, G. and Malmström, M. (1998) Assessing mine water pollution: from laboratory to field scale. In: Herbert, M. and Kovar, K. (editors) *Groundwater Quality: Remediation and Protection,* IAHS Publication No. 250. (ISSN0144-7815), pp. 307–311.

Barton, P. (1978) The acid mine drainage. In: Nriagu, J.O. (Editor) *Sulfur in the Environment. Part II Ecological Impacts*. Wiley, New York, pp. 313–358.

Batty, L.C. (1999) *Metal removal processes in wetlands receiving acid mine drainage*. PhD Thesis, Department of Animal and Plant Sciences, University of Sheffield, UK, 254 pp.

Batty, L.C., Baker, A.J.M., Wheeler, B.D. and Curtis, C.D. (2000) The effect of pH and plaque on the uptake of Cu and Mn in *Phragmites australis* (Cav.) Trin ex. Steudel. *Annals of Botany*, **86**, 647–653.

Batty, L.C., Baker, A.J.M. and Wheeler, B.D. (2001) Metal removal processes in a natural wetland receiving acid mine drainage. *Acta Biotechnologica* (in review).

Batty, L.C., Baker, A.J.M. and Wheeler, B.D. in press. Aluminium and phosphate uptake by *Phragmites australis*: The role of Fe, Mn and Al root plaques. *Annals of Botany*

Bayless, E.R. and Olyphant, G.A. (1993) Acid-generating salts and their relationship to the chemistry of groundwater and storm runoff at an abandoned mine site in southwestern Indiana, USA. *Journal of Contaminant Hydrology*, **12**, 313–328.

Bekendam, R.F. and Pottgens, J.J. (1995) Ground movements over the coal mines of southern Limburg, the Netherlands, and their relation to rising mine waters. In: Proceedings of the 5th International Symposium on Land Subsidence. The Hague, Netherlands 16–20th October, 1995. *International Association of Hydrological Sciences*, Publication no 234, pp. 3–12.

Bell, F.G. (1996) Dereliction: Colliery spoil heaps and their rehabilitation. *Environmental and Engineering Geoscience*, **2**, 85–96.

Benner, S.G., Blowes, D.W. and Ptaceck, C.J. (1997) A full-scale porous reactive wall for prevention of acid mine drainage. *Ground Water Monitoring and Restoration*, **17**, no. 4, 99–107.

Benner, S.G., Blowes, D.W. and Molson, J.W.H. (2001) Modeling preferential flow in reactive barriers: implications for performance and design. *Ground Water*, **39**, 371–379.

Benner, S.G., Blowes, D.W. and Ptacek, C.J. (1997) A full-scale porous reactive wall for prevention of acid mine drainage. *Ground Water Monitoring and Restoration*, **17**, (no. 4), 99–107.

Benner, S.G., Blowes, D.W., Gould, W.D., Herbert, R.B. and Ptaceck, C.J. (1999) Geochemistry of a permeable reactive barrier for metals and acid mine drainage. *Environmental Science and Technology*, **33**, 2793–2799.

Benner, S.G., Blowes, D. and Ptacek, C. (2000) Long-term performance of the Nickel Rim Reactive Barrier: A summary. *Proceedings of the Fifth International Conference on Acid Rock Drainage*, (ICARD 2000), Denver, Colorado, May 21–24, 2000. Volume II, pp. 1221–1225.

Best, G.A. and Aikman, D.I. (1983) The treatment of ferruginous groundwater from an abandoned colliery. *Water Pollution Control*, **82**, 557–566.

Beynon, H., Cox, A. and Hudson, R. (2000) *Digging up trouble. The environment, protest and opencast coal mining*. Rivers Oram Press, London, 306 pp.

Bigham, J.M, Schwertmann, U. and Carlson, L. (1992) Mineralogy of precipitates formed by the bio-geochemical oxidation of Fe(II) in mine drainage, *Catena Supplement*, **21**, 219–232.

Blight, G.E. (1997) Disastrous mudflows as a consequence of tailings dyke failures. *Proceedings of the Institution of Civil Engineers – Geotechnical Engineering*, **125**, 9–18.

Blowes, D.W., Reardon, E.J., Jambor, J.L. and Cherry, J.A. (1991) The formation and potential importance of cemented layers in inactive sulfide mine tailings. *Geochimica et Cosmochima Acta*, **55**, 965–978.

Bochenska, T., Fiszer, J. and Kalisz, M. (2000) Prediction of groundwater inflow into copper mines of the Lubin Glogow Copper District. *Environmental Geology*, **39**, 587–594.

Bond, P.L., Smriga, S.P. and Banfield, J.F. (2000) Phylogeny of microorganisms populating a thick, subaerial, predominantly lithotrophic biofilm at an extreme acid mine drainage site. *Applied And Environmental Microbiology*, **66**, 3842–3849.

Boon, M., Snijder, M., Hansford, G.S. and Heijnen, J.J. (1998) The oxidation kinetics of zinc sulphide with *Thiobacillus ferrooxidans*. *Hydrometallurgy*, **48**, 171–186.

Booth, C.J. and Bertsch, L.P. (1999) Groundwater geochemistry in shallow aquifers above longwall mines in Illinois, USA. *Hydrogeology Journal*, **7**, 561–575.

Booth, C.J. and Spande, E.D. (1992) Potentiometric and aquifer property changes above subsiding longwall mine panels, Illinois Basin Coalfield. *Ground Water*, **30**, 362–368.

Booth, C.J., Spande, E.D., Pattee, C.T., Miller, J.D. and Bertsch, L.P (1998) Positive and negative impacts of longwall mining on a sandstone aquifer. *Environmental Geology*, **34**, 223–233.

Booth, C.J., Curtiss, A.M., Demaris, P.J. and van Roosendaal, D.J. (1999) Anomalous increases in piezometric levels in advance of longwall mining subsidence. *Environmental and Engineering Geoscience*, **5**, 407–417.

Boshoff, G.A. (1999) *Development of integrated biological processing for the biodesalination of sulphate- and metal-rich wastewaters*. PhD Thesis, Rhodes University, South Africa, 164 pp.

Bosman, D.J. (1983) Lime treatment of acid mine water and associated solids/liquid separation. *Water Science and Technology*, **15**, 71–84.

Bowell, R.J., Barta, J., Mansanares, W. and Parshley, J. (1998) Geological controls on pit lake chemistry: Implications for the assessment of water quality in inactive open pits. In: *Proceedings of the International Mine Water Association Symposium on "Mine Water and Environmental Impacts"*, Johannesburg, South Africa, 7–13th September 1998. (Volume II), 375–386.

Brady, B.H.G. and Brown, E.T (1993) *Rock mechanics for underground mining*. (Second edition). Chapman & Hall, New York, 571 pp.

Brant, D.L., Ziemkiewicz, P.F. and Robbins, E.I. (1999) Passive removal of manganese from acid mine drainage at the Shade mining site, Somerset County (PA), USA. In: Goldsack, D., Belzile, N., Yearwood, P., and Hall, G., *Proceedings of "Sudbury '99: Mining and the Environment II"*, held Sudbury, Ontario, Canada. Volume 3, pp. 1241–1249.

Bras, R.L. (1990) *Hydrology. An Introduction to hydrologic science*. Addison-Wesley Publishing Company, Reading, MA. 643 pp.

Brassington, R. (1988) *Field hydrogeology*. (Second edition). Wiley, Chichester, 248 pp.

Bredehoeft, J.D., Papadopolous, S.S. and Cooper, H.H. (1982) Groundwater: The water-budget myth. In: *Scientific basis of water resource management*. Studies in Geophysics. National Academy Press, Washington DC. Chapter 4, pp. 51–57.

Brodie, G.A. (1990) Treatment of acid drainage using constructed wetlands: experiences of the Tennessee Valley Authority. In: Graves, D.H. and Devore, R.W. (Editors), *Proceedings of the 1990 Symposium on Mining, OES Publications,* University of Kentucky, Lexington, Kentucky, USA, pp. 77–83

Brodie, G.A., Hammer, D.A. and Tomljanovich, D.A. (1988) Constructed wetlands for acid control in the Tennessee Valley. *Proceedings of the International Land Reclamation and Mine Drainage Conference and the Third International Conference on the Abatement of Acidic Drainage*. (Pittsburgh, PA, 19–21st April 1988). Volume 1: Mine water and mine waste. US Bureau of Mines Information Circular IC–9183, pp. 325–331.

Brown, M.E. (1997) *The amelioration of contaminated mine water by wetlands*. Unpublished PhD Thesis, Camborne School of Mines, University of Exeter, UK.

Brusseau, M.L. (1996) Evaluation of simple methods for estimating contaminant removal by flushing. *Ground Water*, **34**, 19–22.

Burke S. and Banwart S. (2001) A geochemical model for removal of Fe(II)(aq) from minewater discharges. *Applied Geochemistry*, in press.

Burke, S.P. and Younger, P.L. (2000) Groundwater Rebound in the South Yorkshire Coalfield: A First Approximation using the GRAM Model. *Quarterly Journal of Engineering Geology and Hydrogeology*, **33**, 149–160.

Buckley, J.A. (1992) *The Cornish Mining Industry – A Brief History*. Tor Mark Press, Penryn. 48 pp.

Burrell, R. and Friel, S. (1996) The effect of mine closure on surface gas emissions. In: Proceedings of the IBC UK Conference on *"The Environmental Management of Mining Operations"*, held 23–24th September 1996, London, UK, 15 pp.

Bussière, B., Aubertin, M. and Chapuis, R.P. (2000) An investigation of slope effects on the efficiency of capillary barriers to control AMD. *Proceedings of the Fifth International Conference on Acid Rock Drainage*, (ICARD 2000), Denver, Colorado, May 21–24, 2000. Volume II, pp. 969–977.

Butter, T.J., Evison, L.M., Hancock, I.C., Holland, F.S., Matis, K.A., Philipson, A., Sheikh, A.I. and Zouboulis, A.I. (1998) The removal and recovery of cadmium from dilute aqueous solutions by biosorption and electrolysis at laboratory scale. *Water Research*, **32**, 400–406.

Cain, P., Forrester, D.J. and Cooper, R. (1994) Water inflows at Phalen Colliery in the Sydney Coalfield, and their relation to interaction of workings. In: *Proceedings of the 5th International Minewater Congress*, Nottingham, UK. Volume 1, pp. 83–95.

Cairney, T. and Frost, R.C. (1975) A case study of mine water quality deterioration, Mainsforth Colliery, County Durham. *Journal of Hydrology*, **25**, 275–293.

Calder, J.W (1973) *Inrush at Lofthouse Colliery, Yorkshire. Report on the cause of, and circumstances attending, the inrush which occurred at Lofthouse Colliery, Yorkshire, on 21 March 1973.* Department of Trade and Industry (Cmnd 5419). HMSO, London, 26 pp. plus plans.

Calver, A. (1997) Recharge Response Functions. *Hydrology and Earth Systems Sciences*, **1**, 47–53.

Campbell, C.S. and Ogden, M.H. (1999) *Constructed wetlands in the sustainable landscape.* Wiley, New York, 270 pp.

Canty, M. (2000) Innovative in situ treatment of acid mine drainage using sulfate-reducing bacteria. *Proceedings of the Fifth International Conference on Acid Rock Drainage,* (ICARD 2000), Denver, Colorado, May 21–24, 2000. Volume II, pp. 1139–1147.

Capper, P.L. and Cassie, W.F. (1976) *The mechanics of engineering soils.* (Sixth Edition). E & F N Spon, London, 376 pp.

Carrera, J., Alcolea, A., Bolzicco, J., Bernet, O., Knudby, C., Manzano, M., Saaltink, M., Ayora, C., Domenech, C., de Pablo, J., Cortina, J., Coscera, G., Gilbert, O., Galache, J., Silgado, A. and Mantecon, R. (2001) An experimental geochemical barrier at Aznalcóllar. In: Thornton, S.F., and Oswald, S., (editors) *Preprints of papers for the 3rd International Conference on Groundwater Quality,* (GQ2001, University of Sheffield, UK, 18–21 June 2001), pp. 407–409.

Caruccio, F.T. (1975) Estimating the Acid Potential of Coal Mine Refuse. In: Chadwick, M.J., and Goodman, G.T., (Editors), *The Ecology of Resource Degradation and Renewal.* (Proceedings of the 15th Symposium of the British Ecological Society, 10–12 July 1973). Blackwell Scientific, Oxford, pp. 197–205.

Caruccio, F.T. and Ferm, J.C. (1974) Paleoenvironment – predictor of acid mine drainage problems. *Proceedings of the 5th Coal Mine Drainage Research Symposium*, National Coal Association (USA), Kentucky, pp. 5–9.

Cartwright, A.P (1969) *West Driefontein – ordeal by water.* Gold Fields of South Africa Ltd. J G Ince & Son Ltd, South Africa. 83 pp.

Casey, T.J. (1997) *Unit treatment processes in water and wastewater engineering.* Wiley, Chichester. 280 pp.

Castro, J.M. and Moore, J.N. (2000) Pit lakes: their characteristics and the potential for their remediation. *Environmental Geology*, **39**, 1254–1260.

Catalan, L.J.J., Yanful, E.K., Boucher, J.-F. and Shelp, M.L. (2000) A field investigation of tailings re-suspension in a shallow water cover. *Proceedings of the Fifth International Conference on Acid Rock Drainage,* (ICARD 2000), Denver, Colorado, May 21–24, 2000. Volume II, pp. 921–931.

Chandler, R.J. and Tosatti, G. (1995) The Stava tailings dams failure, Italy. *Proceedings of the Institution of Civil Engineers – Geotechnical Engineering*, **113**, 67–79.

Chang, I.S., Shin, P.K. and Kim, B.H. (2000) Biological treatment of acid mine drainage under sulphate-reducing conditions with solid waste materials as a substrate. *Water Research*, **34**, 1269–1277.

Chanson, H (1999) *The hydraulics of open channel flow.* Arnold, London, 495 pp.

Chen, M., Soulsby, C. and Younger, P.L. (1999) Modelling the evolution of minewater pollution at Polkemmet Colliery, Almond catchment, Scotland. *Quarterly Journal of Engineering Geology*, **32**, 351–362.

Chow, V.T., Maidment, D.R. and Mays, L.W. (1988) *Applied Hydrology.* McGraw-Hill, New York. 572 pp.

Christensen, B., Laake, M. and Lien, T. (1996) Treatment of acid mine water by sulfate reducing bacteria; results from a bench scale experiment. *Water Research*, **30**, 1617–1624.

Clarke, A.M. (1962) Some structural, hydrological and safety aspects of recent developments in south-east Durham. *The Mining Engineer*, **122**, 209–231.

Clarke, L.B. (1995) *Coal mining and water quality.* IEA Coal Research Report No IEACR/80. International Energy Agency, London, 100 pp.

Clarke, L.B. (1996) Environmental aspects of coalbed methane production, with emphasis on water treatment and disposal. *Transactions of the Institution of Mining and Metallurgy*, (Section A), **105**, A105–A113.

Cleary, D. and Thornton, I. (1994) The environmental impact of gold mining in the Brazilian Amazon. In: Hester, R.E., and Harrison, R.M. (Editors), *Mining and its environmental impact.* Issues in Environmental Science and Technology Volume 1. Royal Society of Chemistry, Cambridge, pp. 17–29.

Cliff, M.I. and Smart, P.C. (1988) The use of recharge trenches to maintain groundwater levels. *Quarterly Journal of Engineering Geology*, **31**, 137–145.

Coal Authority (2000) *Underground Coal Gasification Newsletter No. 1.* April 2000. The Coal Authority, Mansfield UK, 4 pp.

Cohen, R.R.H. (1996) The technology and operation of passive mine drainage treatment systems. In: US EPA (editors) Managing environmental problems at inactive and abandoned metals mine sites. EPA Seminar Publication No EPA/625/R-95/007, pp. 18–29.

Cohen, R.R.H. and Staub, M.W. (1992) *Technical manual for the design and operation of a passive mine drainage treatment system.* Report to the US Bureau of Land Reclamation. Colorado School of Mines, Golden, CO. (December 1992), 69 pp.

Cook, N.G.W. (1982) Ground-water problems in open-pit and underground mines. In: Narasimhan, T.N., (Editor), *Recent trends in hydrogeology*. Geological Society of America, Special Paper 189. Boulder, Colorado, pp. 397–405.

Cooper, P. (2001) Constructed wetlands and reed-beds: mature technology for the treatment of wastewater from small populations. *Journal of the Chartered Institution of Water and Environmental Management*, **15**, 79–85.

Cornell University (1999) *Iron Toxicity: What You Don't Know.* Web pages at URL: *http://www.ansci.cornell.edu/plants/toxicagents/iron.html.*

Cousens, R.R.M. and Garrett, W.S. (1970) The flooding at West Driefontein, South Africa. *Proceedings of the Conference "Mining and Petroleum Technology", 9th Commonwealth Mining and Metallurgy Congress.* Institution of Mining and Metallurgy, London. Volume 1: 931–972.

Cranstone, D. (1992) To hush or not to hush – where, when and how? In: Chambers, B. (editor), *Men, mines and minerals of the North Pennines*. The Friends of Killhope, Weardale, UK, pp. 41–48.

Cravotta, C.A. and Trahan, M.K. (1999) Limestone drains to increase pH and remove dissolved metals from acidic mine drainage. *Applied Geochemistry*, **14**, 581–606.

de Marsily, G. (1986) *Quantitative Hydrogeology. Groundwater Hydrology for Engineers*. Academic Press, Orlando. 440 pp.

Demchak, J., Morrow, T. and Skousen, J. (2001) Treatment of acid mine drainage by four vertical flow wetlands in Pennsylvania. *Geochemistry: Exploration, Environment, Analysis*, **1**, 71–80.

Demchak, J., Skousen, J., Bryant, G. and Ziemkiewicz, P. (2000) Comparison of water quality in fifteen underground coal mines in 1968 and 1999. *Proceedings of the Fifth International Conference on Acid Rock Drainage*, (ICARD 2000), Denver, Colorado, May 21–24, 2000. Volume II, pp. 1045–1052.

Dempsey, B.A., Roscoe, H.C., Ames, R., Hedin, R. and Jeon, B-H. (2001) Ferrous oxidation chemistry in passive abiotic systems for the treatment of mine drainage. *Geochemistry: Exploration, Environment, Analysis*, **1**, 81–88.

Desouza, E.M. and Penner, R. (1993) Computer design of underground mine dewatering systems. *CIM Bulletin*, **86**, 55–59.

Diaz, M.A., Monhemius, A.J. and Narayanan, A. (1997) Consecutive hydroxide-sulphide precipitation treatment of acid rock drainage. In: *Proceedings of the 4th International Conference on Acid Rock Drainage,* held Vancouver, British Columbia, May 31st–June 6th 1997. Volume III, 1181–1193.

Dill, S., du Preez, I., Graff, M. and Maree, J. (1994) Biological sulphate removal from acid mine drainage utilising producer gas as a carbon and energy source – process limitations and their resolution. In: *Proceedings of the 5th International Mine water Congress*, Nottingham, UK. Volume 2, pp. 631–641.

Diodato, D.M. and Parizek, R.R. (1994) Unsaturated hydrogeologic properties of reclaimed coal strip mines. *Ground Water*, **32**, pp. 108–118.

Domenico, P.A. and Schwartz, F.W. (1990) *Physical and chemical hydrogeology*. Wiley, New York. 824 pp.

Dooge, J.C.I. (1988) Hydrology in perspective. *Hydrological Sciences Journal*, **33**, 61–85.

Donnelly, L.J. (2000) The reactivation of geological faults during mining subsidence from 1859 to 2000 and beyond. In: *Proceedings of the Conference on "The Legacy of Mineral Extraction"*, organised by the Institution of Mining and Metallurgy and the North of England Institute of Mining and Mechanical Engineers, Newcastle Upon Tyne 18–19th May 2000, pp. 83–121.

Dottridge, J.A. and Keeble, M.R. (1998) Modelling the impact of dewatering in the Lower Greensand aquifer, Surrey. *Quarterly Journal of Engineering Geology*, **31**, 129–135.

Doyle, A. (1997) *Sacrifice, achievement, gratitude. Images of the Great Northern Coalfield in decline*. County Durham Books, Durham, 115 pp.

Driscoll, F.G. (1986) *Groundwater and wells.* (Second edition). Johnson Filtration Systems Inc., St Paul, Minnesota, 1089 pp.

Duckham, H. and Duckham, B.F. (1973) *Great pit disasters: Great Britain 1700 to the present day*. David and Charles, Newton Abbot, 227 pp.

Dudeney, A.W.L. (1997) Removal and utilisation of iron from contaminated waters. *Clay Technology*, **54**, 8–10.

Dudeney, A.W.L., Ball, S. and Monhemius, A.J. (1994) *Treatment processes for ferruginous discharges from disused coal workings*. National Rivers Authority R&D Note 243. National Rivers Authority, Bristol, UK, 60 pp.

Dudgeon, B.A. (1997) Mine water management study for the Lake Cowal gold project, NSW, Australia. In: Veselic, M., and Norton, P.J., (Editors), *Mine Water and the Environment. Proceedings of the 6th International Mine Water Association Congress*, held in Bled, Slovenia, 8–12 September 1997. Volume 2, pp. 273–282.

Dudgeon, C.R. (1985a) Effects of non-Darcy flow and partial penetration on water levels near open-pit excavations. In: *Proceedings of the 18th Congress of the International Association of Hydrogeologists*, Cambridge, England, September 1985, pp. 122–132.

Dudgeon, C.R. (1985b) Unconfined non-Darcy flow near open-pit mines. In: *Proceedings of the 2nd Congress of the International Mine Water Association*, Granada, Spain, September 1985, pp. 443–454.

Dudgeon, C.R. (1997) Problems in separating the effects of drought and mine dewatering in a farming area surrounding a limestone mine. In: Veselič, M. and Norton, P.J. (Editors), *Proceedings of the 6th International Mine Water Association Congress*, held Bled, Slovenia, 8–12th September 1997. Volume 1, pp. 221–233.

Dumpleton, S. and Glover, B.W. (1995) *The impact of colliery closures on water resources with particular regard to the NRA Severn-Trent Region*. British Geological Survey Technical Report WD/95/40. 44 pp.

Durham, A.J.P., Wilson, G.W. and Currey, N. (2000) Field performance of two low infiltration cover systems in a semi arid environment. *Proceedings of the Fifth International Conference on Acid Rock Drainage*, (ICARD 2000), Denver, Colorado, May 21–24, 2000. Volume II, pp. 1319–1326.

Dvorak, D.H. (1996) *The feasibility of using anaerobic water treatment at the Hardin Run Clay Mine*. Report to the Crescent Brick Company, New Cumberland, West Virginia. US Bureau of Mines, Pittsburgh. 33 pp.

Dvorak, D.H., Hedin, R.S., Edenborn, H.M. and McIntyre, P.E. (1992) Treatment of metal-contaminated water using bacterial sulfate reduction: results from pilot-scale reactors. *Biotechnology and Bioengineering*, **40**, 609–616.

Dzombak, D.A. and Morel, F.M.M. (1990) *Surface Complexation Modelling. Hydrous Ferric Oxide*. John Wiley and Sons, New York.

Eden, R.A., Stevensen, I.P. and Edwards, M.A. (1957) Geology of the Country around Sheffield. *Mem. Geol. Surv. Great Britain, Geol. Sheet 100*, H.M.S.O., London.

Edwards, R. (1996) Toxic sludge flows through the Andes. *New Scientist*, 21 Nov 1996, p. 4.

Edwards, P.J. and Maidens, J.B. (1995) Investigations into the impacts of ferruginous minewater discharges in the Pelenna Catchment on salmonid spawning gravels and visual amenity. *National Rivers Authority Welsh Region Internal Report No. PL/EAW/95/6*. Cardiff, UK.

Eger, P. and Wagner, J. (2001) Sulfate reduction – decreases in substrate reactivity and implication for long-term treatment. In: *Proceedings of the 2001 Meeting of the American Association for Surface Mining and Reclamation*, June 3–7, 2001, Albuquerque, New Mexico, USA, pp. 542–557.

Elliott, P., Ragusa, S. and Catcheside, D. (1998) Growth of sulfate-reducing bacteria under acidic conditions in an upflow anaerobic bioreactor as a treatment system for acid mine drainage. *Water Research*, **32**, 3724–3730.

Ellis, D.V. and Robertson, J.D. (1999) Underwater emplacement of mine tailings: case examples and principles. In: Azcue, J.M. (editor) *Environmental impacts of mining activities. Emphasis on mitigation and remedial measures*. Springer, Heidelberg, pp. 123–141.

Elsworth, D. and Liu, J. (1995) Topographic influence of longwall mining on groundwater. *Ground Water*, **33**, 786–793.

Eriksson, N. and Destouni, G. (1997) Combined effects of dissolution kinetics, secondary mineral precipitation, and preferential flow on copper leaching from mining waste rock. *Water Resources Research*, **33**, 471–483.

Eriksson, N., Gupta, A. and Destouni, G. (1997) Comparative analysis of laboratory and field tracer tests for investigating preferential flow and transport in mining waste rock. *Journal of Hydrology*, **194**, 143–163.

Evangelou, V.P. (1995a) *Pyrite oxidation and its control: solution chemistry, surface chemistry, acid mine drainage*. CRC Press, Florida, 293 pp.

Evangelou, V.P. (1995b) Potential microencapsulation of pyrite by artificial inducement of ferric phosphate coatings. *Journal of Environmental Quality*, **24**, 535–542.

Evangelou V.P. and Zhang Y.L. (1995) A review: pyrite oxidation mechanisms and acid mine drainage prevention. *Critical Reviews in Environmental Science and Technology*, **25**, 2, 141–199.

Evenson, C.J. and Nairn, R.W. (2000) Enhancing phosphorus sorption capacity with treatment wetland iron oxyhydroxides. *Proceedings of the 17th National Meeting of the American Society for Surface Mining and Reclamation*, Tampa, FL, June 11–15, 2000.

Ettner, D.C. (1999) Pilot scale constructed wetland for the removal of nickel from tailings drainage, southern Norway. In: *Proceedings of the Congress of the International Mine Water Association*, Sevilla Spain, 13–17 September 1999. Volume I, pp. 207–211.

Ewen, and O'Connell, P.E. (2000) SHETRAN: Distributed river basin flow and transport modelling system. *Journal of Hydrologic Engineering (ASCE)*, **5**, 250–258.

Faulkner, B. and Skousen, J. (1993) Monitoring of passive treatment systems: an update, In: *Proceedings of the Fourteenth Annual West Virginia Surface Mine Drainage Task Force Symposium*, April 27–28, 1993, University of West Virginia, Morgantown, West Virginia, USA.

Faulkner, S.P. and Richardson, C.J. (1989) Physical and chemical characteristics of freshwater wetland soils. In: Hammer, D.A., (editor), *Constructed wetlands for wastewater treatment. Municipal, industrial, agricultural*. Lewis Publishers, Chelsea MI, pp. 41–72.

Fawcett, R.J., Hibberd, S. and Singh, R.N. (1984) An appraisal of mathematical models to predict water inflows into underground workings. *International Journal of Mine Water*, **3**, 33–54.

Fernández-Rubio, R. (1979) Drainage of coal and lignite mines. In: Argall, G.O. and Brawner, C.O. (editors), *Mine Drainage*. (Proceedings of the First International Mine Drainage Symposium, Denver, Colorado, May 1979). M Freeman Publications, San Francisco, pp. 492–506.

Fernández-Rubio, R. (1982) Investigations about the origin of water inflow at Reocín Mine (Santander, Spain). *Proceedings of the First International Mine Water Congress*, Budapest, Hungary, April 19th–24th 1982. International Mine Water Association, pp. 101–120.

Fernández-Rubio, R., Fernández Lorca, S. and Esteban Arlegui, J. (1987) Preventive techniques for controlling acid water in underground mines by flooding. *International Journal of Mine Water*, **6**, 39–52.

Fernández-Rubio, R., León Fábregas, A., Baquero Úbeda, J.C., and Lorca Fernánez, D. (1998) Underground mining drainage: state of the art. In: *Proceedings of the International Mine Water Association Symposium on "Mine Water and Environmental Impacts"*, Johannesburg, South Africa, 7th–13th September 1998. Volume I, pp. 87–112.

Fetter, C.W. (1994) *Applied Hydrogeology*. (Third Edition). Macmillan, 691 pp.

Fetter, C.W. (1999) *Contaminant hydrogeology*. (Second edition). Prentice Hall, New Jersey, 500 pp.

Finkelman, R.B. and Griffin, D.E., 1986, Hydrogen peroxide oxidation: an improved method for rapidly assessing acid-generating potential of sediments and sedimentary rocks. *Reclamation and Revegetation Research*, **5**, 521–534.

Forth, R.A. (1994) Ground settlement and sinkhole development due to the lowering of the water table in the Bank Compartment, South Africa. In: Oliveira, R., Rodrigues, L.F., Coelho, A.G. and Cunha, A.P. (editors), *Proceedings of the 7th Congress of the International Association of Engineering Geologists*, Lisbon, Portugal, 5–9th September 1994. Balkema, Rotterdam, pp. 2933–2940.

Freeze, R.A. and Cherry, J.A. (1979) *Groundwater*. Prentice Hall, Englewood Cliffs, NJ. 604pp.

Frost, R. C. (1979) Evaluation of the rate of decrease in the iron content of water pumped from a flooded shaft mine in County Durham, England. *Journal of Hydrology*, **40**, 101–111.

Fuenkajorn, K. and Daemen, J.J.K. (1996) *Sealing of boreholes and underground excavations in rock*. Chapman & Hall, London, 329 pp.

García-Guinea, J. and Huascar, M. (1997) Mining waste poisons river basin. *Nature*, **387**, 118.

Garritty, P. (1980) *Effects of mining on surface and subsurface water bodies*. Unpublished PhD Thesis. Department of Mining Engineering, University of Newcastle Upon Tyne, UK.

Garritty, P. (1982) Water percolation into fully caved longwall faces. In: Farmer, I.W. (editor), *Strata Mechanics*. (Proceedings of the Symposium on Strata Mechanics held in Newcastle Upon Tyne, 5–7 April, 1982). Developments in Geotechnical Engineering No. 32. Elsevier, Amsterdam, pp. 25–29.

Gatzweiler, R., R. Hähne, M. Eckart, J. Meyer and S. Snagovsky (1997) Prognosis of the flooding of uranium mining sites in east Germany with the help of numerical box-modeling. In: *Proceedings of the*

6th International Mine Water Association Congress, "Minewater and the Environment", held Bled, Slovenia, 8–12th September 1997. Volume 1, pp. 57–64.

Gavaskar, A.R., Gupta, N., Sass, B.M., Janosy, R.J. and O'Sullivan, D. (1998) *Permeable barriers for groundwater remediation: design construction and monitoring.* Battelle Press, Columbus, OH. 176 pp.

Geller, W., Klapper, H. and Salomons, W. (editors) (1998) *Acidic mining lakes. Acid mine drainage, limnology and reclamation.* Springer, Heidelberg. 435 pp.

Germain, D., Tassé, N. and Dufour, C. (2000) A novel treatment for acid mine drainage, using a wood-waste cover preventing sulfide oxidation. *Proceedings of the Fifth International Conference on Acid Rock Drainage,* (ICARD 2000), Denver, Colorado, May 21–24, 2000. Volume II, pp. 987–998.

Girts, M.A., Erickson, P.M. and Kleinmann, R.L.P., (1987) Performance data on Typha and Sphagnum wetlands constructed to treat coal mine drainage. In: *Proceedings of the 8th Annual Surface Mine Drainage Task Force Symposium,* University of West Virginia, Morgantown, West Virginia, USA.

Glover, H.G. (1983) Mine water pollution – an overview of problems and control strategies in the United Kingdom. *Water Science and Technology,* **15**, 59–70.

Gómez de las Heras, J. and Rivadeneira de Vega, J.A. (1999) Dissolution mining and environmental effects in Polanco (Cantabria – Spain). In: *Proceedings of the Congress of the International Mine Water Association,* Sevilla Spain, 13–17 September 1999. Volume II, pp. 807–814.

Goldsmith, J.R. and Graf, D.L. (1957) The system $CaO-MnO-CO_2$: Solid-solution and decomposition relations. *Geochimica et Cosmochimica Acta,* **11**, 310–334.

Goodman, R.E., Moye, D.G., van Schalkwyk, A. and Javandel, I. (1965) Ground water inflows during tunnel driving. *Engineering Geology,* **2**, 39–56.

Gordon, N.D., McMahon, T.A. and Finlayson, B.L. (1992) *Stream Hydrology. An Introduction for Ecologists.* Wiley, Brisbane, 526 pp.

Grapes, T.R. and Connelly, R.J. (1998) Evaluation of mine water inflows and associated impacts at the Olympias Mine, Halkidiki, Greece. In: *Proceedings of the International Mine Water Association Symposium on "Mine Water and Environmental Impacts",* Johannesburg, South Africa, 7th–13th September 1998. (Volume I), pp. 13–23.

Greenhill, P.G. (2000) AMIRA International: AMD research through industry collaboration. *Proceedings of the Fifth International Conference on Acid Rock Drainage,* (ICARD 2000), Denver, Colorado, May 21–24, 2000. Volume I, pp. 13–19.

Greenwell, G.C. (1888) *A glossary of terms used in the coal trade of Northumberland and Durham.* Third edition. Bemrose & Sons, London. (Facsimile edition published 1970 by Frank Graham Publishers, Newcastle Upon Tyne). 92 pp.

Grimalt, J.O., Ferrer, M. and Macpherson, E. (1999) The mine tailing accident in Aznalcóllar. *Science of the Total Environment,* **242**, 3–11.

Grimshaw, P.N., (1992) *Sunshine Miners. Opencast coalmining in Britain, 1942–1992.* British Coal Opencast, Mansfield, Notts. 113 pp.

Grmela, A., and Tylcer, J. (1997) Determination of coal mines water balance. In: Veselic, M. and Norton, P.J. (Editors), *Mine Water and the Environment. Proceedings of the 6th International Mine Water Association Congress,* held in Bled, Slovenia, 8–12 September 1997. Volume 1, pp. 35–42.

Groenewold, G.H. and Rehm, B.W. (1982) Instability of contoured surface-mined landscapes in the northern Great Plains: causes and implications. trip mine spoils – western North Dakota. *Reclamation and Revegetation Research,* **1**, 161–176.

Grosser, J.R., Hagelgans, V., Hentshcel, T. and Priester, M. (1994) Heavy metals in stream sediments – a gold mining area near Los Andes, southern Colombia. *Ambio,* **23**, 146–149.

Gupta, P.K. and Singh, T.N. (1994) Water resource characterization of underground coal working – a case study for part of the Jharia Coal Field. In: *Proceedings of the 5th International Mine Water Congress,* Nottingham, UK. Volume 1: 97–104.

Gusek, J.J. (1998) Three case histories of passive treatment of metal mine drainage. In: *Proceedings of the 19th Annual West Virginia Surface Mine Drainage Task Force Symposium.* West Virginia University, Morgantown, WV, USA.

Gusek, J.J. (2000) Reality check: passive treatment of mine drainage, an emerging technology or proven methodology. In: *Proceedings of the Society of Mining Engineers Annual Meeting,* February 28–March 1, Salt Lake City, Utah, USA.

Gusek, J., Mann, C., Wildeman, T. and Murphy, D. (2000) Operational results of a 1,200 gpm passive bioreactor for metal mine drainage, West Fork, Missouri. In: *Proceedings of the Fifth International Conference on Acid Rock Drainage*, (ICARD 2000), Denver, Colorado, May 21–24, 2000. Volume 2, pp. 1133–1137.

Gustafsson, H.E., Lundgren, T., Lindvall, T., Lindahl, L-E., Eriksson, N., Jönsson, H., Broman, P.G. and Göransson, T. (1999) The Swedish acid mine drainage experience: research development and practice. In: Azcue, J.M. (editor), *Environmental impacts of mining activities. Emphasis on mitigation and remedial measures.* Springer, Heidelberg, pp. 203–228.

Hammack, R.W., Edenborn, H.M. and Dvorak, D.H. (1994) Treatment of water from an open-pit copper mine using biogenic sulfide and limestone: a feasibility study. *Water Research*, **28**, 2321–2329.

Hammer, D.A. (editor) (1989) *Constructed wetlands for wastewater treatment. Municipal, industrial, agricultural.* Lewis Publishers, Chelsea MI. 831 pp.

Hammer, D.A. (1992) *Creating freshwater wetlands*. CRC Press/Lewis Publishers, Boca Raton. 298 pp.

Harries, J.R. and Ritchie, A.I.M. (1985) Pore gas composition in waste rock dumps undergoing pyritic oxidation. *Soil Science*, **140**, 143–152.

Harrison, R., Scott, W.B. and Smith, T. (1989) A Note on the Distribution, Levels and Temperatures of Minewaters in the Northumberland and Durham Coalfield. *Quarterly Journal of Engineering Geology*, **22**, 355–358.

Hartman, H.L., 1987, *Introductory mining engineering*. Wiley, New York. 633 pp.

Hawke, C.J. and José, P.V., (1996) *Reedbed Management for Commercial and Wildlife Interests*. Royal Society for the Protection of Birds, Sandy, Bedfordshire. 212 p.

Hawkins, J.W. (1994) Modeling of a reclaimed surface coal mine spoil aquifer using MODFLOW. In: *Proceedings of the International Land Reclamation and Mine Drainage Conference and the Third International Conference on the Abatement of Acidic Drainage*. Pittsburgh, PA, April 1994. Volume 2 (US Bureau of Mines Special Publication SP 06B–94), pp. 265–272.

Headworth, H.G., Puri, S. and Rampling, B.H. (1980) Contamination of a Chalk aquifer by mine drainage at Tilmanstone, east Kent, U.K. *Quarterly Journal of Engineering Geology*, **13**, 105–117.

Hedin, R.S. (1998) Potential recovery of iron oxides from coal mine drainage. In: *Proceedings of the Nineteenth Annual West Virginia Surface Mine Drainage Task Force Symposium*, West Virginia University, Morgantown, USA.

Hedin, R.S. (1999) *Recovery of Iron Oxides from Polluted Coal Mine Drainage*, Patent No. 5,954,969. U.S. Patent and Trademark Office, Washington D.C.

Hedin, R.S. and Nairn, R.W. (1993) Contaminant removal capabilities of wetlands constructed to treat coal mine drainage. In: Moshiri, G.A. (Editor) *Constructed Wetlands for Water Quality Improvement*. Lewis Publishers, Chelsea, Michigan, USA, pp. 187–197.

Hedin, R.S. and Watzlaf, G.R. (1994) The effects of anoxic limestone drains on mine water chemistry. In: *Proceedings of the International Land Reclamation and Mine Drainage Conference and the Third International Conference on the Abatement of Acidic Drainage*, United States Department of the Interior, Bureau of Mines Special Publication SP 06A–94, Washington DC, pp. 185–195.

Hedin, R.S. and Weaver, T.J., in press, Recovery of iron oxide from an AML site. In: Stuart, B. (editor), *Proceedings of the 2001 National Association of Abandoned Mine Lands Annual Conference*, Aug 19–22, 2001, Athens, Ohio, USA.

Hedin, R.S., Hammack, R.W. and Hyman, D.M. (1988) The potential importance of sulfate reduction in wetlands constructed to treat mine drainage. In: *1988 Mine Drainage and Surface Mine Reclamation, Volume I: Mine Water and Mine Waste*, US Bureau of Mines Information Circular 9183, US Department of the Interior, Washington DC., pp. 382–389.

Hedin, R.S., Nairn, R.W. and Kleinmann, R.L.P. (1994) *Passive treatment of polluted coal mine drainage*. Bureau of Mines Information Circular 9389. United States Department of Interior, Washington DC. 35 pp.

Hedin, R.S., Watzlaf, G.R. and Nairn, R.W. (1994) Passive treatment of acid mine drainage with limestone. *Journal of Environmental Quality*, **23**, 1338–1345.

Hellier, W.W. (1989) Constructed wetlands in Pennsylvania: An overview. In: Salley, J., McCready, R. and Winchlacz, P. (editors), *Proceedings of Biohydrometallurgy 1989*, Jackson Hole, Wyoming, pp. 599–612.

Hellier, W.W., Giovannitti, E.F. and Slack, P.T. (1994) Best professional judgment analysis for constructed wetlands as a best available technology for the treatment of post-mining groundwater seeps. In: Proceedings of the International Land Reclamation and Mine Drainage Conference and the Third

International Conference of the Abatement of Acidic Drainage. Volume 1: Mine Drainage. *US Bureau of Mines Special Publication SP 06A-94,* pp. 60–69.

Henton, M.P. (1979) Abandoned Coalfields: Problem of Pollution. *The Surveyor,* Vol. 153, pp. 9–11.

Henton, M.P. (1981) The problem of water table rebound after mining activity and its effects on ground and surface water quality. In: van Duijvenbooden, W., Glasbergen, P. and van Lelyveld, H. (Editors), *Quality of Groundwater.* (Proceedings of an International Symposium, Noordwijkerhout, The Netherlands, held 23–27th March 1981). Elsevier, The Netherlands, pp. 111–116.

Hering, J. and Stumm, W. (1990) Oxidative and reductive dissolution of minerals. In: *Mineral-Water Interface Geochemistry,* Hochella M.F. and White A.F., Eds, Reviews in Mineralogy, Vol. 23. The Mineralogical Society of America, Washington, D.C.

Herlihy, A.T. and Mills, A.L. (1985) Sulfate reduction in freshwater sediments receiving acid mine drainage. *Applied Environmental Microbiology,* **49,** 179–186.

Herlihy, A.T., Mills, A.L., Hornberger, G.M. and Bruckner, A.E. (1987) The importance of sediment sulfate reduction to the sulfate budget of an impoundment receiving acid mine drainage. *Water Resources Research,* **23,** 287–292.

Herschy, R.W. (1995) *Streamflow Measurement.* (Second Edition). E & F N Spon, London, 524 pp.

Hey, R.D., (1997) Geomorphological basis of river restoration. In: Large, A.R.G. (editor), Floodplain rivers: hydrological processes and ecological significance. *British Hydrological Society Occasional Paper No. 8,* pp. 104–119.

Hobbs, S.L. and Gunn, J. (1998) The hydrogeological effect of quarrying karstified limestone: options for prediction and mitigation. *Quarterly Journal of Enginering Geology,* **31,** 147–157.

Hoek, E., Kaiser, P.K. and Bawden, W.F. (2000) Support of underground excavations in hard rock. Balkema, Rotterdam, 215 pp.

Höglund, L.O. (2000) Mitigation of the environmental impact from mining waste (MiMi) – a Swedish multidisciplinary research programme. *Proceedings of the Fifth International Conference on Acid Rock Drainage,* (ICARD 2000), Denver, Colorado, May 21–24, 2000. Volume I, pp. 3–12.

Holgate, R. (1991) *Prehistoric flint mines.* Shire Publications Ltd, Princes Risborough, UK. 56 pp.

Holmes P.R. and Crundwell F.K. (2000) The kinetics of the oxidation of pyrite by ferric ions and dissolved oxygen: an electrochemical study. *Geochimica et Cosmochimica Acta,* **64,** 263–274.

Hongze, G., Butler, A., Wheater, H. and Vesovic, V. (1997) Modelling the in-situ neutralisation capacity of a karst aquifer for remediation of acid mine drainage. In: Beck, B.F., Stephenson, J.B. and Herring, J.G. (editors), *The engineering geology and hydrology of karst terranes.* Balkema, Rotterdam, pp. 219–224.

Hoover, H.C. and Hoover, L.H. (1950) *De Re Metallica* by Georgius Agricola. Translated from the First Latin Edition of 1556. Dover Publications Inc., New York, 638 pp.

Hubbert, M.K. (1940) The theory of ground-water motion. *Journal of Geology,* **48,** 785–944.

Hughes, D.B. and Clarke, B.G. (2001) The River Aire slope failure at the St Aidans Extension Opencast Coal Site, Wet Yorkshire, United Kingdom. *Canadian Geotechnical Journal,* **38,** 239–259.

Huntsman, B.E., Solch, J.G. and Porter, M.D. (1978) Utilization of Sphagnum species dominated bog for coal acid mine drainage abatement. *Proceedings of the Geological Society of America, 91st Annual Meeting (Abstracts),* Toronto, Ontario, Canada, p. 322.

Hustwit, C.C., Ackman, T.E. and Erickson, P.M. (1991) Role of Oxygen Transfer in Acid Mine Drainage Treatment. *US Bureau of Mines Report RI 9405,* 18 pp.

Huyakorn, P.S. and Pinder, G.F. (1983) *Computational methods in subsurface flow.* Academic Press, Orlando, 473 pp.

ICOLD (2001) *Tailings Dams – Risk of Dangerous Occurrences, Lessons learnt from practical experiences.* International Commission on Large Dams (ICOLD) Bulletin 121. Jointly published by ICOLD, and the United Nations Environmental Programme (UNEP) Division of Technology, Industry and Economics (DTIE). ICOLD, Paris 2001, 144 pp.

Brown, J.M.B. and Ingram, H.A.P. (1988) Changing storage beneath a stationary water-table – an anomaly of certain humified peats. *Quarterly Journal of Engineering Geology,* **21,** 177–182.

Institute For Water Quality Studies (1997) *Status report: Monitoring the impact of saline mine water from the Grootvlei Mine on aspects of the water environmental quality of the Blesbokspruit wetland system.* Government of the Republic of South Africa, Department of of Water Affairs and Forestry (DWAF). DWAF Report No.N/C210/RMQ/1196.

Iribar, V., Izco, F., Tames, P., Antigüedad, I. and da Silva, A. (2000) Water contamination and remedial measures at the Troya abandoned Pb–Zn mine (The Basque Country, northern Spain). *Environmental Geology*, **39**, 800–806.

James, A. (1993) *An Introduction to water quality modelling* (Second Edition). Wiley, New York. 311 pp.

Jarvis, A.P. (2000) *Design, construction and performance of passive systems for the treatment of mine and spoil heap drainage*. Unpublished PhD Thesis, Department of Civil Engineering, University of Newcastle, UK, 231 pp.

Jarvis, A.P. and Younger, P.L. (1997) Dominating chemical factors in mine water induced impoverishment of the invertebrate fauna of two streams in the Durham Coalfield, UK. *Chemistry and Ecology*, **13**, 249–270.

Jarvis, A.P. and Younger, P.L. (1999) Design, construction and performance of a full-scale wetland for mine spoil drainage treatment, Quaking Houses, UK. *Journal of the Chartered Institution of Water and Environmental Management*, **13**, 313–318.

Jarvis, A.P. and Younger, P.L. (2000) Broadening the scope of mine water environmental impact assessment – a UK perspective. *Environmental Impact Assessment Review*, **20**, 85–96.

Jarvis, A.P. and Younger, P.L. (2001) Passive treatment of ferruginous mine waters using high surface area media. *Water Research* (in press).

Johnson, D.B. (1998) Biological abatement of acid mine drainage: the role of acidophilic protozoa and other indigenous microflora. In: Geller, W., Klapper, H. and Salomons, W. (editors). *Acidic Mining Lakes: acid mine drainage, limnology and reclamation*, Springer-Verlag, Berlin, pp. 285–302.

Johnson, D.B., Dziurla, M.A. and Kolmert, A. (2000) Novel approaches for bio-remediation of acidic, metal-rich effluents using indigenous bacteria. *Proceedings of the Fifth International Conference on Acid Rock Drainage*, (ICARD 2000), Denver, Colorado, May 21–24, 2000. Volume 2, pp. 1209–1217.

Johnson, K.L. and Younger, P.L. (2000) Abandonment of Frazer's Grove Fluorspar Mine, North Pennines, UK: Prediction and observation of water level and chemistry changes after closure. *Proceedings of the 7th International Mine Water Association Congress*, Katowice, Poland, September 11th–15th 2000, pp. 271–279.

Johnson, R.H., Blowes, D.W., Robertson, W.D. and Jambor, J.L. (2000) The hydrogeochemistry of the Nickel Rim mine tailings impoundment, Sudbury, Ontario. *Journal of Contaminant Hydrology*, **41**, 49–80.

Johnston, R.H. (1989) The hydrologic responses to development in regional sedimentary aquifers. *Ground Water*, **27**, pp. 316–322.

Johnston, R.H. (1997) Sources of water supplying pumpage from regional aquifer systems of the United States. *Hydrogeology Journal*, **5**, 54–63.

Jones, P.M., Mulvay, S.M. and Fish, D. (1994) The role of sulfate and ionic strength on the shift from acid to alkaline mine drainage in southwest Pennsylvania. In: Proceedings of the International Land Reclamation and Mine Drainage Conference and the Third International Conference of the Abatement of Acidic Drainage. Volume 2: Mine Drainage. *US Bureau of Mines Special Publication SP 06B-94*, 289–295.

Kadlec, R.H. (2000) The inadequacy of first-order treatment wetland models. *Ecological Engineering*, **15**, 105–119.

Kalin, M. (1998) Biological polishing of zinc in a mine waste management area. In: Geller, W., Klapper, H. and Salomons, W. (editors), *Acidic mining lakes. Acid mine drainage, limnology and reclamation*. Springer, Heidelberg, pp. 321–334.

Kelly, M.G. (1988) *Mining and the freshwater environment*. Elsevier Applied Science, London, 231 pp.

Kelly, M.G. (1999) Effects of heavy metals on the aquatic biota. In: Plumlee, G.S. and Logsdon, M.J. (editors), *The environmental geochemistry of mineral deposits. Part A: Processes, techniques and health issues*. Reviews in Economic Geology, Volume 6A. Society of Economic Geologists, Littleton, CO, pp. 363–371.

Kepler, D.A. and McCleary, E.C. (1994) Successive Alkalinity Producing Systems (SAPS) for the Treatment of Acidic Mine Drainage. *Proceedings of the International Land Reclamation and Mine Drainage Conference and the 3rd International Conference on the Abatement of Acidic Drainage*. (Pittsburgh, PA; April 1994). Volume 1: Mine Drainage, pp. 195–204.

Kepler, D.A. and E.C. McCleary, (1997) Passive aluminum treatment successes. In: *Proceedings of the 18th Annual West Virginia Task Force Symposium*, held Morgantown, West Virginia, 15–16 April 1997.

Kesserû, Z. (1994) In-mine sealing of water inrushes. In: *Proceedings of the 5th International Minewater Congress*, Nottingham, UK. Volume 1, 269–280.

Kesserû, Z. (1995) New approaches and results on the assessment of risks due to undermining for mines safety and for protection of water resources. In: *Proceedings of the International Conference on "Water Resources at Risk"*, held by the American Institute of Hydrology and the International Mine Water Association, Denver, Colorado, May 14th–18th 1995, pp. IMWA 53–IMWA 72.

Kesserû, Z. (1997) Argillaceous geological barriers – multi-disciplinary view. In: Veselic, M. and Norton, P.J. (Editors), *Proceedings of the 6th International Mine Water Association Congress*, held Bled, Slovenia, 8–12th September 1997. Volume 2, pp. 397–413.

Kilkenny, W.M. (1968) *A study of the settlement of restored opencast coal sites and their suitability for building development*. Research bulletin No 38, Department of Civil Engineering, University of Newcastle Upon Tyne, 33 pp.

King, J.K., Kostka, J., Frischer, M.E., Saunders, F.M. and Jahnke, R.A. (2001) A quantitative relationship that demonstrates mercury methylation rates in marine sediments are based on the community composition and activity of sulfate-reducing Bacteria. *Environmental Science and Technology*, **35**, 2491–2496.

Kleinmann, R.L.P. (1980) Bactericidal control of acid problems in surface mines and coal refuse. *Proceedings of the Symposium on Surface Mining Hydrology, Sedimentology and Reclamation*, U. of Kentucky, Lexington, Kentucky 40506, December 15, 1980.

Kleinmann, R.L.P. (editor) (2000) *Prediction of water quality at surface coal mines*. Manual prepared by members of the Prediction Workgroup of the Acid Drainage Technology Initiative (ADTI). National Mine Land Reclamation Center, West Virginia University, Morgantown, West Virginia, USA, 241 pp.

Kleinmann, R.L.P. and Crerar, D. (1979) Thiobacillus ferrooxidans and the formation of acidity in simulated coal mine environments. *Geomicrobiology*, **1**, 373–388.

Kleinmann, R.L.P., Crerar, D and Pacelli, R.R. (1981) Biogeochemistry of acid mine drainage and a method to control acid formation. *Mining Engineering*, **33**, 300–306.

Koumantakis, I. and Dimitrakopoulos, D. (1999) Impacts of dewatering of Amynteon open lignite mine of the aquatic environment of Lake Chimatidis, west Macedonia, Greece. In: *Proceedings of the International Congress on Mine, Water and the Environment*, held by the International Mine Water Association, Sevilla, Spain, September 13–17th 1999. Volume 1, pp. 63–67.

Kowalczyk, A., Motyka, J. and Szuwarzynski, M. (2000) Groundwater contamination as a potential result of closing down the Trzebionka Mine, southern Poland. *Proceedings of the 7th International Mine Water Association Congress*, Katowice, Poland, September 11th–15th 2000, pp. 299–308.

Klohn, E.J. (1979) Seepage control for tailings dams. In: Argall, G.O. and Brawner, C.O. (editors), *Mine Drainage*. (Proceedings of the first International Mine Drainage Symposium, Denver, Colorado, May 1979. Miller Freeman Publications, San Francisco, pp. 671–725.

Knight Piésold and Partners (1995) *Wheal Jane Minewater Study. Environmental Appraisal and Treatment Strategy*. Report to the National Rivers Authority South Western Region, Exeter.

Kruseman, G.P. and de Ridder, N.A. (1990) *Analysis and evaluation of pumping test data*. (Second edition). International Institute for Land Reclamation and Improvement (ILRI), Wageningen, The Netherlands. ILRI Publication No. 47, 377 pp.

Kwong, Y.T. J. and Ferguson, K.D. (1997) Mineralogical changes during NP determinations and their implications. In: *Proceedings of the Fourth International Conference on Acid Rock Drainage*, Vol. 1., 435–447, Vancouver, B.C., May 31–June 6, 1997.

Lachmar, T.E. (1994) Application of fracture-flow hydrogeology to acid mine drainage at the Bunker Hill Mine, Kellogg, Idaho. *Journal of Hydrology*, **155**, 125–149.

Ladwig, K.J., Erickson, P.M., Kleinmann, R.L.P. and Posluszny, E.T. (1984) Stratification in water quality in inundated anthracite mines, eastern Pennsylvania. *US Bureau of Mines Report of Investigations RI-8837*. US Bureau of Mines, Pittsburgh, 35 pp.

Laine, D.M. (1997) The treatment of the pumped mine water discharge at Woolley Colliery, West Yorkshire. In: Younger, P.L. (Editor), *Mine water Treatment Using Wetlands*. Proceedings of a National Conference held 5th September 1997, at the University of Newcastle, UK. Chartered Institution of Water and Environmental Management, London, pp. 83–103.

Laine, D.M. (1998) The treatment of pumped and gravity minewater discharges in the UK and an acidic tip seepage in Spain. In: *Proceedings of the International Mine Water Association Symposium on "Mine*

Water and Environmental Impacts", Johannesburg, South Africa, 7th–13th September 1998 (Volume II), pp. 471–490.

Laine, D.M. (2000) Passive water treatment. *World Coal*, **9**(8), 43–46.

Lamb, H.M., Dodds-Smith, M. and Gusek, J. (1998) Development of a long-term strategy for the treatment of acid mine drainage at Wheal Jane. In: Geller, W., Klapper, H. and Salomons, W. (editors), *Acidic mining lakes. Acid mine drainage, limnology and reclamation*. Springer, Heidelberg, pp. 335–346.

Lamont-Black, J., Younger, P.L. and Batty, L.C. (2001) *Factual report of test results for use in designing phosphate removal by passive wetlands*. Report prepared for Cundall Johnson & Partners Ltd by Department of Civil Engineering, University of Newcastle, UK, 11 pp.

Langmuir, D. (1997) *Aqueous Environmental Geochemistry*. Prentice Hall, New Jersey.

Lawrence, R.W. and Wang, Y. (1997) Determination of neutralisation potential in the prediction of acid rock drainage. In: *Proceedings of the Fourth International Conference on Acid Rock Drainage*, Vol. 1, 451–464, Vancouver, B.C., May 31–June 6, 1997.

Lebecka, J., Lukasik, B. and Chalupnik, S. (1994) Purification of saline waters from coal mines from radium and barium. In: *Proceedings of the 5th International Mine Water Congress*, Nottingham, UK. Volume 2, pp. 663–672.

Lebecka, J., Chalupnik, S., Michalik, B., Wysocka, M., Skubacz, K. and Mielnikow, A. (1994) Radioactivity of mine waters in the Upper Silesian Coal Basin and its influence on the natural environment. In: *Proceedings of the 5th International Minewater Congress*, Nottingham, UK. Volume 2, 657–662.

Ledin, M. and Pedersen, K., (1996) The environmental impact of mine wastes – Roles of microorganisms and their significance in treatment of mine wastes. *Earth-Science Reviews*, **41**, 67–108.

Lei, S. (1999) An analytical solution for steady flow into a tunnel. *Ground Water*, **37**, 23–26.

Lemon, R. (1991) Pumping and disposal of deep strata mine water. *Mining Technology*, March 1991, 69–76.

Lerner, D.N., Issar, A.S. and Simmers, I. (1990) Groundwater recharge. A guide to understanding and estimating natural recharge. *International Contributions to Hydrogeology, Volume 8. International Association of Hydrogeologists*. Verlag Heinz Heise, Hannover, 345 pp.

Lewis, P.R. and Jones, G.D.B. (1970) Roman gold-mining in north-west Spain. *Journal of Roman Studies*, **60**, 169–185.

Li, M.G. (1997) Neutralisation potential versus observed mineral dissolution in humidity cell tests for Louvicourt tailings. In: *Proceedings of the Fourth International Conference on Acid Rock Drainage*, Vol. 1, 149–164, Vancouver, B.C., May 31–June 6, 1997.

Li, G.Y. and Zhou, W.F. (1999) Sinkholes in karst mining areas in China and some methods of prevention. *Engineering Geology*, **52**, 42–50.

Li, M., Catalan, L.J.J. and St-Germain, P. (2000) Rates of oxygen consumption by sulphidic tailings under shallow water covers – field measurements and modelling. *Proceedings of the Fifth International Conference on Acid Rock Drainage*, (ICARD 2000), Denver, Colorado, May 21–24, 2000. Volume II, pp. 913–920.

Lind, C.J. and Hem, J.D. (1993) Manganese minerals and associated fine particulates in the streambed of Pinal Creek, Arizona, USA: a mining-related acid drainage problem. *Applied Geochemistry*, Vol. 8, pp. 67–80

Linsley, R.K., Kohler, M.A. and Paulhus, J.L.H. (1988) *Hydrology for engineers*. (SI Metric edition). McGraw-Hill, London, 492 pp.

Liu, J., Elsworth, D. and Matetic, R.J. (1997) Evaluation of the post-mining groundwater regime following longwall mining. *Hydrological Processes*, **11**, 1945–1961.

Llewellyn, K. (1992) *Disaster at Tynewydd. An account of a Rhondda mine disaster in 1877*. (Second Edition). Church in Wales Publications, Penarth, 92 pp.

Lorber, K.E. and Erhart-Schippek, W. (2000) *AURUL-Dam, Baia Mare and NOVAT-Dam, Baia Borsa, Naramures County, Romania. Report on fact-finding mission 16.04.–18.04.2000*. Report prepared for the World Wide Fund for Nature, Danube Carpathian Programme Office, Vienna. Montanuniversität Leoben, Austria, 83 pp.

Lortie, L, Gould, W.D., Stichbury, M., Blowes, D.W. and Thurel, A. (1999) Inhibitors for the prevention of acid mine drainage (AMD). *Proceedings of the Sudbury '99 Conference "Mining and the Environment II"*, held Sudbury Ontario, Canada. Volume 3, pp. 1191–1198.

Machemer, S.D. and Wildeman, T.R. (1992) Adsorption compared with sulfide precipitation as metal removal processes from acid mine drainage in a constructed wetland. *Journal of Contaminant Hydrology*, **9**, 115–131.

Macklin, M.G., Payne, I., Preston, D. and Sedgwick, C. (1996) *Review of the Porco mine tailings dam burst and associated mining waste problems, Pilcomayo basin, Bolivia*. University of Wales, Aberystwyth. Report to the UK Overseas Development Administration, 33 pp.

Malmström, M., Destouni, G, Banwart, S. and Strömberg, B. (2000) Resolving the scale-dependence of mineral weathering rates. *Environmental Science and Technology*, **34**, No. 7, 1375–1377.

Manzano, M., Ayora, C., Domenech, C., Navarette, P., Garralon, A. and Turrero, M.-J. (1999) The impact of the Aznalcóllar mine tailing spill on groundwater. *Science of the Total Environment*, **242**, 189–209.

Marsden, M., Holloway, D. and Wilbraham, D. (1997) The Position in Scotland. In: Bird, L. (Editor), *Proceedings of the UK Environment Agency Conference on "Abandoned Mines: Problems and Solutions"*. Held at Tapton Hall, University of Sheffield, 20th–21st March 1997, pp. 76–84.

Mason, E. (editor) (1951) *Practical coal mining for miners*. (Second edition). Virtue and Company Ltd, London. Volumes I and II. 787 pp.

McCready, R.G.L. (1982) Bacterial oxidation of sulfur as a means of reclaiming Solonetzic soil. *Solonetzic Soils in Alberta*, J.C. Hermans, Ed, 13–31, Alta Agric.

McCready, R.G.L. and Krouse, H.R. (1982b) Sulfur isotope fractionation during the oxidation of elemental sulfur by Thiobacilli in a Solonetzic Soil. *Can. J. Soil Science*, **62**, 105–110.

McGregor, R., Blowes, D., Ludwig, R., Pringle, E., Choi, M. Booth, R., Duchene, M. and Pomeroy, M. (1999) Remediation of heavy-metal contaminated groundwater using a porous reactive wall within an urban setting. *Proceedings of the Sudbury '99 Conference "Mining and the Environment II"*, held Sudbury Ontario, Canada. Volume 2, pp. 645–653.

McIntire, P.E. and Edenborn, H.M. (1990) The use of bacterial sulfate reduction in the treatment of drainage from coal mines. In: *Proceedings of the 1990 Conference Mining and Reclamation*, University of West Virginia, Morgantown, West Virginia, USA, pp. 409–415.

McKibben, M.A. and Barnes, H.L. (1986) Oxidation of pyrite in low temperature solutions: rate laws and surface textures. *Geochim. Cosmochim. Acta*, **50**, 1509–1520.

McRae, C.W.T., Blowes, D.W. and Ptacek, C.J. (1999) *In situ* removal of arsenic from groundwater using permeable reactive barriers: a laboratory study. *Proceedings of the Sudbury '99 Conference "Mining and the Environment II"*, held Sudbury Ontario, Canada. Volume 2, pp. 601–609.

Meech, J.A., Veiga, M.M. and Tromans, D. (1997) Emission and stability of mercury in the amazon. *Canadian Metallurgical Quarterly*, **36**, 231–239.

Metcalf and Eddy Inc. (1991) *Wastewater engineering. Treatment, disposal, reuse*. (Third edition). McGraw-Hill, New York, 1334 pp.

Metcalfe, B. (1994) Reclamation of markedly acidic minestone waste tips using sewage sludge as a soil forming material. *Proceedings of the 5th International Mine Water Congress*, Nottingham, UK. Volume 2, pp. 727–757.

Miller, G.C., Lyons, W.B. and Davis, A. (1996) Understanding the water quality of pit lakes. *Environmental Science and Technology*, **30**(3), 118A–123A.

Miller, J.T. and Thompson, D.R. (1974) *Seepage and mine barrier width*. 5th Symposium – Coal Mine Drainage Research. National Coal Association, Kentucky, pp. 103–127.

Miller, S., Robertson, A. and Donohue, T. (1997) Advances in acid drainage prediction using the net acid generation (NAG) test. In: *Proceedings of the Fourth International Conference on Acid Rock Drainage*, Vol. 2, 535–549, Vancouver, B.C., May 31–June 6, 1997.

Minett, S.T. (1987) *The hydrogeology of parts of the Northumberland and Durham Coalfield related to opencast mining operation*. Unpublished PhD Thesis, Newcastle Upon Tyne.

Minett, S.T., Blythe, D.A., Hallam, G.D. and Hughes, D.B. (1986) Analysis of an Advanced Dewatering Scheme at an Opencast Coal Site in Northumberland. In: Cripps, J.C., Bell, F.G. and Culshaw, M.G. (editors), *Groundwater in Engineering Geology*. Engineering Geology Special Publication No 3, pp. 347–352. Geological Society, London.

Mitchell, P., Potter, C. and Watkins, M. (2000) Treatment of acid rock drainage: field demonstration of silica micro encapsulation technology and comparison with an existing caustic soda-based system. *Proceedings from the Fifth International Conference on Acid Rock Drainage* (ICARD 2000), Vol. 2, 1035–1044.

Mitsch, W.J. and Gosselink, J.G. (2000) *Wetlands* (Third Edition). Wiley, New York, 920 pp.

Morgan-Jones, M., Bennett, S. and Kinsella, J.V. (1984) The hydrological effects of gravel-winning in an area west of London, United Kingdom. *Ground Water*, **22**, 154–161.

Motyka, J. and Postawa, A. (2000) Influence of contaminated Vistula River water on the groundwater entering the Zakrzówek limestone quarry, Cracow region, Poland. *Environmental Geology*, **39**, 398–404.

Mueller, R.F. (2001) Microbially mediated thallium immobilization in bench scale systems. *Mine Water and the Environment*, **20**, 17–29.

Muggli, D.L., Pelletier, C.A., Poling, G.W. and Schwamberger, E.C. (2000) Injected ARD plume behaviour in a pit lake utilizing in situ dye studies. *Proceedings of the Fifth International Conference on Acid Rock Drainage*, (ICARD 2000), Denver, Colorado, May 21–24, 2000. Volume I, pp. 305–318.

Nairn, R.W., Griffin, B.C., Strong, J.D. and Hatley, E.L. (2001) Remediation challenges and opportunities at the Tar Creek Superfund Site, Oklahoma. In: *Proceedings of the 2001 Meeting of the American Association for Surface Mining and Reclamation*, June 3–7, 2001, Albuquerque, New Mexico, USA, pp. 579–584

Narayan, K. (1998) Groundwater contamination: the Omai Case. In: Wheater, H. and Kirby, C. (eds.), *Hydrology in a Changing Environment*. (Proceedings of the International Symposium organised by the British Hydrological Society, 6th–10th July 1998, Exeter, UK). Wiley, Chichester. Volume II, pp. 41–46.

National Research Council (1981) Coal mining and ground-water resources in the United States. Report of the Committee on Ground-Water Resources in Relation to Coal Mining, Board on Mineral and Energy Resources, Commission on Natural Resources, National Research Council. National Academy Press, Washington DC, 197 pp.

NCB (1973) *Spoil heaps and lagoons. Technical handbook*. National Coal Board, Mining Department, London. 232 pp.

NCB (1975) *The subsidence engineers' handook*. National Coal Board, Mining Department, London. 111 pp.

NCB (1977) *Surveying practice. NCB (Mining) Codes and Rules*. National Coal Board, Mining Department, London. 41 pp. plus figures.

NCB (1982) *Technical management of water in the coal mining industry*. Mining Department, National Coal Board, London, 129 pp.

Newman, L.L., Herasymuik, G.M., Barbour, S.L., Fredlund, D.G. and Smith, T. (1997) The hydrogeology of waste rock dumps and a mechanism for unsaturated preferential flow. In: *Proceedings of the Fourth International Conference on Acid Rock Drainage* (held Vancouver, Canada, May 31–June 6, 1997). Volume III, pp. 551–565.

Ngah, S.A., Reed, S.M. and Singh, R.N. (1984) Groundwater problems in surface mining in the United Kingdom. *International Journal of Mine Water*, **3**, 1–12.

Nicholson, R.V., Gillham, R.W., Cherry, J.A. and Reardon, E.J. (1989) Reduction of acid generation in mine tailings through the use of moisture-retaining cover layers as oxygen barriers. *Canadian Geotechnical Journal*, **26**, 1–8.

Nordstrom, D.K., Alpers, C.N., Ptacek, C.J. and Blowes, D.W. (2000) Negative pH and extremely acidic mine waters from Iron Mountain, California. *Environmental Science & Technology*, **34**, 254–258.

Nordstrom, D.K. and Southam, G. (1997) Geomicrobiology of sulfide mineral oxidation. In: *Reviews in Mineralogy*, Vol. 35, pp. 361–390. Mineralogical Society of America, Washington, D.C.

Norris, P.R. and Johnson, D.B. (1998) Acidophilic Microorganisms. In: Horikoshi, K. and Grant, W.D. (editors), *Extremophiles: Microbial Life in Extreme Environments*. John Wiley, New York, pp. 133–154.

Norton, P.J. (1983) *A study of groundwater control in British surface coal mining*. Unpublished PhD Thesis, Nottingham University, UK.

Novak, P. (1994) Improvement of water quality in rivers by aeration at hydraulic structures. In: Hino, M. (editor), *Water Quality and its Control*. Balkema, Rotterdam, pp. 147–168.

Nuttall, C.A. and Younger, P.L. (1999) Reconnaissance hydrogeochemical evaluation of an abandoned Pb–Zn orefield, Nent Valley, Cumbria, UK. *Proceedings of the Yorkshire Geological Society*, **52**, 395–405.

Nuttall, C.A. and Younger, P.L. (2000) Zinc removal from hard circum-neutral mine waters using a novel closed-bed limestone reactor. *Water Research*, **34**, 1262–1268.

O'Brien, W. (1996) *Bronze age copper mining in Britain and Ireland*. Shire Publications Ltd, Princes Risborough, UK, 64 pp.

Olem, H. and Unz, R.F. (1977) Acid mine drainage treatment with rotating biological contactors. *Biotechnology and Bioengineering*, **19**, 1475–1491.

Olem, H. and Unz, R.F. (1980) Rotating-disc biological treatment of acid mine drainage. *Journal of the Water Pollution control Federation*, **52**, 257–269.

O'Shay, T.A., Hossner, L.R. and Dixon, J.B. (1990) A modified hydrogen peroxide oxidation method for determination of potential acidity in pyritic overburden. *J. Environ. Quality*, **19**, 778–782.

Orchard, R.J. (1975) Working under bodies of water. *The Mining Engineer*, **170**, 261–270.

Orchard, W.G. (1991) *A glossary of mining terms*. (ISBN 185022 053 0). Dyllansow Truran Publishers, Redruth, 42 pp.

Pamely, C. (1904) *The Colliery Manager's Handbook. A comprehensive treatise on the laying-out and working of collieries, designed as a book of reference for colliery managers and for the use of coal-mining students preparing for first-class certificates*. Crosby Lockwood & Son, London, 1178 pp.

Parkin, G. and Adams, R. (1998) Using catchment models for groundwater problems: evaluating the impacts of mine dewatering and groundwater abstraction. In: Wheater, H. and Kirby, C. (editors), *Hydrology in a Changing Environment*. Volume II. Wiley, Chichester, 269–279.

Peters, T.W. (1978) Mine drainage problems in north Derbyshire. *Mining Engineer*, **137**, 463–474.

Pfeifer, S. and Shubert, I. (1999) The oxidation of pyrite at pH 7 in the presence of reducing and nonreducing Fe(III)-chelators. *Geochim. et Cosmochim. Acta*, 63:19/20, 3171–3182.

Price, W.A. (1997) *DRAFT Guidelines and Recommended Methods for the Prediction of Metal Leaching and Acid Rock Drainage at Minesites in British Columbia*. British Columbia Ministry of Employment and Investment, Energy and Minerals Division, Smithers, B.C.

Phillips, P., Bender, J., Simms, R., Rodriguez-Eaton, S. and Britt, C. (1995) Manganese removal from acid coal-mine drainage by a pond containing green algae and microbial mat. *Water Science and Technology*, Vol. **31**, No. 12, pp. 161–170.

Pirrie, D. and Camm, G.S. (1999) The impact of mining on sedimentation in the coastal zone of Cornwall. In: Scourse, J.D. and Furze, M.F.A. (editors), *The Quaternary of West Cornwall. Field Guide*. Quaternary Research Association, London, pp. 62–73.

Pirrie, D., Camm, G.S., Sear, L.G. and Hughes, S.H. (1997) Mineralogical and geochemical signature of mine waste contamination, Tresillian River, Fal Estuary, Cornwall, UK. *Environmental Geology*, **29**, 58–65.

Plumlee, G.S. and Edelmann, P. (1995) *The Summitville Mine and its downstream effects*. US Geological Survey Open-File Report 95–23. Denver, Colorado, 11 pp.

Plumptre, J.H. (1959) Underground waters of the Kent Coalfield. *Transactions of the Institution of Mining Engineers*, 119, pp. 155–164 plus plates.

Poulin, R., Hadjigeorgiou, J. and Lawrence, R.W. (1996) Layered mine waste co-mingling for mitigation of acid rock drainage. *Transactions of the Institution of Mining and Metallurgy*, (Section A), **105**, pp. A55–A62.

Preene, M., Roberts, T.O.L., Powrie, W. and Dyer, M.R. (2000) *Groundwater control – design and practice*. CIRIA Report C515. Construction Industry Research and Information Association (CIRIA), London. 204 pp.

Pretorius, P.J. and Linder, P.W. (2001) The adsorption characteristics of δ–manganese dioxide: a collection of diffuse double layer constants for the adsorption of H^+, Cu^{2+}, $Ni2+$, Zn^{2+}, $Cd2+$ and Pb^{2+}. *Applied Geochemistry*, **16**, 1067–1082.

Puente, C. and Atkins, J.T. (1987) Simulation of rainfall-runoff response in mined and unmined catchments in coal areas of West Virginia. U.S. *Geological Survey Water Supply Paper 2298*, 48 pp.

Pulles, W. (2000) Development of passive mine water treatment technology. In: *Proceedings of the BIOY2K Conference, (Water Institute of South Africa Conference on "The Biotechnology of Acid Mine Drainage Wastewater Treatment")*, Grahamstown, South Africa, January 23–28th 2000. Invited paper. [Abstract]. BIOY2K Programme and Abstracts Volume, pp. 600–601.

Ranson, C.M. and Edwards, P.J. (1997) The Ynysarwed experience: active intervention, passive treatment and wider issues. In: Younger, P.L. (Editor), *Mine water Treatment Using Wetlands*. Proceedings of a National Conference held 5th September 1997, at the University of Newcastle, UK. Chartered Institution of Water and Environmental Management, London, pp. 151–164.

Ratsep, A. and Liblik, V. (2000) Technogenic waterflows generated by oil shale mining: impact on Purtse catchment rivers. *Oil Shale*, **17**, 95–112.

Reddish, D.J., Yao, X.L., Benbia, A., Cain, P. and Forrester, D.J. (1994) Modelling of caving over the Lingan and Phalen mines in the Sydney Coalfield Cape Breton. In: *Proceedings of the 5th International Minewater Congress*, Nottingham, UK. Volume 1, pp. 105–124.

Reddy, M.R., Raju, N.J., Reddy, Y.V. and Reddy, T.V.K. (2000) Water resources development and management in the Cuddapah district, India. *Environmental Geology*, **39**, 342–352.

Rimstidt, J.D. and Newcomb, W.D. (1993) Measurement and analysis of rate data: the rate of reaction of ferric iron with pyrite. *Geochim. Cosmochim. Acta*, **57**, 1919–1934.

Robins, N.S. (1990) *Hydrogeology of Scotland*. HMSO for British Geological Survey, London. 90 pp.

Robins, N.S. and Younger, P.L. (1996) Coal abandonment – mine water in surface and near-surface environment: Some historical evidence from the United Kingdom. *Proceedings of the Conference "Minerals, Metals and the Environment II"*, Prague, Czechoslavakia, 3–6 September 1996. Institution of Mining and Metallurgy, London, pp 253–262.

Robinson, K.E. and Toland, G.C. (1979) Case histories of different seepage problems for nine tailings dams. In: Argall, G.O. and Brawner, C.O. (editors) *Mine Drainage*. (Proceedings of the first International Mine Drainage Symposium, Denver, Colorado, May 1979. Miller Freeman Publications, San Francisco, pp. 781–800.

Rose, P.D., Boshoff, G.A., van Hille, L.C.M., Dunn, K.M. and Duncan, J.R. (1998) An integrated algal sulphate reducing high rate ponding process for the treatment of acid mine drainage wastewaters. *Biodegradation*, **9**, 247–257.

Rowley, M.V., Warkentin, D.D., Yan, V.T. and Piroshco, B.M. (1994) The Biosulfide Process: integrated biological/chemical acid mine drainage treatment. In: *Proceedings of the International Land Reclamation and Mine Drainage Conference, and the Third International Conference on the Abatement of Acidic Drainage*. Held Pittsburgh, PA, 24–29 April 1994. US Bureau of Mines Special Publication SP 06A-94, pp. 205–213.

Rózkowski, A. and Rózkowski, J. (1994) Impact of mine waters on river water quality in the Upper Silesian Coal Basin. In: *Proceedings of the 5th International Minewater Congress*, Nottingham, UK. Volume 2: 811–821.

Roy, S. and Worrall, F. (1999) Pyrite oxidation in a coal-bearing strata: the use of P-blocking techniques. In: *Proceedings of the International Congress on Mine, Water and Environment*, Vol. II, R.F. Rubio, Ed., 547–550, held in Sevilla, Spain, September 13–17, 1999. International Mine Water Association.

Rieuwerts, J.H. (1998) *Glossary of Derbyshire lead mining terms*. Peak District Mines Historical Society, Matlock Bath, Derbyshire, 192 pp.

Rimmer, D.L. and Younger, A. (1997) Land reclamation after coal-mining operations. In: Hester, R.E. and Harrison, R.M. *Contaminated land and its reclamation*. Thomas Telford, London, pp. 73–90.

Ripley, E.A., Redmann, R.E. and Crowder, A.A. (1995) *Environmental Effects of Mining*. St Lucie Press. 356 pp.

Rogoz, M. (1994) Computer simulation of the process of flooding up a group of mines. *Proceedings of the 5th International Mine Water Congress*, Nottingham, UK. Volume 1, pp. 369–377.

Rózkowski, A. (1997) Impact of coal mining on quality and quantity of fresh groundwater in the Upper Silesian Coal Basin, Poland. In: Veselic, M. and Norton, P.J. (Editors), *Proceedings of the 6th International Mine Water Association Congress*, held Bled, Slovenia, 8–12th September 1997. Volume 1, pp. 95–103.

Runkel, R.L. (1998) One-dimensional transport with inflow and storage (OTIS): a solute transport model for streams and rivers. *US Geological Survey Water Resources Investigations Report 98-4018*, 73 pp.

Rytuba, J.J., Enderlin, D., Ashley, R., Seal, R. and Hunerlach, M.P. (2000) Evolution of the McLaughlin Gold Mine Pit Lakes, California. *Proceedings of the Fifth International Conference on Acid Rock Drainage*, (ICARD 2000), Denver, Colorado, May 21–24, 2000. Volume I, pp. 367–375.

Sadler, P. and Rees, A. (1998) Use of probabilistic methods for sizing a wetland to treat mine water discharging from the former Polkemmet Colliery, West Lothian, Scotland. In: *Proceedings of the International Mine Water Association Symposium on "Mine Water and Environmental Impacts"*, Johannesburg, South Africa, 7th–13th September 1998. (Volume II), pp. 401–411.

Saul, H. (1936) Outcrop water in the South Yorkshire Coalfield. *Transactions of the Institution of Mining Engineers*, **93**, 64–94

Saul, H. (1948) Mine water. *Transactions of the Institution of Mining Engineers*, **107**, 294–310.

Saul, H. (1949) Mines drainage. *Transactions of the Institution of Mining Engineers*, **108**, 359–370.

Saul, H. (1959) Water problems in the coalfields of Great Britain. *Colliery Guardian*, **199**, 191–199; 229–234.

Saul, H. (1970) Current mine drainage problems. *Transactions of the Institution of Mining and Metallurgy* (Section A), **79**, A63–A80.

Schmolke, C.M.R. (1998) Leadhills, Scotland–Flood and pollution threat from disused lead workings. In: Fox, H.R., Moore, H.M. and McIntosh, A.D. (editors), *Land Reclamation: Achieving Sustainable Benefits*. Balkema, Rotterdam, pp. 517–523.

Schnoor, J.L. and Stumm, W. (1985) Acidification of aquatic and terrestrial systems. In: *Chemical Processes in Lakes*, W. Stumm (Ed.). John Wiley and Sons, New York.

Schwartz, F.W. and Crowe, A.S. (1985) Simulation of changes in ground-water levels associated with strip mining. *Geological Society of America Bulletin*, **96**, 253–262.

Scullion, J. and Edwards, R.W. (1980) The effect of pollutants from the coal industry on the fish fauna of a small river in the South Wales Coalfield. *Environmental Pollution (Series A)*, **21**, 141–153.

Sengupta, M. (1993) *Environmental impacts of mining: monitoring, restoration and control*. Lewis Publishers, Boca Raton, Fl. 508 pp.

Shaw, 1994, *Hydrology in practice*. (Third Edition). Chapman & Hall, London. 569 pp.

Shea, P. (2000) The U.S. Interior Department's Abandoned Mine Lands Program. *Proceedings of the Fifth International Conference on Acid Rock Drainage*, (ICARD 2000), Denver, Colorado, May 21–24, 2000. Volume I, pp. 21–27.

Shepherd, R. (1993) *Ancient Mining*. Institution of Mining and Metallurgy/Elsevier Applied Science, London, 494 pp.

Sherwood, J.M. (1997) *Modelling minewater flow and quality changes after coalfield closure*. Unpublished PhD Thesis, Department of Civil Engineering, University of Newcastle, UK, 241 pp.

Sherwood, J.M. and Younger, P.L. (1997) Modelling groundwater rebound after coalfield closure. In: Chilton, P.J., *et al.*, (Editors), *Groundwater in the urban environment, Volume 1: Problems, processes and management*. (Proceedings of the XXVII Congress of the International Association of Hydrogeologists, Nottingham, UK, 21–27 September 1997). A.A. Balkema Publishers, Rotterdam, pp. 165–170.

Shevenell, L. (2000a) Analytical method for predicting filling rates of mining pit lakes: example from the Getchell Mine, Nevada. *Mining Engineering*, **52**(3), 53–60.

Shevenell, L. (2000b) Evaporative concentration in pit lakes: Example calculations for the Getchell Pit Lakes, Nevada. *Proceedings of the Fifth International Conference on Acid Rock Drainage*, (ICARD 2000), Denver, Colorado, May 21–24, 2000. Volume I, pp. 337–345.

Shevenell, L.A. (2000c) Water quality in pit lakes in disseminated gold deposits compared to two natural, terminal lakes in Nevada. *Environmental Geology*, **39**, 807–815.

Sibrell, P.L., Watten, B.J., Friedrich, A.E. and Vinci, B.J. (2000) ARD remediation with limestone in a CO_2 pressurized reactor. *Proceedings of the Fifth International Conference on Acid Rock Drainage*, (ICARD 2000), Denver, Colorado, May 21–24, 2000. Volume II, pp. 1017–1026.

Sidle, R.C., Kamil, I., Sharma, A. and Yamashita, S. (2000) Stream response to subsidence from underground coal mining in central Utah. *Environmental Geology*, **39**, 279–291.

Singer P.C. and Stumm W. (1970) Acid mine drainage: the rate limiting step. *Science*, **167**, 1121–1123.

Singh, R.N. (1986) Mine water inundations. *International Journal of Mine Water*, **5**(2), 1–28.

Singh, R.N. and Atkins, A.S. (1983) Design considerations for mine workings under accumulations of water. *International Journal of Mine Water*, **4**, 35–56.

Skarzynska, K.M. and Michalski, P. (1999) Environmental effects of the deposition and re-use of colliery spoils. In: Azcue, J.M. (editor), *Environmental impacts of mining activities. Emphasis on mitigation and remedial measures*. Springer, Heidelberg, pp. 179–200.

Skousen, J. (1991) Anoxic limestone drains for acid mine drainage treatment. *Green Lands*, **21**(4), 30–35.

Skousen, J., Politan, K., Hilton, T. and Meek, A. (1990) Acid mine drainage treatment systems: chemicals and costs. *Green Lands*, **20**(4), 31–37.

Smedley, P.L., Edmunds, W.M. and Pelig-Ba, K.B. (1996) Mobility of arsenic in groundwater in the Obuasi gold-mining area of Ghana: some implications for human health. In: Appleton, J.D., Fuge, R. and McCall, G.J.H. (Editors), Environmental geochemistry and health. *Geological Society Special Publication No. 113*, London, pp. 163–181.

Smethurst, G. (1988) Basic water treatment for application world-wide. (Second edition). Thomas Telford, London. 216 pp.

Smit, J.P. (1999) The treatment of polluted mine water. In: *Proceedings of the Congress of the International Mine Water Association*, Sevilla Spain, 13–17 September 1999. Volume II, pp. 467–471.

Smith, D.I., Atkinson, T.C. and Drew, D.P. (1976) The hydrology of limestone terrains. In: Ford, T.D. and Cullingford, C.H.D. (Editors), *The Science of Speleology*. Academic Press, London, pp. 179–212.

Smith, E.S. and Connell, D.H. (1979) The role of water in the failure of tailings dams. In: Argall, G.O. and Brawner, C.O. (editors) *Mine Drainage*. Proceedings of the first International Mine Drainage Symposium, Denver, Colorado, May 1979. Miller Freeman Publications, San Francisco, pp 627–650.

Smith, J.A., and Colls, J.J. (1996) Groundwater rebound in the Leicestershire Coalfield. *Journal of the Chartered Institution of Water and Environmental Management*, **10**, 280–289.

Smith, S. (1923) *Lead and zinc ores of Northumberland and Alston Moor*. Memoirs of the Geological Survey, Special Reports on the Mineral Resources of Great Britain. Volume XXV. London, HMSO, 110 pp.

Sopwith, T. (1833) *An account of the mining district of Alston Moor, Weardale and Teesdale*. W. Davison, Alnwick. (Facsimile edition published 1984, Davis Books, Newcastle Upon Tyne), 183 pp.

Spratt, A.K. and Wieder, R.K. (1988) Growth responses and iron uptake in Sphagnum plants and their relation to acid mine drainage. In: *Proceedings of the International Land Reclamation and Mine Drainage Conference and the Third International Conference on the Abatement of Acidic Drainage*. (Pittsburgh, PA, 19–21st April 1988). Volume 1: Mine water and mine waste. US Bureau of Mines Information Circular IC-9183, pp. 279–286.

Stark, L., Stevens, E., Webster, H. and Wenerick, W. (1990) Iron loading, efficiency and sizing in a constructed wetland receiving mine drainage. In: Skousen, J., Sencindiver, J. and Samuel, D. (editors), *Proceedings of the 1990 Mining Reclamation Conference and Exhibition*, (West Virginia University Publications Service), volume II, pp. 393–401.

Stark, L.R., Williams, F.M., Stevens, S.E. and Eddy, D.P. (1994) Iron retention and vegetative cover at the Simco constructed wetland: an appraisal through year eight of operation. In: *Proceedings of the International Land Reclamation and Mine Drainage Conference and the Third International Conference on the Abatement of Acidic Drainage*. Pittsburgh, PA, April 1994. Volume 1 (US Bureau of Mines Special Publication SP 06A-94), pp. 89–98.

St-Arnaud, L. (1994) Water covers for the decommissioning of sulfidic mine tailings impoundments. In: *Proceedings of the International Land Reclamation and Mine Drainage Conference and the Third International Conference on the Abatement of Acidic Drainage*. Pittsburgh, PA, April 1994. Volume 1 (US Bureau of Mines Special Publication SP 06A-94), pp. 279–287.

Starr, R. C. and Cherry, J.A. (1994) *In situ* remediation of contaminated ground water: the funnel-and-gate system. *Groundwater*, **32**, 465–476.

Staub, M.W. and Cohen, R.H. (1992) A passive mine drainage treatment system as a bioreactor: treatment efficiency, pH increase and sulfate reduction in two parallel reactors. *Proceedings of the 1992 National Meeting of the American Society for Surface Mining Reclamation*, Duluth, MN.

Stichbury, M., Béchard, G., Lortie, L. and Gould, W.D. (1995) Use of inhibitors to prevent acid mine drainage. In: Hynes, T.P. and Blanchette, M. (editors) *Conference Proceedings, Vol. II Sudbury '95, Mining and the Environment,* pp. 613–622.

Strangeways, I. (2000) Measuring the environment. Cambridge University Press, Cambridge. 365 pp.

Streetley, M. (1998) Dewatering and environmental monitoring for the extractive industry. *Quarterly Journal of Engineering Geology*, **31**, 125–127.

Strömberg, B. (1997) *Weathering Kinetics of Sulphidic Mining Waste*. PhD Thesis, TRITA–OOK–1043, Dept. of Inorganic Chemistry, The Royal Institute of Technology, Stockholm.

Strömberg, B. and Banwart, S.A. (1994) Kinetic modelling of geochemical processes at the Aitik mining waste rock site in northern Sweden. *Applied Geochemistry*, **9**, 583–595.

Stumm, W. and Morgan, J.J. (1996) Aquatic Chemistry (3rd Edition). J. Wiley and Sons, New York.

Suter, G.W., Lixmoore, R.J. and Smith, E.D. (1993) Compacted soil barriers at abandoned landfill sites are likely to fail in the long term. *Journal of Environmental Quality*, **22**, 217–226.

Sverdrup, H.U. (1990) *The Kinetics of Base Cation Release due to Chemical Weathering*. Lund University Press.

Tarutis, W.J., Stark, L.R. and Williams, F.M. (1999) Sizing and performance estimation of coal mine drainage wetlands. *Ecological Engineering*, **12**, 353–372.

Tassé, N. (2000) Efficient prevention of sulphide oxidation by an organic cover: for hoe long can a reactive barrier be reactive? *Proceedings of the Fifth International Conference on Acid Rock Drainage*, (ICARD 2000), Denver, Colorado, May 21–24, 2000. Volume II, pp. 979–986.

Taylor, J., Fowell, R.J. and Wade, L. (2000) Effects of abandoned shallow bord and pillar workings on surface development. *Transactions of the Institution of Mining and Metallurgy (Section A: Mining Technology)*, **109**, A140–A145

Taylor, T.J. (1858) The Archaeology of the Coal Trade. In: 'Memoirs chiefly illustrative of the antiquities of Northumberland communicated at the annual meetings of the Archaeological Institute of Great Britain and Ireland held at Newcastle-on-Tyne in August 1852'. (Reprinted in facsimile by Frank Graham Publishers, Newcastle Upon Tyne, 1971), 76 pp.

Taylor, R.K. (1988) Coal measures mudrocks: composition, classification and weathering processes. *Quarterly Journal of Engineering Geology*, **21**, 85–99.

Tenenbein, M. (1998) Toxicokinetics and toxicodynamics of iron poisoning. *Toxicology Letters*, **103**, 653–656.

Theis, C.V. (1935) The relation between the lowering of the piezometric surface and the rate and duration of discharge of a well using ground-water storage. *Transactions of the American Geophysical Union*, **16**, 519–524.

Theis, C.V. (1940) The source of water derived from wells: Essential factors controlling the response of an aquifer to development. *Civil Engineering*, **10**, 277–280.

Thomas, R.C., Romanek, C.S., Coughlin, D.P. and Crowe, D.E. (1999) Treatment of acid mine drainage using anaerobic constructed treatment wetlands: predicting longevity with substrate neutralization potential. In: Goldsack, D., Belzile, N., Yearwood, P. and Hall, G. (editors), *Proceedings of the Sudbury '99 Conference "Mining and the Environment II"*, held Sudbury Ontario, Canada. Volume 2, pp. 449–458.

Thompson, L.C. (1996) Cyanide biotreatment and metal biomineralization in spent ore and process solutions. In: US EPA (editors) *Managing environmental problems at inactive and abandoned metals mine sites*. EPA Seminar Publication No EPA/625/R-95/007, pp. 18–29.

Thornton, F.C. (1995) Manganese removal from water using limestone-filled tanks. *Ecological Engineering*, **4**, 11–18.

Tòtaro, E.A., Lucadamo, L., Coppa, T., Turano, C. and Gervasi, R. (1992) Effects of iron pollution on macroinvertebrates promoting organic-matter transformation in soils of Presila-Cosentina (Italy). *Biology and Fertility of Soils*, **14**, 223–229.

Tremblay, G.A. (2000) The Canadian Mine Environment Neutral Drainage 2000 (MEND 2000) Program. *Proceedings of the Fifth International Conference on Acid Rock Drainage*, (ICARD 2000), Denver, Colorado, May 21–24, 2000. Volume I, pp. 33–30.

Trexler, B.D. (1979) Hydrogeology of a lead-zinc mine. In: Argall, G.O. and Brawner, C.O. (editors), *Mine Drainage*. (Proceedings of the First International Mine Drainage Symposium, Denver, Colorado, May 1979). Miller Freeman Publications Inc., San Francisco, CA, pp. 552–570.

Tsukamoto, T.K. and Miller, G.C. (1999) Methanol as a carbon source for microbiological treatment of acid mine drainage. *Water Research*, **33**, 1365–1370.

Tuttle, L.H., Dugan, P.R. and Randles, C.I. (1969) Microbial sulfate reduction and its potential utility as an acid mine water pollution abatement procedure. *Applied Microbiology*, **17**, 297–302.

UNEP (1996) *Environmental and safety incidents concerning tailings dams at mines. Results of a survey for the years 1980–1996*. Prepared by Mining Journal Research Services for the United Nations Environment Programme, Paris, 130 pp.

UNEP (2000) *Cyanide spill at Baia Mare, Romania. Spill of liquid and suspended waste at the Aural S.A. retreatment plant in Baia Mare. Assessment Mission Report*. United Nations Environment Programme, Office for the Co-ordination of Humanitarian Affairs. UNEP, Geneva, March 2000, 60 pp.

US EPA (1983) Design Manual: Neutralization of Acid Mine Drainage. United States Environmental Protection Agency, Office of Research and Development, Environmental Research Laboratory, 231 pp.

Vandiviere, M.M. and Evangelou, V.P. (1998) Comparative testing between conventional and micro-encapsulation approaches in controlling pyrite oxidation. *Journal of Geochemical Exploration*, **64**, 161–176.

van Schalkwyk, L.G. and Bellamy, R.E.S. (1994) Sealing of high pressure water fissures in South African mines. In: *Proceedings of the 5th International Mine water Congress*, Nottingham, UK. Volume 1, 299–316.

van Tonder, G.J., Maree, J.P. and Millard, P. (1994) Neutralisation of acid coal mine water with dolomite in a fluidised-bed reactor. In: *Proceedings of the 5th International Mine water Congress*, Nottingham, UK. Volume 2, pp. 587–597.

Vick, S.G. (1983) *Planning, design and analysis of tailings dams*. Wiley, New York. 369 pp.

Vidal, J. (1998) Filipino gold. *Guardian Weekend*, 28–2–1998, pp. 10–19.

Vile, M.A. and Wieder, R.K. (1993) Alkalinity generation by Fe(III) reduction versus sulfate reduction in wetlands constructed for acid-mine drainage treatment. *Water Air And Soil Pollution*, **69**, 425–441.

Vuori, K.M. (1995) Direct and indirect effects of iron on river ecosystems. *Annales Zoologici Fennici*, **32**, 317–329.

Waltham, A.C. (1994) *Foundations of engineering geology*. Blackie, Academic and Professional, London. 88 pp.

Walton-Day, K. (1999) Geochemistry of the processes that attenuate acid mine drainage in wetlands. In: Plumlee, G.S. and Logsdon, M.J. *The environmental geochemistry of mineral deposits. Part A: Processes, techniques and health issues*. Reviews in Economic Geology, Volume 6A, Society of Economic Geologists, Littleton, CO, pp. 215–228.

Walton-Day, K., Runkel, R.L., Kimball, B.A. and Bencala, K.E. (2000) Application of the solute-transport models OTIS and OTEQ and implications for remediation in a watershed affected by acid mine drainage, Cement Creek, Animas River Basin, Colorado. *Proceedings of the Fifth International Conference on Acid Rock Drainage, (ICARD 2000), Denver, Colorado*, May 21–24, 2000. Volume I, pp. 389–399.

Wang, Y. and Reardon, E.J. (2001) A siderite/limestone reactor to remove arsenic and cadmium from wastewaters. *Applied Geochemistry*, **16**, 1241–1249.

Wardrop, D.R., Leake, C.C. and Abra, J. (2001) Practical techniques that minimize the impact of quarries on the water environment. *Transactions of the Institution of Mining and Metallurgy (Section B: Applied Earth Sciences)*, **110**, B5–B14.

Warhurst, A.C. (1999) Best available practices in mining on land of minerals. In: Nath, B., Hens, L., Compton, P., and Devuyst, D. (Editors), *Environmental Management in Practice: Volume 2. Compartments, Stressors, Sectors*. Routledge, London, 1999, 145–165.

Watkins, G. (1979) The Steam Engine in Industry. Mining and the Metal Trades. Moorland Publishing, Ashbourne, UK. (Reprinted in paperback 1994). 128 pp.

Watzlaf, G.R. (1997) Passive treatment of acid mine drainage in down-flow limestone systems. In: *Proceedings of the 1997 National Meeting of the American Society for Surface Mining and Reclamation*, (ASSMR, Princeton, WV), pp 611–622.

Watzlaf, G.R., Schroeder, K.T. and Kairies, C. (2000) Long-term performance of alkalinity-producing passive systems for the treatment of mine drainage. *Proceedings of the 2000 National Meeting of the American Society for Surface Mining and Reclamation*, Tampa, Florida, June 11–15, 2000.

Watzlaf, G.R., Schroeder, K.T. and Kairies, C.L. (2001) Modeling of iron oxidation in a passive treatment system. In: *Proceedings of the 2001 Meeting of the American Association for Surface Mining and Reclamation*, June 3–7, 2001, Albuquerque, New Mexico, USA, pp. 626–638

Waybrant, K.R., Blowes, D.W. and Ptacek, C.J. (1998) Selection of reactive mixtures for use in permeable reactive walls for treatment of mine drainage. *Environmental Science and Technology*, **32**, 1972–1979.

Wegelin, M. (1996) *Surface water treatment by roughing filters. A design, construction and operation manual*. Swiss Centre for Development Cooperation in Technology and Management (SKAT). IT Publications, London, 184 pp.

Wheater, H. and Kirby, C. (editors) (1998) *Hydrology in a Changing Environment*. (3 volumes). Wiley, Chichester.

Wherli, B. (1990) Redox reactions of metal ions at mineral surfaces. In: *Aquatic Chemical Kinetics*, W. Stumm (Ed.), 311–336. J. Wiley and Sons, New York.

White, A.F. and Brantley, S.L. (Eds.) (1995) *Chemical Weathering Rates of Silicate Minerals, Reviews in Mineralogy, Vol. 31*. The Mineralogical Society of America, Washington, D.C.

White, A.F. and Petersen, M.L. (1990) Role of reactive-surface area characterization in geochemical kinetic models. In: Chemical Modelling of Aqueous Systems II. (Eds. D.L. Melchior and R.L. Bassett), Chap. 35, pp. 461–475. American Chemical Society Symposium Series 416.

Whittaker, B.N. and Reddish, D.J. (1989) Subsidence – occurrence, prediction and control. *Developments in Geotechnical Engineering*, **56**, Elsevier, Amsterdam.

Whittaker, B.N., Singh, R.N. and Neate, C.J. (1979) Effect of longwall mining on ground permeability and subsurface drainage. In: Argall, G.O. and Brawner, C.O. (editors), *Mine Drainage. (Proceedings of the First International Mine Drainage Symposium*, Denver, Colorado, May 1979). M. Freeman Publications, San Francisco, pp. 161–183.

Whittaker, D.R. (1968) An investigation of ground movement patterns around stopes in hard rock. Unpublished M.Sc. Thesis, Department of Mining Engineering, University of Newcastle, UK. 59 pp. plus figures.

Whitworth, K.R. (1982) Induced changes in the permeability of Coal Measures strata as an indicator of the mechanics of rock deformation above a longwall face. In: Farmer, I.W. (editor), *Strata Mechanics*.

(Proceedings of the Symposium on Strata Mechanics held in Newcastle Upon Tyne, 5–7 April, 1982). Developments in Geotechnical Engineering No. 32. Elsevier, Amsterdam, pp. 18–24.

Wieder, R.K. (1989) A survey of constructed wetlands for acid coal mine drainage treatment in the eastern United States. *Wetlands*, **9**, 299–315.

Wieder, R.K. and Lang, G.E. (1982) Modification of acid mine drainage by a freshwater wetland. In: McDonald, B.R. (Editor), *Proceedings of the Symposium on Wetlands of the Unglaciated Appalachian Region*, West Virginia University, Morgantown, West Virginia, USA.

Wieder, R.K., Lang, G.E. and Whitehouse, A.E. (1985) Metal removal in Sphagnum-dominated wetlands: experience with a man-made wetland system. In: Brooks, R.P., (Editor), *Wetlands and Water Management on Mined Lands*, Pennsylvania State University, University Park, Pennsylvania, USA, pp. 353–364.

Wieder, R.K., Linton, M.N. and Heston, K.P. (1990) Laboratory mesocosm studies of Fe, Al, Mn, Ca and Mg dynamics in wetlands. *Water, Air and Soil Pollution*, **51**, 181–196.

Wiersma, C.L. and Rimstidt, J.D. (1984) Rate of reaction of pyrite and marcasite with ferric iron at pH 2. *Geochim. Cosmochim. Acta*, **48**, 85–92.

Williamson, M.A. and Rimstidt, J.D. (1994) The kinetics and electrochemical rate-determining step of aqueous pyrite oxidation. *Geochim. Cosmochim. Acta*, **58**:24, 5443–5454.

Wilby, R.L. (1997) *Contemporary Hydrology. Towards holistic environmental science*. Wiley, Chichester. 354 pp.

Wildeman, T.R., Brodie, G. and Gusek, J.J. (1993) *Wetlands design for mining operations*. BiTech Publishers, Richmond, British Columbia.

Willow, M.A. and Cohen, R.R.H. (1998) pH and dissolved oxygen as factors controlling treatment efficiencies in wet substrate, bio-reactors dominated by sulfate-reducing bacteria. In *"Tailings and Mine Waste '98"* (Symposium Volume of the American Society for Surface Mining and Reclamation).

Wills, B. A. (1992) *Mineral processing technology: an introduction to the practical aspects of ore treatment and mineral recovery*. (Fifth edition). Pergamon, New York, 855 pp.

Wilmoth, R.C. (1973) *Applications of reverse osmosis to acid mine drainage treatment*. US Environmental Protection Agency, Report EPA-670/2-73-100.

Wilson, D. and Brown, D. (1997) Fast-track provision of St Aidan's Reverse Osmosis Water-Treatment Plant. *Journal of the Chartered Institution of Water and Environmental Management*, **11**, 271–276.

Wilson, J.A., Wilson, G.W. and Fredlund, D.G. (2000) Numerical modeling of vertical and inclined waste rock layers. *Proceedings of the Fifth International Conference on Acid Rock Drainage*, (ICARD 2000), Denver, Colorado, May 21–24, 2000. Volume I, pp. 257–266.

Winczewski, L.M. (1979) An overview of western North Dakota lignite strip mining processes and resulting subsurface characterisation. In: Wali, M.K., (editor), *Ecology and Coal Resource Development*. Volume II, pp. 677–684.

Wing, R. and Gee, G. (1994) Quest for the perfect cap. *Civil Engineering*, (October 1994), pp. 38–41.

Winland, R.L., Trainer, S.J. and Bigham, J.M. (1991) Chemical composition of aqueous precipitates from Ohio coal mine drainage. *Journal of Environmental Quality*, **20**, 452–460.

Winter, T.C., Harvey, J.W., Franke, O.L. and Alley, W.M. (1998) Ground water and surface water: a single resource. *United States Geological Survey Circular 1139*. Denver, Colorado, 79 pp.

Wolersdorfer, C. (2002) Mine water tracing. In: Younger, P.L. and Robins, N.S. (editors), *Mine water hydrogeology and geochemistry*. Special Publication of the Geological Society, London (in press).

Wood, A. and Reddy, V. (1998) Acid mine drainage as a factor in the impacts of underground minewater discharges from Grootvlei Gold Mine. In: *Proceedings of the International Mine Water Association Symposium on "Mine Water and Environmental Impacts"*, Johannesburg, South Africa, 7th–13th September 1998. (Volume II), pp. 387–398.

WWF International (1998) *Analysis and evaluation of the clean-up activities of the toxic spill in Guadiamar river*. Report prepared for WWF International, Copenhagen, by Buser & Finger (Zurich) and Roth & Partner GmbH (Karlsruhe), 30th July 1998. 19 pp.

Young, C.A. and Jordan, T.S. (1996) Cyanide Remediation: Current and Past Technologies. In: *Proceedings of the 10th Annual Conference on Hazardous Waste Research*, pp. 104–129.

Younger, P.L. (1993a) Possible Environmental Impact of the Closure of Two Collieries in County Durham. *Journal of the Institution of Water and Environmental Management*, **7**, 521–531.

Younger, P.L. (1993b) Simple Generalised Methods for Estimating Aquifer Storage Parameters. *Quarterly Journal of Engineering Geology*, **26**, 127–135.

Younger, P.L. (1997a) The Longevity of Minewater Pollution: A Basis for Decision-Making. *Science of the Total Environment*, 194/195, 457–466.

Younger, P.L. (Editor) (1997b) *Minewater Treatment Using Wetlands*. Proceedings of a National Conference held 5th September 1997, at the University of Newcastle, UK. Chartered Institution of Water and Environmental Management, London, 189 pp.

Younger, P.L. (1998a) Adit hydrology in the long-term: observations from the Pb–Zn mines of northern England. In: *Proceedings of the International Mine Water Association Symposium on "Mine Water and Environmental Impacts"*, Johannesburg, South Africa, 7th–13th September 1998. (Volume II), pp. 347–356.

Younger, P.L. (1998b) Coalfield abandonment: geochemical processes and hydrochemical products. In: Nicholson, K., *Energy and the Environment. Geochemistry of Fossil, Nuclear and Renewable Resources*. Society for Environmental Geochemistry and Health. McGregor Science, Aberdeen, pp. 1–29.

Younger, P.L. (1998c) Hydrological consequences of the abandonment of regional mine dewatering schemes in the UK. In: Arnell, H., and Griffin, J. (Editors), Hydrology in a Changing Environment. *Poster Papers and Index, International Symposium organised by the British Hydrological Society*, 6th–10th July 1998, Exeter, UK). British Hydrological Society Occasional Paper No 9, pp. 80–82.

Younger, P.L. (1998d) Design, construction and initial operation of full-scale compost-based passive systems for treatment of coal mine drainage and spoil leachate in the UK. In: *Proceedings of the International Mine Water Association Symposium on "Mine Water and Environmental Impacts"*, Johannesburg, South Africa, 7th–13th September 1998. (Volume II), pp. 413–424.

Younger, P.L. (1998e) Long term sustainability of groundwater abstraction in north Northumberland. In: Wheater, H. and Kirby, C. (eds.), Hydrology in a Changing Environment. (*Proceedings of the International Symposium organised by the British Hydrological Society*, 6th–10th July 1998, Exeter, UK). Volume II, pp. 213–227.

Younger, P.L. (1999) Pronóstico del ascenso del nivel freático en minas subterráneas y sus consecuencias medio-ambientales. *Boletín Geológico y Minero*, **110**, 407–422.

Younger, P.L. (2000a) Predicting temporal changes in total iron concentrations in groundwaters flowing from abandoned deep mines: a first approximation. *Journal of Contaminant Hydrology*, **44**, 47–69.

Younger, P.L. (2000b) Nature and practical implications of heterogeneities in the geochemistry of zinc-rich, alkaline mine waters in an underground F–Pb mine in the UK. *Applied Geochemistry*, **15**, 1383–1397.

Younger, P.L. (2000c) Mine water pollution in the long-abandoned Cleveland Ironstone field, north-east England. *Proceedings of the 7th National Hydrology Symposium, British Hydrological Society*. Newcastle Upon Tyne, 6–8th September 2000.

Younger, P.L. (2000d) Holistic remedial strategies for short- and long-term water pollution from abandoned mines. *Transactions of the Institution of Mining and Metallurgy* (Section A), **109**, A210–A218.

Younger, P.L. (2000e) The adoption and adaptation of passive treatment technologies for mine waters in the United Kingdom. *Mine Water and the Environment*, **19**, 84 –97

Younger, P.L. (2001a) Mine water pollution in Scotland: nature, extent and preventative strategies. *Science of the Total Environment*, **265**, 309–326.

Younger, P.L. (2001b) Passive treatment of European mine waters: the European Commission's 'PIRAMID' project. In: *Proceedings of the International Mine Water Association Symposium*, April 24th–28th 2001, Belo Horizonte, Brasil.

Younger, P.L. and Adams, R. (1999) *Predicting mine water rebound*. Environment Agency R&D Technical Report W179. Bristol, UK. 108 pp.

Younger, P.L. and Harbourne, K.J. (1995) "To pump or not to pump": Cost-benefit analysis of future environmental management options for the abandoned Durham Coalfield. *Journal of the Chartered Institution of Water and Environmental Management*, **9**(4), 405–415.

Younger, P.L. and LaPierre, A.B. (2000) "Uisge Mèinne": Mine water hydrogeology in the Celtic lands, from *Kernow* (Cornwall) to *Ceapp Breattain* (Cape Breton, Nova Scotia). In: Robins, N.S. and Misstear, B.D.R. (editors), *Groundwater in the Celtic regions: Studies in Hard-Rock and Quaternary Hydrogeology*. Special Publication No 182, Geological Society of London, pp. 35–52.

Younger, P.L., Banwart, S.A., Jarvis, A.P. and Burke, S.P. (2001) Optimisation of Ochre Accretion at Source (GR/L54790 GR/L55421): Final Report to the Engineering and Physical Sciences Research Council. Universities of Newcastle and Sheffield, UK. May 2001, 6 pp.

Younger, P.L., Barbour, M.H. and Sherwood, J.M. (1995) Predicting the Consequences of Ceasing Pumping from the Frances and Michael Collieries, Fife. In: Black, A.R. and Johnson, R.C. (Editors), *Proceedings of the Fifth National Hydrology Symposium, British Hydrological Society*. Edinburgh, 4–7th September 1995, pp. 2.25–2.33

Younger, P.L., Curtis, T.P., Jarvis, A.P. and Pennell, R. (1997) Effective passive treatment of aluminium-rich, acidic colliery spoil drainage using a compost wetland at Quaking Houses, County Durham. *Journal of the Chartered Institution of Water and Environmental Management*, **11**, 200–208.

Younger, P.L., Large, A.R.G. and Jarvis, A.P. (1998) The creation of floodplain wetlands to passively treat polluted minewaters. In: Wheater, H. and Kirby, C. (eds.), *Hydrology in a Changing Environment*. Volume I, pp. 495–515.

Zaluski, M., Trudnowski, J., Canty, M. and Harrington Baker, M.A. (2000) Performance of field-bioreactors with sulfate-reducing bacteria to control acid mine drainage. *Proceedings of the Fifth International Conference on Acid Rock Drainage*, (ICARD 2000), Denver, Colorado, May 21–24, 2000. Volume 2, pp. 1169–1175.

Ziemkiewicz, P.F., Skousen, J.G., Brant, D.L., Sterner, P.L. and Lovett, R.J. (1997) Acid mine drainage treatment with armored limestone in open channels. *Journal of Environmental Quality*, **26**, 1017–1024.

Zinck, J.M. and Griffith, W.F. (2000) An assessment of HDS-type lime treatment processes – efficiency and environmental impact. *Proceedings of the Fifth International Conference on Acid Rock Drainage*, (ICARD 2000), Denver, Colorado, May 21–24, 2000. Volume 2, pp. 1027–1034.

Zipper, C., Balfour, W., Roth, R. and Randolph, J. (1997) Domestic water supply impacts by underground coal mining in Virginia, USA. *Environmental Geology*, **29**, 84–93.

Zou, D.H.S. and Huang, Y. (1999) A proposed method of underwater capping for tailings disposal. In: Goldsack, D., Belzile, N., Yearwood, P. and Hall, G. (editors), *Proceedings of the Sudbury '99 Conference "Mining and the Environment II"*, held Sudbury Ontario, Canada. Volume 2, pp. 671–679.

INDEX